The Liver

The Organ of Flow, Conversion, Storage, and Survival

First Edition: April 2026
ISBN: 978-1-966035-16-9
Published by Elite Nutrition Research
EliteNutritionResearch.com

Disclaimer: This book is for informational purposes only. The author and publisher make no representations or warranties with respect to the accuracy or completeness of the contents of this book and disclaim all warranties, including without limitation warranties of fi tness for a particular purpose. The advice and strategies contained herein may not be suitable for every individual. The author and publisher shall not be liable for any loss or damage, including but not limited to special, incidental, consequential, or other damages.

Medical Disclaimer:
This book is intended for informational purposes only and is not a substitute for professional medical advice, diagnosis, or treatment. The content provided is based on the author's research, knowledge, and personal opinions. Always seek the advice of your physician or other qualifi ed health provider with any questions you may have regarding a medical condition. Never disregard professional medical advice or delay in seeking it because of something you have read in this book.

The author and publisher are not responsible for any health issues or consequences resulting from the use or misuse of the information contained in this book. Individual results may vary, and the suggestions herein may not be suitable for everyone. Readers are encouraged to consult with healthcare professionals before making any changes to their diet, exercise, or lifestyle routines.

ELITE

NUTRITION RESEARCH

Elite Nutrition Research is a science-driven health and education company dedicated to understanding and optimizing human physiology at its most fundamental level. Our work is built on an architecture-first approach—examining how nutrients interact with cellular systems, membranes, energy production, and signaling pathways to shape overall health and performance. Rather than following trends, we focus on first principles, combining biochemical insight, formulation precision, and practical application to create both advanced nutritional products and in-depth educational systems.

Through our books, courses, and formulations, Elite Nutrition Research aims to give individuals direct access to the mechanisms that govern their biology empowering them to think clearly, act intentionally, and build lasting health from the inside out.

To explore our full system of products, courses, and research, visit:
www.elitenutritionresearch.com

Ian Miller is an independent health writer, researcher, publisher, and founder of Elite Nutrition Research, focused on human biology, digestion, metabolism, and first-principles nutrition. His work is built on a simple conviction: the body can only be understood clearly when food is judged through anatomy, physiology, and biological function rather than through diet trends, food marketing, or cultural nutrition narratives. Across his books and research projects, he aims to make complex physiology readable without flattening it into slogans.

Through his books, Miller continues to build a body of work aimed at restoring clarity to modern nutrition by grounding it in the material realities of the human body.

Table of Contents

Part I

Why the Liver Sits Where It Sits

Chapter 1

The Organ Downstream from the Intestine

The liver sits where it sits because digestion is not finished when food leaves the intestinal lumen. The lumen is the hollow interior of the gut, the space where acid, enzymes, bile, and mechanical breakdown turn food into smaller chemical units. Even after that work, what crosses the intestinal wall is not automatically safe, balanced, or ready for general circulation. It is a raw post-digestive stream. Glucose arrives in surges. Amino acids arrive in uneven waves. Fructose comes in needing immediate handling. Lipid fragments, microbial byproducts, bacterial cell wall fragments, histamine, alcohol, drug molecules, bilirubin derivatives, and whatever else survived digestion or was produced during fermentation all begin moving toward the blood. The body does not send that first chemical flood directly to the brain, heart, kidneys, muscles, and endocrine organs. It routes it through the liver first. That placement is not incidental anatomy. It is control.

The intestine is an absorptive surface, not a final judge. Its job is to move usable compounds across the epithelial barrier, the thin cellular lining that separates the gut contents from the body. But absorption is permissive. It allows entry. It does not finish the accounting. A molecule can cross the intestinal wall and still need to be stored, converted, detoxified, neutralized, conjugated, packaged, buffered, or excreted. Conjugated means chemically modified so it becomes easier to transport or eliminate. The liver is positioned immediately downstream because the body needs a second organ to interpret what the intestine has allowed through. The portal vein exists for exactly this reason. It collects blood from the stomach, intestines, pancreas, and spleen and delivers that blood straight to the liver before it joins the wider circulation. This is called portal circulation. It is a separate vascular route built to ensure that the chemical results of digestion are reviewed before they are distributed.

That review process has to happen fast and at scale. After a meal, the intestine can deliver a heavy carbohydrate load, nitrogen from amino acids, water-soluble vitamins, minerals, bacterial metabolites, and a large number of foreign compounds all at once. If that blood bypassed the liver, systemic circulation would take the full hit in real time. Glucose would spike harder. Ammonia from nitrogen handling would travel farther before being neutralized. Drugs and alcohol would move into tissues with less first-pass metabolism, meaning less early modification during the first trip through the liver. Hormone breakdown would be less controlled. Microbial fragments from the gut would circulate more freely. The body would lose its main buffering chamber. The liver exists in that downstream position because absorbed chemistry needs to be slowed, sorted, and reshaped before it becomes everybody else's problem.

The organ is built for this workload. It receives blood from two sources at once. The portal vein brings nutrient-rich, chemically crowded blood from the digestive tract. The hepatic artery brings oxygen-rich arterial blood to support the liver's own high metabolic demand. Those two inflows meet in specialized vascular channels called sinusoids. Sinusoids are not ordinary capillaries. They are wide, slow-flow exchange spaces lined in a way that allows close contact between blood contents and hepatocytes, the liver's main working cells. This matters because the liver is not simply filtering particles out of blood like a screen. It is chemically rewriting the blood. Hepatocytes take in glucose and store some of it as glycogen. They convert ammonia into urea so the kidneys can excrete it safely. They process fructose, package lipids into lipoproteins, modify hormones, conjugate bilirubin, transform drugs, produce plasma proteins, and secrete bile components. The blood entering from the gut is chemically different from the blood leaving the liver. That is the point of the organ's position.

This placement also reveals something deeper about digestion itself. Digestion is not only the breakdown of food. It is also the management of exposure. The intestine absorbs. The liver decides consequences. A meal is not just nutrition entering the body. It is a biochemical event that can destabilize blood composition if it is not handled properly. That is especially clear with compounds the body cannot simply use as-is.

Fructose is a good example. Much of the absorbed fructose from the intestine is extracted rapidly by the liver because peripheral tissues are not meant to be the first site of its large-scale handling. Ammonia is another. It is generated from amino acid metabolism and gut microbial activity, and it is toxic enough that it has to be trapped quickly in the urea cycle before it builds up in the bloodstream. Bacterial endotoxin, a structural component of certain bacterial cell walls, can also enter portal blood in small amounts from the gut. The liver is positioned to intercept that traffic through Kupffer cells, its resident macrophages, before that inflammatory burden spreads. The organ sits downstream because the gut is productive, but it is not clean.

There is another reason the liver occupies this position, and it is easy to miss if the subject is taught too abstractly. The body needs a place where incoming abundance can be converted into usable timing. Absorption happens according to when food arrives. Physiology cannot operate on that schedule alone. The brain, muscle, blood, endocrine system, and kidneys need steadier conditions than the intestine can provide. The liver absorbs the irregularity of eating and turns it into managed release. It stores glycogen when glucose is abundant, then releases glucose later when intake stops. It converts excess amino acid nitrogen into excretable form. It builds plasma proteins continuously instead of only when food is present. It produces bile so fat digestion can occur in the next meal. It recycles, stores, modifies, and redistributes. In other words, the liver is anatomically downstream from the intestine because the intestine works in episodes and the body has to survive continuously.

Once you see that, the familiar phrase that the liver is a detox organ starts to feel too small to be useful. The liver is the organ that stands between digestion and distribution. It is placed at the border where food stops being matter in the gut and starts becoming chemistry in the body. That border is dangerous. Too much arrives too fast. Some of it is useful only after conversion. Some of it is harmful unless neutralized. Some of it needs to be stored for later. Some of it needs to be sent back out in bile. The liver sits there because no organism can afford to let the raw output of digestion pour straight into the general bloodstream without review. The gut brings in the outside world. The liver decides what form of that world the rest of the body will be allowed to see.

The Portal Vein as the Direct Line from Digestion to Chemical Processing

The portal vein is the vessel that makes the liver's position matter. Without it, the liver would still be a large metabolic organ, but it would not stand in immediate command of the post-digestive stream. The portal system changes that. It collects venous blood from the stomach, small intestine, large intestine, pancreas, and spleen and delivers that blood straight into the liver before it is allowed to join the wider circulation. This is not ordinary return flow on its way back to the heart. It is targeted delivery. The body builds a dedicated vascular route because digestion creates a chemical surge that needs to be handled before it is distributed. The portal vein is the pipe that carries that surge.

What travels through it is not just nutrition. It is the first dissolved aftermath of eating. Glucose absorbed from carbohydrate digestion enters portal blood in high concentration. Amino acids released from protein digestion enter with it. Water-soluble vitamins, minerals, organic acids, microbial fermentation products, bacterial cell wall fragments, ammonia precursors, bilirubin derivatives, alcohol, plant compounds, drugs, and hormone metabolites all ride the same route. The intestine does not sort these into clean categories of useful and harmful. It transfers what crosses. The portal vein then delivers that mixed chemical cargo to the liver in concentrated form. That is why the portal vein should be understood as a functional structure, not just an anatomical term. It is the direct line between absorption and judgment.

This arrangement lets the liver perform first-pass handling, which means the first major round of extraction, conversion, buffering, and detoxification before the rest of the body sees the blood. Fructose is one of the clearest examples. It is absorbed by the intestine, enters portal blood, and is taken up heavily by the liver, where it is phosphorylated and pushed into hepatic metabolism. The same logic applies to many drugs. A compound absorbed from the gut may lose a substantial fraction of its original form during this first passage through hepatocytes before it ever reaches general circulation in meaningful quantity. The portal vein exists so that this first-pass event can happen at the correct place and at the correct intensity. It prevents the body from treating every absorbed molecule as if it were already system-ready.

Nitrogen handling makes the necessity even clearer. Protein digestion releases amino acids, but amino acid traffic is never just about building tissue. Some amino acids are deaminated, meaning their nitrogen group is removed during metabolism. Gut microbes also generate nitrogenous compounds that can enter portal blood. Ammonia sits downstream of this chemistry, and ammonia is dangerous because it diffuses easily and disrupts brain function when it accumulates. The liver, receiving portal blood first, captures much of this nitrogen burden and channels it into the urea cycle, converting ammonia into urea, a water-soluble waste that the kidneys can excrete. The portal vein therefore is not merely supplying nutrients to the liver. It is bringing unstable chemistry to the only organ positioned to neutralize it fast enough.

The same vessel also makes the liver the first immune filter for the gut. The intestine contains an enormous bacterial population, and even in health small amounts of microbial products can cross into portal blood. That means the liver is continuously exposed to lipopolysaccharide, peptidoglycan fragments, metabolites produced by fermentation, and antigenic material that came from food or microbes. Kupffer cells, the resident macrophages of the liver, sit in the sinusoids ready to sample that incoming blood. They clear, tolerate, or respond depending on the nature of the signal. The portal vein is what forces the liver into that role. It makes the organ the immune buffer between the intestine and the rest of the body. The gut sends chemical traffic. The portal vein delivers it. The liver decides how much of that traffic becomes systemic reality.

Once you see the portal vein correctly, the whole architecture of digestion sharpens. Absorption is not completion. It is transfer into a controlled internal channel. The portal vein is that channel. It takes what the intestine has liberated and moves it to the liver for storage, conversion, packaging, conjugation, detoxification, immune screening, and redistribution. Blood does not leave the gut as finished nutrition. It leaves as raw material, potential fuel, possible toxin, signaling burden, and microbial report. The portal vein is the direct line that carries all of it to chemical processing before the rest of the body is asked to live with the result.

Why Food Does Not Enter the Body as Finished Nutrition

Food enters the mouth as structure, not as nourishment already delivered. Muscle tissue arrives as bundled proteins, connective tissue, fat droplets, minerals held inside cellular architecture, and water bound within a physical matrix. Plant tissue arrives locked inside cellulose walls, vacuoles, fiber networks, starch granules, sugars, acids, polyphenols, mineral-binding compounds, and water compartments that digestion has to break open before any serious extraction can occur. The body cannot absorb a steak, a potato, an egg, or an apple in the form those foods are eaten. It has to dismantle them first. Chewing reduces particle size and increases exposed surface area. Gastric acid denatures proteins, meaning it unfolds them so enzymes can get at peptide bonds that were previously buried inside folded structure. Pepsin begins protein cleavage in the stomach. Pancreatic enzymes continue that breakdown in the small intestine. Bile emulsifies fat, breaking large lipid masses into smaller suspended droplets so lipases can act at the surface. Carbohydrases split larger carbohydrates into smaller sugars. Brush border enzymes on the intestinal surface finish the job. By the time the intestine is ready to absorb, food is no longer functioning as food in the everyday sense. It has been reduced to chemical fragments, and even those fragments are not all equally usable.

Absorption is selective, conditional, and incomplete. A molecule has to survive digestion in an absorbable form, remain soluble in the gut lumen, approach the intestinal surface, cross the mucus layer, pass through or between epithelial cells, and then survive the first round of handling after entry. The lumen is the hollow interior of the intestine where digestion continues and where liberated compounds still remain outside the body in a meaningful physiological sense. That distinction matters. A nutrient present in the lumen is not yet nourishment inside tissue. Calcium can be present in a meal and still fail to enter meaningfully if it is trapped in an insoluble complex. Protein can be listed on a label and still arrive at the absorptive surface in reduced usable yield if digestion is impaired or if part of that protein remains physically inaccessible inside the food matrix. Fat-soluble compounds can be present in abundance and still enter poorly if bile flow is weak and micelle formation is poor. A micelle is a tiny transport structure formed by bile acids that helps lipid digestion products stay suspended in the watery environment of the intestine long enough to reach the absorptive membrane. The body does not take in nutrients according to food labels. It takes in what digestion can actually liberate and what transport systems can actually move.

Even once molecules cross the intestinal wall, they do not arrive as finished tissue fuel ready for universal use. Many of them still need sorting, conversion, packaging, or neutralization. Glucose may be absorbed efficiently, but that does not mean it can simply flood the circulation without control. Amino acids may enter portal blood, but their nitrogen load still has to be managed. Fructose may cross the intestinal membrane, but much of its burden falls immediately on the liver because peripheral tissues are not designed to be the first site of large fructose handling. Lipids illustrate the point even more clearly. Long-chain fatty acids and monoglycerides do not move directly into portal blood in their original digestive state and then circulate as finished nutrition. Inside intestinal cells they are re-esterified, meaning rebuilt into triglycerides, then packaged with cholesterol, phospholipids, and apolipoproteins into chylomicrons, which are large transport particles that enter lymph before eventually reaching the bloodstream. The body is constantly changing the form of what it absorbs because raw digestive fragments are often not stable, not soluble, not targetable, or not safe enough to circulate unprocessed.

Some absorbed compounds are even less finished than they appear because they carry a processing cost rather than a ready benefit. Ammonia precursors, microbial fermentation products, bacterial cell wall fragments, alcohol, xenobiotics, plant defense compounds, and drug molecules can all cross into the body's internal channels after digestion, but none of them count as nutrition in the functional sense. They are chemical burdens that have entered the same route as nutrients. Xenobiotics are foreign compounds, substances the body does not make for its own structural use. The intestine does not possess the full authority to decide the fate of all this incoming material. It transfers what crosses. The liver then receives much of that traffic through the portal vein and begins the second stage of judgment: extraction, storage, conjugation, packaging, detoxification, secretion into bile, or release into the wider circulation. This is why food does not enter the body as finished nutrition. What leaves the intestine is still mixed chemistry. It is partly fuel, partly building material, partly waste burden, partly signaling load, and partly potential injury.

The deeper point is that nutrition is not what a food contains. Nutrition is what remains biologically available after structure has been broken, digestion has acted, inhibitors have bound what they can bind, transporters have moved what they can move, and downstream organs have processed what the intestine has passed forward. Food enters as architecture. Digestion dismantles that architecture. Absorption selects from the debris. The liver then decides what form of that absorbed chemistry the rest of the body will actually be allowed to use. A meal is therefore not a packet of finished nourishment entering the bloodstream intact. It is a staged physiological event that begins with disassembly and ends, if all goes well, with a much smaller set of compounds that the body has successfully extracted, transformed, and made safe enough to call nourishment.

The Liver as the First Major Checkpoint After Intestinal Absorption

Once a molecule crosses the intestinal wall, it has entered the body in one sense, but it has not yet been cleared for general use. That distinction is where most shallow nutrition writing breaks down. Intestinal absorption is a transfer event. It moves compounds out of the gut lumen and into the internal circulation, but it does not finish the work of sorting value from burden. The small intestine is built to absorb. It is lined with enterocytes, the absorptive cells whose apical surface faces the lumen and whose basolateral surface releases absorbed material toward blood or lymph. Those cells digest, transport, and package, but they do not possess the full metabolic authority needed to stabilize the chemical aftermath of a meal. Glucose can cross. Amino acids can cross. Water-soluble vitamins, minerals, microbial metabolites, drug molecules, alcohol, and a range of other compounds can cross as well. Crossing is not the same as being ready. The body therefore routes much of this post-absorptive stream through the liver first because a second, more powerful checkpoint is required immediately after entry.

That checkpoint exists because the intestine is permissive by design. It decides whether a compound can pass a membrane, not whether the whole organism is ready for the consequences. After a meal, portal blood can carry a dense and irregular chemical mix. Glucose may rise sharply and require buffering before it spreads through the wider circulation. Fructose often needs heavy first-pass hepatic handling because large peripheral exposure would be metabolically disordered. Amino acids arrive with nitrogen that may later need to be stripped and converted into urea. Bilirubin derivatives, microbial fermentation products, bacterial cell wall fragments, and foreign compounds absorbed from food or drugs can enter that same flow. The liver is the first major checkpoint because it is the first organ with enough metabolic range, synthetic capacity, enzymatic machinery, and excretory connection to decide what happens next. It can extract, store, convert, neutralize, package, or send back out in bile. That is a very different level of control than simple membrane transport.

The portal circulation makes this checkpoint physically unavoidable. Blood draining the stomach, intestines, pancreas, and spleen enters the portal vein and is delivered directly into the hepatic sinusoids, the liver's specialized low-pressure exchange channels. There, hepatocytes and other liver cells meet the incoming chemistry before it reaches the rest of the body in full force. Hepatocytes can pull in glucose and convert part of it into glycogen, a storage polymer that reduces the immediate glycemic load reaching systemic blood. They can capture ammonia and channel it into the urea cycle so nitrogen waste becomes a water-soluble compound the kidneys can excrete. They can modify hormones, conjugate bilirubin, build plasma proteins, repackage lipids, and begin biotransformation of foreign compounds. Conjugation means attaching chemical groups to a molecule so its solubility, reactivity, or excretability changes. This is not passive screening. The liver does not stand there like a gate guard checking labels. It rewrites the chemical form of what just entered.

This position also makes the liver the first major immunologic checkpoint after absorption. The intestine is not sterile, and the blood leaving it does not carry nutrients alone. Small amounts of bacterial products and antigenic material can pass from the gut into portal blood even under ordinary conditions. The liver is built to receive that traffic without letting every exposure become systemic inflammation. Kupffer cells, the liver's resident macrophages, sit in the sinusoids where they can sample portal blood and clear microbial debris. Sinusoidal endothelial cells and hepatocytes participate in this filtering environment as well. The checkpoint therefore is not only metabolic. It is inflammatory, immunologic, and structural. The liver has to decide what gets neutralized quietly, what gets tolerated, what gets excreted, and what demands a defensive response. The organ stands downstream from the intestine because the intestine opens a door. Someone has to control what walks through.

The phrase first major checkpoint matters because it explains why liver dysfunction changes the meaning of absorption itself. A nutrient may be absorbed normally and still be mishandled if the liver cannot store, convert, or distribute it properly. A toxin or drug may be absorbed in modest quantity and still create exaggerated effects if first-pass metabolism is impaired. Microbial byproducts that would normally be contained may circulate farther. Ammonia that would normally be trapped may reach the brain. Bilirubin that would normally be conjugated and excreted may accumulate in blood and stain tissue yellow.

In other words, intestinal absorption tells you what entered the portal stream. Liver function tells you what the body was actually able to do with it. That is why the liver deserves to be taught not as a vague detox organ but as the first major checkpoint after intestinal absorption. The gut opens the border. The liver decides what form of that absorbed chemistry the rest of the body will be forced to live with.

Why the Liver Cannot Be Understood as Only a Detox Organ

Calling the liver a detox organ shrinks the organ down to one public-relations function and hides the actual architecture of its work. Detox is not wrong, but it is radically incomplete. The liver does transform and clear foreign compounds, endogenous waste products, excess hormones, bilirubin, ammonia-derived nitrogen, and a range of reactive intermediates the body does not want left circulating freely. But that is only one lane in a much larger traffic system. The liver is also a storage organ, a synthetic organ, a secretory organ, a fuel-buffering organ, an immune interface, an endocrine modifier, and a flow organ built around the management of incoming portal blood and outgoing bile. If you teach it only as a detox organ, the reader sees a filter. What is actually present is a chemical command center attached directly to the intestine.

Detox itself is too vague to do real explanatory work unless it is broken into mechanisms. The liver does not remove bad things by some general cleansing force. It takes specific molecules and changes their chemistry. A fat-soluble compound may need to be oxidized, reduced, or hydrolyzed in Phase I biotransformation, which means it is chemically altered, often through enzyme systems such as cytochrome P450. It may then need Phase II conjugation, where the liver attaches another chemical group, such as glucuronide, sulfate, glycine, or glutathione, to make that compound more water-soluble or easier to pump into bile or urine. That is real detoxification. But even here the word detox can mislead, because the liver often makes a molecule more reactive before it makes it safer. The work is hazardous chemistry, not spa language. And while that chemistry is happening, the same hepatocytes are also storing glycogen, assembling lipoproteins, producing albumin, secreting bile components, handling bilirubin, converting ammonia into urea, and modifying steroid and thyroid hormones. The organ is processing far more than toxins.

The liver's central task is not simply to clear threats. It is to keep blood chemically usable. After a meal, glucose enters portal blood in a surge.

The liver takes a substantial fraction of that load and stores part of it as glycogen so the systemic circulation does not receive the full spike unbuffered. Amino acids arrive and are sorted between protein synthesis, energy metabolism, and nitrogen disposal. Fructose is taken up heavily by the liver and routed into hepatic metabolism because the body is not built to let large amounts of it wash indiscriminately through peripheral tissues first. Lipid remnants return from intestinal fat handling and are repackaged, stored, oxidized, or exported. This is not detox work in the narrow sense. This is post-absorptive governance. The liver sits downstream from the intestine because food does not enter the body as finished nutrition and absorbed chemistry is too unstable, too irregular, and too mixed to be released into general circulation without control.

Bile formation alone is enough to break the detox-only model. The liver continuously manufactures bile, a fluid containing bile acids, phospholipids, cholesterol, conjugated bilirubin, electrolytes, and water. That bile is sent through the biliary tree and later delivered into the intestine, where bile acids emulsify fat and help form micelles, small transport structures that keep lipid digestion products suspended long enough to reach the absorptive surface. In other words, the liver is not only cleaning up after digestion. It is upstream of digestion too. It prepares one of the essential chemical tools required for fat absorption in the next meal. At the same time, bile serves as an excretory route for compounds the body wants removed. The same outgoing stream helps digest food and eliminate waste. That dual role is typical of the liver. It is always doing more than one thing at once.

Its synthetic function is equally incompatible with the detox label. Hepatocytes produce albumin, the major plasma protein that helps hold fluid inside the vascular space and bind many circulating compounds during transport. They produce clotting factors, anticoagulant proteins, acute-phase proteins involved in inflammation, apolipoproteins needed for lipid transport, and signaling molecules such as hepcidin and thrombopoietin that influence iron traffic and platelet production. The liver also stores glycogen, vitamin A, vitamin B12, iron, and copper. It participates in vitamin D metabolism, thyroid hormone handling, steroid modification, and the control of ammonia through the urea cycle. Once you start listing the organ's actual outputs, the detox framing collapses. A failing liver does not merely allow toxins to build up. It stops building the blood correctly, stops stabilizing fuel supply correctly, stops supporting digestion correctly, and stops managing whole-body chemistry correctly.

The immune role pushes the point even further. Portal blood coming from the intestine carries not only nutrients but also bacterial fragments, metabolites made by gut microbes, food antigens, and inflammatory signals. The liver receives this traffic first and has to decide what gets cleared quietly, what gets tolerated, what gets excreted, and what deserves an immune response. Kupffer cells, the resident macrophages of the liver, sit in the sinusoids and sample that incoming blood. Liver sinusoidal endothelial cells and hepatocytes participate in the same interface. This makes the liver an immune buffer between the gut and the rest of the body. It is not a janitor cleaning up after the fact. It is a border organ deciding how much of the intestine's chemical and microbial reality reaches the wider organism.

So the problem with calling the liver only a detox organ is not that detoxification is unimportant. It is that the phrase makes the organ sound secondary, as if its main job were to take out the trash after the real physiology has already happened. The opposite is true. The liver is part of the core physiology that makes usable blood, usable nutrients, usable hormones, usable fats, and usable waste disposal possible in the first place. Detox is one function inside a larger design. The liver does not merely remove what the body does not want. It determines what form of incoming chemistry the body will be allowed to live on.

The Five Core Verbs of Liver Physiology: Receive, Convert, Store, Build, Excrete

The liver becomes easier to understand when its endless list of functions is collapsed into five verbs. Receive, convert, store, build, excrete. Those five actions are not a simplification that flattens the organ. They are the underlying pattern that lets the complexity stay coherent. Everything the liver does can be read through them. Blood arrives from the portal vein carrying the dissolved consequences of digestion. The liver receives that flow. The chemistry inside that blood is then altered, buffered, neutralized, repackaged, or activated. The liver converts. Some of what arrives is set aside for later use. The liver stores. Some of what the body needs cannot simply be absorbed from food in finished form and has to be manufactured inside the organ itself. The liver builds. Some compounds are too reactive, too abundant, too insoluble, or too spent to remain in circulation and must be sent out through bile or prepared for renal excretion through the kidneys. The liver excretes. Once those verbs are fixed in the reader's mind, the organ stops looking like a chaotic catalog of unrelated jobs and starts reading like a single system built around chemical control.

Receive comes first because the liver is positioned downstream from the intestine to take custody of incoming blood before the rest of the body has to deal with it. Portal blood is not clean fuel. It is mixed post-digestive traffic. Glucose comes in after carbohydrate absorption. Amino acids come in after protein digestion. Fructose, lactate, microbial metabolites, bacterial cell wall fragments, bilirubin derivatives, drugs, alcohol, hormones returning for breakdown, and a range of foreign compounds travel in the same stream. The liver receives that blood through the portal vein and slows it through the sinusoids, the specialized exchange channels where hepatocytes and other liver cells can sample what has arrived. Receiving in this context does not mean passively allowing blood to pass through the organ. It means taking first possession of the chemical aftermath of eating, intestinal absorption, splenic drainage, and part of endocrine turnover. The liver is the first major organ asked to make sense of that flow.

Convert is the verb that gives the liver its chemical authority. Much of what arrives cannot remain in its incoming form. Ammonia is converted into urea so it can be carried safely in blood and then removed in urine. Bilirubin is conjugated, meaning chemically modified to become water-soluble enough to move into bile. Glucose can be converted into glycogen for storage or derived from noncarbohydrate substrates through gluconeogenesis when food is absent. Fatty acids can be oxidized, triglycerides can be assembled, cholesterol can be turned into bile acids, and hormones can be activated, inactivated, or prepared for clearance. Drugs and xenobiotics, foreign compounds the body does not make for its own structure, may be oxidized in Phase I metabolism and then conjugated in Phase II metabolism so their solubility and excretability change. This is why the word detox is too small. The liver is not merely neutralizing poisons. It is constantly changing the form of incoming and circulating molecules so they become usable, storable, exportable, or disposable.

Store is the verb that explains why the liver protects continuity rather than merely reacting to intake. Eating is intermittent. Physiology cannot be. The liver stores glycogen so blood glucose can be supported between meals. It stores vitamins, especially vitamin A and vitamin B12, along with iron and copper. It stores energy and chemical reserve in ways that allow the body to survive the gap between intake and demand. Storage is not passive accumulation. It is controlled holding. A useful molecule is retained in a form that can be mobilized later without destabilizing the present. Glycogen is a clear example. Free glucose in excess would distort blood chemistry if left circulating unchecked. Converted into glycogen, it becomes compact, osmotically manageable, and retrievable. The same logic applies to mineral and vitamin reserves. Storage turns temporary abundance into delayed availability. Without that function, every missed meal, every shortfall in intake, and every shift in demand would hit the body with much greater force.

Build is the least appreciated verb in public discussion of the liver and one of the most important. The liver is not simply handling what arrives. It is manufacturing major components of the internal environment. Hepatocytes build albumin, the main plasma protein that helps hold fluid inside the vasculature and bind many molecules during transport. They build clotting factors, anticoagulant proteins, complement proteins, apolipoproteins for lipid transport, and acute-phase proteins that reshape plasma chemistry during inflammation. The liver also builds bile, which means it creates part of the digestive machinery required for fat absorption in the intestine. It produces urea. It produces ketone bodies during fasting. It generates signaling molecules such as hepcidin and thrombopoietin that influence iron movement and platelet production. Once the verb build is seen clearly, liver failure becomes more intelligible. The problem is not only that toxins accumulate. The body also begins losing the proteins, transport systems, digestive chemistry, and blood-stabilizing outputs that the liver had been supplying every hour of the day.

Excrete completes the sequence because chemistry that cannot remain in circulation has to leave the body in an orderly way. The liver plays a central role in that removal, but it often does so by changing compounds first and then directing them toward the appropriate exit. Some compounds are sent into bile and enter the intestine for elimination in stool. Bilirubin is a classic example. Cholesterol can also leave through bile, directly or after conversion into bile acids. Other compounds are made more water-soluble and released into blood so the kidneys can filter them into urine. Urea is the clearest example of that route. Excretion therefore is not separate from the earlier verbs. The liver receives a compound, converts it into a safer or more transportable form, may transiently store or bind related intermediates, builds the transport chemistry needed to move it, and then excretes it through bile or prepares it for renal loss. Nothing about this process is casual. Excretion is the disciplined end of a longer metabolic judgment.

These five verbs also explain why liver disease spreads its effects so widely. If receiving is distorted because blood cannot move through the organ properly, portal pressure rises and upstream flow becomes dangerous. If conversion fails, ammonia, bilirubin, hormones, drugs, and reactive intermediates begin behaving differently in blood and tissue. If storage fails or reserves are consumed, the body loses buffering capacity. If building fails, albumin falls, clotting destabilizes, bile production suffers, and plasma composition shifts. If excretion fails, compounds meant to leave accumulate and start injuring tissue or altering signaling. The liver's importance is not hidden in some obscure specialty function. It is visible in the fact that these five verbs touch digestion, circulation, endocrinology, immunity, renal burden, and brain function all at once.

So if the reader wants one durable frame for the rest of the book, this is it. The liver receives the consequences of digestion and circulation. It converts raw chemistry into safer or more useful forms. It stores reserve. It builds critical components of blood and bile. It excretes what cannot stay. Every chapter ahead is a more detailed version of one or more of those verbs. Together they form a single argument. The liver keeps the internal environment livable by deciding what enters the blood, what changes form, what is held back, what is manufactured, and what is pushed out.

Chapter 2

Gross Anatomy of the Liver

The liver is large because the workload is large. In the adult body it is the biggest solid internal organ, not counting the skin, and one of the heaviest organs in the body, usually weighing roughly three to four pounds in a healthy adult, though body size, sex, blood volume, and metabolic state shift that number. That mass is not decorative reserve. It reflects constant throughput. The liver has to receive a huge volume of blood, process nutrients and toxins arriving from the gut, manufacture plasma proteins, build bile, store glycogen, hold vitamins and minerals, and maintain this work without pauses. A small organ could not do that job at organism scale. The liver is broad, heavy, and densely vascular because it is built for high-volume chemical handling. When you hold that first fact clearly, size stops feeling like anatomy trivia and starts reading like function made visible.

Its location is equally purposeful. The liver sits high in the right upper abdomen, tucked directly under the diaphragm, which is the large muscular sheet that separates the chest from the abdominal cavity and drives breathing. Most of the liver is protected by the lower right rib cage, with part of it extending toward the midline and a smaller portion crossing into the left upper abdomen. That protected placement matters. The organ is soft, highly vascular, and constantly filled with moving blood. It is safer shielded beneath the ribs than exposed lower in the abdomen. It also sits close to the stomach, duodenum, gallbladder, pancreas, and the incoming portal circulation that drains the digestive tract. The liver is not simply in the abdomen. It is positioned at the anatomical crossroads between digestion, circulation, and bile delivery. Superiorly it molds against the diaphragm. Inferiorly and posteriorly it relates to the stomach, right kidney, duodenum, colon, and great vessels. The organ occupies premium territory because the traffic it handles is central to survival.

From the outside, the liver has a smooth capsule and a firm but not rigid consistency, with a dark reddish-brown color produced by its dense blood content and rich vascular supply. Its general shape is often described as wedge-like, broader on the right and tapering toward the left, but the shape is better understood as molded by its neighbors. The diaphragm presses its superior surface into a broad convex curve, while the organs beneath it create impressions on the undersurface. This gives the liver two major exterior faces. The diaphragmatic surface is the smooth upper and anterior surface that lies against the diaphragm. The visceral surface is the undersurface that faces the abdominal organs and contains more landmarks, grooves, and entry points. These are not abstract categories. The diaphragmatic surface reflects protection and contour. The visceral surface reflects traffic, attachment, and exchange with the rest of the abdominal system.

The right lobe dominates the organ's exterior mass. It is much larger than the left lobe and occupies most of the right upper abdomen. The left lobe is thinner and extends across the midline toward the left. On gross inspection that right-left division is the first thing most people notice, but the external lobes do not tell the full functional story. Still, the surface landmarks matter because they show how the organ is organized in real space. On the visceral surface you can identify the gallbladder fossa, which is the shallow bed where the gallbladder sits against the liver. You can identify the fissures and grooves associated with the round ligament and venous ligament, remnants of fetal circulation that still mark the adult liver's surface. You can identify the caudate lobe and quadrate lobe, smaller named regions on the inferior surface that help orient anatomy even if they are less important to a beginner than the larger right and left divisions. The liver's surface is therefore not featureless smooth tissue. It is mapped by its developmental history, its vascular entry points, and its contact with surrounding organs.

One of the most important external landmarks is the porta hepatis, the transverse gateway on the visceral surface where major vessels and ducts enter or leave the organ. This is the liver's hilum, meaning its main entry and exit zone. The portal vein enters here carrying nutrient-rich venous blood from the digestive tract. The hepatic artery enters carrying oxygen-rich arterial blood. The hepatic ducts leave carrying bile that hepatocytes have secreted into the biliary tree. Lymphatics and nerves travel through this region as well. If you want one gross anatomical feature that announces the liver's function without a word of physiology, it is this one. The porta hepatis shows that the liver is an exchange organ built around flow. Blood in, bile out, signals in, processed chemistry out. The organ's surface is organized around movement, not just mass.

Its external attachments reinforce that same idea. The liver is held in place by peritoneal reflections, folds of the abdominal lining that connect it to the diaphragm and anterior abdominal wall. The falciform ligament is the most familiar of these, a sickle-shaped fold that helps divide the right and left anatomical lobes on the anterior surface. There are also coronary and triangular ligaments that anchor the liver to the diaphragm. These attachments stabilize a heavy organ that moves with respiration. Every breath shifts the liver slightly because the diaphragm descends and rises against it. That respiratory movement matters clinically because the liver edge may descend with deep inspiration and become easier to feel on physical exam. Even here, surface anatomy is tied to behavior. The liver is not bolted in place. It rides the mechanics of breathing while staying suspended under the rib cage in a way that protects vascular inflow, venous outflow, and biliary drainage.

The surface anatomy also explains why enlargement changes the physical exam in recognizable ways. A healthy liver may sit mostly concealed beneath the ribs, with only the edge sometimes palpable in thin individuals or with deep inspiration. But when the liver enlarges, whether from congestion, fat accumulation, inflammation, infiltration, or tumor burden, it can extend farther below the right costal margin, which is the lower edge of the rib cage. The edge may feel smooth, rounded, firm, tender, or nodular depending on the process. That surface change is not superficial information. It reflects tissue events underneath. A smooth enlarged liver suggests one class of problem. A hard irregular liver suggests another. Gross anatomy becomes the first visible layer of pathology because disease eventually distorts shape, size, and surface.

So before the reader ever reaches histology, enzyme systems, or disease mechanisms, the gross liver already teaches the logic of the organ. It is large because the chemical workload is massive. It is high and protected because it is soft, vascular, and indispensable. It is broad on the right because that is where most of its bulk is needed. It presents a smooth diaphragmatic surface above and a more complex visceral surface below because one side is molded by protection and the other by traffic. Its main gateway, the porta hepatis, announces that the organ is built around controlled flow. Even on the surface, the liver does not look like a passive gland. It looks like what it is: a heavy, central processing organ placed under the diaphragm at the point where digestion, blood chemistry, and biliary drainage converge.

Right Lobe, Left Lobe, Caudate Lobe, Quadrate Lobe

At the gross anatomical level, the liver is divided into named lobes so the reader can begin orienting the organ in physical space before moving into its deeper vascular logic. The two largest external divisions are the right lobe and the left lobe. The right lobe makes up most of the liver's visible mass. It occupies the greater part of the right upper abdomen, rises high under the right hemidiaphragm, and extends posteriorly toward the kidney and inferior vena cava, the large vein that returns blood from the lower body to the heart. The left lobe is smaller, flatter, and narrower. It stretches across the upper abdomen toward the left side and lies closer to the stomach and distal esophagus. On the anterior surface, the falciform ligament helps mark the visible separation between these two large lobes. That external division is useful for orientation, for imaging, and for physical description. But it is still a surface division. The liver's true functional organization runs deeper through vessels and ducts, and later that deeper map will matter more than the superficial lobe names.

The right lobe is large because it houses most of the organ's bulk, most of the blood handling volume, and much of the tissue that gives the liver its heavy right-sided presence beneath the ribs. Its diaphragmatic surface is broad and convex, pressing upward against the diaphragm. Its visceral surface is more complex, molded by contact with the gallbladder, right kidney, hepatic flexure of the colon, and duodenum. The gallbladder sits in a fossa, a shallow depression, on the inferior surface of the right lobe. This relationship matters because bile made by hepatocytes ultimately drains through the biliary tree and can be stored in the gallbladder before release into the small intestine. The right lobe therefore is not just the bigger side of the organ. It is the dominant mass through which a large share of hepatic blood flow, bile production, metabolic handling, and clinical pathology are appreciated on exam and imaging.

The left lobe narrows and thins as it extends toward the left upper abdomen, but it should not be thought of as a minor accessory flap. It is still fully liver, fully hepatocellular, fully vascular, and fully involved in the organ's metabolic workload. It lies closer to the stomach and contributes to the liver's cross-midline reach, reminding the reader that the organ is not confined to the right side even though its weight is concentrated there. On imaging or dissection, the left lobe often appears more elegant and blade-like compared with the heavier right lobe. That visual difference reflects geometry, not lesser significance. The left lobe still receives portal inflow, arterial inflow, and participates in the same continuous work of receiving, converting, storing, building, and excreting. The body does not assign one lobe to metabolism and the other to decoration. The whole organ shares the burden.

The caudate lobe and quadrate lobe are smaller named regions seen mainly on the visceral surface, and they matter because they train the eye to notice that the liver's underside is not a blank plane. The caudate lobe sits posteriorly, near the inferior vena cava and above the porta hepatis, the main gateway where major vessels and ducts enter or leave the organ. Because of that position, the caudate lobe occupies a strategically interesting piece of anatomy. It lies close to major inflow and outflow structures and is often discussed in relation to venous drainage and cross-connections within the liver. The quadrate lobe sits more anteriorly on the inferior surface, between the gallbladder fossa and the fissure for the round ligament. It is quadrangular in outline, which is where the name comes from, and it helps complete the basic surface map of the liver's underside. These smaller lobes are less important to the beginner than the right and left lobes, but they stop the reader from oversimplifying the organ into a clean two-part wedge.

What makes these lobe names both useful and limited is that they describe what the outside of the liver looks like, not the full logic of how blood and bile move through it. A person can memorize right lobe, left lobe, caudate lobe, and quadrate lobe and still not understand the liver well. Functional anatomy depends much more on how branches of the portal vein, hepatic artery, hepatic veins, and bile ducts divide the organ into segments with their own inflow, outflow, and drainage patterns. That is why surgeons and radiologists often rely more heavily on segmental anatomy than on the older gross lobe names alone. Still, the external lobes matter as a first map. They let the reader see mass distribution, organ orientation, the underside landmarks, and the way the porta hepatis, gallbladder bed, and venous grooves organize the visceral surface.

So these four lobe names should be learned as the surface language of the liver. The right lobe is the dominant bulk. The left lobe reaches across the midline and balances the organ's upper abdominal spread. The caudate lobe occupies a posterior position near major vascular structures. The quadrate lobe marks part of the inferior surface near the gallbladder and round ligament. Together they teach the reader to stop seeing the liver as a single featureless red mass. It has contour, sidedness, named regions, and a visible external architecture. That external architecture is the first layer. Beneath it sits the more important truth: the liver is divided not only by shape, but by flow.

Ligaments and Anatomical Relations to Diaphragm, Stomach, Gallbladder, Pancreas, and Intestine

The liver is held in place by more than weight and crowding. It is suspended, draped, and stabilized by peritoneal reflections, folds of the serous membrane that lines the abdominal cavity and covers many of its organs. These folds are commonly called ligaments, but they are not ligaments in the orthopedic sense. They do not behave like dense fibrous straps binding bone to bone. They are sheets and folds of abdominal lining that help anchor the liver to surrounding structures, maintain orientation, and preserve the geometry of a heavy organ that rises and falls with every breath. The falciform ligament runs from the anterior abdominal wall to the liver and helps mark the external division between the right and left anatomical lobes on the front surface. The coronary ligaments and right and left triangular ligaments secure the liver to the underside of the diaphragm. On the inferior surface, the lesser omentum connects the liver to the lesser curvature of the stomach and the proximal duodenum. Together these structures do not immobilize the liver so much as suspend it in controlled relationship to the organs and spaces around it. That distinction matters because the liver must remain stable without becoming rigid. Its blood vessels, bile ducts, and venous outflow all depend on preserved orientation during constant respiratory motion.

The diaphragm is the liver's nearest superior neighbor and one of the main structures shaping its surface. The broad upper surface of the liver, called the diaphragmatic surface, is smooth and convex because it is molded directly against the underside of the diaphragm. Every inhalation drives the diaphragm downward, and the liver moves with it.

This is why the liver edge can descend below the costal margin during deep inspiration and become easier to feel on physical examination. The relationship is intimate enough that the liver carries the physical imprint of diaphragmatic contact. This superior placement under the right dome of the diaphragm also explains why the liver is protected by the lower rib cage and why disease involving the liver can sometimes produce symptoms that feel mechanically related to breathing, pressure, or fullness in the right upper abdomen. The liver does not simply sit below the diaphragm. It is shaped by it, moves with it, and depends on that sheltered subdiaphragmatic position for protection.

Inferiorly and to the left, the liver relates closely to the stomach. The left lobe extends toward the stomach and overlies part of it, particularly near the lesser curvature. This matters anatomically because the stomach is upstream in the digestive chain, breaking food down mechanically and chemically before chyme enters the duodenum. Chyme is the semi-fluid mixture of partially digested food and gastric secretions that leaves the stomach for the small intestine. The liver does not directly digest that material, but through bile production it prepares one of the essential chemical tools needed once gastric contents enter the small intestine and fat digestion begins in earnest. The lesser omentum helps connect liver and stomach, forming a structural bridge that also transmits vessels, nerves, and biliary relationships in the upper abdomen. The proximity is not random. The stomach begins the digestive event. The liver stands nearby to handle the consequences that follow once absorption and portal transport begin.

The gallbladder sits directly against the undersurface of the liver in a shallow depression called the gallbladder fossa. This is one of the clearest examples of a true functional anatomical relationship. The gallbladder is not an independent digestive organ operating on its own schedule. It is attached to the liver's biliary system and stores bile that the liver has already made. Hepatocytes secrete bile into canaliculi, tiny channels between liver cells, and that bile flows through progressively larger ducts until it can either move toward the intestine or be diverted into the gallbladder for storage and concentration. So the physical contact between the liver and gallbladder reflects direct continuity of function. One organ manufactures the fluid. The other stores and concentrates it. When fat reaches the duodenum and hormonal signals trigger gallbladder contraction, the bile delivered into the intestine is still liver-made chemistry.

The pancreas lies deeper and more posterior than the liver, but it is part of the same regional machinery of digestion, especially through its close relationship with the duodenum and biliary tree. The head of the pancreas rests within the curve of the duodenum, and the common bile duct descends through this region before entering the small intestine. This means liver bile flow and pancreatic enzyme flow converge near the same entry point into the duodenum. Anatomically, this makes the upper abdominal digestive system read less like a row of separate organs and more like a coordinated cluster around a single processing zone. The liver produces bile. The pancreas produces bicarbonate and digestive enzymes. The stomach supplies acidic chyme. The duodenum receives and integrates these streams. The relationship is tight enough that obstruction, inflammation, or compression in one part of this region can disturb the others. The liver's anatomy therefore cannot be taught honestly without placing it in this shared upper abdominal crossroads.

Its relation to the intestine is even more important functionally. The liver is not attached to the full length of the intestine in a direct surface-to-surface way, but it is inseparably tied to the intestine through portal circulation and bile delivery. The duodenum lies just inferior to the liver and receives bile entering through the biliary tract. Farther along, the jejunum and ileum absorb nutrients, water, electrolytes, and a wide range of chemical compounds, and much of that absorbed material is then carried back to the liver through the portal vein. So the intestine feeds the liver chemically even when it is not pressed directly against it anatomically. This is why the liver should be understood as downstream from the intestine even though part of that downstream relationship is vascular rather than purely physical. The intestine breaks apart the meal and transfers compounds across its lining. The liver receives the result. Their anatomical relationship is therefore a combination of proximity, ductal communication, and dedicated blood flow.

When all these relations are seen together, the logic of the upper abdomen becomes much sharper. The diaphragm shelters the liver and shapes its superior surface. The stomach sits nearby as the first major chamber of digestion. The gallbladder lies tucked against the liver as its bile reservoir. The pancreas shares the same regional outlet into the duodenum, contributing enzymes and alkaline fluid to the digestive stream. The intestine receives bile, absorbs the digested products of the meal, and sends much of that absorbed chemistry back to the liver through portal blood. The ligaments and peritoneal folds hold these relationships in disciplined spatial order. The liver is not hanging in the abdomen as an isolated block of tissue. It is suspended inside a structural web that places it exactly where digestion turns into circulation, where bile enters the gut, and where the chemical consequences of eating can be brought under control.

The Dual Blood Supply: Portal Vein and Hepatic Artery

The liver is unusual because it does not live on one blood stream. It is built to receive two at once, and that fact tells you almost everything about its role in the body. Most organs are supplied mainly by arterial blood, which brings oxygen under pressure from the heart and supports the metabolic needs of the tissue. The liver does receive arterial blood through the hepatic artery, but that is only part of the story. The larger share of its inflow comes through the portal vein, which carries venous blood draining the stomach, intestines, pancreas, and spleen. Venous blood usually means blood returning after tissues have already extracted oxygen and nutrients, but portal venous blood is different. It is relatively lower in oxygen than arterial blood, yet it is rich in newly absorbed nutrients, microbial products, gut-derived signaling molecules, hormones, drug compounds, and metabolic debris coming straight from digestion and splenic breakdown. The liver therefore is not supplied like an ordinary organ because it is not doing an ordinary job. It needs one inflow to feed the tissue and another inflow to deliver the chemistry it has been built to handle.

The portal vein is the dominant functional stream. It brings the post-absorptive consequences of the meal. Glucose enters from the intestine after carbohydrate digestion. Amino acids enter after protein breakdown. Fructose, lactate, water-soluble vitamins, minerals, bacterial metabolites, bilirubin-related compounds, alcohol, medications, and a wide range of foreign molecules travel in that same blood. This is chemically busy blood. It is not clean fuel on its way back to the heart. It is cargo blood. The liver receives it first because the body cannot allow all of that absorbed chemistry to pass unbuffered into systemic circulation. Hepatocytes need first access to the load so they can extract glucose, store some as glycogen, convert ammonia into urea, conjugate bilirubin, modify hormones, repackage lipids, clear microbial fragments, and begin biotransformation of compounds the rest of the body should not see in their incoming form. The portal vein therefore is not just nourishment for the liver. It is delivery of work.

The hepatic artery provides the second requirement. It brings oxygen-rich blood from the systemic circulation, and that oxygen is essential because the liver is metabolically expensive tissue. Hepatocytes run intense synthetic and oxidative chemistry all day. They build albumin and clotting factors, maintain ion gradients, power the urea cycle, carry out gluconeogenesis, assemble lipoproteins, produce bile components, and operate large enzyme systems in the endoplasmic reticulum and mitochondria. None of that can happen on nutrient-rich venous inflow alone. The portal vein delivers chemical substrate and incoming burden, but the hepatic artery helps keep the tissue alive enough to process it. This is the key distinction. Portal blood tells the liver what has arrived from the gut. Arterial blood gives the liver the oxygen needed to deal with it.

These two inflows do not remain separate for long. They enter the liver through branches that ultimately feed the sinusoids, the specialized vascular channels where blood slows and exchange begins. In the sinusoids, portal venous blood and hepatic arterial blood mix. That mixed blood then passes along hepatocyte plates toward the central veins. The arrangement is elegant because it solves two problems at once. It exposes hepatocytes to gut-derived chemistry in high concentration while also supplying enough oxygen to sustain the tissue's own processing power. The liver is therefore not choosing between perfusion and information. It receives both together. One stream tells it what has entered the body from the digestive tract. The other stream helps power the response.

This dual supply also explains why injury patterns inside the liver are not uniform. Blood enters from the portal tract side, where branches of the portal vein, hepatic artery, and bile duct travel together. As blood moves through the sinusoid toward the central vein, oxygen tension gradually falls. That means hepatocytes near the portal region, often called periportal cells or zone 1 cells, live in a more oxygen-rich environment. Hepatocytes nearer the central vein, often called centrilobular cells or zone 3 cells, see blood later, after some oxygen has already been extracted. This gradient matters because different metabolic jobs favor different parts of the lobule. It also means some injuries strike one zone harder than another. Low oxygen states and congestion often hit the downstream, more oxygen-vulnerable regions first. Some toxin handling is also concentrated there because the enzyme profile changes across the lobule. The liver's dual blood supply does not just nourish the organ. It creates an internal geography of metabolism and vulnerability.

The balance between portal venous flow and hepatic arterial flow is also physiologically important. The portal vein usually carries the larger volume of blood into the liver, while the hepatic artery contributes a smaller but crucial fraction with higher oxygen content. If portal flow falls, arterial compensation can increase to a degree, helping preserve hepatic oxygenation. If arterial supply is compromised, the portal stream still brings volume and nutrient-rich blood, but the tissue becomes more vulnerable because chemical traffic without adequate oxygen support is a dangerous bargain. The liver can tolerate a great deal, but it is still bound by perfusion. It cannot process what it cannot energize.

This arrangement reveals the true nature of the organ. The liver is not simply a gland that secretes bile or a filter that cleans blood after the fact. It is a receiving station built at the exact border where absorbed chemistry enters internal circulation. It needs the portal vein because digestion sends it raw material, metabolic load, microbial signals, and potential toxins. It needs the hepatic artery because none of that can be processed without oxygen-rich support. The two vessels together define the organ's role. One brings the message from the gut. The other brings the power to answer it.

The Hepatic Veins and Return to the Systemic Circulation

Blood does not stay in the liver after it has been processed. It has to leave in an orderly way, and the route out matters just as much as the route in. The portal vein and hepatic artery bring blood into the organ, but the hepatic veins are what carry it back into the general circulation after the liver has done its work. This is the exit side of the system. Once portal venous blood and arterial blood have entered the sinusoids, mixed, slowed, and passed along the hepatocyte plates, they collect in central veins, then in larger venous channels, and finally drain into the major hepatic veins, which empty directly into the inferior vena cava. The inferior vena cava is the large vein that returns blood from the lower body to the heart. This means the liver sits between two vascular worlds. One world brings in chemically crowded blood from the gut. The other carries blood away after the liver has extracted, converted, stored, built, and prepared compounds for release. The hepatic veins therefore are not an afterthought in liver anatomy. They are the final outflow path by which processed blood is returned to the rest of the body.

This return flow tells you something important about what the liver is actually doing. The blood leaving through the hepatic veins is not the same blood that entered through the portal vein. It may carry less glucose because some has been taken up and stored as glycogen. It may carry less ammonia because nitrogen has been converted into urea. It may carry altered hormone levels because steroid and thyroid-related compounds have been modified. It may carry newly synthesized plasma proteins, packaged lipids, glucose released during fasting, ketone bodies, or conjugated compounds that have been diverted toward bile instead of being allowed to remain in circulation. The hepatic veins therefore do not simply drain used blood from an organ that consumed oxygen. They carry the output of hepatic judgment. What reaches systemic circulation after the liver has handled portal inflow is a chemically revised version of what the gut and spleen originally sent forward.

The path that creates this outflow is built around the microscopic architecture of the lobule. Blood enters near the portal triads, where branches of the portal vein and hepatic artery sit alongside bile ducts, and then moves through the sinusoids toward the center of the lobule. As it moves, hepatocytes lining the sinusoidal space remove and add compounds continuously. By the time that blood reaches the central vein, it has already passed through a field of intense metabolic decision-making. The central veins then gather this processed blood and merge into larger venous channels that eventually form the right, middle, and left hepatic veins. These major hepatic veins empty into the inferior vena cava on the posterior aspect of the liver. The direction matters. Blood enters the liver from one side of the microscopic unit, moves across it, and leaves through another. Bile flows in the opposite direction. This opposing arrangement keeps inflow, chemical processing, and waste secretion organized rather than chaotic. Blood is being cleaned, buffered, and reformulated while bile is simultaneously being secreted away from that blood toward the duct system.

The hepatic veins also reveal how dependent the liver is on unobstructed flow. Inflow alone is not enough. Blood has to be able to leave. If venous outflow becomes impaired, pressure builds backward into the liver, sinusoids become congested, hepatocytes receive less efficient oxygen delivery, and the tissue begins to suffer. The liver is soft and densely vascular, so obstruction on the outflow side can quickly distort its internal environment. This is one reason venous congestion from right-sided heart failure or direct hepatic venous obstruction can produce a large tender liver and a dark congested appearance on pathology. The problem is not only that too much blood is present. It is that the liver cannot clear its own vascular traffic. A processing organ built around continuous flow becomes compromised when the exit narrows or backs up. The organ may still be perfused, but perfusion without drainage is unstable. Blood has to arrive, pass through, and leave on schedule.

This outflow pathway also helps explain why liver disease becomes systemic so quickly. The hepatic veins empty straight into the central venous circulation, which means whatever the liver releases after processing enters the wider blood supply with direct consequence for the rest of the body. If the liver is maintaining order, the blood leaving through the hepatic veins helps stabilize glucose, nitrogen balance, plasma protein composition, lipid transport, and hormone handling. If the liver is failing, the blood leaving it may carry excess bilirubin, altered drug chemistry, poorly controlled metabolites, and a reduced supply of proteins the body depends on. The liver does not work in isolation and then file its output away internally. It returns its final product directly to circulation. The hepatic veins are the route by which the results of liver function become the chemical environment of the organism.

So the vascular story of the liver is not complete until the hepatic veins are understood clearly. The portal vein brings the consequences of digestion. The hepatic artery brings oxygen-rich support. The sinusoids create the exchange field where incoming blood is examined and altered. The central veins collect that revised blood. The hepatic veins then deliver it back to the heart through the inferior vena cava. This is the completed arc of hepatic blood flow. Blood arrives carrying raw post-digestive chemistry. It leaves carrying the liver's decision about what the rest of the body will be allowed to receive.

The Extrahepatic Biliary Tree and Why Bile Flow Must Be Taught Alongside Blood Flow

The liver does not only receive and process blood. It also manufactures an outgoing fluid stream that has to leave the organ in an organized way. That stream is bile, and once bile exits the microscopic canaliculi inside the liver and enters larger ducts beyond the hepatic tissue itself, it moves through what is called the extrahepatic biliary tree. Extrahepatic simply means outside the liver. This tree is the duct system that carries bile from the liver to the gallbladder and the small intestine. It includes the right and left hepatic ducts, which drain bile from the right and left sides of the liver, the common hepatic duct formed by their union, the cystic duct connecting the gallbladder to the main channel, and the common bile duct that descends toward the duodenum, the first segment of the small intestine. Near its terminal portion, that duct usually joins the pancreatic duct or runs close beside it before bile enters the intestine through the major duodenal papilla under control of the sphincter of Oddi, a muscular valve that regulates flow. This is not plumbing detail added for completeness. It is one of the liver's main exit routes, and without it the organ cannot finish several of its most important jobs.

Bile has to move because the liver uses it for two kinds of work at once. First, bile is digestive chemistry. It carries bile acids and phospholipids that help emulsify dietary fat in the intestinal lumen, the hollow interior of the gut where digestion occurs before absorption. Emulsify means breaking large fat masses into much smaller suspended droplets so lipases can attack the surface more efficiently. Without that step, fat digestion remains physically incomplete and lipid absorption falls apart. Second, bile is an excretory stream. It carries conjugated bilirubin, excess cholesterol, bile acids, and a range of compounds that the liver has modified and decided to push out of the body through the intestine. Conjugated means chemically altered into a more transportable form, usually more water-soluble or more compatible with secretion. So the extrahepatic biliary tree is not only a digestive duct system. It is also part of the organ's waste-handling architecture. That dual role is exactly why bile flow cannot be treated as secondary.

If blood flow tells you what is arriving at the liver, bile flow tells you what the liver is sending away. These are parallel stories and they have to be taught together because they describe opposite directions of the same organ's labor. Blood enters through the portal vein and hepatic artery, passes through the sinusoids, and is altered as it moves toward the hepatic veins. Bile begins at the hepatocyte, the liver's main working cell, moves into canaliculi between adjacent hepatocytes, then into progressively larger bile ducts, and eventually out through the extrahepatic biliary tree toward the gallbladder or duodenum. Blood brings nutrient load, hormone traffic, microbial products, drugs, alcohol, nitrogen burden, and everything absorbed from the intestine that requires review. Bile carries away part of the result of that review. When these two streams are separated in teaching, the liver starts to look like it only takes things in. In reality it is constantly taking blood in, chemically rewriting it, and sending selected compounds out through bile at the same time.

The gallbladder makes this even clearer. It sits on the undersurface of the liver and connects to the biliary tree through the cystic duct. When the intestine is empty, a substantial fraction of bile leaving the liver can be diverted into the gallbladder for storage and concentration. Water and electrolytes are removed there, making the bile denser and more potent. When fat enters the duodenum, hormonal and neural signals trigger gallbladder contraction and relaxation of the sphincter of Oddi so bile can be expelled into the intestine. This means the extrahepatic biliary tree is not just an always-open drainage channel. It is a controlled flow system with timing, storage, concentration, and release. The liver produces bile continuously. The intestine needs it in pulses around meals. The biliary tree and gallbladder solve that timing problem.

Teaching bile flow alongside blood flow also matters because the failure patterns are different, and the reader needs to see both to understand liver disease properly. Blood-flow problems distort delivery, oxygenation, exchange, and venous return. Portal hypertension, congestion, and ischemic injury belong mainly to that side of the story. Bile-flow problems create cholestasis, which means impaired bile formation or impaired bile movement. Once bile slows or backs up, the chemistry that was supposed to leave begins to accumulate in liver tissue and blood. Bilirubin rises. Itching develops because retained biliary compounds circulate abnormally. Stool may become pale because less bilirubin pigment reaches the intestine. Urine may darken because conjugated bilirubin spills into the blood and is filtered by the kidneys. Fat digestion worsens because bile acids are not reaching the duodenum effectively. That means the same obstruction can create both an excretory failure and a digestive failure at once. A person who only learned liver blood flow will miss half the mechanism.

This is why the extrahepatic biliary tree has to be taught early and treated as part of the organ's core design. The liver is built around inflow and outflow, but not only in the vascular sense. Blood comes in carrying the consequences of digestion. Blood leaves carrying the liver's revised chemical decision. Bile leaves carrying digestive tools and excreted burden. The ducts outside the liver are the visible continuation of that second stream. Right and left hepatic ducts join, the common hepatic duct forms, the gallbladder connects through the cystic duct, the common bile duct descends, and the final output enters the intestine where digestion resumes. This is the second half of liver physiology made anatomical. The liver does not only manage what enters the body. It also determines what must leave it, and bile is one of the main ways that decision becomes real.

Chapter 3

Blood In, Bile Out

The liver makes no sense until the reader sees that it is running two streams at once. Blood comes in carrying the chemical aftermath of digestion. Bile goes out carrying both digestive tools and selected waste. Those streams do not move in the same direction because they are not solving the same problem. Blood is the incoming report. It brings glucose, amino acids, hormones, microbial fragments, drug compounds, bilirubin precursors, nitrogen burden, and the broader consequences of intestinal absorption. Bile is the outgoing decision. It carries bile acids, phospholipids, cholesterol, conjugated bilirubin, and other compounds the liver has chosen either to use in digestion or to push toward elimination. One stream brings raw chemistry into the organ for review. The other stream carries the result of that review away from the organ. That is the basic logic. The liver is built where incoming and outgoing chemical traffic must be kept separate if the organ is going to control either one.

At the microscopic level, that separation is built directly into the architecture of the hepatocyte. A hepatocyte is not a featureless cell floating in tissue. It has polarity, which means different sides of the same cell are specialized for different kinds of exchange. One side faces the sinusoid, the specialized blood channel where portal venous blood and hepatic arterial blood mix and pass slowly enough for exchange to occur. The opposite side contributes to the bile canaliculus, the tiny channel formed between adjacent hepatocytes where bile begins its movement out of the liver. This arrangement matters because the same cell that receives compounds from blood is also secreting other compounds into bile, but it is doing those two jobs across different membrane domains.

The liver cannot afford to let incoming blood cargo and outgoing biliary cargo dissolve into one confused internal pool. The organ maintains directional order at the cellular level.

Blood therefore moves one way through the liver's functional units. It enters from branches of the portal vein and hepatic artery located in the portal triads, then flows through the sinusoids toward the central vein. As it moves, hepatocytes remove some compounds, modify others, release selected products back into the blood, and hold or redirect still others. Bile begins inside that same field of hepatocytes but travels in the opposite direction, away from the central region and back toward the bile ducts located near the portal triads. This means blood and bile are not simply separate fluids. They are separate directional systems crossing the same tissue for opposite reasons. Blood is moving inward through the logic of delivery and sampling. Bile is moving outward through the logic of secretion and disposal. The geometry is elegant because it allows the liver to inspect incoming chemistry while simultaneously exporting compounds that need to leave through the biliary route.

This opposite-direction traffic is what allows the liver to connect digestion to clearance without collapsing the two into the same act. Take bilirubin as one example. Unconjugated bilirubin arrives in blood bound to albumin after heme breakdown elsewhere in the body. Hepatocytes take it up from the blood-facing side, conjugate it so it becomes more water-compatible and transportable, then secrete it out the bile-facing side into the canalicular system. The same logic applies to bile acids synthesized from cholesterol. The liver builds them inside hepatocytes, then pumps them into bile so they can travel to the intestine and assist with fat digestion. The incoming side and the outgoing side are linked, but they are not interchangeable. A molecule may arrive through blood and leave through bile. Another may arrive through blood and be released back into blood in altered form. Another may be stored. Another may be sent toward renal excretion instead. The liver stays coherent because it preserves direction while making those decisions.

The digestive consequence of this design is easy to miss if bile is taught as a minor accessory fluid. Bile is not an afterthought. It is one of the liver's main outgoing products, and its direction matters because fat digestion depends on delivery into the intestinal lumen, not into the bloodstream. Bile acids have to leave hepatocytes, pass through ducts, and reach the duodenum so they can emulsify dietary fat and help form micelles, which are small transport assemblies that keep lipid digestion products suspended in the watery environment of the gut. If bile flowed the wrong way, or leaked into the wrong compartment, digestion would fail and tissue injury would begin. Blood cannot do bile's job, and bile cannot do blood's job. Their separation is functional discipline made visible.

The pathological consequence is just as important. Problems on the blood side and problems on the bile side do not produce the same disease pattern. If blood flow is impaired, the organ may suffer ischemia, congestion, reduced exchange, or portal hypertension. If bile flow is impaired, retained bile acids and biliary constituents begin damaging hepatocytes and bile ducts, bilirubin accumulates, stool pigment changes, urine darkens, itching develops, and fat absorption worsens. One traffic system fails through inflow or outflow of blood. The other fails through impaired secretion or obstruction of bile. They intersect inside the same organ, but they must be read separately if the reader is going to understand liver disease with any real precision.

So the phrase blood in, bile out is more than a tidy summary. It is the governing logic of the organ. The liver receives chemical burden through blood, judges it, transforms it, stores part of it, releases selected products back into circulation, and sends other compounds away through bile. Opposite-direction traffic is how the organ keeps arrival and removal from becoming the same event. Incoming blood says what the body has absorbed, broken down, or returned for handling. Outgoing bile says what the liver has chosen to use for digestion or push toward disposal. The liver works because those two statements remain distinct.

Blood Flow Through the Liver Versus Bile Flow Out of the Liver

Blood and bile move through the same organ, but they are not participating in the same event. Blood flow is about delivery, exchange, and return. Bile flow is about secretion, concentration, and exit. Blood enters the liver carrying what has been absorbed, broken down, mobilized, or returned for handling. It arrives through the portal vein and hepatic artery, passes through the sinusoids, and leaves through the hepatic veins. Bile begins inside hepatocytes, enters tiny canaliculi formed between adjacent cells, moves into progressively larger ducts, and then leaves the liver through the biliary tree on its way to the gallbladder or small intestine. One stream is bringing chemical material to the liver for judgment. The other is carrying selected compounds away from the liver after that judgment has been made. If the reader confuses those two directions, the organ collapses into a blur. If the reader keeps them distinct, liver physiology begins to read clearly.

The direction of blood flow is shaped by the need to expose incoming chemistry to hepatocytes before that chemistry reaches the wider circulation. Portal venous blood arrives rich in nutrient load and digestive aftermath. Glucose, amino acids, microbial metabolites, hormone fragments, drug compounds, alcohol, bilirubin precursors, and a range of gut-derived signals enter the sinusoids alongside oxygen-rich arterial blood. These sinusoids are low-pressure exchange channels, not ordinary capillaries. They slow the flow enough for hepatocytes to pull compounds in, release others back out, and alter the chemistry of the passing blood before it collects into central veins and exits through the hepatic veins toward the inferior vena cava. Blood flow therefore crosses the liver as a processing route. It is delivering substrates, burdens, and signals to the organ, then carrying away a revised internal mixture after the liver has buffered glucose, handled ammonia, modified hormones, built proteins, or initiated biotransformation of foreign compounds.

Bile flow begins from a different side of the same cell. Hepatocytes are polarized, which means one membrane domain faces blood and another faces bile. That polarity is one of the liver's great architectural achievements. A hepatocyte can take bilirubin up from blood, conjugate it so it becomes more transportable, then secrete it out into bile. It can synthesize bile acids from cholesterol, assemble bile components, and pump them into canaliculi without letting those compounds spill back into the sinusoidal blood space. The canaliculi then drain into bile ductules and larger ducts that converge toward the extrahepatic biliary tree. Bile is not traveling through the liver to deliver oxygen or nutrients to tissue. It is leaving because the liver has produced it as an outgoing fluid. That fluid contains bile acids for fat digestion, phospholipids to stabilize the mixture, cholesterol for disposal or recycling dynamics, conjugated bilirubin for excretion, and other compounds selected for biliary removal. Blood flow feeds hepatic work. Bile flow expresses hepatic output.

This difference becomes even sharper when the reader asks what each stream is trying to accomplish physically. Blood has to remain within a vascular circuit. It is carrying dissolved compounds in a fluid built to perfuse tissue and then return to the heart. Bile is not part of that circulatory loop. It is a secreted fluid destined for the gut lumen, where it will either help digest fat or carry waste out toward fecal elimination. Blood is recirculating transport. Bile is one-way export with partial recycling later through enterohepatic circulation, the return of reabsorbed bile acids from the intestine back to the liver. Even when bile acids are recovered and reused, bile itself is still functioning as an outward stream. It leaves hepatocytes on purpose because fat digestion and certain forms of excretion require a route out of the vascular space and into the intestinal tract.

The pathology of the liver also changes depending on which stream is disturbed. When blood flow through the liver is impaired, the problems tend to center on delivery, exchange, pressure, and venous drainage. Portal hypertension, congestion, ischemic injury, and altered sinusoidal exchange belong mainly to the blood side of the story. When bile flow out of the liver is impaired, the problem shifts toward retention. Bile acids accumulate where they should not. Conjugated bilirubin backs up into blood. Stool loses pigment because less bilirubin reaches the intestine. Urine darkens because conjugated bilirubin is now being filtered by the kidneys. Fat digestion weakens because less bile reaches the duodenum. The tissue consequence is different because the traffic failure is different. A blood-flow problem deprives or distorts the liver's processing field. A bile-flow problem traps the liver's outgoing chemistry inside the wrong compartments.

So blood flow through the liver and bile flow out of the liver should never be taught as parallel versions of the same thing. Blood enters to be reviewed. Bile leaves because the review has produced something that must either aid digestion or be removed. Blood carries incoming burden and substrate. Bile carries outgoing digestive chemistry and excreted load. Blood moves through sinusoids toward venous return. Bile moves through canaliculi and ducts toward the intestine. The liver sits between those streams and keeps them from contaminating each other. That is one of the deepest structural truths of the organ. It survives by receiving chemistry through blood and answering with bile.

Why This Counter-Direction Design Matters

The liver has to do two opposite jobs in the same piece of tissue without letting them collapse into each other. It has to receive blood carrying the chemical aftermath of digestion, and at the same time it has to secrete bile carrying digestive tools and waste products away from that blood. That is why the counter-direction design matters. Blood enters through the portal tracts, moves through the sinusoids, and drains toward the central veins. Bile begins at the canalicular surface of hepatocytes and moves the other way, from the center of the working field back toward the bile ducts in the portal region. Those opposite directions keep intake and output from becoming chemically confused. The liver can extract compounds from blood, modify them, and then choose whether to release them back into circulation, store them, or push them into bile. Without directional separation, the organ would lose one of its most important powers, which is the ability to decide where a molecule belongs next.

This matters first at the level of membrane control. A hepatocyte does not face blood on all sides. One surface is built to exchange with sinusoidal blood. Another surface is built to secrete into bile. That polarity lets the cell do different chemistry at different edges. Bilirubin shows the logic cleanly. Unconjugated bilirubin arrives from blood bound to albumin. The hepatocyte takes it in on the blood-facing side, conjugates it so it becomes more transportable, then pumps it out the bile-facing side into the canalicular stream. The molecule crosses the cell, but it does not drift randomly through tissue. The architecture forces direction. The same basic logic applies to bile acids, cholesterol, phospholipids, and many conjugated compounds selected for biliary excretion. The liver is not only modifying molecules. It is assigning them to the correct exit route.

The counter-direction design also protects digestion itself. Bile is not a disposable side fluid. It is required for fat handling in the intestinal lumen. Bile acids and phospholipids have to leave the hepatocyte, enter the duct system, and reach the duodenum if the body is going to emulsify dietary fat and build micelles that can carry lipid digestion products to the absorptive surface. If bile flow were not kept distinct and directed, the chemistry needed for fat digestion would not reliably reach the intestine. At the same time, compounds that should be eliminated through bile, such as conjugated bilirubin and certain cholesterol-derived products, could leak into the wrong compartment. The organ would stop being a disciplined traffic controller and start behaving like a mixing chamber. That would be disastrous. Digestion would weaken, excretion would blur, and blood chemistry would become harder to stabilize.

It also matters because the liver is handling chemically incompatible worlds at once. Blood is an aqueous transport medium delivering glucose, amino acids, hormones, microbial fragments, drugs, and nitrogen burden in continuous flow. Bile is a detergent-rich secretion built to carry hydrophobic compounds, meaning compounds that do not move comfortably in water, toward the gut. These are different chemical environments with different purposes. The counter-direction design keeps those environments adjacent but separate. Hepatocytes can pull compounds from blood, transform them, and then place them into bile without exposing the circulation to inappropriate concentrations of biliary contents. That separation is especially important because bile acids are useful in the intestine but damaging when retained inside hepatocytes or leaked abnormally into blood. The body needs them in the lumen, not washing backward through the wrong spaces.

The clinical importance is just as large. When blood-side flow fails, the pattern of injury reflects impaired delivery, pressure handling, exchange, and venous drainage. When bile-side flow fails, the pattern reflects retention. Bilirubin rises. Bile acids accumulate. Itching develops. Stool loses pigment. Fat absorption worsens. Hepatocytes and bile ducts are exposed to compounds that should have moved out. The counter-direction design is what makes these syndromes distinguishable in the first place. Blood problems and bile problems are not interchangeable because the organ was built with separate directional systems for separate physiological purposes. Once one side backs up, the consequences follow the traffic pattern that has been broken.

So the deeper reason this design matters is that the liver is not merely processing chemistry. It is sorting chemistry into destinations. Some compounds need to stay in blood in altered form. Some need to be held in reserve. Some need to be sent to the kidneys after conversion. Some need to be secreted into bile and removed through the intestine. The counter-direction design gives the liver the physical architecture required to make those distinctions real. Blood brings the incoming problem. Bile carries part of the answer away.

Portal Blood as Nutrient-Rich but Potentially Hazardous Blood

Portal blood is rich because it is the first blood leaving the digestive tract after absorption, but that richness is exactly why it is dangerous. It carries the chemical results of the meal in concentrated form before the rest of the body has had a chance to dilute, store, or neutralize them. Glucose can be high after carbohydrate absorption. Amino acids arrive in waves after protein digestion. Fructose enters needing immediate hepatic handling. Water-soluble vitamins, minerals, organic acids, microbial metabolites, bilirubin-related compounds from splenic breakdown, drug molecules, alcohol, and hormone fragments can all appear in the same stream. This is not ordinary venous return. It is chemically crowded blood. The liver receives it first because the body cannot afford to let that post-digestive surge enter the general circulation without inspection. Portal blood is nutrient-rich in the sense that it carries absorbed substrate, but it is hazardous in the deeper physiological sense that the same stream carries excess, instability, and burden.

The hazard begins with concentration and timing. Nutrients do not leave the intestine in a slow even trickle designed around the needs of every tissue. They arrive according to the mechanics of gastric emptying, enzymatic digestion, transport capacity, meal composition, and intestinal transit. That means portal blood can deliver sharp surges rather than balanced supply. Glucose is a clear example. If large amounts of absorbed glucose were allowed to flow straight into systemic blood without first-pass hepatic uptake, blood sugar excursions would be harder to control and peripheral tissues would see a more violent metabolic hit. The liver acts as a buffer by extracting part of that load, converting some into glycogen, and releasing glucose later when needed. The same general logic applies to amino acids and fructose. Portal blood is therefore not dangerous because nutrients are bad. It is dangerous because unbuffered abundance is chemically disruptive.

The second hazard is that portal blood carries compounds that are not finished nutrition at all. It carries the consequences of digestion, not just the rewards. Gut microbes generate metabolites that can enter portal flow. Small amounts of bacterial structural material can cross from the intestine and travel to the liver. Ammonia precursors and other nitrogenous compounds can enter from intestinal and metabolic activity. Alcohol absorbed from the gut reaches the liver rapidly in portal blood. Drugs taken by mouth do the same. Plant compounds, food additives, and xenobiotics, foreign compounds the body does not make for its own structural use, can all arrive in this same stream. This is why portal blood should never be imagined as a clean nutrient river. It is a mixed report from the intestine. Some of its contents are useful, some need conversion, some need storage, some need to be excreted, and some need to be neutralized before they circulate further.

Its nutrient richness can even become hazardous by forcing the liver into metabolically expensive chemistry. Amino acids illustrate this well.

They are valuable building material, but when intake exceeds immediate synthetic need or when amino acids are used as fuel, nitrogen has to be removed. That pushes ammonia generation downstream, and ammonia is neurotoxic enough that the liver has to capture it quickly and channel it into the urea cycle. Fructose presents a different kind of burden. It is absorbed from the intestine and delivered heavily to the liver, where much of its initial handling occurs because the liver is the main first-pass site equipped to absorb that load. In both cases the same truth holds. A compound can arrive as nutrient and still create a processing burden if it arrives in excess, in the wrong pattern, or in a form that requires immediate biochemical control.

The immune hazard is just as important. The intestine is not sterile, and the barrier separating gut contents from blood is highly active rather than absolute. That means portal blood regularly carries signals that tell the liver something about the microbial life of the gut. Kupffer cells, the resident macrophages of the liver, sit in the sinusoids where portal blood first arrives and sample this incoming stream. They clear debris, respond to danger, and help maintain tolerance so the body does not launch full inflammatory war against every meal and every microbial fragment. But that tolerance has limits. If the intestinal barrier becomes more permeable, if microbial load shifts, or if inflammatory signaling rises, the hazard carried by portal blood increases. The liver then bears more of the cost. Portal blood is therefore not only a nutrient delivery route. It is an immunologic challenge line arriving from the gut every hour of the day.

This is why the liver's position downstream from the intestine is so exact. Portal blood has to be rich, because life depends on moving absorbed nutrients inward. But it cannot be trusted, because the intestine does not send forward only what the body wants in finished form. The liver receives this blood first so it can extract glucose, sort amino acids, process fructose, conjugate bilirubin, modify hormones, screen microbial traffic, and begin the handling of alcohol, drugs, and other foreign compounds before the chemistry of the meal becomes the chemistry of the whole organism. Portal blood carries value and threat in the same fluid. The liver exists where it does because those two cannot be separated upstream. They have to be separated there.

Arterial Blood as Oxygen Support for Constant Metabolic Work

The portal vein brings the liver its workload, but the hepatic artery helps make that workload survivable. Portal blood is rich in absorbed nutrients, gut-derived metabolites, microbial products, hormones, drugs, and the broader chemical aftermath of digestion, yet it is venous blood. That means it carries less oxygen than arterial blood arriving directly from the systemic circulation. The liver cannot run on chemical substrate alone. It is not enough for glucose, amino acids, and other compounds to arrive at the hepatocyte if the cell does not also receive the oxygen needed to extract energy from them and power its own internal machinery. This is why the hepatic artery matters. It delivers oxygen-rich blood to an organ that is performing continuous high-cost metabolic work. The portal vein tells the liver what needs to be done. The hepatic artery helps supply the oxygen required to do it.

That oxygen demand is enormous because hepatocytes are not passive storage cells. They are running dense, uninterrupted chemistry. They maintain membrane gradients, which are the differences in ion concentration and electrical charge that let the cell transport molecules in and out with control. They synthesize proteins such as albumin and clotting factors. They convert ammonia into urea through an energy-consuming pathway that spans both mitochondria and cytosol. They carry out gluconeogenesis, the building of glucose from noncarbohydrate substrates during fasting. They oxidize fatty acids, assemble triglycerides and lipoproteins, synthesize cholesterol and bile acids, conjugate bilirubin, operate detoxification systems in the smooth endoplasmic reticulum, and secrete bile components into canaliculi. None of this is metabolically cheap. The liver is chemically busy even when the person is sitting still and has not eaten for hours. It remains busy after meals, during fasting, during illness, during hormonal shifts, and during drug exposure. A tissue doing this much work cannot be perfused with low-oxygen blood alone and remain intact.

The liver's dual inflow solves that problem by letting portal venous blood and hepatic arterial blood meet in the sinusoids. These are specialized low-pressure vascular channels where exchange occurs between blood and hepatocytes. Portal blood brings the incoming substrate and burden. Arterial blood raises the oxygen content of the mixed sinusoidal flow so hepatocytes can sustain oxidative metabolism while processing what the portal stream has delivered. Oxidative metabolism means ATP generation through oxygen-dependent pathways inside mitochondria. ATP, adenosine triphosphate, is the main immediately usable energy currency of the cell. Without adequate ATP production, hepatocytes cannot maintain transporters, cannot hold ion gradients, cannot perform conjugation and synthesis efficiently, and cannot preserve the membrane polarity that separates blood-facing work from bile-facing secretion. The organ would still be receiving chemistry, but its ability to impose order on that chemistry would begin to fail.

This oxygen support also creates an internal gradient across the liver lobule. Blood enters from the portal region, where branches of the portal vein and hepatic artery lie together, and moves toward the central vein. The hepatocytes closest to the portal region see blood first and therefore live in a more oxygen-rich environment. Hepatocytes closer to the central vein see blood later, after some oxygen has already been consumed upstream. This means the liver is not oxygenated uniformly. Its own architecture creates zones with different metabolic conditions. The more oxygen-rich periportal cells are especially suited to certain oxidative and synthetic tasks, while the more downstream centrilobular cells operate in a relatively lower oxygen environment and carry different enzymatic profiles. This matters because oxygen delivery is not simply about keeping the organ alive in the abstract. It shapes which parts of the liver perform which jobs best and which parts fail first when perfusion drops or congestion develops.

The importance of arterial oxygen becomes most obvious when it is lost. A hepatocyte can be surrounded by nutrient-rich blood and still become dysfunctional if oxygen delivery is inadequate. Substrate without oxygen is not metabolic support. It is incomplete supply. The liver's core functions depend heavily on aerobic energy production, meaning energy production that uses oxygen. When oxygenation falls, ATP generation falls with it, transporter systems weaken, membrane integrity becomes harder to maintain, bile secretion can falter, and the tissue becomes more vulnerable to necrosis, which is uncontrolled cell death from injury. This is why ischemia, congestion, shock, or reduced arterial perfusion can produce characteristic patterns of liver injury even when blood is technically still present in the organ. Presence of flow is not enough. The quality of that flow matters, and oxygen content is one of the main reasons.

So arterial blood should not be thought of as a secondary backup because the portal vein already brings so much volume. The hepatic artery is the oxygen arm of liver support. It helps sustain the mitochondria, the energy-intensive synthetic work, the detoxification systems, the transport machinery, and the constant biochemical sorting that define the organ. The liver is not only a receiver of post-digestive chemistry. It is a high-output metabolic engine. Portal blood delivers the message from the gut. Arterial blood keeps the engine lit while the organ answers it.

The Liver as a High-Flow Exchange Organ Rather Than Passive Tissue

The liver is often pictured as a dense reddish mass sitting quietly under the ribs, as if its importance comes from what it contains rather than from what is constantly moving through it. That picture is wrong at the level that matters. The liver is not passive tissue waiting for chemistry to drift into it. It is a high-flow exchange organ built around continuous traffic. Huge volumes of blood are delivered to it every day through the portal vein and hepatic artery, and that blood is not arriving for simple nourishment of the tissue itself. It is arriving because the liver is positioned to inspect, alter, store, synthesize, and redirect what the rest of the body cannot safely absorb or circulate unfiltered. This is why its vascular architecture is so elaborate. The organ is designed less like a static reservoir and more like a controlled processing field where blood slows, spreads, meets specialized cells, and leaves chemically changed.

That distinction matters because exchange is the real event. Portal blood brings the aftermath of digestion in concentrated form. Arterial blood brings oxygen to support the metabolic cost of handling it. These two streams mix in the sinusoids, which are not ordinary capillaries. They are broad, low-pressure vascular channels built to expose flowing blood to hepatocytes with unusual efficiency. The sinusoidal endothelial cells are fenestrated, meaning they contain pores that allow plasma and dissolved compounds to move more freely toward the hepatocyte surface than they could across a tighter capillary wall. Between the sinusoid and the hepatocyte lies the space of Disse, a narrow exchange compartment where nutrients, hormones, metabolites, lipoproteins, and signaling molecules can move between blood and liver cells. This is not tissue sitting behind a wall. This is tissue built to make contact with passing chemistry at scale.

A passive organ could not do what the liver does because the liver's function depends on rapid decisions being made against moving blood. Glucose arrives after a meal and has to be buffered before systemic levels swing too far. Amino acids arrive and have to be routed toward protein synthesis, oxidation, or nitrogen disposal. Fructose arrives and is taken up heavily by hepatocytes. Microbial metabolites and bacterial fragments arriving from the gut have to be sampled and often cleared. Bilirubin has to be taken up from blood, conjugated, and moved toward bile. Hormones return for modification. Drugs arrive for first-pass metabolism. Lipid remnants are repackaged. Plasma proteins are synthesized and released. None of this resembles passive storage. These are active exchanges occurring while blood is moving, not after it has stopped. The liver lives in the stream. Its physiology depends on contact with flow.

The cell types inside the liver reinforce this same truth. Hepatocytes do the large-scale metabolic work, but they are not alone. Kupffer cells sit inside the sinusoidal space and sample portal blood for microbial debris and inflammatory signals. Liver sinusoidal endothelial cells regulate permeability and maintain the exchange environment. Stellate cells occupy the space of Disse and respond to injury, vitamin A storage demands, and matrix remodeling. All of these cells are positioned around blood flow, not away from it. Their jobs make sense only in a tissue that is constantly receiving, screening, and reformulating incoming chemistry. Even the damage patterns of the liver reflect this high-flow design. Congestion, ischemia, toxin injury, cholestatic retention, and portal hypertension all distort the organ by disturbing movement, gradients, and exposure, not just by harming a static piece of flesh.

This is also why the liver can fail dramatically when flow is impaired even if much of the tissue is still physically present. A liver deprived of proper inflow or trapped under poor outflow is not simply underfed. Its entire exchange architecture begins to collapse. Oxygen gradients shift. Sinusoids congest. Hepatocytes lose efficient access to the passing blood they are supposed to process. Bile secretion may falter because the energy and membrane polarity required for directional transport begin to weaken. The problem is not merely injury to cells in isolation. The problem is breakdown of an organ whose normal state is controlled throughput. A high-flow exchange organ can look anatomically present and still become functionally unavailable if circulation through its microarchitecture is distorted.

So the reader should stop picturing the liver as a quiet detox sponge. That image hides the mechanism. The liver is a dense, vascular, continuously working exchange organ built to stand in the path of incoming blood and decide what happens next. It receives chemistry in motion, not in theory. It extracts from that stream, adds to it, subtracts from it, redirects part of it into bile, and releases the rest back into circulation in altered form. Its importance comes from traffic, contact, and transformation. The liver is not defined by stillness. It is defined by what can happen there while blood is moving.

Chapter 4

The Microscopic Liver

What a Lobule Is

A liver lobule is the basic repeating unit of liver tissue, the smallest large-scale pattern you can recognize when you stop looking at the organ as one heavy red mass and start looking at how blood, bile, and cells are arranged for work. It is not a single cell type and it is not a little sac filled with liver fluid. It is a structural field. Within that field, hepatocytes are arranged in plates, sinusoids run between those plates carrying mixed portal venous and arterial blood, bile begins in tiny canaliculi formed between adjacent hepatocytes, and all of this traffic is organized around a directional logic. Blood tends to enter from the periphery of the lobule, where portal tracts sit, and move inward toward a central vein. Bile begins at the level of the hepatocyte and moves outward in the opposite general direction toward bile ducts located near the portal regions. So when you say lobule, you are naming a microscopic traffic pattern, not just a shape. The liver is built by repeating this pattern thousands of times.

Classically, the lobule is described as roughly hexagonal in outline when viewed in cross-section, with a central vein in the middle and portal triads at the corners. A portal triad contains three key structures traveling together: a branch of the portal vein, a branch of the hepatic artery, and a bile duct. Sometimes lymphatics and nerves are present nearby as well, but those three structures are the core teaching set because they announce the logic of the tissue. Blood comes in from portal and arterial branches, spreads through the sinusoids, and drains into the central vein. Bile is secreted by hepatocytes into canaliculi and eventually reaches the bile ducts located back near the portal regions. The hexagon description is useful because it gives the reader a first map, but the map matters more than the geometry. The point is not to memorize a six-sided figure. The point is to see that the microscopic liver is organized around inflow, exchange, and outflow.

Inside the lobule, hepatocytes are arranged in plates or cords, usually one cell thick in the healthy liver, with sinusoidal blood flowing in the spaces between them. This matters because the hepatocyte is placed close enough to the blood to sample what arrives quickly and close enough to neighboring hepatocytes to create bile canaliculi between cells. The sinusoidal lining is fenestrated, meaning it contains pores that allow plasma and dissolved compounds to move out of the flowing blood and toward the hepatocyte surface more easily than they could across a tight ordinary capillary wall. Between the sinusoid and the hepatocyte sits the space of Disse, a narrow exchange compartment where much of this transfer occurs. The lobule therefore is not a random clump of liver cells. It is a carefully arranged exchange zone that lets hepatocytes take up glucose, amino acids, bilirubin, hormones, microbial products, and drug compounds from blood while also releasing proteins back into blood and secreting other compounds into bile.

That arrangement is what makes the lobule such an important teaching unit. It is the first place where the reader can see that liver physiology depends on position. A hepatocyte near the incoming portal side of the lobule does not live in the same chemical environment as a hepatocyte near the central vein. The blood reaching those two cells differs in oxygen content, nutrient concentration, and exposure history because blood has already moved through part of the sinusoidal network by the time it reaches the central region. This is why the lobule is more than a histology diagram. It creates gradients. It creates zones. It creates the conditions that let different parts of the same tissue specialize in somewhat different tasks and become vulnerable to somewhat different injuries. The microscopic design is already preparing the reader for later concepts such as zonation, toxin patterns, ischemic injury, and metabolic compartmentation.

So the safest way to understand a lobule is this: it is the liver's local processing district. Blood enters at the edges, moves through an exchange network lined with working cells, and drains at the center. Bile begins in those same working cells and moves back toward the edge through a different route. Hepatocytes sit in the middle of both decisions. The lobule is the smallest unit in which that full logic can be seen clearly. Once the reader understands the lobule, the liver stops looking like a bag of chemistry and starts looking like organized tissue built to control moving blood, moving bile, and the molecular consequences of both.

What an Acinus Is and Why Function Is Not Identical Across the Tissue

The lobule gives the reader a clean structural map, but the acinus gives the reader a more useful functional map. A hepatic acinus is a way of looking at liver tissue through the logic of blood supply and metabolic exposure rather than through the neat symmetry of a hexagon. Instead of centering the picture on a central vein, the acinus centers it on the vascular entry point where terminal branches of the portal vein and hepatic artery deliver blood into the surrounding parenchyma, the working tissue of the organ. From that entry zone, blood spreads outward through the sinusoids and moves toward the central veins of neighboring lobules. The acinus therefore is the piece of tissue being fed by one small portal vascular unit and drained by central veins downstream. This matters because hepatocytes do not all encounter the same blood at the same time. Some cells see blood first, rich in oxygen and newly delivered substrates. Others see that same blood later, after some oxygen has already been consumed and some chemistry has already been altered. The liver is one organ, but it is not one uniform chemical environment.

That is why the acinus is so important. It makes the gradients visible. The tissue closest to the incoming portal blood is called zone 1, often referred to as the periportal region. These hepatocytes are first in line. They see the highest oxygen tension, the freshest nutrient delivery, and the earliest arrival of hormones, microbial products, and absorbed compounds returning from the intestine. Farther downstream sits zone 2, the intermediate region. Farthest from the incoming supply and closest to the central vein lies zone 3, often called the centrilobular region. These cells receive blood last. By the time blood reaches them, oxygen has fallen, nutrient proportions have shifted, and some upstream extraction has already taken place. This means the liver is not flat in function. It is graded. The acinus teaches that what a hepatocyte does depends partly on where it sits inside the flow path.

Once that gradient is seen, the idea that liver tissue behaves identically everywhere becomes impossible to maintain. Zone 1 cells, with better oxygen access, are well positioned for energy-intensive oxidative metabolism, gluconeogenesis, and early handling of blood arriving from the portal circulation. Gluconeogenesis is the synthesis of glucose from noncarbohydrate substrates such as lactate, glycerol, and certain amino acids. Zone 3 cells live in a more oxygen-poor environment, but they are also the cells exposed last to the downstream consequences of sinusoidal flow and are especially important in pathways such as glycolysis, lipogenesis, and much of cytochrome P450 mediated drug metabolism. Lipogenesis means the synthesis of fat, particularly fatty acids and triglyceride components. Cytochrome P450 refers to a major enzyme family used in chemical modification of drugs, toxins, and endogenous compounds. These regional differences are not academic fine print. They explain why one part of the liver is better at one kind of work while another part is more vulnerable to a different kind of injury.

The injury patterns make the point even more sharply. When oxygen delivery falls, zone 3 hepatocytes often suffer first because they already live at the downstream edge of the oxygen gradient. A fall in perfusion does not hit a perfectly equal tissue field. It strikes a tissue whose inner geography is already unequal by design. The same logic applies to toxic injury. If a toxin or drug becomes dangerous only after enzymatic activation, and those enzymes are concentrated more heavily in downstream hepatocytes, then the damage will tend to cluster there. In other cases, compounds arriving in highest concentration from portal inflow may strike periportal cells first because those cells meet the incoming chemistry before anyone else does. The acinus therefore turns pathology into something intelligible. Injury patterns are not random splashes on a microscope slide. They often follow the architecture of flow.

This is why function is not identical across the tissue. The liver is built around movement, and movement creates gradients. Gradients create zones. Zones create different workloads, different enzyme profiles, different oxygen availability, and different vulnerabilities. The acinus is the framework that lets the reader see all of that at once. The lobule teaches where blood enters and leaves and where bile travels. The acinus teaches that the same liver tissue changes character along the route of that blood. One organ. One repeating structure. Multiple local environments. Once the reader understands the acinus, the liver stops looking like a uniformly red factory and starts looking like a living metabolic terrain shaped by flow.

Portal Triads, Central Veins, Sinusoids, Canaliculi

The microscopic liver is easiest to read when four structures are fixed clearly in the mind: portal triads, sinusoids, central veins, and canaliculi. Those four are the traffic architecture of the tissue. Portal triads sit at the edge of the classical lobule and mark the incoming side of the system. Each triad contains a branch of the portal vein, a branch of the hepatic artery, and a bile duct, usually accompanied by lymphatics and nerves in the surrounding connective tissue. The portal vein branch brings nutrient-rich but relatively oxygen-poor blood from the digestive organs. The hepatic artery branch brings oxygen-rich arterial blood to support the metabolic cost of hepatic work. The bile duct stands there as the outgoing conduit for bile that has already been secreted by hepatocytes and is now leaving the microscopic field. So a portal triad is not just a cluster of tubes. It is the local entry point for blood and the local exit point for bile. That combination tells you immediately that the liver's smallest working units are organized around opposing streams.

From those portal inflow branches, blood enters the sinusoids. Sinusoids are the liver's exchange channels, and they are not ordinary capillaries. They are wider, lower-pressure vascular spaces built to slow blood down and expose it to hepatocytes with unusual efficiency. Their endothelial lining is fenestrated, meaning it contains pores that allow plasma and dissolved molecules to pass more freely toward the hepatocyte surface. Between that sinusoidal lining and the hepatocyte lies the space of Disse, the narrow exchange compartment where much of the actual transfer occurs. Glucose, amino acids, hormones, bilirubin, lipoprotein remnants, microbial metabolites, drug compounds, and other dissolved cargo move out of the flowing blood toward hepatocytes. At the same time, hepatocytes release newly synthesized proteins, glucose, ketone bodies, urea precursors and other processed outputs back toward the blood. Kupffer cells, the resident macrophages of the liver, sit in the sinusoidal space as well, sampling portal blood for bacterial fragments and other inflammatory signals. The sinusoid therefore is not a passive pipe. It is the place where incoming chemistry is inspected, altered, cleared, and redistributed while the blood is still moving.

As blood travels through the sinusoids, it moves inward toward the central vein. The central vein sits at the center of the classical lobule and functions as the local outflow channel. It collects the mixed blood that has already passed through the hepatocyte plates and sinusoidal network, then drains into larger veins that ultimately become the hepatic veins and return blood to the inferior vena cava. The central vein is therefore the downstream end of the exchange field. What reaches it is not chemically identical to what entered through the portal triads. By the time the blood arrives there, the liver may have extracted glucose, handled ammonia, modified hormones, conjugated bilirubin, repackaged lipids, cleared microbial debris, and released newly built plasma proteins or metabolites back into circulation. The central vein marks the completion of that local pass through the tissue. In simple terms, portal triads are where blood enters the working zone. Central veins are where that revised blood leaves it.

Canaliculi are the bile side of the same system. They are tiny channels formed between adjacent hepatocytes at the apical, bile-facing membrane of the cells. Apical here means the specialized surface of the hepatocyte built for secretion into bile, as opposed to the blood-facing surface that interacts with the sinusoid. Canaliculi are not independent tubes lined by separate duct cells at their origin. They are narrow grooves created by the hepatocytes themselves, sealed by tight junctions so bile stays in its proper compartment. Into these canaliculi hepatocytes secrete bile acids, phospholipids, cholesterol, conjugated bilirubin, and other biliary constituents. From there bile moves outward toward progressively larger ductules and eventually into the bile ducts of the portal triads. This is the crucial directional reversal in liver tissue. Blood enters near the portal triad and flows toward the central vein. Bile begins between hepatocytes deeper in the tissue and flows back toward the portal region. The same hepatocyte is participating in both systems at once, taking compounds from blood on one side and secreting compounds into bile on the other.

Once these four structures are seen together, the microscopic liver stops looking like random histology and starts reading like engineered tissue. Portal triads deliver inflow and collect biliary outflow. Sinusoids create the exchange field where blood moves slowly enough for hepatocytes and immune cells to act on it. Central veins drain the processed blood away. Canaliculi begin the outgoing bile stream at the level of the hepatocyte itself. Everything the liver does depends on the discipline of that arrangement. It can take up bilirubin from blood and send it into bile because the blood route and bile route are distinct. It can receive glucose, amino acids, and microbial products from portal blood while simultaneously secreting bile acids and conjugated waste in the opposite direction because its traffic architecture prevents confusion of compartments. Portal triads, central veins, sinusoids, and canaliculi are therefore not just landmarks to memorize. They are the physical grammar of liver function.

The Space of Disse and Why Exchange Happens There

The space of Disse is the narrow interface where blood-facing liver physiology actually happens. It lies between the sinusoidal endothelium, the thin cellular lining of the sinusoid, and the hepatocyte surface. The sinusoid is the specialized vascular channel carrying mixed portal venous and hepatic arterial blood through the liver. The hepatocyte is the main working cell that has to take up glucose, amino acids, bilirubin, hormones, lipoprotein remnants, microbial metabolites, and drug compounds from that blood while also releasing newly made proteins and other processed molecules back into it. Those two structures cannot simply be pressed tightly together like sealed plumbing because exchange would be strangled. The liver solves that problem by creating a narrow but open transfer compartment between them. That compartment is the space of Disse. It is small in width, but functionally it is one of the most important spaces in the organ because it is where blood-borne chemistry leaves the moving vascular stream and becomes available to the hepatocyte.

Exchange happens there because the sinusoidal wall is built to permit it. Unlike the tight continuous capillaries found in tissues that need a stronger barrier, the liver sinusoid is lined by fenestrated endothelial cells. Fenestrated means they contain pores. In the healthy liver these cells also lack the kind of thick continuous basement membrane that would otherwise create a more restrictive filter. So plasma, dissolved nutrients, hormones, lipoprotein particles of the right class, and many other solutes can pass out of the sinusoidal blood and enter the space of Disse with relative ease. Red blood cells remain within the vascular lumen, but the fluid phase of blood and the compounds dissolved or suspended in it can approach the hepatocyte closely. The hepatocyte surface facing this space is not smooth and indifferent. It projects microvilli, tiny membrane folds that increase surface area and improve contact with the exchanged fluid. More surface area means more transport opportunity. The liver therefore builds exchange into both sides of the interface. The sinusoid opens the gate. The hepatocyte expands the receiving surface.

This arrangement matters because the liver's workload depends on rapid high-volume transfer rather than slow occasional sampling. After a meal, glucose arrives in portal blood and has to move across this interface so hepatocytes can buffer part of the load and store some as glycogen. Amino acids move across it for protein synthesis or nitrogen handling. Unconjugated bilirubin reaches the hepatocyte from this blood-facing side before it can be conjugated and pushed toward bile. Hormones returning from circulation, microbial products arriving from the gut, and drugs absorbed from the intestine all meet the liver here first. At the same time, hepatocytes release albumin, clotting factors, lipoproteins, glucose during fasting, and many other products back across the same interface into the circulating blood. The space of Disse is therefore the liver's local trading floor. Compounds do not simply diffuse through solid tissue at random. They enter this narrow perisinusoidal compartment, meet a dense microvillous hepatocyte surface, and are then taken up, modified, stored, or released according to the cell's transporters and metabolic state.

The space also matters because it houses more than traffic alone. Stellate cells reside here in the healthy liver, sitting quietly in the perisinusoidal environment while storing vitamin A in lipid droplets. Under chronic injury they can transform into collagen-producing cells and begin laying down extracellular matrix, the structural material that accumulates during fibrosis. Once that matrix expands, the exchange space thickens and stiffens. The sinusoidal wall becomes less permissive, hepatocyte access to flowing plasma worsens, and the liver begins to lose one of its key advantages, which is fast low-resistance contact between blood chemistry and working cells. This is one reason fibrosis is so destructive even before the liver becomes grossly cirrhotic. The problem is not only scar in the abstract. The problem is that the exchange architecture is being buried. The blood can still arrive, but access to the hepatocyte becomes less clean, less efficient, and less immediate.

The space of Disse also contributes to lymph formation. Some of the fluid filtered out of the sinusoid enters this compartment and then moves onward into lymphatic drainage pathways. That detail matters because it shows again that the liver is a high-flow exchange organ, not a passive block of tissue. Blood is moving, plasma is filtering, solutes are crossing, hepatocytes are sampling, proteins are being secreted, and fluid is being redistributed continuously. The space of Disse sits inside that motion as the actual zone of contact. If the reader wants one microscopic location where liver blood handling becomes real, this is it.

The sinusoid carries the stream. The hepatocyte performs the work. The space of Disse is the place where those two can meet without losing their separate identities.

How Microscopic Architecture Determines Macroscopic Function

The liver does not perform large jobs because it is merely large. It performs large jobs because its microscopic structure has been arranged so that blood can be exposed, sampled, altered, and returned at extraordinary scale without losing directional order. That is the first principle the reader has to keep. The macroscopic functions people recognize, glucose buffering, ammonia control, bile production, bilirubin clearance, lipid packaging, hormone modification, plasma protein synthesis, immune filtering, all begin as microscopic placement problems. Which surface of the hepatocyte faces blood. Which surface faces bile. How close the flowing plasma can get to the cell. Where oxygen enters. How pressure is handled. Where outflow collects. Where bile exits. These are not histology details sitting underneath the real story. They are the real story in small form. The liver works as a whole because its smallest repeating structures have already solved the traffic problem.

Take blood handling first. The liver can buffer a meal because portal blood does not simply wash through a sealed capillary bed and disappear. It enters sinusoids, which are wide, low-pressure exchange channels lined by fenestrated endothelial cells, meaning cells with pores that allow plasma and dissolved compounds to move more freely toward hepatocytes. Between the sinusoid and the hepatocyte sits the space of Disse, the narrow exchange compartment where incoming chemistry meets the cell surface. Hepatocytes extend microvilli into that space, increasing surface area and improving contact with the passing plasma. That microscopic design is why the liver can strip glucose from portal blood after a meal, store some of it as glycogen, release glucose later during fasting, take up amino acids for synthesis or nitrogen handling, capture bilirubin, modify hormones, and begin first-pass drug metabolism before the rest of the body sees the full load. The macroscopic function, stabilization of blood chemistry, depends on microscopic openness, closeness, and controlled exposure.

The same principle governs bile formation. A hepatocyte is polarized, meaning one side is specialized for blood exchange and another side is specialized for bile secretion. That polarity is not decorative cell biology.

It is the reason the liver can take a compound in from blood and send it out toward the intestine without mixing the two streams. Bilirubin can be taken up from the sinusoidal side, conjugated so it becomes more transportable, then secreted across the canalicular side into bile. Bile acids synthesized from cholesterol can be directed into canaliculi, the tiny channels formed between adjacent hepatocytes, and then move toward bile ducts while blood continues toward the central vein. On the scale of the whole organ, that architecture becomes bile flow, fat digestion support, and biliary excretion. If hepatocyte polarity breaks down, if canaliculi are injured, or if the exchange between blood-facing and bile-facing surfaces becomes disordered, the liver does not merely lose a microscopic feature. It loses one of its largest visible functions. Jaundice, cholestasis, fat malabsorption, and retained biliary compounds all begin with small directional failures.

Microscopic architecture also determines metabolic specialization across the tissue. Blood does not enter the liver evenly everywhere at once. It arrives at portal regions and flows toward central veins, creating gradients in oxygen, substrate delivery, and chemical exposure. That means hepatocytes living near the portal tract encounter a different environment from hepatocytes living near the central vein. Some cells see higher oxygen and earlier nutrient delivery. Others see downstream blood with less oxygen and a different chemical profile. This is why liver function is zoned rather than flat. One region is better positioned for certain oxidative and synthetic tasks. Another region is more involved in downstream chemical handling, including much of the cytochrome P450 activity that modifies drugs and toxins. On the scale of the whole organ, this zonation becomes functional diversity. On the scale of disease, it becomes patterned vulnerability. Hypoxia, congestion, toxin injury, and fatty change do not strike a featureless field. They follow the internal geography created by sinusoidal flow.

The immune role of the liver is built the same way. Kupffer cells are placed inside the sinusoidal space so they can sample portal blood the moment gut-derived microbial products arrive. Sinusoidal endothelial cells help shape what crosses and how inflammatory contact is managed. Stellate cells sit in the space of Disse where they can support normal matrix conditions, store vitamin A, and then, under chronic injury, begin laying down collagen that stiffens the exchange field and degrades hepatic function. What the body later experiences as portal hypertension, fibrosis, poor exchange, synthetic decline, or impaired detoxification starts as microscopic changes in where cells are sitting and what kind of surface they are maintaining. Scar does not simply appear as a big structural problem. It first thickens the exchange interface, narrows sinusoidal freedom, and begins strangling the intimacy between blood and hepatocyte that whole-organ function depends on.

So when the liver is described macroscopically as a glucose buffer, a nitrogen-clearing organ, a bile-forming organ, an endocrine modifier, an immune filter, or a plasma-protein factory, each of those statements is really a compressed description of tissue architecture doing its job correctly. The large function is the visible outcome. The microscopic arrangement is the cause. The liver can manage the chemistry of the whole body because, at small scale, it has already arranged blood inflow, bile outflow, membrane polarity, oxygen gradients, immune surveillance, and exchange surfaces with ruthless precision. Macroscopic function is microscopic order multiplied across the organ.

Why the Reader Must Learn This Structure Before Learning Disease

Disease in the liver is not a collection of labels floating above the tissue. It is damage happening to an existing architecture. If the reader does not first understand what a lobule is, what an acinus is, where blood enters, where bile exits, what a sinusoid does, why the space of Disse exists, how hepatocytes face two different compartments, and how oxygen and substrate change as blood moves across the tissue, then every later disease chapter turns into memorization without mechanism. Jaundice becomes a yellow symptom instead of failed bilirubin handling across a polarized cell. Cholestasis becomes a diagnosis word instead of impaired bile formation or bile movement through a directional secretion system. Cirrhosis becomes "severe liver disease" instead of collapse of exchange architecture, distortion of sinusoidal flow, rising vascular resistance, and loss of organized hepatocyte function. The structure has to come first because the structure is what disease is acting on.

The liver especially punishes shallow learning because so many of its diseases look different on the surface while damaging the same underlying traffic system. One process injures hepatocytes directly. Another obstructs bile flow. Another congests venous outflow. Another overloads the tissue with fat. Another activates stellate cells and lays down collagen in the space of Disse. Another alters the portal blood arriving from the gut. To a reader without the structural map, these seem like separate stories. To a reader who knows the architecture, they become variations on a smaller set of failures. Blood cannot enter cleanly. Blood cannot cross the exchange field efficiently. Oxygen gradients become unstable. Hepatocytes lose membrane polarity. Bile cannot leave in the proper direction. Outflow pressure rises. Scar thickens the interface. The point of learning structure first is not academic discipline for its own sake. It is compression. The architecture reduces dozens of disease names to a smaller number of intelligible failure patterns.

This is also what lets the reader understand laboratory findings and symptoms later without treating them like trivia. Elevated aminotransferases mean more when the reader knows hepatocytes are being injured inside a tissue built around blood-facing and bile-facing domains. A cholestatic pattern means more when the reader can picture bile supposed to be moving through canaliculi and ducts but instead being retained. Hyperbilirubinemia means more when the reader sees bilirubin taken up from blood, conjugated inside the hepatocyte, then secreted into bile, and understands which part of that route has broken. Portal hypertension means more when the reader knows portal blood is supposed to move through a low-resistance sinusoidal field before draining into central and hepatic veins. Without structure, tests and symptoms sit on the page as disconnected facts. With structure, they become readable consequences.

There is another reason the order matters. The liver is easy to misteach because it is easy to summarize badly. People are told it detoxifies, stores glycogen, and makes bile, then they are hurried into hepatitis, fatty liver, cirrhosis, liver enzymes, and jaundice. That sequence produces recognition without understanding. The reader can repeat names but cannot see the organ. A better sequence begins with the tissue because tissue imposes limits on what disease can be. A hepatocyte can only fail in the compartments it actually has. Blood can only congest in channels that actually exist. Fibrosis can only injure function by distorting the real exchange surfaces and flow routes the organ depends on. Once the reader sees the microscopic liver as a built environment, disease stops being mysterious. It becomes damage to placement, direction, permeability, gradient, flow, and cell behavior.

So the reader must learn this structure first because the liver is an architectural organ before it is a disease list. Its illnesses are distortions of organized blood handling, bile secretion, exchange, zonation, and outflow. Teach the disease names before the structure, and the book becomes a glossary with anecdotes. Teach the structure first, and every later chapter has somewhere real to stand. The reader is no longer memorizing that the liver can fail. The reader can see exactly what kind of order had to break for that failure to occur.

Part II

The Cells That Make the Liver Work

Chapter 5

Hepatocytes: The Main Working Cell

What Hepatocytes Are

Hepatocytes are the main working cells of the liver, the cells that carry most of the organ's chemical burden and make most of its visible functions possible. When people say the liver stores glycogen, makes bile, converts ammonia into urea, conjugates bilirubin, builds albumin, modifies hormones, packages lipids, or performs first-pass handling of drugs and alcohol, they are usually talking about hepatocyte labor. The liver contains other essential cells, Kupffer cells, stellate cells, sinusoidal endothelial cells, cholangiocytes, but hepatocytes are the dominant parenchymal cells, meaning the principal functional tissue of the organ rather than its support framework. They make up most of the liver's cellular mass and most of its actual throughput. If the liver is imagined as a processing field, hepatocytes are the workers covering nearly all of that field.

What makes a hepatocyte different from a generic metabolic cell is that it is built for simultaneous traffic in more than one direction. One side of the hepatocyte faces the sinusoidal blood space, where portal venous blood and hepatic arterial blood bring nutrients, hormones, microbial products, bilirubin, lipoprotein remnants, drugs, and waste precursors into close contact with the cell. Another side faces the bile canaliculus, the tiny channel formed between neighboring hepatocytes where bile begins its outward path. That polarity matters because the hepatocyte is not simply soaking in blood and reacting to it. It is taking compounds in from one side, changing them internally, then deciding whether to release the result back into blood, store it, or secrete it into bile. Bilirubin may come in from blood and leave in bile. Glucose may come in from portal blood and leave later in fasting as newly released glucose. Cholesterol may be used for membrane needs, packaged into lipoproteins, or converted into bile acids. The hepatocyte is therefore not just the liver's main cell. It is the organ's main sorting cell.

Structurally, hepatocytes are large polygonal epithelial cells, meaning they are tightly arranged cells with organized membrane surfaces and specialized directional function. They usually sit in plates one cell thick, with sinusoidal blood flowing along one side and neighboring hepatocytes pressed alongside them. Their cytoplasm is dense with the machinery needed for heavy biochemical work. Mitochondria generate ATP, adenosine triphosphate, the immediately usable energy currency of the cell, to power urea synthesis, transport processes, and other energy-expensive reactions. Rough endoplasmic reticulum supports protein synthesis. Smooth endoplasmic reticulum is heavily involved in lipid metabolism and chemical biotransformation, including much of the cytochrome P450 activity used in drug handling. The Golgi apparatus helps process and package secreted products. Peroxisomes contribute to lipid oxidation and other specialized reactions. Glycogen granules and lipid droplets may be present depending on feeding state and metabolic load. A hepatocyte therefore looks like what it is. Not a quiet structural cell. A chemically crowded workstation.

Their job is to keep the chemistry of the internal environment usable from hour to hour. After a meal, hepatocytes take up portal glucose and convert some of it into glycogen so blood sugar does not rise as violently in the wider circulation. During fasting, they break glycogen back down and later build new glucose through gluconeogenesis, the synthesis of glucose from noncarbohydrate substrates such as lactate, glycerol, and certain amino acids. They receive amino acid traffic and help manage nitrogen by converting ammonia into urea. They synthesize bile acids from cholesterol, secrete bile components, conjugate bilirubin so it can be excreted, assemble lipoproteins for lipid transport, build plasma proteins such as albumin and clotting factors, and modify steroid, thyroid, and other hormone-related compounds. This is why hepatocytes have to be taught early and taught properly. They are not one liver cell type among many equivalent options. They are the central biochemical labor force of the organ.

They are also a good example of how liver function depends on architecture rather than on isolated chemistry alone. A hepatocyte can only do this amount of work because it sits in the right place, close enough to sinusoidal blood to sample incoming compounds rapidly, close enough to neighboring hepatocytes to form canaliculi for bile flow, and close enough to the space of Disse, the narrow exchange compartment between sinusoidal blood and the hepatocyte surface, to maintain high-volume transfer. Its function depends on contact, polarity, and flow. When fibrosis expands the extracellular matrix around the sinusoid, that contact worsens. When membrane polarity breaks down, bile secretion suffers. When oxygen delivery falls, ATP production falls and synthetic work weakens. When toxic burden rises beyond handling capacity, intracellular stress accumulates and hepatocyte injury begins. So to understand a hepatocyte is already to understand a large part of liver disease. A failing liver is often, at its core, a field of hepatocytes losing the structural conditions required to keep blood, bile, and metabolism in order.

The cleanest way to think about hepatocytes is this. They are the liver's principal decision-making cells, but they do not make decisions in the abstract. They make them chemically. They take in what blood brings. They alter it. They store part of it. They build what the body needs from it. They secrete what must leave by bile. Almost every major function of the liver passes through them in one form or another. If the organ is a border between digestion and circulation, hepatocytes are the cells standing on that border, deciding what form of incoming chemistry the body will be allowed to live with.

Their Polarity and Why One Side Faces Blood While Another Faces Bile

A hepatocyte is not built like a generic round cell floating in fluid. It is polarized, which means different parts of its membrane are assigned to different jobs and are physically organized to keep those jobs separate. One surface faces the sinusoid, where blood is moving past carrying glucose, amino acids, bilirubin, hormones, microbial products, lipoprotein remnants, drugs, and countless other compounds arriving from the portal and arterial circulation. Another surface faces the bile canaliculus, the narrow channel formed between adjacent hepatocytes where bile begins its outward route toward the bile ducts and eventually the intestine. That separation is not a minor structural detail. It is the reason the liver can take something in from blood, change it inside the cell, and then choose whether the result belongs back in blood or out in bile. Without polarity, the hepatocyte would lose directional authority. It could still perform chemistry, but it would struggle to assign products to the correct compartment.

The blood-facing side exists because the liver has to interrogate moving plasma continuously. Compounds arriving in the sinusoid are not all equal and they are not all destined for the same fate. Some need immediate uptake for storage, like glucose after a meal. Some need biochemical conversion, like ammonia-derived nitrogen being pushed toward urea synthesis or bilirubin being prepared for biliary excretion. Some need to be modified and returned to circulation, such as newly released glucose during fasting, plasma proteins, lipoproteins, or hormone metabolites. The sinusoidal membrane is therefore equipped for exchange. It sits adjacent to the space of Disse, the narrow compartment between sinusoidal blood and the hepatocyte, and it carries transporters and receptors that let the hepatocyte sample, import, and export compounds rapidly. Microvilli on this surface increase area and improve contact with the passing plasma. The blood-facing side is built for intake, sensing, and selective release into the internal circulation.

The bile-facing side exists because some liver work cannot end in blood without becoming harmful or useless. Bile acids, phospholipids, cholesterol, and conjugated bilirubin need to be secreted into a separate stream that will leave the liver through the biliary tree. Conjugated means chemically modified into a more transportable form. Bilirubin is a clean example of why the two surfaces must remain distinct. It arrives at the blood-facing side largely in an unconjugated form bound to albumin. The hepatocyte takes it up, modifies it internally so it becomes more water-compatible and better suited for controlled secretion, then pumps it across the bile-facing membrane into the canaliculus. The canaliculus is not blood. It is the beginning of an excretory and digestive route. If bilirubin were sent back toward the wrong side, clearance would fail. The same is true for bile acids synthesized from cholesterol. They are useful in the gut lumen for fat digestion, but damaging if retained in excess inside hepatocytes or misdirected into the wrong compartment. The hepatocyte needs a dedicated membrane domain for secretion into bile because the body needs a route that is not the bloodstream.

This membrane polarity also lets the liver run two opposite traffic systems through the same tissue without letting them contaminate each other. Blood is an incoming review stream. It brings the chemical aftermath of digestion and systemic metabolism to the liver for handling. Bile is an outgoing decision stream. It carries compounds away from the hepatocyte toward the intestine for fat digestion or elimination. The hepatocyte sits between those streams like a two-sided customs station. One side receives the incoming load. The other side dispatches selected cargo outward. Tight junctions between adjacent hepatocytes help seal the canalicular space so bile does not leak backward into the blood-facing environment. That barrier is crucial. Bile is a detergent-rich secretion. It is useful where it belongs and injurious where it does not. Polarity is what keeps the blood compartment and the biliary compartment from dissolving into one confused chemical field.

The consequence is that a hepatocyte can perform sequence, not just reaction. It can take in a molecule from blood, alter it through enzymatic work, sort it internally, then send it either back to circulation, into storage pathways, or across the apical canalicular membrane into bile. Apical here means the bile-facing membrane domain. Basolateral refers to the blood-facing and lateral surfaces involved in exchange with the sinusoid and neighboring structures. That sequence is what makes the liver an organ of routing rather than simple metabolism. A cell without polarity can still change chemistry. A cell with polarity can change chemistry and direct the outcome. That is a much more powerful design.

This is also why polarity failure is a real disease mechanism rather than a microscopic curiosity. If the hepatocyte loses membrane organization, transporters are misplaced, tight junctions weaken, canalicular secretion suffers, and biliary compounds can accumulate or move in the wrong direction. The result may be cholestasis, which means impaired bile formation or flow, or more diffuse breakdown of hepatic handling. A cell that can no longer keep its blood side and bile side distinct cannot maintain the liver's most basic promise, which is that incoming chemistry will be judged and routed correctly. The hepatocyte works because one side faces blood and another faces bile. That split is not an inconvenience the cell has to manage. It is the reason the organ can do its job at all.

Their Massive Metabolic Workload

A hepatocyte is busy in a way most cells are not. It does not serve one narrow tissue purpose and then wait for instructions. It sits in the path of continuous chemical traffic and has to keep acting on that traffic minute by minute without losing directional control. Portal blood arrives carrying glucose after a meal, amino acids after protein digestion, microbial metabolites from the gut, bilirubin from heme breakdown, hormone fragments returning for modification, lipoprotein remnants, alcohol, drugs, and a wide range of endogenous and foreign compounds that the body cannot simply leave unexamined. At the same time the hepatocyte is already maintaining ion gradients across its membranes, already synthesizing plasma proteins, already generating bile components, already storing or mobilizing glycogen depending on feeding state, and already handling part of the body's nitrogen burden through urea production. This is why the hepatocyte has to be taught as a high-throughput cell. It is not occasionally metabolic. It is constitutively metabolic, meaning the workload is built into its normal existence.

Its energy demand reflects that burden. The hepatocyte has to make adenosine triphosphate, ATP, in large amounts because so much of its work depends on active transport and enzyme-driven conversion. Molecules do not simply drift into the right compartment on their own. Bile acids have to be secreted into canaliculi against gradients. Bilirubin has to be taken up, conjugated, and exported. Ammonia has to be trapped and converted into urea through reactions spanning the mitochondria and cytosol. Gluconeogenesis, the synthesis of glucose from noncarbohydrate substrates such as lactate, glycerol, and certain amino acids, is energetically expensive. Protein synthesis is expensive. Lipoprotein assembly is expensive. Even maintaining membrane polarity is expensive, because transporters have to stay in the correct domains and ionic differences across the membrane have to be preserved continuously. The hepatocyte therefore is not merely surrounded by chemistry. It is paying an energy cost to keep that chemistry sorted.

The workload is also massive because the cell has to switch functions without changing identity. In the fed state, hepatocytes extract portal glucose and convert part of it into glycogen, synthesize fatty acids and triglyceride components from excess substrate when appropriate, and process incoming amino acids and fructose. In fasting, that same cell shifts toward glycogenolysis, which is glycogen breakdown, then toward gluconeogenesis as glycogen reserves fall. It increases fatty acid oxidation, supports ketone production, and keeps blood glucose available for tissues that depend on it. During inflammation it changes the profile of proteins it produces, increasing acute-phase reactants while other priorities shift. During drug exposure it may increase biotransformation load through cytochrome P450 systems in the smooth endoplasmic reticulum. During cholestatic stress it may be forced to handle retained bile constituents that are themselves injurious to membranes. The same hepatocyte is therefore not running one static program. It is rebalancing pathways according to feeding, fasting, infection, endocrine signals, toxic exposure, and overall organism need.

What makes this even harder is that the hepatocyte is often working on compounds that can damage it while it is processing them. Ammonia is dangerous before it becomes urea. Bilirubin is not yet ready for excretion before conjugation. Bile acids are useful in the intestinal lumen but irritating and potentially toxic when retained inside hepatic tissue at the wrong concentration. Drug metabolism can generate reactive intermediates, which are chemically aggressive intermediate products formed during transformation. Fatty acid overload can swell the cell with lipid droplets and distort organelle function. Oxidative stress can rise when mitochondrial throughput, detoxification demand, and redox burden begin outrunning the cell's protective systems. Redox refers to the balance between oxidation and reduction reactions that determines how electrons are being handled inside the cell. The hepatocyte therefore does not simply process raw materials into useful outputs. It often has to survive the chemistry of transition while doing so.

This is why hepatocyte failure carries such wide consequences. When a cell with this much workload begins to lose ATP, lose membrane polarity, accumulate fat, swell with oxidative stress, or fall behind on transport and synthesis, the liver does not just become slightly less efficient. Blood chemistry changes. Bile flow changes. Nitrogen disposal changes. Hormone handling changes. Plasma protein production changes. Drug clearance changes. The body loses one of its main chemical stabilizers. The hepatocyte earns its status as the liver's main working cell because it is doing the labor that keeps the post-digestive world from spilling into systemic chaos. Its workload is massive because the body keeps sending it problems that cannot be postponed.

Why Hepatocytes Are Built for Synthesis, Packaging, Conversion, and Secretion

A hepatocyte is built the way it is because the liver cannot survive as a mere receiving surface. Blood arrives carrying raw material, but raw material is not enough. The body needs proteins that do not arrive ready-made from the intestine, transport particles that must be assembled before lipids can circulate safely, bile constituents that have to be manufactured and secreted into a separate compartment, and waste products that often need chemical modification before they can leave the body without injuring tissue on the way out. The hepatocyte therefore is not designed as a passive absorber. It is designed as a biochemical factory cell. It has to take in heterogeneous incoming chemistry, pull it apart, rebuild it into new forms, direct it toward blood or bile, and do all of that continuously while maintaining its own membrane gradients, redox balance, and polarity. Redox refers to the cell's management of electron transfer, one of the core determinants of how well it can run energy production and detoxification without drifting into oxidative stress.

Synthesis is central because the liver does not simply redistribute nutrients that came from food. It manufactures essential internal products the rest of the body depends on. Hepatocytes synthesize albumin, clotting factors, complement proteins, acute-phase reactants, apolipoproteins, bile acids, cholesterol, phospholipids, glucose during fasting, and many signaling-related proteins and intermediates that shape systemic physiology. This is why the cytoplasm of the hepatocyte is crowded with rough endoplasmic reticulum, smooth endoplasmic reticulum, Golgi apparatus, mitochondria, and other organelles needed for heavy biosynthetic work. Rough endoplasmic reticulum supports protein production. Smooth endoplasmic reticulum supports lipid metabolism and chemical transformation. The Golgi modifies, sorts, and packages products before they are secreted. Mitochondria supply ATP, adenosine triphosphate, the immediately usable energy currency required to keep the entire system moving. A cell that has to build this much cannot be architecturally simple.

Packaging is equally important because many compounds cannot move through the body in their raw digestive form. Lipids are the clearest example. Fat cannot simply dissolve into aqueous blood and drift toward tissues in a stable usable way. Hepatocytes therefore package triglycerides, cholesterol, phospholipids, and apolipoproteins into lipoprotein particles so hydrophobic cargo can travel through plasma without precipitating into chaos. Hydrophobic means poorly compatible with water. The same logic appears in protein secretion and bile formation. The liver has to place compounds into the correct transport format if they are going to move through blood, bile, or tissue safely. Packaging is not decorative wrapping. It is structural permission. Without it, circulation would receive compounds in the wrong physical form, with poor solubility, poor targeting, poor stability, or direct toxicity.

Conversion is what gives the hepatocyte its authority over the post-digestive stream. Glucose can be converted into glycogen for storage. Glycogen can be broken back down. Lactate, glycerol, and amino acid carbon skeletons can be converted into glucose through gluconeogenesis. Ammonia can be converted into urea so the kidneys can excrete nitrogen safely. Bilirubin can be conjugated so it becomes compatible with biliary secretion. Cholesterol can be converted into bile acids. Drugs, steroid hormones, and foreign compounds can be oxidized, reduced, hydrolyzed, conjugated, and otherwise transformed so their behavior inside the body changes. This is why hepatocytes need such dense enzymatic machinery. The body is not asking them to hold chemistry still. It is asking them to change the fate of that chemistry. A molecule arriving from blood may be unusable, unstable, excessive, or dangerous in its incoming form. The hepatocyte exists to make it something else.

Secretion completes the logic because much of what the hepatocyte builds or modifies must leave the cell in a disciplined direction. Albumin and clotting factors are secreted into blood. Bile acids, phospholipids, cholesterol, and conjugated bilirubin are secreted into canaliculi, the tiny bile channels formed between adjacent hepatocytes. Urea is released back into blood so the kidneys can eliminate it. Lipoproteins are secreted into circulation so lipids can reach distant tissues. This is why hepatocyte polarity matters so much. One membrane domain faces the sinusoidal blood side. Another faces the bile canalicular side. The same cell can therefore receive compounds from blood, transform them internally, then secrete the result either back into blood or outward into bile depending on what physiology demands. A nonpolar cell could still perform chemistry. It could not perform routing with this degree of precision.

The deeper reason hepatocytes are built for these four jobs is that the liver sits at the point where digestion ends and internal chemistry begins. What comes from the intestine is not ready-made tissue support. It is mixed substrate, burden, signal, and waste precursor. The body needs a cell type capable of turning that unstable incoming mixture into organized biological outputs. That is exactly what the hepatocyte is. It synthesizes what the body lacks, packages what cannot circulate raw, converts what cannot stay as it arrived, and secretes what must move on. The liver works because hepatocytes are built not merely to receive chemistry, but to re-author it.

The Organelles That Matter: Mitochondria, Endoplasmic Reticulum, Peroxisomes, Lysosomes, Golgi

A hepatocyte only makes sense when the reader stops seeing it as one cell and starts seeing it as a crowded internal district built for heavy throughput. The liver's major functions are too chemically expensive and too mechanically organized to be carried by an empty sac of cytoplasm. A hepatocyte has to generate large amounts of energy, synthesize proteins and lipids, modify hormones and xenobiotics, package molecules for secretion, digest damaged internal components, and keep reactive chemistry from tearing the cell apart while all of that is happening. Xenobiotics are foreign compounds, substances the body does not make for its own structure or signaling. That workload is distributed across organelles, the specialized internal compartments of the cell. Mitochondria, endoplasmic reticulum, peroxisomes, lysosomes, and Golgi are not background details in a textbook cell diagram. They are the working hardware of hepatocyte physiology. When the liver fails, these compartments are often where the failure begins.

Mitochondria matter first because the hepatocyte is an energy-hungry cell running constant transport, synthesis, and conversion. Mitochondria are the organelles that generate most of the cell's ATP, adenosine triphosphate, through oxidative phosphorylation, the oxygen-dependent process that turns fuel-derived electrons into usable chemical energy. The hepatocyte needs that ATP to maintain membrane gradients, to run transporters that move bile acids and other compounds into the correct compartment, to support urea cycle chemistry that traps nitrogen in excretable form, and to sustain gluconeogenesis, the synthesis of glucose from noncarbohydrate substrates such as lactate, glycerol, and certain amino acids. Mitochondria also handle major parts of fatty acid oxidation, redox balance, and apoptotic signaling, the internal pathways that can commit a damaged cell to controlled death when injury becomes too great. A hepatocyte with weak mitochondria may still be present under the microscope, but it begins losing the energetic authority to remain a liver cell in full working order. Bile secretion weakens. Nitrogen handling weakens. Oxidative stress rises. Fat accumulates more easily because fuel cannot be oxidized cleanly. The whole organ starts drifting toward backlog.

The endoplasmic reticulum is the next major compartment because the hepatocyte is both a factory and a chemical transformation site. The rough endoplasmic reticulum is studded with ribosomes and specializes in protein synthesis, especially proteins destined for secretion or membrane insertion. This is where the hepatocyte begins producing albumin, clotting factors, complement proteins, transport proteins, and many other exported products that shape plasma behavior across the body. The smooth endoplasmic reticulum handles a different class of labor. It is deeply involved in lipid synthesis, cholesterol metabolism, bile acid related chemistry, calcium handling, and biotransformation of drugs, hormones, and other foreign or endogenous compounds. Much of the cytochrome P450 system sits here. Cytochrome P450 enzymes oxidize compounds, often making them more reactive first so they can later be conjugated into a more water-compatible or excretable form. This is why hepatocytes need so much smooth endoplasmic reticulum. The liver is not merely storing incoming chemistry. It is rewriting it. When endoplasmic reticulum stress rises, which means protein folding, lipid handling, or detoxification demand begins exceeding the compartment's capacity, the hepatocyte starts losing precision. Misfolded proteins accumulate. Membrane lipid composition shifts. Toxic intermediates can build. A factory organelle under chronic strain becomes a source of injury.

Peroxisomes matter because not all oxidative work belongs inside mitochondria. These small organelles handle specialized oxidation reactions, especially very-long-chain fatty acid metabolism, certain steps in bile acid synthesis, and detoxification of hydrogen peroxide through catalase and related enzymes. Hydrogen peroxide is a reactive oxygen species, a chemically active oxygen-containing molecule that can damage proteins, lipids, and DNA if it accumulates in the wrong place. Peroxisomes therefore help the hepatocyte perform oxidation without letting every oxidative reaction become a diffuse cellular hazard. They also contribute to plasmalogen synthesis and other lipid pathways that matter for membrane structure. In liver disease, peroxisomal strain or insufficiency can distort lipid handling and raise oxidative burden. The important principle is that oxidation inside the hepatocyte is compartmentalized on purpose. The cell does not dump all reactive chemistry into one shared interior soup. It assigns different oxidative tasks to different organelles so fuel handling, detoxification, and membrane biology can proceed without excessive collateral damage.

Lysosomes are the degradative side of hepatocyte order. They are acidic digestion compartments filled with enzymes that break down damaged proteins, worn-out organelles, endocytosed material, and other cellular debris. The hepatocyte depends on them because a cell doing this much synthesis, transport, oxidation, and chemical conversion is constantly generating components that need to be cleared or recycled. Lysosomes participate in autophagy, the process by which the cell captures and degrades parts of itself to remove damaged structures and recover usable building blocks. This is especially important when mitochondria are injured, proteins misfold, or lipid burden rises. Without lysosomal function, damage accumulates faster than the cell can reset its own interior. Debris that should have been cleared remains in circulation inside the cytoplasm, and the hepatocyte becomes progressively more disordered from within. In that sense lysosomes protect not by building or converting, but by preventing internal congestion. A hepatocyte can only keep processing the body's chemistry if it can also digest the wreckage created by that processing.

The Golgi apparatus is the finishing and sorting station. Proteins and lipids made in the endoplasmic reticulum do not simply emerge in final usable form. They often require modification, glycosylation, concentration, and packaging before they can be secreted or inserted into the correct membrane domain. Glycosylation means the controlled addition of carbohydrate groups to proteins or lipids, often changing stability, recognition, and trafficking. The Golgi takes what the cell has made and decides where it goes next. In a hepatocyte that means directing proteins toward secretion into blood, placing transporters into the correct membrane domain, and supporting the polarized architecture that keeps the blood-facing side and bile-facing side distinct. This is one reason Golgi function matters so much in the liver. The hepatocyte is not only making products. It is routing them. A secretion organ without a reliable sorting system would lose direction. Albumin might still be synthesized, but secretion would falter. Transport proteins might still be made, but membrane polarity would blur. Bile-facing and blood-facing identity would begin to degrade.

Taken together, these organelles explain why the hepatocyte can do so much and why it fails so widely when internal order breaks. Mitochondria supply energy and redox control. Endoplasmic reticulum builds proteins and lipids while transforming hormones and xenobiotics. Peroxisomes carry specialized oxidation and help restrain reactive oxygen damage. Lysosomes clear internal wreckage and recycle material so the cell does not drown in its own workload. The Golgi modifies, sorts, and dispatches what the cell has made. None of these organelles are optional support cast. They are the internal architecture that lets a hepatocyte synthesize, package, convert, and secrete at the scale required by the liver. When the reader understands them, liver disease becomes easier to read. The failing organ is often a field of hepatocytes whose internal districts are no longer holding their posts.

Why Hepatocyte Injury Creates System-Wide Consequences

Hepatocyte injury spreads beyond the liver because hepatocytes are not servicing the liver alone. They are servicing the blood. They sit in the path of portal inflow, receive the chemical aftermath of digestion, reshape it, and then send altered blood back into the systemic circulation. That means damage to hepatocytes does not remain trapped inside a local tissue compartment the way damage to a structural cell might. It changes what leaves the organ. If the hepatocyte cannot take up bilirubin properly, bilirubin rises in blood. If it cannot conjugate and secrete bile constituents properly, biliary compounds accumulate in tissue and circulation. If it cannot trap ammonia in the urea cycle, nitrogen burden escapes hepatic control and reaches the brain. If it cannot build albumin and clotting factors, the composition of plasma changes. The consequence is immediate in principle even when it is gradual in appearance. Hepatocyte injury becomes systemic because the hepatocyte is one of the main cells deciding what kind of blood the rest of the body receives.

This is easiest to see when the reader stops thinking of liver injury as inflammation in the abstract and starts thinking in terms of lost tasks. After a meal, hepatocytes normally pull glucose out of portal blood and buffer the post-absorptive surge. During fasting, they release glucose from glycogen and later build new glucose through gluconeogenesis. They receive amino acid traffic, remove nitrogen when necessary, and convert ammonia into urea so the kidneys can excrete it safely. They take up unconjugated bilirubin from blood, alter it into a more transportable form, and secrete it into bile. They synthesize albumin, clotting factors, apolipoproteins, complement proteins, and many other plasma components that shape vascular behavior, lipid transport, and inflammatory response. They modify steroid hormones, thyroid-related compounds, drugs, alcohol, and foreign chemicals. Once those tasks are seen clearly, hepatocyte injury stops reading like damage to one more organ cell. It becomes loss of blood chemistry control, fuel timing control, nitrogen control, biliary control, and plasma-building capacity all at once.

The metabolic consequences are broad because hepatocytes sit at the center of post-digestive handling. If they are swollen, oxidatively stressed, ATP-depleted, fat-loaded, or dying, they stop imposing order on incoming substrate. ATP, adenosine triphosphate, is the cell's immediately usable energy currency, and without enough of it transport, synthesis, and directional secretion begin to fail. Glucose handling becomes less stable. Glycogen storage and mobilization become less reliable. Lipid traffic becomes harder to package cleanly, and triglycerides may accumulate inside the cells rather than being oxidized or exported efficiently. Hormone modification shifts. Drug handling shifts. A person with hepatocyte injury is therefore not dealing with a liver problem in isolation. They are dealing with altered systemic metabolism because one of the main organs that rewrites incoming nutrient traffic has begun to fall behind.

Nitrogen handling makes the danger even more visible. Ammonia is generated from amino acid metabolism and gut microbial activity, and it becomes dangerous quickly if it is not converted into urea. Hepatocytes are the cells that perform that conversion. When they lose enough functional capacity, ammonia is no longer contained at the liver. It enters systemic circulation and begins to disrupt brain function. Confusion, slowed thinking, sleep reversal, asterixis, and hepatic encephalopathy sit downstream of this failure. The brain does not care that the original problem began in a hepatocyte. It experiences the consequence as toxic chemistry reaching neural tissue. This is one of the clearest examples of why liver injury cannot be treated as local pathology. The cell that failed was hepatic. The organ that begins to malfunction visibly may be the brain.

The same outward spread happens with bile and bilirubin. Hepatocytes are polarized cells. One membrane domain faces blood. Another faces bile canaliculi, the tiny channels between adjacent hepatocytes where bile begins its outward route. If injury disrupts membrane polarity, transporter placement, or canalicular secretion, compounds that should have been sent into bile are retained or misdirected. Bilirubin accumulates in blood and tissues. Bile acids and related compounds can accumulate where they do not belong. Fat digestion worsens because the intestine receives less of the bile chemistry it depends on. That means hepatocyte injury can create both a systemic blood chemistry problem and a digestive failure at the same time. The same damaged cell that failed to clear bilirubin also failed to support the next meal.

The plasma consequences are just as serious. Hepatocytes do not simply clean blood. They help build it. Albumin is made in hepatocytes and released into circulation, where it helps maintain oncotic pressure, the protein-driven force that keeps fluid inside the vasculature. Clotting factors are also made largely by hepatocytes. When synthesis drops, fluid begins to misplace more easily, edema and ascites become more likely, and coagulation becomes unstable. Drug binding can change as albumin changes. Inflammatory signaling can change as hepatic protein output changes. The blood itself becomes a less stable medium. So even when hepatocyte injury is silent at first, the organ is already beginning to lose one of its core systemic roles, which is maintaining a plasma environment compatible with organized circulation.

This is the deeper reason hepatocyte injury creates system-wide consequences. Hepatocytes are not local support cells buried inside an organ with narrow responsibilities. They are high-throughput interface cells standing between the intestine and the bloodstream, between nitrogen release and renal excretion, between bilirubin production and biliary removal, between nutrient arrival and metabolic timing, between raw chemical load and blood worth circulating. Damage enough of them, and the body does not merely lose liver tissue. It loses one of the main cellular systems that keeps the internal environment coherent.

Chapter 6

Sinusoids and the Exchange Surface

A sinusoid is not just a wider capillary with a different name. It is a specialized exchange channel built for a different class of work. Ordinary capillaries are usually designed to serve tissue in a more limited way. They deliver oxygen, carry away carbon dioxide and waste, and allow controlled exchange of water and small solutes according to the needs of the surrounding organ. Many capillary beds are built to preserve tighter boundaries because the tissue they serve cannot tolerate indiscriminate exposure. The liver cannot operate under that kind of narrow-contact design. It sits downstream from the intestine and has to inspect, extract, modify, package, store, and redirect a dense chemical stream arriving from portal blood. That requires a vascular channel that does more than nourish the tissue. It has to expose the blood to the tissue in a sustained and permissive way. The sinusoid exists for that reason.

The first difference is structural openness. Ordinary continuous capillaries have a more restrictive endothelial lining and a clearer basement membrane, which is the thin supportive layer beneath the endothelial cells that helps define the vascular barrier. Liver sinusoids are different. Their endothelial cells are fenestrated, meaning they contain pores, and the barrier between flowing plasma and hepatocyte is far more permissive. That looseness is not a flaw. It is the design. Plasma and dissolved compounds can leave the flowing blood more freely and enter the space of Disse, the narrow exchange compartment between the sinusoidal wall and the hepatocyte surface. That means glucose, amino acids, bilirubin, hormones, lipoprotein remnants, microbial metabolites, drugs, and many other compounds do not have to fight through a tightly guarded barrier before the hepatocyte can act on them. The sinusoid is built to make contact easier because the liver's job depends on contact.

The second difference is the flow environment itself. Sinusoids are wider and more irregular than ordinary capillaries, and blood moves through them more slowly and at lower pressure. That slower flow matters because exchange requires time. The liver is not trying to rush blood past a tissue that only needs oxygen delivered. It is trying to hold incoming chemistry in a usable field long enough for hepatocytes to sample it, transport it, alter it, and sometimes release products back into circulation before the blood exits through the central vein. Slow flow also gives Kupffer cells, the resident macrophages of the liver, time to inspect portal blood for bacterial fragments, debris, and inflammatory signals coming from the gut. A standard high-barrier capillary would starve the liver of the very exposure it needs to do its work. The sinusoid slows the stream so the organ can read it.

A third difference is what the sinusoid is for. Ordinary capillaries mainly support the tissue they perfuse. Sinusoids support the tissue, but they also deliver the liver's actual workload. Portal blood is not merely feeding hepatocytes. It is bringing them jobs. Glucose after a meal must be buffered. Amino acids must be sorted. Ammonia precursors must be managed. Bilirubin must be taken up and prepared for biliary excretion. Hormones must be modified. Drugs and xenobiotics may need first-pass handling. Xenobiotics are foreign compounds, substances the body does not make for its own structural use. In other words, the sinusoid is not simply a nutrient supply line. It is an intake channel for metabolic burden. That changes what kind of vascular surface the liver needs. It needs a channel that serves as an interface, not merely a pipe.

The hepatocyte side of the equation is built to match this. The hepatocyte does not present a flat indifferent membrane to the sinusoidal space. Its blood-facing surface carries microvilli, tiny membrane projections that increase area and improve contact with the fluid in the space of Disse. More area means more transport opportunity. That allows high-volume uptake and release. The hepatocyte can take in glucose, bilirubin, hormones, and lipoprotein remnants from the blood-facing side while also releasing albumin, glucose during fasting, urea-related products, and other synthesized or modified compounds back toward circulation. This makes the sinusoid and hepatocyte a matched pair. The vascular channel is permissive and slow. The cell surface is expanded and chemically active. Together they create an exchange field much more intimate than an ordinary capillary bed.

This architecture also explains why sinusoidal damage carries such heavy consequences. If the endothelial surface loses its fenestration, if the exchange space thickens with fibrosis, if congestion raises pressure and slows or distorts passage abnormally, or if inflammatory injury alters the sinusoidal lining, then the liver begins losing the very interface that makes it a liver. Blood may still be present in the tissue, but presence is not enough. The problem is not whether blood arrived. The problem is whether blood could meet hepatocytes under the right structural conditions. Once that meeting becomes harder, glucose buffering weakens, bilirubin handling weakens, nitrogen control weakens, protein secretion into blood becomes less efficient, and the whole organ starts to function more like poorly perfused scarred tissue than an exchange organ.

So what makes sinusoidal blood channels different from ordinary capillaries is not one feature but a whole design logic. They are wider. They are slower. They are more permeable. They sit beside hepatocytes across a specialized exchange space rather than behind a tight barrier. They carry workload, not just nourishment. They invite scrutiny by metabolic cells and immune cells at the same time. A capillary usually serves tissue. A sinusoid exposes blood to a processing field. That is why the liver can do what other organs cannot. It does not merely receive circulation. It opens circulation up.

Fenestrated Endothelium and Rapid Movement of Solutes

The sinusoidal endothelium is built to let chemistry move. That is what makes it different from a vascular lining whose main job is to defend a barrier. In many tissues the endothelial wall is tighter because the surrounding cells cannot afford broad exposure to whatever is passing through the bloodstream. The liver has the opposite problem. It cannot do its job unless the blood is made available to hepatocytes quickly and at high volume. The endothelial cells lining the sinusoid therefore are fenestrated, which means they contain small pores that allow plasma and dissolved compounds to pass out of the flowing blood far more easily than they could across an ordinary tight capillary wall. These pores are not random leaks. They are part of the organ's exchange design. They make the sinusoid into an access channel rather than a sealed tube.

That structural openness matters because the liver is not trying to exchange only oxygen and carbon dioxide. It is handling nutrient surges, bilirubin, hormones, amino acids, glucose, microbial metabolites, lipoprotein remnants, drug molecules, alcohol, and a large number of other compounds arriving from the portal circulation. Those compounds have to leave the central blood stream and enter the perisinusoidal space fast enough for hepatocytes to act on them before the blood exits through the central vein. The perisinusoidal space, also called the space of Disse, is the narrow interface between the fenestrated endothelium and the hepatocyte surface. Once plasma and solutes pass through the fenestrations, they are brought into close contact with hepatocyte microvilli, the tiny membrane projections that increase surface area and improve transport opportunity. The exchange event therefore is not abstract diffusion through liver tissue in general. It is structured movement through a porous endothelial wall into a specialized transfer compartment designed for rapid cellular contact.

Rapid movement of solutes is essential because the liver is being asked to impose order on moving blood, not on a stored sample. After a meal, portal glucose may rise sharply and hepatocytes need access to that glucose quickly if they are going to buffer the load and store part of it as glycogen. Amino acids have to reach the hepatocyte for protein synthesis or nitrogen handling. Unconjugated bilirubin has to be taken up from blood before it can be conjugated and secreted toward bile. Lipoprotein remnants returning from intestinal fat handling have to reach the hepatocyte surface for processing. Hormones and xenobiotics have to be exposed to hepatic transporters and enzymes early enough to matter. Xenobiotics are foreign compounds, substances the body does not make for its own structural use. If the endothelial wall were too restrictive, all of this work would slow, backlog would develop, and the liver would lose one of its defining abilities, which is rapid first-pass handling of the post-digestive stream.

The word solute matters here because much of what the liver handles is traveling dissolved or suspended in plasma rather than packaged as solid particles crossing whole. A solute is a substance dissolved in a fluid. Glucose, amino acids, lactate, urea precursors, bile salt related intermediates, many drugs, and many hormone forms are all examples of compounds that can move in this way. The fenestrated sinusoidal endothelium gives those compounds a relatively direct route out of the vascular lumen and toward the hepatocyte. Larger cellular elements such as red blood cells remain within the main blood channel. That separation is important. The liver wants access to plasma chemistry, not collapse of the vascular compartment altogether. Fenestration is therefore a controlled permeability design. It opens the door to what the hepatocyte needs to sample while preserving the basic integrity of blood flow.

This arrangement also explains why sinusoidal injury is so serious. If fenestrations are lost, if the endothelial lining becomes less permissive, or if fibrosis thickens the exchange pathway between blood and hepatocyte, solute movement slows and the liver begins to act less like an exchange organ and more like a tissue cut off from its own workload. Glucose reaches hepatocytes less cleanly. Bilirubin uptake becomes less efficient. Lipoprotein handling suffers. Hormone and drug exposure at the hepatocyte surface changes. The injury may begin microscopically, but the functional consequences spread upward fast because the whole organ depends on rapid access to the chemical contents of blood. A liver without a permissive sinusoidal interface still has hepatocytes, but it no longer lets them meet the stream properly.

So fenestrated endothelium is not a minor histology fact. It is one of the reasons the liver can be the liver. The pores in the sinusoidal lining let plasma chemistry move out of flowing blood and into immediate contact with hepatocytes at the speed the organ requires. The liver does not merely receive blood. It opens blood up to inspection. Fenestration is part of how that opening is made real.

The Importance of Blood Slowing Down Inside the Liver

The liver cannot do high-volume chemical review if blood moves through it like blood moves through a transit pipe. Speed would defeat the organ's purpose. Portal blood arrives carrying the dissolved aftermath of digestion, glucose, amino acids, bilirubin precursors, microbial metabolites, hormone fragments, drug molecules, alcohol, lipoprotein remnants, and a wide range of other compounds that the body cannot afford to distribute blindly. The liver is supposed to intercept that stream, expose it to hepatocytes, and decide what gets extracted, altered, stored, released back into circulation, or redirected toward bile. That kind of work requires contact time. Blood has to slow down enough for exchange to happen. A fast stream nourishes poorly when the task is not just perfusion but interpretation.

This is why the sinusoid is built as a low-pressure exchange channel rather than a tight high-speed capillary. As blood enters from terminal branches of the portal vein and hepatic artery, it spreads into these broader vascular spaces and loses velocity. That slowing matters because hepatocytes are not working on the blood from a distance. Plasma has to leave the main flowing column, pass through the fenestrated sinusoidal endothelium, enter the space of Disse, and reach the microvillous surface of the hepatocyte. Fenestrated means the endothelial lining contains pores that make plasma exchange easier. The space of Disse is the narrow transfer compartment between the sinusoid and the hepatocyte. Microvilli are small membrane projections that increase the cell's effective surface area. All of this architecture is built around one requirement: the chemistry in the blood must have enough time to meet the chemistry of the cell.

That time requirement becomes obvious the moment you think about what the hepatocyte is being asked to do. After a meal it may need to remove a substantial fraction of incoming glucose from portal blood and convert some of it into glycogen. It may need to take up bilirubin and prepare it for biliary excretion. It may need to absorb amino acids, sort their carbon skeletons toward synthesis or fuel use, and begin trapping their nitrogen burden through urea cycle activity. It may need to take up drug molecules for first-pass metabolism, modify hormone forms, receive lipoprotein remnants, or clear microbial signals arriving from the gut. None of those events are instantaneous in the physiological sense. They require membrane transport, intracellular handling, enzymatic work, and often directional secretion. Slow sinusoidal flow does not make the liver lazy. It gives the organ the physical condition required for precision.

The immune function of the liver also depends on this deceleration. Kupffer cells, the resident macrophages sitting in the sinusoidal space, need access to the passing stream if they are going to clear bacterial fragments, debris, and other danger signals coming from the intestine. If portal blood raced through the liver without slowing, the immune interface would be blunted. The organ would receive the gut's chemical and microbial report too quickly and too shallowly. A slower stream gives metabolic cells time to extract and transform, and gives immune cells time to inspect and respond. The liver is therefore not only slowing blood for nutrient handling. It is slowing blood so the gut's burden can be screened before that burden becomes systemic.

Slowing also helps preserve gradients across the tissue. As blood moves through the sinusoid, hepatocytes nearer the portal region see it first, with higher oxygen and a fresher nutrient profile. Hepatocytes closer to the central vein see blood later, after some oxygen and substrates have already been extracted upstream. That sequential exposure is one of the reasons liver function is zoned rather than identical across the whole lobule. If blood moved too rapidly, the organ would lose some of the controlled gradient that helps organize distinct metabolic roles across the tissue. The liver is not built as a uniform bath. It is built as a directed field where timing, concentration, and position all matter.

The importance of slowing becomes even clearer in disease. When sinusoidal flow is distorted by congestion, fibrosis, inflammation, or vascular pressure changes, the problem is not simply that blood is present or absent. The problem is that the conditions of contact are wrong. Blood may move too poorly, pool abnormally, or face increased resistance. Exchange becomes less efficient. Hepatocytes lose clean access to the passing plasma they are supposed to govern. A liver can therefore be full of blood and still function badly if the blood is not moving through the exchange surface in the right way. The organ depends on disciplined slowness, not stagnation and not haste.

So blood slowing down inside the liver is not a passive hemodynamic side effect. It is part of the organ's intelligence. The liver needs flow, but it needs processed flow, blood moving slowly enough to be read, altered, and judged before it is released back to the body. Speed is useful in transport. The liver is doing more than transport. It is trying to turn raw portal chemistry into blood the rest of the organism can live on.

How the Liver Samples Incoming Chemistry

The liver does not sample incoming chemistry by waiting for blood to announce what it contains in some abstract signaling language. It samples by exposing the blood physically to the right cells in the right order. Portal blood enters carrying glucose after absorption, amino acids after protein digestion, bilirubin-related compounds, microbial metabolites, bacterial fragments, hormone traffic, lipoprotein remnants, drugs, alcohol, and a range of other dissolved burdens and substrates. That blood is slowed in the sinusoids, the specialized exchange channels of the liver, and the plasma portion of that blood is allowed to move across a fenestrated endothelial lining into the space of Disse, the narrow compartment lying directly against the hepatocyte surface. Fenestrated means the lining contains pores. The result is that the liver does not sample from a distance. It places the chemistry of the incoming blood within immediate reach of the cells built to judge it.

The first layer of that sampling is structural. Hepatocytes face the sinusoid with a membrane domain specialized for uptake, sensing, and controlled release. Microvilli project from this blood-facing surface into the space of Disse, increasing contact area so more of the passing plasma can be interrogated at once. This matters because the liver is not looking for one molecule at a time. It is receiving a moving mixture. Glucose transporters allow hepatocytes to pull in glucose when portal levels rise. Other transport systems bring in amino acids, bilirubin, bile acid intermediates, lipoprotein remnants, and many classes of drugs and endogenous compounds. Some of this sampling is simple uptake driven by concentration and transporter availability. Some of it is receptor-mediated, meaning the cell reads the presence of a molecule through binding and signaling before deciding how strongly to respond. But all of it depends on the same physical truth. The chemistry has to reach the membrane.

The second layer is metabolic sampling. A hepatocyte does not merely detect what is present. It detects what the incoming chemistry demands. Rising portal glucose shifts intracellular metabolism toward glycogen synthesis and post-meal buffering. Amino acid influx changes protein synthesis, nitrogen handling, and enzyme activity. Fructose arrival produces a different hepatic handling pattern from glucose because the pathways and constraints are different. Hormone delivery changes the cell's transcriptional and enzymatic priorities. Xenobiotics, foreign compounds the body does not make for its own structural use, activate biotransformation machinery such as cytochrome P450 systems. Microbial metabolites and inflammatory signals arriving from the gut can alter acute-phase responses and other hepatic outputs. In other words, the liver samples not only by identifying molecules, but by reading the metabolic pressure those molecules create inside the cell. The incoming blood tells the hepatocyte what has happened upstream in the gut and what kind of internal response is now required.

The third layer is immune sampling. Portal blood does not carry nutrients alone. It also carries evidence of microbial life in the intestine. Kupffer cells, the resident macrophages of the liver, sit within the sinusoidal space where they can inspect the incoming stream for bacterial fragments, damaged-cell debris, and inflammatory cues. Liver sinusoidal endothelial cells also participate in this screening environment. This means the liver is reading two reports at once every time portal blood arrives. One report is nutritional and metabolic. The other is microbial and immunologic. The organ has to decide whether the incoming material represents ordinary digestive traffic that should be tolerated and processed quietly or whether it represents a threat that requires a defensive response. That is why the liver is such an important immune buffer between the gut and the rest of the body. It samples the intestine's chemical output before systemic circulation is forced to live with it.

Timing matters too. The liver samples chemistry as it is moving, not after it has settled. Blood enters from the portal region and travels through the sinusoid toward the central vein. Hepatocytes nearer the incoming side see fresher, more oxygen-rich, less previously altered blood. Hepatocytes farther downstream see blood later, after some extraction and oxygen use have already occurred. This means the organ is sampling along a gradient, not in one uniform snapshot. One region is first contact. Another is downstream interpretation. The architecture of the acinus, the functional unit defined by blood supply gradients, makes that possible. This is part of why liver function is zoned and why certain injuries and metabolic tasks cluster in different regions of the tissue.

So the liver samples incoming chemistry through exposure, transport, metabolism, and immune surveillance all at once. Blood is slowed. Plasma is opened to the exchange surface. Hepatocytes take in and interpret compounds according to transporters, receptors, and intracellular metabolic state. Kupffer cells and endothelial cells read the microbial and inflammatory content of the same stream. The organ does not passively receive information. It creates the conditions under which the information in blood becomes readable. That is how the liver knows what has just come out of the intestine and what the rest of the body should be allowed to receive next.

The Exchange Surface Between Portal Blood and Hepatocytes

The exchange surface between portal blood and hepatocytes is where digestion becomes body chemistry in a controlled way. Portal blood arrives carrying the immediate chemical results of intestinal absorption, glucose, amino acids, water-soluble vitamins, minerals, microbial metabolites, hormone fragments, drug molecules, bilirubin-related compounds, and a range of other substances that have crossed the gut barrier but have not yet been fully judged by the organism. That blood does not remain sealed inside a distant vascular tube while hepatocytes guess at its contents. The liver opens the stream. Portal blood enters the sinusoids, the specialized low-pressure vascular channels of the liver, and the plasma portion of that blood is allowed to move across a fenestrated endothelial lining into the space of Disse, the narrow compartment between sinusoidal blood and the hepatocyte surface. Fenestrated means the endothelial cells contain pores. The result is an exchange field built for immediate contact rather than guarded separation.

That contact surface matters because hepatocytes are not trying to extract one or two simple nutrients from blood. They are handling mixed post-digestive traffic. Glucose may need to be removed rapidly and stored as glycogen. Amino acids may need to be used for protein synthesis, oxidized for energy, or stripped of nitrogen so ammonia can be converted into urea. Bilirubin has to be taken up from blood and prepared for secretion into bile. Hormones may need modification. Lipoprotein remnants returning from intestinal fat handling may need to be processed. Drugs and xenobiotics, foreign compounds the body does not make for its own structural use, may need first-pass metabolism. None of that works well if the blood-facing surface of the hepatocyte is held behind a tight barrier. The liver needs a broad, permissive, high-contact interface because the workload is chemically dense and time-sensitive.

The hepatocyte side of this surface is built to match that demand. The blood-facing membrane projects microvilli into the space of Disse, increasing surface area and giving transporters, receptors, and membrane enzymes more working frontage against the exchanged plasma. More surface area means more opportunity to move molecules in and out. Transporters can pull in glucose, bilirubin, bile acid intermediates, amino acids, and many other compounds according to concentration gradients, membrane specialization, and cellular need. Receptors can detect hormonal and signaling conditions in the blood and shift intracellular priorities accordingly. Newly made products can also move the other way. Albumin, clotting-related proteins, lipoproteins, glucose released during fasting, and other compounds synthesized or processed by the hepatocyte can be returned to the blood across this same interface. The exchange surface is therefore not an intake port only. It is a bidirectional control zone where the liver edits the chemistry of the passing circulation.

This is one reason the space of Disse cannot be treated as empty filler between vessel and cell. It is the working margin where plasma leaves the main vascular lumen and becomes available to the hepatocyte without losing the overall direction of blood flow. Red blood cells remain inside the sinusoid, but the dissolved chemical contents of plasma move into this narrow exchange environment and meet the microvillous hepatocyte surface at close range. That arrangement gives the liver something most organs do not need to the same degree, a way to review large amounts of blood chemistry quickly without collapsing the distinction between blood compartment and cell interior. Portal blood keeps moving. The hepatocyte keeps sampling. Exchange happens in the middle.

The immune side of this surface matters too. Portal blood does not carry nutrients alone. It also carries microbial signals from the gut, bacterial structural fragments, inflammatory cues, and metabolites produced by intestinal organisms. Kupffer cells, the resident macrophages of the liver, sit in the sinusoidal space and participate in sampling this stream before it reaches the wider circulation. Liver sinusoidal endothelial cells also contribute to the behavior of this interface. So the exchange surface is metabolic and immunologic at the same time. It is where the liver reads not only what has been absorbed from food, but also what kind of microbial and inflammatory burden is arriving from the intestine.

When this exchange surface is damaged, liver function declines long before the organ disappears anatomically. If fenestrations are lost, if fibrosis thickens the space between blood and hepatocyte, if sinusoidal pressure changes distort flow, or if inflammation alters endothelial behavior, the contact between portal blood and hepatocytes becomes less efficient. Glucose buffering worsens. Bilirubin uptake worsens. Hormone and drug handling change. Lipoprotein processing suffers. The liver can still be full of cells and blood yet fail because the interface between them is no longer built for rapid controlled exchange. That is the deeper point. The liver does not govern portal blood merely because blood passes through it. It governs portal blood because the tissue has constructed an exchange surface intimate enough to make that governance real.

What Happens When Sinusoidal Architecture Is Damaged

Sinusoidal architecture matters because the liver works through contact, and sinusoids are the structure that makes that contact possible. Blood is supposed to enter the liver, slow down, spread through a low-pressure exchange field, and expose its plasma chemistry to hepatocytes across a fenestrated endothelial lining and the space of Disse. Fenestrated means the endothelial cells contain pores that allow plasma and dissolved compounds to pass toward the hepatocyte surface with unusual ease. The space of Disse is the narrow transfer compartment between the sinusoid and the hepatocyte. If that architecture is damaged, the problem is not just that a vessel wall looks abnormal under the microscope. The problem is that the liver begins losing its ability to let blood and hepatocytes meet under the right conditions. Once that meeting deteriorates, nearly every major hepatic function starts to lose efficiency.

One of the first things that can happen is capillarization of the sinusoid. This means the sinusoidal lining begins to behave less like a permissive exchange surface and more like an ordinary tighter capillary. Fenestrations are reduced or lost, the endothelial barrier becomes less open, and the space between blood and hepatocyte becomes functionally harder to cross. Fibrosis often worsens this by depositing extracellular matrix in the space of Disse, thickening the interface and physically separating flowing plasma from the hepatocyte membrane. The hepatocyte is still present. The blood is still present. But the contact surface that made the liver a rapid exchange organ is no longer clean. Glucose reaches the cell less efficiently. Bilirubin uptake becomes less efficient. Lipoprotein remnants, hormones, and drug molecules no longer meet the hepatocyte with the same immediacy. The organ begins to lose throughput without necessarily losing bulk.

This damage also changes hemodynamics, which means the behavior of blood flow itself. Healthy sinusoids are low-resistance channels. Scar deposition, endothelial dysfunction, swelling, inflammatory distortion, and architectural collapse all increase resistance to portal flow. Blood has a harder time crossing the tissue, so portal pressure begins to rise upstream. That is the beginning of portal hypertension. Once portal blood cannot move cleanly through the sinusoidal bed, the problem escapes the microscopic scale and becomes a whole-organ and then whole-body problem. The spleen enlarges. Collateral vessels may form. Ascites, the accumulation of fluid in the abdominal cavity, becomes more likely. In other words, sinusoidal damage is one of the bridges between cellular liver injury and vascular liver disease. The microarchitecture hardens and the circulation pays for it.

Oxygen handling worsens too. Sinusoids do not merely deliver nutrients and metabolic burden. They create the oxygen gradient that supports zonation across the liver, meaning different regions of the tissue receive blood in different sequence and with different oxygen content.

When sinusoidal flow is distorted, congested, or shunted abnormally, that gradient becomes less stable. Hepatocytes that already live at the downstream edge of oxygen delivery become more vulnerable to hypoxic injury, which is injury caused by inadequate oxygen availability. ATP production falls. ATP, adenosine triphosphate, is the cell's immediately usable energy currency. Without enough of it, hepatocytes cannot maintain membrane transporters, cannot preserve polarity cleanly, cannot support bile secretion, and cannot sustain the full load of synthesis, conversion, and detoxification. So damage to sinusoidal architecture is not just about access to substrates. It is also about access to oxygen and the energetic authority to keep being a hepatocyte.

The immune environment shifts as well. Kupffer cells and liver sinusoidal endothelial cells are positioned around the sinusoid because the liver has to sample portal blood for microbial fragments and inflammatory signals arriving from the gut. When architecture becomes distorted, immune signaling tends to change with it. Endothelial dysfunction, altered shear forces, retained gut-derived inflammatory cues, and hepatocyte stress all feed into a more activated local environment. Stellate cells, which normally sit relatively quiet in the space of Disse storing vitamin A, are more likely to activate under these conditions and begin producing collagen. That creates a feedback loop. Injury distorts the sinusoid. Distorted sinusoids promote inflammation and stellate activation. Stellate activation deposits more matrix. More matrix further damages the sinusoidal exchange surface. A temporary injury can therefore become a self-reinforcing remodeling process if the underlying burden persists.

The consequences spread into bile handling too, even though bile does not flow through the sinusoid itself. Hepatocytes depend on adequate sinusoidal exchange and oxygen delivery to maintain the polarity and ATP-dependent transport needed for canalicular secretion. If sinusoidal architecture is damaged enough, hepatocytes become metabolically weaker and less able to take up bilirubin from blood, conjugate it, and secrete it into bile. Bile acids and other biliary constituents may accumulate within tissue if secretion and flow begin to falter. So the damage starts on the blood-facing side of the cell and then spills into the bile-facing side. The liver stops behaving like a coordinated two-stream organ. Blood handling worsens first or most visibly, but biliary competence eventually follows it downward.

This is why sinusoidal injury is one of the most important turning points in liver disease. A liver can survive a surprising amount of hepatocyte stress if the exchange field remains open, the gradients remain intact, and blood can still meet the cell surface properly. But once sinusoidal architecture is damaged, the organ begins losing the physical conditions that made its chemical power possible. Blood still arrives, but it is no longer presented correctly. Hepatocytes still exist, but they are no longer being supplied or exposed correctly. Fibrosis, congestion, endothelial dysfunction, portal hypertension, impaired uptake, reduced oxygenation, and downstream synthetic and biliary failure all begin to connect. The liver does not fail only because its cells are hurt. It fails because the exchange surface that gave those cells access to the moving chemistry of life starts to close.

Chapter 7

Kupffer Cells, Stellate Cells, and Endothelial Cells

Kupffer cells are the liver's resident macrophages, the built-in phagocytic immune cells stationed inside the sinusoidal blood channels where portal blood first arrives. Resident means they live in the tissue as part of its standing architecture rather than appearing only after an injury. Macrophage means a large scavenger cell built to engulf, digest, and interpret foreign material, damaged cellular debris, and inflammatory signals. In the liver this role carries unusual importance because portal blood is not clean vascular traffic. It comes directly from the intestine and spleen carrying not only nutrients, but also bacterial fragments, microbial metabolites, aged blood-cell breakdown products, endotoxin, immune stimuli, and all the low-level chemical noise that digestion and gut microbial life continuously generate. Kupffer cells sit in that stream on purpose. They are positioned inside the sinusoidal lumen, attached to or embedded within the endothelial environment, so the first major organ downstream from the intestine can screen what has just crossed into internal circulation.

Their job is often described too loosely as immune defense, but their real role is more exact. Kupffer cells are part scavenger, part sensor, part traffic controller. They engulf bacteria and particulate debris that entered portal blood from the gut. They clear remnants of damaged red blood cells and other circulating material arriving for disposal. They detect inflammatory patterns through receptors that recognize microbial structures and injury signals, then decide whether the situation calls for quiet clearance, cytokine release, or escalation into a stronger immune response. Cytokines are signaling proteins used by cells to communicate inflammatory and immune states. This matters because the liver cannot behave as if every meal is an infection. Portal blood constantly carries small amounts of microbial evidence from the intestine. The liver therefore needs immune cells that can clear routine burden without igniting unnecessary systemic inflammation. Kupffer cells are part of that solution. They defend, but they also restrain.

Their location explains why they matter so much. A macrophage sitting deep in a solid tissue can only react after something has already arrived and spread. Kupffer cells are stationed inside the exchange stream itself. Blood slows in the sinusoids, and as that blood passes, Kupffer cells are able to sample what it contains before the chemistry reaches the wider circulation. This makes them one of the main reasons the liver functions as an immune buffer between the gut and the rest of the body. Bacterial lipopolysaccharide, also called endotoxin, is one example. It is a structural component of the outer membrane of certain bacteria and a powerful immune stimulus when it appears in the wrong place. Small amounts of endotoxin can enter portal blood even under ordinary conditions. Kupffer cells help contain that burden. Without them, the intestine's microbial reality would reach the systemic circulation with far less screening.

They also help determine whether liver stress stays local or begins spreading into a larger inflammatory program. When Kupffer cells detect danger signals strongly enough, they release cytokines and other mediators that affect neighboring hepatocytes, stellate cells, endothelial cells, and recruited immune cells. That can be protective in the short term. A true infection or major tissue injury needs coordination, not silence. But the same system becomes dangerous when stimulation is chronic. Persistent gut-derived inflammatory traffic, ongoing hepatocyte injury, alcohol exposure, fat accumulation, toxin burden, or cholestatic stress can keep Kupffer cells activated for too long. Then the cell that once protected the liver begins helping drive tissue damage. Cytokines rise. Oxidative stress rises. Stellate cells are more likely to activate and begin producing collagen. Endothelial behavior changes. Hepatocyte survival becomes harder. The immune cell that was supposed to clear the stream now starts thickening the consequences of what the stream contains.

This is why Kupffer cells have to be understood as part of liver architecture rather than as a generic immune add-on. They are not just there because every organ gets some immune coverage. They are there because the liver occupies a uniquely exposed position. The organ sits downstream from the gut, upstream from the systemic circulation, and directly in the path of blood carrying nutrients mixed with microbial and inflammatory burden. Kupffer cells are one of the structural reasons that arrangement is tolerable. They help remove what should not circulate, interpret what should and should not trigger alarm, and shape the inflammatory climate of the tissue around them. In health they make portal exposure survivable. In chronic stress they can become part of the machinery that turns exposure into fibrosis and progressive liver disease. That dual role is exactly what the liver requires. An immune cell soft enough to tolerate normal gut traffic. Sharp enough to respond when the traffic turns dangerous.

The Liver's Constant Exposure to Bacterial Fragments from the Gut

The liver is exposed to bacterial fragments all day because the intestine is not a sterile pipe and the gut barrier is not an absolute wall. The intestinal lumen contains a vast microbial population living directly above a thin epithelial lining, a mucus layer, and an immune surface that has to absorb nutrients without opening the body to uncontrolled invasion. That arrangement works, but it is never the same thing as perfect exclusion. Small amounts of microbial material constantly enter the portal circulation. This includes lipopolysaccharide, a structural component of the outer membrane of many gram-negative bacteria, along with peptidoglycan fragments from bacterial cell walls, flagellin from bacterial motility structures, bacterial DNA, and a large number of metabolites produced during fermentation and microbial breakdown of food. The gut therefore does not send the liver nutrients alone. It sends a running microbial report.

That report is normal physiology, not automatically a pathology. The liver is built with the expectation that portal blood will contain low-level bacterial evidence. Kupffer cells, the resident macrophages sitting in the sinusoids, sample this incoming stream and help clear what should not circulate farther. Liver sinusoidal endothelial cells and hepatocytes also participate in reading and responding to microbial signals. This is one of the main reasons the liver sits where it does. It is not only downstream from absorption. It is downstream from the intestine's bacterial reality. The organ has to decide, continuously, whether what arrived is routine traffic that should be tolerated and cleared quietly or a stronger danger signal that requires escalation. That decision is being made on moving blood, not in a sterile laboratory setting. The liver lives in contact with the microbiological edge of digestion.

This constant exposure creates a difficult balance. If the liver reacted to every bacterial fragment as if it were a bloodstream infection, systemic inflammation would become a normal feature of eating and intestinal life. If it ignored those signals completely, microbial burden would spread too freely into the wider circulation. So the hepatic immune environment is tuned for discrimination. Low-level portal exposure is often met with containment, scavenging, and controlled immune tolerance. Tolerance in this setting does not mean indifference. It means the liver clears and interprets microbial material without converting every encounter into tissue-damaging inflammation. That is one of the organ's quieter achievements. It protects the rest of the body from gut-derived microbial traffic while avoiding constant overreaction to the very traffic it was built to receive.

The problem begins when the quantity, frequency, or quality of that incoming burden changes. If the intestinal barrier becomes more permeable, if microbial overgrowth or dysbiosis shifts the chemical output of the gut, if alcohol or inflammatory injury weakens epithelial integrity, or if stagnant intestinal contents change bacterial behavior, the liver begins receiving a heavier inflammatory load. Then the same signals that were manageable at low level become chronic stimulation. Kupffer cells release more cytokines. Endothelial behavior shifts. Hepatocytes work under a higher inflammatory burden. Stellate cells are more likely to activate and produce collagen, thickening the exchange architecture and moving the tissue toward fibrosis. What began as ordinary portal exposure can therefore become one of the engines of chronic liver injury when upstream control in the gut starts to break.

This is why the liver should be understood as an immune buffer organ as much as a metabolic one. It is constantly exposed to bacterial fragments not because the body failed to defend itself, but because the digestive system is built around intimate contact with a microbial world. The intestine absorbs from that world. The portal vein carries part of its evidence forward. The liver receives the first pass. Its job is to keep that evidence from becoming systemic chaos.

How the Liver Clears Debris Without Overreacting to Every Meal

The liver clears debris by treating portal blood as something that must be screened constantly but not dramatized constantly. That distinction is essential. Every meal sends more than nutrients toward the liver. Portal blood carries microbial metabolites, bacterial cell wall fragments, endotoxin, damaged host molecules, worn-out blood components, food-derived antigens, and the low-level chemical disorder that naturally follows digestion in a microbe-rich intestine. If the liver responded to each of those arrivals as if the body were under open microbial invasion, eating itself would become an inflammatory event severe enough to damage tissue daily. If it ignored them completely, the rest of the circulation would be left exposed to the intestine's particulate and microbial burden. The liver survives between those two failures by building a filtering system that is active without being hysterical.

Kupffer cells are central to this restraint. They sit in the sinusoids, directly in the path of portal blood, and act as resident macrophages, meaning long-lived scavenger immune cells embedded within the tissue itself. They engulf particulate debris, clear bacterial fragments, and remove damaged cellular material before it spreads farther downstream. But clearance is only part of their role. They also interpret context. A low-level trickle of microbial products arriving with otherwise ordinary portal blood does not mean the same thing as a large disruptive burst arriving alongside tissue injury, endotoxin excess, cholestasis, or hepatocyte stress. The Kupffer cell is not just eating debris. It is reading the meaning of that debris in relation to the rest of the stream. That is one reason the liver can remain tolerant during ordinary digestion while still being capable of alarm when conditions shift.

This tolerance is not passive weakness. It is an actively maintained immune discipline. Liver sinusoidal endothelial cells, Kupffer cells, hepatocytes, and other immune-active cells in the hepatic environment help shape a local setting in which low-grade exposure from the gut is often met with containment, scavenging, and signal dampening rather than full inflammatory escalation. Signal dampening means the tissue limits how strongly inflammatory pathways are amplified when the incoming burden is familiar, low-level, and structurally consistent with ordinary gut traffic. That matters because the gut is never sending the liver a perfectly sterile stream. The liver has to live with constant microbial evidence without turning that evidence into constant tissue damage. In that sense, immune tolerance in the liver is not neglect. It is a survival strategy built for a portal blood supply that is inherently contaminated in small ways by the biology of digestion.

The architecture helps make that restraint possible. Blood slows in the sinusoids instead of racing through a tight capillary bed. That slower flow gives Kupffer cells and sinusoidal endothelial cells time to inspect the stream. The endothelial surface is fenestrated, meaning porous, so plasma chemistry is exposed broadly enough for metabolic handling, but the immune response remains compartmentalized within a tissue built to interpret gut-derived traffic. Hepatocytes themselves also contribute to this environment by handling incoming molecules rapidly and by participating in signaling that can either support tolerance or amplify stress. The liver therefore does not clear debris through brute inflammatory force. It clears through layered review. Structural slowing. Local phagocytosis. Signal interpretation. Selective escalation only when the incoming burden stops looking like ordinary digestion and starts looking like injury, overload, or invasion.

The problem comes when that distinction begins to fail. If intestinal permeability rises, if microbial output becomes more inflammatory, if alcohol weakens barrier integrity, if fat accumulation stresses hepatocytes, or if chronic injury keeps inflammatory signals high, then the same system that once cleared quietly begins to react more aggressively. Kupffer cells release more cytokines. Endothelial behavior shifts. Stellate cells are more likely to activate and begin laying down collagen. Cytokines are signaling proteins that tell nearby cells the tissue is under threat. Once that message becomes chronic, the liver can no longer treat each incoming meal as routine traffic. The immune setting hardens. The stream coming from the gut has not become fully foreign, but it is no longer being interpreted as harmlessly ordinary either. That is one of the key transitions from healthy hepatic tolerance to chronic liver inflammation.

So the liver clears debris without overreacting to every meal by refusing two extremes. It does not pretend the portal stream is clean, and it does not behave as if every bacterial fragment is a crisis. It screens, engulfs, degrades, and interprets. It maintains a filtered tolerance toward the ordinary microbial noise of digestion while keeping enough defensive capacity in reserve to respond when the burden becomes dangerous. That balance is one of the quiet achievements of liver physiology. The organ lives downstream from the gut and manages to keep ordinary eating from becoming systemic inflammation.

Stellate Cells as Vitamin A Storage Cells in Health

In the healthy liver, stellate cells are quiet cells living in the space of Disse, the narrow exchange compartment between the sinusoidal blood channel and the hepatocyte surface. They are not the main metabolic workers of the liver and they are not part of the incoming blood stream itself. They sit just outside that stream, embedded in the perisinusoidal environment where exchange, structural support, and future tissue response are all being quietly prepared. In health, one of their defining jobs is storage of vitamin A. More specifically, they store retinoids, the family of compounds built around vitamin A activity, in lipid droplets within their cytoplasm. This gives the liver one of its major reserve functions. The body does not have to depend on immediate daily intake for every moment of vitamin A need because stellate cells are holding a buffered supply inside the organ that stands at the center of nutrient handling.

That storage role matters because vitamin A is not a trivial nutrient moving casually through the blood. It is involved in epithelial integrity, immune function, visual chemistry, gene regulation, growth, and differentiation, which means the controlled development and maintenance of specialized cell identity. A compound with that much biological authority cannot simply be left ungoverned in circulation. Too little creates dysfunction. Too much creates toxicity. The body therefore benefits from a storage system that can hold vitamin A in a protected form and release it in a regulated way when needed. Stellate cells provide that system. They are strategically placed to receive retinoid traffic coming through the liver and to retain it as part of the organ's broader reserve function. In this sense they are not decorative support cells. They are nutrient custodians embedded directly into the liver's exchange architecture.

Their location helps explain why the job belongs to them. The space of Disse is where plasma leaving the sinusoid comes into close contact with hepatocytes, and it is also where stellate cells reside under normal conditions. That puts them near incoming blood-derived nutrients without placing them inside the main metabolic lane that hepatocytes occupy. They are close enough to the chemical traffic to receive and store retinoid material, but they are not burdened with the full hepatocyte workload of glycogen buffering, urea synthesis, bile secretion, and plasma protein production. This division of labor is efficient. The hepatocyte handles the major throughput of blood chemistry. The stellate cell helps preserve structural and nutritional reserve in the background. Healthy liver tissue is full of these kinds of specialized placements. Each cell sits where its role makes the most physiological sense.

The storage form also matters. Vitamin A is not being held there as loose abstract "nutrition." It is stored in lipid droplets as retinyl esters, chemically stabilized forms suited for reserve. That detail helps the reader understand the deeper logic of hepatic storage. Useful compounds are often not held in the same form they traveled in. They are packaged into a state that is safer, denser, and more controllable. The liver does this with glycogen for glucose, with certain minerals, and here through stellate cells with vitamin A related compounds. Storage is not passive accumulation. It is structured retention. The body is turning transient intake into future availability.

This quiet storage role is also one of the reasons stellate cells are so important later in disease. A healthy stellate cell storing retinoid is a cell living in reserve mode. It is holding nutrient capital and helping maintain the normal perisinusoidal environment. When chronic injury changes the local signals, that same cell can abandon its storage identity, lose those lipid droplets, and shift toward collagen production and tissue remodeling. That later transformation matters enormously in fibrosis, but it is easier to understand if the reader first sees what the stellate cell is in health. It begins as a quiet vitamin A reservoir cell situated in the exchange margin of the liver, helping preserve nutritional reserve and tissue stability. The scar-building cell of disease is not a foreign invader arriving from nowhere. It is a local cell that has changed its job.

So in the healthy liver, stellate cells should be understood first as reserve cells, especially vitamin A storage cells, living in the space between moving blood and working hepatocyte. They are part of the organ's quieter intelligence. The liver does not only process what is arriving now. It also holds what may be needed later. Stellate cells are one of the clearest expressions of that principle.

Stellate Cells as Scar-Building Cells in Injury

In the healthy liver, stellate cells sit quietly in the space of Disse storing retinoids, helping preserve the exchange environment without drawing attention to themselves. Injury changes that identity. When hepatocytes are damaged, when Kupffer cells release inflammatory cytokines, when oxidative stress rises, when bile acids are retained, when alcohol, metabolic overload, viral injury, or chronic gut-derived inflammatory traffic keep the tissue under pressure, stellate cells stop behaving like reserve cells and begin behaving like wound-response cells. They activate. That word needs to be taken literally. The cell changes its phenotype, meaning it changes its functional identity. It loses much of its quiet vitamin A storage character, sheds its lipid droplets, proliferates, contracts more like a myofibroblast, and begins producing extracellular matrix proteins, especially collagen. This is the hinge point between liver stress and liver scarring. The cell that once held reserve now begins building restraint around injury.

That shift makes sense from the tissue's point of view, at least at first. An injured organ needs a way to reinforce damaged regions, contain disruption, and stabilize a threatened microenvironment. Stellate cells are well positioned for that response because they live exactly where the exchange surface is most vulnerable, between sinusoidal blood and hepatocytes. When the local environment becomes dangerous, they are already there, already sensing inflammatory mediators, already exposed to oxidized lipids, reactive oxygen species, apoptotic debris from injured hepatocytes, and cytokines released by Kupffer cells and endothelial cells. Reactive oxygen species are chemically aggressive oxygen-containing molecules that can damage membranes, proteins, and DNA when they accumulate excessively. Under those conditions the stellate cell begins laying down extracellular matrix, the structural protein scaffold outside the cell, as if trying to patch and reinforce the wounded territory. In acute injury that response may help preserve tissue integrity. In chronic injury it becomes one of the main engines of architectural failure.

The scar-building role becomes destructive because the liver depends on openness. Sinusoids are supposed to remain low-resistance exchange channels. The space of Disse is supposed to remain a narrow permissive interface where plasma chemistry can reach hepatocyte microvilli quickly. Once activated stellate cells begin depositing collagen and other matrix components into that perisinusoidal space, the exchange surface thickens and stiffens. Blood is still present. Hepatocytes are still present. But the intimacy between them starts to disappear. Fenestrated endothelial behavior changes. Solute movement becomes less efficient. Oxygen handling worsens. Portal resistance rises. This is why fibrosis is not merely scar as a visual finding. It is scar placed in the worst possible location for liver function. The injury is architectural before it is merely cosmetic.

Stellate cells also contribute contractile force, which means they do not simply lay down matrix and stand back. Once activated, they can constrict the sinusoidal environment and help increase intrahepatic vascular resistance. That matters because portal blood is supposed to move through the liver under relatively low resistance before draining into the central and hepatic veins. If activated stellate cells narrow that microvascular space while also stiffening it with matrix, portal pressure rises upstream. This is one of the early bridges between cellular injury and portal hypertension. The reader should see the logic clearly. Stellate activation does not only damage exchange. It changes flow. A scar-building response inside the exchange surface becomes a hemodynamic problem for the whole organ.

The tragedy of chronic liver disease is that this response can become self-reinforcing. Hepatocyte injury activates Kupffer cells and inflammatory signaling. Inflammatory signaling activates stellate cells. Activated stellate cells deposit matrix and distort sinusoids. Distorted sinusoids worsen exchange, oxygen delivery, and hepatocyte stress. More hepatocyte stress creates more inflammatory signaling. The original insult may have been alcohol, steatotic overload, viral hepatitis, cholestatic retention, toxin exposure, or chronic gut-derived inflammatory burden. But once stellate activation is established, the tissue begins participating in its own remodeling. The liver is no longer only suffering an insult. It is rebuilding itself incorrectly.

This is why stellate cells have to be understood in both states. In health they are reserve cells, especially vitamin A storage cells, helping preserve the quiet architecture of the perisinusoidal space. In injury they become scar-building cells, converting that same strategic position into a site of collagen deposition, vascular resistance, and exchange failure. The important point is not that the liver contains cells that make scar. Many tissues do. The important point is where these cells sit and what their transformation does to function. When stellate cells activate, the liver begins losing the very openness that made it a high-flow exchange organ in the first place. Scar is not laid over the work. It is laid into the work.

Why Fibrosis Begins as a Cellular Response Before It Becomes Structural Collapse

Fibrosis does not begin as scar in the gross anatomical sense. It begins as cells changing their behavior under pressure. That distinction matters because readers often imagine fibrosis as if the liver were slowly being wrapped in passive scar from the outside, the way a wound on skin leaves a visible patch behind. The liver does not fail that way. It fails from inside the exchange architecture outward. Chronic injury alters signaling in hepatocytes, Kupffer cells, endothelial cells, and especially stellate cells long before the organ looks shrunken, nodular, or cirrhotic on imaging or gross inspection. The first event is not structural collapse. The first event is that cells stop maintaining a low-resistance exchange field and begin behaving as if the tissue is under continuous wound threat. Fibrosis is therefore not scar appearing out of nowhere. It is wound logic that refused to turn off.

The process usually starts with persistent injury rather than one dramatic insult. Hepatocytes may be stressed by fat overload, alcohol, viral injury, cholestatic bile retention, toxin exposure, hypoxia, or chronic inflammatory traffic arriving from the gut. Injured hepatocytes release danger signals, oxidative byproducts, and fragments of dying cellular material. Kupffer cells sense that damage and respond with inflammatory cytokines. Endothelial cells alter their behavior as the sinusoidal environment becomes more hostile. Stellate cells, which normally sit in the space of Disse storing vitamin A and helping maintain the perisinusoidal environment, are exposed to all of this at close range. They receive the message that the tissue is no longer in a state of quiet metabolic work. It is in a state of chronic threat. This is where fibrosis begins. Not in the collagen itself, but in the decision of the cell to switch roles.

Once stellate cells activate, the entire logic of the local tissue changes. The quiet storage cell becomes a wound-response cell. It loses its retinoid-rich lipid droplets, proliferates, becomes more contractile, and starts producing extracellular matrix proteins such as collagen. Extracellular matrix means the structural material outside cells that gives tissue shape, support, and mechanical restraint. In an acute injury, some matrix deposition can be adaptive. A damaged exchange field may need temporary reinforcement. But the liver depends on openness more than on rigidity. Its function requires permissive sinusoids, a thin space of Disse, rapid movement of solutes, and intimate contact between plasma chemistry and hepatocyte microvilli. The moment stellate cells begin building matrix persistently in that space, the wound response has started to interfere with the physiology it was supposed to protect.

That is why fibrosis is cellular before it is structural. Long before the liver becomes grossly nodular, the exchange surface is already being thickened. Fenestrated sinusoidal endothelium begins losing its open character. The space of Disse becomes less permissive as collagen and other matrix proteins accumulate. Hepatocytes are still present, blood is still present, but the distance between them functionally increases. Solutes do not move as cleanly. Oxygen gradients become less favorable. Transport becomes less efficient. Bile-facing work begins to suffer because the hepatocyte is metabolically weaker and less well supplied from the blood-facing side. Portal blood meets more resistance as the once-soft exchange field becomes mechanically tighter. So fibrosis does not wait for cirrhosis to begin harming function. The harm starts when the cells begin rebuilding the tissue in a way that no longer suits a high-flow exchange organ.

The vascular consequences are part of the same story. Activated stellate cells are not only collagen-producing cells. They also become more contractile, which means they can narrow the sinusoidal environment and help increase resistance to portal flow. The liver is supposed to receive portal blood into a low-pressure exchange network, not force it through a progressively stiffening maze. Once cellular activation begins tightening and reinforcing that network, portal pressure starts to rise upstream. This can happen before the reader would think of the organ as structurally collapsed. The collapse begins microscopically as increasing resistance, increasing matrix deposition, and worsening distortion of flow. The larger consequences, portal hypertension, congestion, collateral vessel formation, ascites, and eventually cirrhosis, are later expressions of a process that started as a cellular shift in identity and behavior.

This is also why fibrosis can be difficult to appreciate early if the reader is looking only for visible scar. The liver may still be large. The lobes may still be present. The overall organ may still look like a liver. But the internal operating surface is changing. Cells that once preserved exchange are now obstructing it. Cells that once stored reserve are now laying down mechanical restraint. Cells that once signaled repair briefly are now maintaining a chronic inflammatory program. Structural collapse comes later, when enough of these microscopic changes accumulate and begin to distort the larger architecture into nodules, vascular rerouting, and widespread exchange failure. By then the disease is easier to see. It was already active long before it became obvious.

So fibrosis begins as a cellular response because the liver first interprets chronic injury as a wound that needs stabilization. The problem is that the organ being stabilized is an organ that survives by permeability, flow, and contact. A response that might protect another tissue can slowly disable this one. Stellate cells activate. Cytokine signaling persists. Matrix accumulates. Sinusoids narrow. Exchange worsens. Portal resistance rises. The scar appears later as a structural fact, but the real beginning was cellular. The liver starts to fail when its own cells begin rebuilding it for the wrong kind of survival.

Chapter 8

Zonation: Not Every Hepatocyte Does the Same Job

Zone 1, Zone 2, Zone 3

The liver is one organ, but it is not one flat field of identical cells doing identical work. Blood does not arrive everywhere at once, and that single fact changes the chemistry each hepatocyte experiences. Portal venous and hepatic arterial blood enter at the portal tracts, then move through the sinusoids toward the central vein. As that blood travels, oxygen falls, nutrient proportions change, signaling molecules are partly extracted, and exposure to toxins or metabolites can shift from one part of the lobule to another. The result is zonation. Hepatocytes are arranged along a gradient, and that gradient divides the tissue into functional regions often called zone 1, zone 2, and zone 3. This is not an arbitrary teaching diagram. It is the metabolic geography created by flow. A hepatocyte near the incoming blood supply is living in a different environment from a hepatocyte farther downstream. The cell's position changes what it sees, what it is best equipped to do, and what is most likely to injure it.

Zone 1 is the periportal zone, meaning the region closest to the portal triads where blood first enters from branches of the portal vein and hepatic artery. These hepatocytes see the freshest blood, the highest oxygen tension, and the earliest delivery of absorbed nutrients, hormones, gut-derived signals, and other compounds arriving from the intestine. Because they live in the richest oxygen environment, they are well suited to energy-intensive oxidative work. This includes gluconeogenesis, the synthesis of glucose from noncarbohydrate substrates such as lactate, glycerol, and certain amino acids, along with much of the early handling of ammonia through the urea cycle and other tasks that benefit from strong oxidative support. They are first in line for the incoming stream. That makes them powerful cells for extraction and first-contact processing, but it also means they are the first to meet anything harmful arriving in high concentration from portal blood.

Zone 2 sits between the periportal and centrilobular extremes. It is the transitional zone, neither first-contact nor last-contact, neither highest oxygen nor lowest oxygen, neither the cleanest view of incoming blood nor the most downstream chemically altered view. This middle territory is easy to under-teach because it lacks the dramatic identity of zones 1 and 3, but that would be a mistake. Zone 2 is part of how the liver preserves continuity across the gradient rather than splitting into two disconnected physiological worlds. It carries intermediate metabolic behavior and often helps the tissue absorb shifts in demand, exposure, and injury pattern. When the organ is healthy, this zone helps smooth the gradient. When the organ is stressed, it can become part of the transitional burden between upstream oxidative support and downstream biochemical vulnerability.

Zone 3 is the centrilobular zone, the region nearest the central vein where blood arrives last. By the time blood reaches this area, some oxygen has already been consumed upstream, and some of the chemical content has already been partly extracted or altered. These hepatocytes therefore live in a relatively lower-oxygen environment. That difference matters. Zone 3 cells are more vulnerable to hypoxic injury, which is injury caused by inadequate oxygen availability, because they already sit at the thinner end of the oxygen gradient. At the same time, this region is heavily involved in glycolysis, lipogenesis, and much of the cytochrome P450 system that handles drugs and other xenobiotics. Lipogenesis is the synthesis of fat. Xenobiotics are foreign compounds, substances the body does not make for its own structural use. This means zone 3 is a chemically important but more vulnerable territory. It often carries a high detoxification burden while living closer to the edge of oxygen sufficiency. That is a bad combination when blood flow drops or toxin load rises.

Once the reader sees these three zones clearly, many liver injury patterns stop looking random. A drop in oxygen delivery will often strike zone 3 hardest because those hepatocytes are already downstream in the oxygen gradient. A toxin that arrives in high concentration from the portal blood may damage zone 1 more heavily because those cells meet it first. A compound that becomes dangerous only after enzymatic activation may create more downstream centrilobular injury if the activating machinery is concentrated there. Fat accumulation also often shows preferential patterns because lipid handling is not identical across the gradient. Zonation therefore is not a fine point added for specialists. It is one of the main reasons the liver behaves like a complex organ rather than a red sponge. Position changes chemistry. Chemistry changes function. Function changes vulnerability.

So zone 1, zone 2, and zone 3 should be understood as a living flow map written into the tissue itself. Zone 1 is the high-oxygen first-contact territory near the portal supply. Zone 2 is the intermediate bridge. Zone 3 is the lower-oxygen downstream territory near the central vein, carrying different metabolic priorities and different risks. The liver keeps one organ-wide purpose, but it distributes that purpose across a gradient. Not every hepatocyte does the same job because not every hepatocyte is standing in the same blood.

Why Oxygen and Substrate Gradients Matter

Oxygen and substrate gradients matter because the liver is not bathing all of its hepatocytes in the same blood at the same time. Blood enters near the portal triads carrying oxygen from the hepatic artery and absorbed chemistry from the portal vein, then moves through the sinusoids toward the central vein. As it travels, oxygen is consumed, nutrients are extracted, hormones and signaling molecules are partly handled, and the chemical profile of that blood changes from one end of the lobule to the other. That means a hepatocyte near the portal region is not living in the same physiological world as a hepatocyte near the central vein. One sees fresher, more oxygen-rich, less previously handled blood. The other sees blood later, after upstream cells have already taken their share and after some of the oxygen reserve has been spent. The gradient is not incidental. It is the reason the liver can divide labor across the tissue instead of forcing every hepatocyte to do every job under identical conditions.

Oxygen matters first because different kinds of hepatic work carry different energetic demands. Oxidative metabolism, gluconeogenesis, and urea cycle activity depend heavily on a strong oxygen-supported ATP supply. ATP, adenosine triphosphate, is the cell's immediately usable energy currency. Hepatocytes in the periportal region, the cells closest to incoming blood, live where oxygen is most available, which makes them better suited for these energy-intensive oxidative tasks. Farther downstream, in the centrilobular region, hepatocytes live with less oxygen and a different metabolic pressure. They are still fully functional liver cells, but they are operating in a shifted environment, one that favors different pathways such as glycolysis, lipogenesis, and much of the cytochrome P450 mediated handling of drugs and toxins. So the oxygen gradient is not just about survival. It is part of the organ's specialization map. It helps determine which chemistry is best performed where.

Substrate gradients matter for the same reason. Portal blood enters carrying newly absorbed glucose, amino acids, microbial metabolites, hormone signals, xenobiotics, and many other compounds that have just arrived from the intestine or associated digestive organs. Xenobiotics are foreign compounds, substances the body does not make for its own structural use. Hepatocytes near the inflow side meet these compounds first and in higher early concentration. That gives them a first-contact role in extraction, buffering, and conversion. As blood moves downstream, the concentration of some substrates falls because upstream hepatocytes have already removed or altered part of the load. Other compounds may rise in relative importance because they are products of upstream handling or because they persist longer in the stream. This means the downstream hepatocyte is not merely under lower oxygen. It is also working on a chemically edited version of portal blood. The substrate gradient therefore helps assign which parts of the liver are first responders to incoming chemistry and which parts are dealing with what remains after that first pass.

These gradients matter most when the tissue is stressed. If oxygen delivery falls because of shock, congestion, or impaired perfusion, the centrilobular cells are often injured first because they already live at the downstream edge of the oxygen supply. If a toxin arrives in high concentration from portal blood and is intrinsically damaging before much processing occurs, the periportal zone may take the first hit because it sees that load earliest. If a drug becomes dangerous only after enzymatic activation, and that activating system is concentrated more heavily downstream, the injury may cluster in the centrilobular region instead. In other words, the gradient explains patterned injury. It tells you why one form of damage does not distribute evenly across the liver. The tissue is not uniform, so the injury is not uniform.

The gradients also protect the organ by making it more flexible. The liver does not need every hepatocyte to behave identically if different parts of the blood stream are already presenting different problems along the route. One region can be optimized for high oxygen oxidative work. Another can be optimized for downstream chemical handling. A middle zone can buffer transitions between the two. That is a more efficient design than forcing one metabolic profile onto every hepatocyte regardless of position. The gradient turns flow into specialization. Specialization turns one organ into a more adaptable system.

So oxygen and substrate gradients matter because they are the reason zonation is real rather than theoretical. They shape what each hepatocyte sees, what each hepatocyte can do most effectively, and what each hepatocyte is most vulnerable to when physiology starts to fail. The liver is not organized as equal cells in equal blood. It is organized as a moving landscape, and the gradient is what gives that landscape its logic.

Which Zones Handle What Kinds of Chemistry

Once the reader understands that blood enters near the portal triads and moves toward the central vein, zonation stops being an abstract labeling system and becomes a way to predict where different chemistry will be handled. Zone 1, the periportal zone, sees blood first. It receives the highest oxygen tension and the freshest delivery of absorbed nutrients, gut-derived signals, and arterial support. That makes it especially well suited for oxidative and energy-intensive tasks that depend on strong oxygen availability. Gluconeogenesis belongs here in large part, because building glucose from lactate, glycerol, and amino acid carbon skeletons is costly work. The urea cycle also leans heavily into this territory, which makes sense because periportal hepatocytes are in a strong position to capture incoming ammonia and other nitrogen-related burden early, before those compounds move farther downstream. Early bile acid related synthesis, cholesterol handling, and some of the liver's first-contact work with incoming portal chemistry also fit the logic of zone 1. This is the front line for high-oxygen extraction, conversion, and stabilization.

Zone 3, the centrilobular zone near the central vein, lives in a different chemical world. By the time blood reaches it, oxygen has already fallen and some of the more immediately extracted substrates have already been partly handled upstream. That lower-oxygen environment shifts what this region tends to do well. Glycolysis, the breakdown of glucose for energy in a way that can proceed under relatively lower oxygen conditions, is more prominent here. Lipogenesis, the synthesis of fatty acids and triglyceride components, is also strongly associated with this downstream region. Much of the cytochrome P450 system is concentrated here as well, and that matters enormously because these enzymes carry a large share of drug metabolism and xenobiotic transformation. Xenobiotics are foreign compounds, substances the body does not make for its own structure or normal signaling. This makes zone 3 chemically powerful but also more vulnerable. It is asked to perform significant transformation work, including activation of some compounds into more reactive intermediates, while living closer to the edge of oxygen sufficiency. That combination explains why centrilobular injury is such a common pattern in toxic and hypoxic liver damage.

Zone 2 sits between these extremes and often gets described too vaguely, but it is not metabolically empty. It is the transitional field where the upstream and downstream chemistries meet and blur. The blood reaching zone 2 is not as oxygen-rich as in zone 1 and not as oxygen-poor as in zone 3. The substrate profile has already been partly edited, but not exhausted. That means zone 2 participates in intermediate handling and helps create continuity across the lobule rather than an abrupt metabolic cliff. It is part of how the liver distributes work fluidly across the gradient. The cleanest way to think about zone 2 is not as a third dramatic specialty, but as the tissue that helps the gradient remain usable. It absorbs transition. It shares burden. It prevents the liver from splitting into only an upstream oxidative zone and a downstream detoxification zone with nothing between them.

This division of chemistry matters because it explains why the same liver can perform mutually different tasks at the same time without forcing every hepatocyte into the same intracellular state. Periportal cells are better placed to handle early incoming substrates and oxygen-demanding chemistry. Centrilobular cells are better positioned for downstream chemical transformation, lipogenic handling, and lower-oxygen metabolic programs. The organ uses blood flow itself to assign labor. That is efficient. It means the liver does not have to build one universal hepatocyte equally optimized for every possible job. It builds a gradient and lets position change responsibility.

It also means certain diseases and exposures create patterned damage because they are colliding with this preexisting chemistry map. A toxin arriving in high concentration from portal blood may damage zone 1 more if its injury happens before much enzymatic conversion. A compound that becomes more dangerous only after cytochrome P450 processing may create zone 3 injury because the activation machinery is concentrated there. Hypoxia often punishes zone 3 first because it already lives at the thinner end of the oxygen gradient. Fat accumulation often appears prominently in zone 3 because lipogenic pathways are strong there and oxidative reserve is less generous. The liver is not assigning chemistry randomly, and it is not receiving injury randomly either.

So when the question is which zones handle what kinds of chemistry, the answer begins with flow. Zone 1 handles much of the high-oxygen first-contact work, especially gluconeogenesis, early ammonia capture through the urea cycle, and other oxidative tasks suited to rich inflow. Zone 3 handles more of the downstream lower-oxygen work, especially glycolysis, lipogenesis, and much of drug and xenobiotic metabolism through cytochrome P450 systems. Zone 2 bridges the two and preserves continuity across the gradient. The chemistry changes because the blood changes. The job changes because the position changes. That is the real logic of zonation.

Why Certain Toxins Damage Specific Zones

Toxins do not damage the liver in a perfectly even wash because the liver is not a perfectly even tissue field. Blood enters near the portal triads and moves toward the central vein. As it moves, oxygen falls, substrate concentrations shift, enzyme exposure changes, and one zone of hepatocytes inherits a different chemical environment from the next. This means a toxin is not arriving at one homogeneous mass of liver cells. It is entering a gradient. Some hepatocytes see that toxin first and in higher incoming concentration. Other hepatocytes see it later, under lower oxygen conditions, after it has already been partly transformed by upstream handling or after it has reached the region where activating enzymes are more abundant. The injury pattern therefore follows the route of blood and the geography of metabolism. A toxin damages the zone that is most exposed to its dangerous form, not necessarily the zone that is physically closest to where the liver sits in the abdomen.

Zone 1, the periportal zone, is the first territory blood reaches after entering from branches of the portal vein and hepatic artery. These hepatocytes live in the highest oxygen environment and meet portal-borne compounds first. That matters for toxins that are intrinsically damaging in the form they arrive in, before much hepatic conversion has occurred. If a harmful compound enters from the gut in high concentration and can injure cells directly on contact or early uptake, periportal hepatocytes are more likely to take the first hit because they stand at the front edge of exposure. They are the earliest readers of portal blood, which makes them powerful for first-pass handling but also vulnerable when the incoming stream itself is immediately hostile. The same logic applies to severe portal-borne inflammatory burdens. First contact can become first injury.

Zone 3, the centrilobular zone near the central vein, is vulnerable for almost the opposite reason. These hepatocytes receive blood later, after oxygen has already been partly extracted upstream, so they live in a relatively lower-oxygen environment from the start. At the same time, zone 3 carries much of the cytochrome P450 machinery that modifies drugs and xenobiotics. Xenobiotics are foreign compounds, substances the body does not make for its own normal structure or signaling. Cytochrome P450 enzymes often protect the body by changing the chemistry of these compounds so they can later be conjugated and excreted, but that first transformation can also generate reactive intermediates, unstable chemically aggressive forms that are more dangerous than the original molecule. This is one reason centrilobular toxicity is so common. Zone 3 is both enzymatically active in biotransformation and metabolically closer to the edge of oxygen sufficiency. It is being asked to run hazardous chemistry with less respiratory margin.

That combination is especially damaging because detoxification is not a clean subtraction problem. A hepatocyte may take up a drug or toxin and, through oxidation, reduction, or hydrolysis, produce a reactive intermediate that now needs rapid containment through glutathione conjugation or other defensive pathways. Glutathione is a major intracellular antioxidant and conjugating molecule used to neutralize electrophilic stress, meaning chemically reactive compounds that would otherwise attack proteins, lipids, and DNA. If that secondary defense capacity is overwhelmed, the cell that created the reactive intermediate becomes the cell exposed to it. Zone 3 is often the place where this happens because much of that activating machinery is concentrated there. A toxin that looked relatively harmless on entry can therefore become dangerous only after it reaches the region most equipped to transform it. The liver injures itself at the site of activation.

Oxygen gradients deepen the pattern. A zone 3 hepatocyte already works in a lower-oxygen setting than a zone 1 hepatocyte. If a toxin impairs mitochondrial function, disrupts oxidative phosphorylation, increases redox stress, or forces the cell into metabolically expensive damage control, centrilobular cells are more likely to fail first because they had less oxygen reserve to begin with. Oxidative phosphorylation is the main oxygen-dependent process by which mitochondria generate ATP, adenosine triphosphate, the cell's immediately usable energy currency. A periportal cell and a centrilobular cell may encounter the same toxic burden, but the centrilobular cell is starting from a more precarious energetic position. That is why hypoxia, venous congestion, and many forms of toxic injury tend to converge on zone 3. It is not the weakest zone in some moral sense. It is the zone living closest to the threshold where high-risk chemistry becomes unsustainable.

The reverse pattern also makes sense once this architecture is clear. If the toxic event depends less on enzymatic activation and more on raw incoming concentration, then zone 1 may be the more exposed region. The first cells that encounter the portal stream can absorb the largest early burden. If the injury is driven by immediate contact with a caustic or strongly reactive incoming compound, periportal damage becomes easier to understand. First delivery becomes first exposure. The liver is not choosing a random zone to injure. It is revealing where the dangerous form of the insult became physiologically real.

So certain toxins damage specific zones because the liver distributes risk along the same gradients it uses to distribute work. Zone 1 receives blood first, with more oxygen and higher incoming concentration of portal-borne compounds. Zone 3 receives blood later, with less oxygen and more of the enzyme systems that can convert a relatively harmless compound into a reactive one. A toxin injures where its chemistry meets the wrong conditions: highest early exposure, greatest activation potential, thinnest oxygen reserve, weakest defensive margin, or some combination of these. The pattern of injury is therefore not a mystery laid on top of the liver. It is the liver's own architecture making toxic logic visible.

Why Ischemia and Hypoxia Create Different Injury Patterns from Drug Toxicity

Ischemia and hypoxia injure the liver through failure of supply. Drug toxicity usually injures it through chemical conversion, reactive burden, or direct molecular interference. Those are not the same kind of assault, so they do not strike the tissue in the same pattern. Ischemia means inadequate blood flow. Hypoxia means inadequate oxygen availability at the tissue level. A drug or toxin, by contrast, may arrive in adequate blood, with adequate oxygen still present, and yet injure hepatocytes because the compound itself is intrinsically harmful or because hepatic enzymes convert it into a more dangerous intermediate. The liver's zonation makes these differences visible. Hepatocytes are not living under equal oxygen conditions across the lobule. Zone 1 near the portal tract sees blood first and receives the richest oxygen supply. Zone 3 near the central vein sees blood last and lives at the thinner end of the oxygen gradient. That means oxygen failure and chemical toxicity begin from different premises before either one has damaged a single cell.

In ischemia and hypoxia, the decisive variable is not what specific compound arrived. It is whether the hepatocyte has enough oxygen to maintain ATP production and keep doing the work that defines it. ATP, adenosine triphosphate, is the cell's immediately usable energy currency, and hepatocytes spend it constantly. They need it for membrane gradients, transporter function, bile secretion, protein synthesis, urea cycle activity, and every other energy-dependent reaction that keeps blood chemistry stable. Zone 3 hepatocytes already receive blood after oxygen has been partly extracted upstream. They are therefore operating with less reserve before any injury begins. When perfusion falls or oxygen delivery drops, they are usually the first cells pushed past the threshold where oxidative phosphorylation, the oxygen-dependent mitochondrial process that generates most ATP, can no longer support function. This is why ischemic and hypoxic injury often cluster in the centrilobular region. The damage follows the preexisting oxygen gradient.

Drug toxicity follows a different map because the danger often depends on chemistry rather than supply. Some drugs are directly toxic in the form they arrive. Others become dangerous only after hepatic metabolism changes them. Much of that activating chemistry is concentrated in zone 3, especially through cytochrome P450 systems in the smooth endoplasmic reticulum. Cytochrome P450 enzymes protect the body by modifying xenobiotics, foreign compounds the body does not make for its own normal structure or signaling, so they can later be conjugated and eliminated. But that first step can produce reactive intermediates, unstable chemically aggressive forms that are more dangerous than the parent compound. Zone 3 therefore often becomes the site of drug injury not because blood flow failed there first, but because the local enzyme profile converted the compound into its damaging form there. The pattern may look superficially similar to hypoxic injury because both often involve centrilobular hepatocytes, but the mechanism is different. In one case the cell dies because oxygen support collapses. In the other it may die because it manufactured its own poison while still being perfused.

That difference matters because the surrounding tissue state is different too. In ischemia and hypoxia, the cell is starving energetically. Ion gradients fail. Membrane integrity weakens. Transport collapses. Bile secretion falters. Urea synthesis and other ATP-intensive pathways fall behind. The injury is global at the level of cellular energy. In drug toxicity, the cell may initially have adequate flow and oxygen yet still accumulate covalent protein damage, oxidative stress, mitochondrial injury, membrane lipid damage, or endoplasmic reticulum stress because of the specific compound or its metabolite. Covalent protein damage means reactive intermediates are chemically binding to cellular proteins in ways that distort their function. Endoplasmic reticulum stress means folding, synthesis, or detoxification burden inside that compartment is exceeding the cell's capacity to manage it cleanly. So the histologic pattern may overlap, but the physiology underneath it does not. Hypoxia removes the fuel for the machinery. Drug toxicity jams or corrodes the machinery while the fuel may still be present.

The periportal region can help show the contrast even more clearly. Zone 1 cells see incoming blood first and in the highest oxygen state. They are comparatively protected against pure oxygen failure for that reason. But they are also the first cells exposed to compounds arriving in high concentration from portal blood. If a toxin is intrinsically damaging before much metabolism has occurred, periportal injury can make sense because the earliest exposure happened there. That pattern is much easier to explain with chemical toxicity than with ischemia. Blood flow problems do not strike zone 1 first under most ordinary hepatic conditions because zone 1 is where oxygen delivery is strongest. Portal-borne toxic burden, however, can strike there first because that is where concentration is highest before the compound has been diluted, extracted, or transformed.

So ischemia and hypoxia create different injury patterns from drug toxicity because they are injuring different vulnerabilities. Oxygen failure follows the flow gradient and punishes the cells living closest to the edge of energetic sufficiency. Drug toxicity follows compound concentration, enzyme distribution, metabolite generation, and intracellular defense capacity. One is fundamentally a perfusion and oxygen problem. The other is fundamentally a molecular burden problem, even when oxygen is still present. The liver's zonation makes those differences readable. It shows whether the cell died because it could no longer power itself or because the chemistry passing through it became incompatible with survival.

How Expert Liver Thinking Begins with Location Inside the Lobule

Expert liver thinking begins when the reader stops asking only what the liver does and starts asking where inside the lobule a given event is happening. That shift changes everything. The novice treats the liver as one sheet of red tissue and assumes a toxin, a hypoxic insult, a metabolic burden, or a cholestatic stress is hitting a uniform field of cells. The expert knows the opposite is true. Blood enters near the portal triad, moves through the sinusoids, and leaves through the central vein. Oxygen falls along that path. Substrate exposure changes along that path. Enzyme distribution changes along that path. The hepatocyte nearest the portal inflow is not living in the same chemical world as the hepatocyte near the central vein. So when injury appears, the first useful question is not simply what kind of liver disease this is. The first useful question is where the tissue is being asked to fail.

That question works because location inside the lobule is already a map of physiology. Periportal hepatocytes in zone 1 see incoming blood first. They live in the richest oxygen environment and meet the highest early concentrations of portal-borne compounds. They are well positioned for oxidative work, early ammonia handling through the urea cycle, and other first-contact processing tasks. Centrilobular hepatocytes in zone 3 see blood later, after oxygen has already been consumed upstream and after some chemistry has already been altered. They carry more of the downstream drug-metabolizing burden through cytochrome P450 systems and more of the lower-oxygen metabolic program. Zone 2 bridges those extremes. So when an expert sees centrilobular necrosis, steatosis, cholestatic stress, periportal inflammation, or mixed zonal injury, that pattern is not decorative pathology. It is a clue about flow, oxygen, enzyme exposure, and the form in which the insult became dangerous.

This is why the lobule is not just a histology concept. It is a reasoning tool. Hypoxia usually strikes the cells already living closest to the oxygen edge, which means zone 3 often suffers first. A portal-borne toxin that is dangerous in its incoming form may hit zone 1 harder because those cells encounter the highest early concentration. A drug that becomes damaging only after enzymatic activation may injure zone 3 because that is where more of the activating machinery is concentrated. Fat accumulation often favors downstream regions because lipid handling, oxygen reserve, and metabolic burden are not distributed evenly. The expert does not memorize these as disconnected associations. The expert asks what blood looked like when it entered, what it looked like by the time it arrived downstream, and which hepatocytes were structurally positioned to pay the cost.

Location also helps prevent shallow interpretation of symptoms, labs, and pathology. Elevated transaminases alone do not tell you enough. Jaundice alone does not tell you enough. Even the word hepatitis does not tell you enough. The organ has to be read through its microgeography. Is the injury centered where oxygen is lowest. Is it centered where bile is being secreted poorly. Is it centered where portal inflow first contacts the tissue. Is the exchange surface being thickened. Are central veins congested. Are portal tracts inflamed. These are location questions, and they are the beginning of mechanism. Without them, liver reasoning collapses into category labels. With them, the pathology starts explaining itself.

So expert liver thinking begins with location inside the lobule because the lobule is where flow becomes function and where function becomes vulnerability. The liver is not injured in the abstract. It is injured in a place, under a gradient, inside a built architecture. Once the reader starts there, the organ stops looking mysterious. A pattern of damage becomes a pattern of position. A pattern of position becomes a pattern of mechanism. That is the beginning of real hepatology.

Part III

The Liver as the Keeper of Blood Chemistry

Chapter 9

Glucose Traffic After a Meal

What Happens to Absorbed Carbohydrate Before It Reaches the General Circulation

Carbohydrate does not leave the intestine as finished blood sugar already assigned to the rest of the body. It leaves as a post-digestive glucose load that has to pass through hepatic control first. That is the first correction the reader needs to make. Digestion in the small intestine breaks starches and disaccharides down into absorbable sugars, mainly glucose, galactose, and fructose. Those sugars cross the intestinal epithelium through transporters on the enterocyte, the absorptive cell lining the small intestine, then exit the cell on the blood-facing side and enter the portal circulation. At that point they are inside the body, but they are not yet part of the final systemic blood chemistry the brain, muscle, kidney, adipose tissue, and endocrine system will see. They are in transit toward the liver. The portal vein is carrying them to the organ that will decide how much of that carbohydrate load is extracted immediately, how much is stored, how much is converted, and how much is released onward into the general circulation.

This matters because intestinal absorption is episodic and often abrupt. A carbohydrate-rich meal can deliver a sharp rise in portal glucose concentration that would be metabolically destabilizing if it were allowed to strike the systemic circulation without buffering. The body solves that problem by putting the liver in the middle of the route. Portal blood rich in absorbed glucose enters the sinusoids, the liver's low-pressure exchange channels, where hepatocytes meet that incoming sugar before it reaches the rest of the body at full force. Hepatocytes take up glucose through transport systems that allow high-volume traffic when concentrations rise. They do not merely trap sugar blindly. They read the fed state through insulin, portal substrate load, and intracellular metabolic conditions, then shift toward uptake and storage. Some of that incoming glucose is immediately phosphorylated, which means a phosphate group is attached to it, helping keep it inside the cell and committing it to further metabolism. From there it may be burned for energy, routed into glycogen synthesis for storage, or used as substrate for other biosynthetic pathways. The key point is that the first destination of absorbed carbohydrate is not the whole body equally. It is hepatic review.

The liver also uses the meal to decide timing. The intestine delivers according to what was eaten, how quickly the stomach empties, how fast enzymes work, and how rapidly transporters move sugar across the gut wall. The rest of the body cannot operate on that raw schedule. The brain, red blood cells, endocrine tissues, and peripheral organs need something steadier than the chemistry of a single meal arriving in pulses. The liver acts as the first buffer by absorbing part of the incoming glucose surge and converting some of it into glycogen, the branched storage polymer that lets large amounts of glucose be packed in a compact and osmotically manageable form. Osmotically manageable matters because free glucose in excess would pull water with it and distort intracellular balance if it were stored without conversion. Glycogen solves that storage problem. So before absorbed carbohydrate becomes general blood sugar, some of it has already been delayed, compacted, and held in reserve by the liver.

Not all absorbed carbohydrate is handled identically, and that difference matters to blood chemistry. Glucose and galactose enter portal blood and are heavily subject to hepatic first-pass handling, but fructose is even more liver-centered in its early fate. Much of the fructose absorbed from the intestine is taken up rapidly by hepatocytes because peripheral tissues are not built to be the main first-pass site for large fructose loads. That means the liver is not only buffering quantity. It is sorting forms. Different sugars arrive through the same portal route but impose different metabolic demands once they reach the hepatocyte. Expert understanding begins there. Carbohydrate is not one undifferentiated fuel blob entering the bloodstream. It is a class of absorbed sugars entering a gatekeeper organ that decides what chemical form, what concentration, and what timing the rest of the body will be allowed to experience.

This first-pass control also protects the systemic circulation from the intestine's raw irregularity. If absorbed carbohydrate poured directly into general blood without hepatic extraction, post-meal glucose excursions would be steeper, insulin signaling would be forced into a more reactive pattern, and tissues not designed to receive the first concentrated wave would be hit harder and faster. The liver dampens that rise. It does not eliminate it, and it does not act alone, because insulin secretion, peripheral uptake, gastric emptying, meal composition, and muscle activity all matter. But anatomically and chemically the liver is the first large buffer the absorbed carbohydrate load meets. This is why liver physiology belongs at the center of glucose physiology. Blood sugar is not only a pancreatic story and not only a peripheral tissue story. It is a portal traffic story first.

So what happens to absorbed carbohydrate before it reaches the general circulation is not passive transport. It is staged control. Digestion reduces food carbohydrate to absorbable sugars. The intestine moves those sugars into portal blood. The portal vein carries them to the liver. Hepatocytes extract a substantial fraction, phosphorylate, store, burn, or redirect that incoming substrate according to the fed state. Only after this first round of hepatic handling does the remaining carbohydrate load move onward through the hepatic veins into the wider circulation. Blood sugar after a meal is therefore not simply intestinal output. It is intestinal output after liver judgment.

Hepatic Uptake of Glucose

Hepatic uptake of glucose begins the moment portal blood leaves the intestine carrying the first serious sugar load from a meal. This is where carbohydrate stops being a digestive event and becomes a blood chemistry event. Glucose absorbed across the intestinal epithelium enters the portal vein and reaches the liver in concentrations that are often far higher than what the rest of the circulation should see unbuffered. The hepatocyte is positioned to intercept that load before it disperses system-wide. Its blood-facing membrane meets portal plasma across the sinusoid and the space of Disse, and from there glucose moves into the cell through transport systems built for high-capacity exchange rather than tight scarcity control. That matters because the liver is not trying to protect itself from glucose. It is trying to absorb enough of the incoming surge to prevent the bloodstream from becoming the raw extension of the meal.

Once glucose enters the hepatocyte, the first important step is phosphorylation. The cell adds a phosphate group to glucose, converting it into glucose-6-phosphate. That change matters because it traps the molecule inside the cell and commits it to metabolism rather than allowing it to drift back out into blood unchanged. In the liver this phosphorylation step is carried out largely by glucokinase, an enzyme suited for handling high post-meal glucose loads without saturating too early. That design is important. The liver is not supposed to hoard glucose aggressively during fasting when blood levels are low. It is supposed to respond strongly when portal glucose rises after a meal. Glucokinase fits that logic. It becomes especially useful when glucose is abundant, which makes hepatic uptake meal-responsive rather than indiscriminate. The hepatocyte therefore does not merely absorb glucose. It chemically claims it.

What happens next depends on the fed state, insulin signaling, and the broader energetic condition of the cell. Some of that glucose-6-phosphate is routed into glycogen synthesis, where glucose units are linked together into a large branched storage polymer that can be held in compact form without the osmotic burden that free glucose would impose. Some enters glycolysis, the pathway that breaks glucose down toward pyruvate and downstream energy production. Some may be diverted into the pentose phosphate pathway, which helps generate reducing power and biosynthetic intermediates, or into lipogenic pathways when carbohydrate availability stays high enough to push the liver beyond immediate storage and oxidation needs. The important point is that hepatic uptake is not one mechanical sink. It is the opening move in a sorting process. The liver is deciding whether incoming glucose should be stored, burned, redirected into biosynthesis, or held in a form that can later be released back into circulation when feeding stops.

Insulin shapes this uptake, but not in the simplified way it is often taught. Peripheral tissues such as skeletal muscle and adipose tissue rely heavily on insulin-responsive glucose transporters to increase glucose entry after a meal. The liver works differently. Hepatic glucose entry is driven strongly by the concentration gradient created by portal delivery and by the activity of glucokinase once glucose is inside. Insulin still matters, but more as a metabolic signal telling the hepatocyte what to do with the glucose it is receiving than as the sole gatekeeper of entry itself. It promotes glycogen synthesis, suppresses hepatic glucose output, and shifts the cell toward storage and fed-state metabolism. This is a crucial distinction. The liver is anatomically positioned to take first possession of post-meal glucose because of where portal blood goes, not merely because insulin waves a membrane gate open. Position comes first. Hormonal instruction refines the response.

This is why hepatic uptake of glucose is one of the body's main anti-chaos mechanisms after a meal. The intestine absorbs according to what was eaten and how fast the meal was digested. The liver receives that sugar load and converts part of it from immediate blood burden into intracellular substrate, glycogen reserve, or downstream metabolic material. Without that uptake step, systemic blood glucose would reflect the intestine too directly. The bloodstream would become a more violent mirror of gastric emptying and carbohydrate absorption. The liver interrupts that. It takes the first shock. It turns a meal into something the circulation can live with.

Glycogen Synthesis and Short-Term Storage

Once glucose has entered the hepatocyte and been phosphorylated into glucose-6-phosphate, the liver has to decide whether that carbon should be burned now or held for later. One of its main answers is glycogen synthesis. Glycogen is the liver's short-term carbohydrate reserve, a large branched polymer made by linking glucose units together into a compact stored form that can be mobilized quickly when blood glucose begins to fall. This matters because free glucose is osmotically active. If the hepatocyte tried to store large amounts of sugar simply as free intracellular glucose, water movement and cellular balance would become unstable. By converting many individual glucose molecules into one large branched storage structure, the cell can hold substantial carbohydrate reserve without paying that same osmotic price. Glycogen is therefore not just stored sugar. It is sugar structurally reorganized into a storable state.

The synthesis begins after glucose-6-phosphate has been generated inside the hepatocyte. That molecule is shifted into glucose-1-phosphate and then activated further into uridine diphosphate glucose, the immediate donor form used to extend the glycogen chain. Glycogen synthase then adds glucose units onto the growing polymer, while branching enzyme creates side branches that make the final structure highly branched and compact. Those branches are not decorative architecture. They matter because they create many terminal ends where glucose can later be added or removed rapidly. The liver is not building a static starch-like lump. It is building a readily accessible reserve designed for quick turnover. Glycogen synthesis is therefore a timing operation. The organ is taking part of the meal and converting it into near-future blood stability.

This storage role is called short-term for a reason. Hepatic glycogen is not the body's long-horizon calorie vault in the way adipose tissue stores triglyceride. It is a rapid-access glucose reserve built to buffer the gap between meals, overnight fasting, and other short intervals when the bloodstream still needs glucose but the intestine is no longer supplying it. The liver stores in the fed state so it can release in the post-absorptive state. That is why glycogen synthesis belongs in the same chapter as blood chemistry control rather than in some isolated storage chapter. The liver is not hoarding carbohydrate for itself. It is holding carbohydrate in trust for the circulation.

Insulin is part of the signal that pushes the hepatocyte in this direction. After a meal, rising portal glucose and rising insulin tell the liver that incoming substrate is abundant and immediate systemic release is not the priority. Glycogen synthase activity is favored. Hepatic glucose output is suppressed. The cell shifts toward storage rather than release. But the deeper logic is anatomical before it is hormonal. The liver sees the sugar surge first because portal blood reaches it first. That gives the organ first possession of the meal's glucose burden, and glycogen synthesis is one of the main ways it converts that burden into order. Instead of letting all absorbed carbohydrate continue downstream as immediate blood sugar, the liver delays part of it by packing it into glycogen.

This short-term reserve also protects the rest of the body from the irregularity of eating. Meals do not arrive continuously. Digestion is pulsed. The brain, red blood cells, renal medulla, and other glucose-dependent tissues still require a more continuous glucose environment than the intestine can provide. Glycogen gives the liver a way to smooth that discontinuity. A carbohydrate-rich meal becomes partly immediate fuel and partly stored future support. The bloodstream does not have to live at the mercy of every feeding gap because the liver has built a temporary reserve in advance.

So glycogen synthesis is best understood as a structural solution to a timing problem. The liver takes incoming glucose, chemically traps it inside the hepatocyte, then reorganizes it into a branched polymer that can be stored compactly and released quickly later. It is short-term storage because it is designed for near-future use, not for indefinite energy banking. The meal arrives in pulses. Glycogen lets the liver turn part of that pulse into continuity.

How the Liver Dampens Post-Meal Glucose Spikes

A meal can deliver glucose into portal blood far faster than the rest of the body should be forced to absorb it. That is the basic problem the liver solves after eating. The intestine is an absorptive organ, not a precision blood-sugar regulator. Once carbohydrate has been digested down to absorbable sugars and moved across the intestinal lining, portal glucose can rise sharply depending on meal composition, gastric emptying speed, intestinal transit, and transporter activity. If that incoming load poured straight into the general circulation without interception, systemic glucose excursions would be steeper, insulin demand would be harsher, and peripheral tissues would be exposed to the intestine's raw timing rather than to controlled metabolic release. The liver sits between those two events so the chemistry of absorption does not become the chemistry of systemic shock.

It dampens that rise first by taking first possession of portal glucose. Blood from the intestine reaches the liver before it reaches the wider circulation, which means hepatocytes see the meal's glucose surge in concentrated form and get the first opportunity to remove part of it. Glucose enters hepatocytes and is phosphorylated, trapping it inside the cell as glucose-6-phosphate. That step matters because once the glucose has been chemically claimed in this way, it is no longer just passing through as free sugar. The liver can now decide its fate. Some is burned immediately according to local energy demand. A substantial fraction is routed toward glycogen synthesis, which converts glucose into a compact branched storage polymer. Some may move into other metabolic pathways depending on the broader fed state. The important point is that hepatic uptake interrupts the direct line between intestinal absorption and systemic glycemia. The liver does not stop glucose from entering the body. It slows the speed at which that glucose is allowed to become everybody else's problem.

Insulin reinforces this buffering state, but the liver's anatomical position is what makes the first-pass control possible in the first place. After a meal, rising insulin signals that the body is in a fed state and supports hepatic glycogen synthesis while suppressing hepatic glucose output. But insulin is not acting on an organ that discovered glucose late. It is acting on an organ already standing in front of the main incoming stream. That means the liver can absorb and store part of the glucose burden at the exact moment the burden appears, rather than trying to correct the spike only after systemic blood sugar has already risen more dramatically. This is one reason blood sugar control cannot be reduced to pancreatic insulin release alone. The liver is the first major hydraulic buffer in the system. It receives the wave before the rest of the body does.

The dampening effect also comes from timing. The intestine delivers in pulses. The body needs steadier conditions than that. By storing part of the incoming glucose as glycogen, the liver converts an abrupt carbohydrate event into delayed availability. Instead of allowing all absorbed glucose to continue into systemic circulation immediately, it captures part of the surge and holds it in reserve for later release when the meal is over and blood glucose begins to fall. This is the real anti-spike logic of hepatic glycogen synthesis. It is not storage for storage's sake. It is storage used as a blood chemistry stabilizer. The meal arrives now. The full burden does not have to.

There is another layer to this buffering that often gets missed. The liver is not only reducing the peak. It is shaping the profile of the whole post-meal curve. By extracting glucose early, storing some of it, and suppressing its own glucose production in the fed state, the liver lowers the amplitude of the rise and delays the need for later correction. A sharper spike usually demands a sharper fall. A dampened rise is metabolically quieter. Peripheral tissues are exposed to a more controlled systemic glucose pattern, and the body avoids having post-meal chemistry driven entirely by the raw speed of intestinal absorption. The liver therefore behaves like a metabolic shock absorber. It does not erase the meal. It prevents the bloodstream from feeling the full force of it all at once.

This is why hepatic dysfunction can destabilize glucose handling even when the pancreas is still present and insulin is still being produced. A liver that cannot take up glucose cleanly, cannot synthesize glycogen effectively, or continues inappropriately releasing glucose during the fed state loses part of its buffering capacity. The result is not just poor storage. It is a bloodstream forced to reflect the intestine more directly. Post-meal spikes become easier to generate because the first major interceptive organ is no longer imposing enough order on the incoming load.

So the liver dampens post-meal glucose spikes by standing in the path of portal glucose, taking up a substantial fraction early, phosphorylating and trapping that glucose inside hepatocytes, converting part of it into glycogen, and suppressing its own glucose output while the meal is still arriving. The intestine absorbs. The liver buffers. Blood sugar after a meal is not simply the result of what was eaten. It is what remains after hepatic first-pass control has taken its share.

Why the Liver Is Central to Fed-State Stability

The fed state is not simply the condition of having nutrients present in the gut or rising in the blood. It is the condition in which the body has to absorb an incoming nutrient load without allowing that load to destabilize the internal environment. That is why the liver is central. The intestine can digest and absorb, but it does not impose whole-body order on what it absorbs. It moves glucose, amino acids, fructose, vitamins, minerals, and countless other compounds across the epithelial barrier and into portal blood. From there the chemical problem becomes immediate. Too much glucose arriving too fast will distort blood sugar. Too much amino acid traffic without controlled handling will increase nitrogen burden. Fructose requires heavy first-pass hepatic management. Lipid processing has to be coordinated with bile flow, lipoprotein handling, and downstream tissue delivery. The fed state therefore is not secured by absorption alone. It is secured by hepatic first-pass control.

This is easiest to see with glucose, but the principle is broader than glucose. After a meal, the liver receives portal blood before the rest of the circulation does, which gives hepatocytes first possession of the post-digestive surge. They can extract part of the incoming glucose, phosphorylate it so it stays in the cell, convert some into glycogen, divert some into immediate energy metabolism, and suppress hepatic glucose release while external supply is abundant. That alone stabilizes the fed state by preventing the bloodstream from becoming a direct mirror of intestinal absorption. But the liver is also handling amino acids at the same time, sorting them toward protein synthesis, oxidative use, or nitrogen disposal. It is taking in fructose and routing it through hepatic metabolism. It is repackaging lipid traffic, synthesizing and secreting bile components, and adjusting hormone handling in response to the meal. Fed-state stability is therefore not one pathway staying calm. It is many pathways being coordinated at once, and the liver is one of the few organs placed to do that coordination in real time.

The liver also matters because the fed state is a signal, not just a substrate load. Rising insulin, increased portal nutrient delivery, changing gut hormones, and the metabolic consequences of absorption tell the liver that the body is in intake mode rather than fasting mode. Hepatocytes respond by shifting program. Glycogen synthesis is favored. Gluconeogenesis is suppressed. Hepatic glucose output falls. Lipid handling changes. Protein synthesis priorities shift. Bile production continues in a context now supported by incoming nutrients rather than reserve mobilization. The organ is therefore not simply acting as a passive sponge for what the intestine delivers. It is reading the fed state and reorganizing internal chemistry around that information. This is one reason the liver is so central to metabolic coherence. It does not just receive nutrients. It changes the meaning of nutrient arrival for the whole body.

Fed-state stability also depends on timing, and timing is one of the liver's core talents. The meal arrives in bursts. Blood chemistry cannot be allowed to live in bursts of equal violence. The liver smooths this by turning part of immediate abundance into delayed availability. Glycogen is the clearest example, but the same logic applies more broadly. Incoming nutrients are buffered, stored, transformed, or repackaged so the rest of the body experiences a more disciplined curve rather than a raw digestive surge. The liver does not eliminate the fed state. It civilizes it. That is the deeper physiological point. A stable fed state is not one in which nutrients simply appear in blood. It is one in which nutrient appearance has already been filtered through an organ capable of absorbing irregularity and returning order.

This is why fed-state instability appears so quickly when hepatic handling is impaired. If the liver cannot take up glucose cleanly, cannot store glycogen appropriately, cannot suppress its own glucose output, cannot process fructose efficiently, or cannot coordinate lipid and amino acid handling under fed-state signals, then the bloodstream begins reflecting the intestine too directly. Post-meal glucose excursions become harsher. Nitrogen handling becomes less controlled. Lipid traffic becomes less orderly. Hormonal and metabolic transitions become rougher. The person may still be eating, digesting, and absorbing, but the fed state is no longer being governed properly. Intake is happening. Stability is not.

So the liver is central to fed-state stability because it stands between the meal and the organism. It receives the first flood of absorbed chemistry, interprets the hormonal meaning of that flood, extracts part of it, stores part of it, transforms part of it, suppresses pathways that belong to fasting, and releases onward only what the circulation can handle more safely. The intestine creates entry. The liver creates control. That is the difference between being fed and being metabolically stable while fed.

The Relationship Between Insulin, Portal Blood, and Hepatic Handling

The liver does not respond to insulin in isolation, and it does not respond to portal blood as if substrate and hormone were separate stories. They arrive as one fed-state message. After a meal, carbohydrate is digested and absorbed, glucose enters the portal vein, and that portal blood delivers a concentrated sugar load directly to the liver before the rest of the body sees it in full. At the same time, rising blood glucose stimulates insulin secretion from the pancreas, and because the pancreas also drains into the portal circulation, the liver is exposed early to relatively high portal insulin concentrations during the post-meal state. That arrangement matters. The hepatocyte is not being asked to interpret glucose alone or insulin alone. It is receiving substrate abundance and hormonal instruction together, at the same anatomical checkpoint. Portal blood says the meal has arrived. Insulin says store, restrain output, and handle this as fed-state chemistry rather than fasting chemistry.

This is why hepatic glucose handling after a meal is so different from glucose handling in muscle or adipose tissue. In peripheral tissues, insulin plays a stronger gate-opening role for cellular glucose entry through insulin-responsive transport systems. The liver works from a different design. Hepatocytes are already positioned in the path of the incoming portal glucose surge, and glucose can enter them efficiently when concentrations rise. Insulin still matters enormously, but less as a simple permission signal for entry and more as a signal that changes intracellular fate. It pushes the hepatocyte toward glycogen synthesis, favors glycolytic and lipogenic pathways when substrate abundance is high enough, and suppresses hepatic glucose production by inhibiting glycogen breakdown and gluconeogenesis. Gluconeogenesis is the synthesis of glucose from noncarbohydrate substrates such as lactate, glycerol, and certain amino acids. So insulin in the liver is not merely about pulling glucose in. It is about telling the cell what not to release and what to build instead.

Portal blood gives that instruction a chemical context. The liver is not being told to enter fed mode based on hormone alone. It is being hit simultaneously by the physical arrival of glucose, fructose, amino acids, gut-derived signals, and post-digestive substrate traffic. That is why portal blood and insulin have to be taught together. If insulin rises but substrate does not arrive, the hepatic response is not the same as when insulin rises in the presence of a dense portal nutrient stream. If glucose arrives without the proper insulin response, hepatic extraction, storage, and suppression of glucose output become less coherent. The liver is reading both the message and the material. Hormone without substrate is incomplete instruction. Substrate without hormone is incomplete control.

This partnership is what lets the liver act as the first stabilizer of the fed state. Rising portal glucose gives hepatocytes the burden to manage. Rising portal insulin gives them the command to shift toward storage and away from endogenous glucose release. Together those signals help the liver take part of the incoming glucose load into the cell, phosphorylate it, and route it toward glycogen rather than allowing the full burden to continue directly into the systemic circulation. At the same time, insulin suppresses hepatic glucose export, which means the liver stops adding internally produced glucose to a bloodstream that is already being supplied from the intestine. That double effect matters. The liver is not only taking more in. It is also sending less out. Post-meal blood sugar stability depends heavily on both halves of that shift.

The same logic extends beyond glucose. Amino acid handling changes in the fed state. Lipid handling changes. Fructose is taken up and processed in a strongly liver-centered way. Hormonal metabolism shifts. Protein synthesis priorities shift. The liver is therefore not merely a sugar sponge responding to insulin. It is an organ reading portal blood and insulin together as a systems-level state change. Fed state is not just an increase in fuel. It is a reprogramming of hepatic traffic. Incoming abundance is being converted into storage, synthesis, packaging, and restrained output rather than continued endogenous release.

When this relationship breaks down, the consequences spread quickly. If portal substrate delivery is high but insulin signaling is impaired, the liver may fail to suppress glucose production properly even while glucose is arriving from the intestine. It may continue acting partly as if the body were fasting. If insulin is present but hepatocyte handling is metabolically disordered, glycogen synthesis may be blunted, glucose buffering may weaken, and downstream lipid production may become more chaotic. The bloodstream then begins reflecting the meal too directly, because the organ meant to absorb the first shock is no longer reading hormone and substrate in a coordinated way.

So the relationship between insulin, portal blood, and hepatic handling is best understood as a three-part control system. Portal blood delivers the meal. Insulin declares the fed state. The hepatocyte translates both into action. It takes up incoming substrate, suppresses glucose release, shifts toward storage and synthesis, and helps prevent intestinal absorption from becoming immediate systemic instability. The intestine provides the surge. Insulin provides the command. The liver turns the two into order.

Chapter 10

Fasting, Glycogen, and Gluconeogenesis

The moment eating stops, the body does not enter a void. It enters a transition problem. Nutrient entry from the intestine falls away, but tissue demand does not. The brain still needs glucose. Red blood cells still require glucose because they lack mitochondria and cannot oxidize fat directly for their own energy. The kidneys, the endocrine system, immune cells, and every other tissue continue drawing on the circulation even though the gut is no longer sending a fresh post-meal stream through the portal vein. This is why fasting has to be understood as a controlled metabolic handoff rather than as simple absence of food. The liver sits at the center of that handoff. In the fed state it receives incoming substrate and buffers the meal. In fasting it has to reverse the direction of support. Instead of absorbing the first shock of absorption, it now has to keep blood chemistry stable from internal reserves and internally generated output.

The first phase of that support is hepatic glycogen breakdown. During the fed state the liver stored part of incoming glucose as glycogen, a branched polymer compact enough to hold large amounts of carbohydrate without the osmotic instability that free intracellular glucose would create. Once external intake stops and insulin falls while glucagon rises, the liver begins dismantling that glycogen reserve. Glycogen phosphorylase cleaves glucose units from the glycogen chain, and those units are processed toward glucose-6-phosphate. The liver has a special advantage here because it contains glucose-6-phosphatase, an enzyme that allows glucose-6-phosphate to be converted into free glucose and released back into the bloodstream. Muscle stores glycogen too, but muscle glycogen is mainly for local use. Liver glycogen is for the circulation. That difference matters. The liver is not just feeding itself during fasting. It is feeding the blood.

This glycogen phase is a bridge, not a permanent solution. The reserve is useful precisely because it is rapid and temporary. It buys time while the body shifts into the deeper fasting program. As glycogen stores begin to fall, the liver increasingly turns to gluconeogenesis, the synthesis of new glucose from noncarbohydrate substrates. Lactate returning from peripheral tissues, glycerol released from triglyceride breakdown in adipose tissue, and amino acid carbon skeletons, especially from alanine and other glucogenic amino acids, begin feeding this pathway. This is where fasting reveals its real architecture. The bloodstream is no longer being maintained by what just arrived from the intestine. It is being maintained by the liver's ability to rebuild glucose from materials already inside the organism. The body does not merely coast on stored sugar. It starts manufacturing replacement.

That manufacturing is expensive and tightly organized. Gluconeogenesis is not glycolysis run backward in a casual way. It requires distinct enzymes, mitochondrial participation, ATP, adenosine triphosphate, as the cell's immediately usable energy currency, and coordinated control over substrate flow. Pyruvate can be converted into oxaloacetate in the mitochondria, then moved through a sequence of reactions that eventually regenerate glucose. Lactate arriving from anaerobic metabolism in peripheral tissues can be recycled in this way. Glycerol from fat breakdown enters the pathway through a different route. Amino acid carbon skeletons join through still other entry points. The liver is therefore rebuilding blood glucose from fragments of prior metabolism. That is why fasting is not simple depletion. It is a highly managed reconstruction of circulating fuel.

At the same time, the body is shifting away from total dependence on glucose where it can. Fatty acids released from adipose tissue begin supplying more energy to many peripheral tissues, and the liver increases fatty acid oxidation to support its own energy needs and later ketone production. That point is essential because gluconeogenesis costs energy. The liver cannot manufacture glucose for the rest of the body unless it is powering that work through other fuels. Fat oxidation becomes one of the main energetic supports for hepatic glucose maintenance during fasting. This is also where the division of labor becomes clearer. The liver uses fat heavily to spare glucose for tissues that need it most. It is protecting the circulation by changing its own substrate strategy.

The hormonal setting makes this possible. Insulin falls as nutrient inflow declines. Glucagon rises and signals the liver to mobilize glycogen and support glucose output. Catecholamines and cortisol can reinforce this broader fasting shift depending on timing and stress conditions. But hormones do not act on an abstract metabolic space. They act on an organ already structured for reserve storage, blood glucose release, and reconstruction of sugar from noncarbohydrate material. The liver is responsive because it was built for this exact state change. In the fed state it stores the excess. In fasting it turns reserve and recycled substrates into continuity.

This is why the body does not crash the moment a meal ends. The liver prevents that. It absorbs the end of feeding the same way it absorbed the beginning of feeding, by imposing order on a transition the intestine cannot manage once intake stops. Glycogen provides the short early bridge. Gluconeogenesis carries the deeper fasting burden. Fat oxidation helps power the process. The bloodstream remains livable because the liver changes from a first-pass buffer of incoming carbohydrate into a producer of outgoing glucose. Fasting is therefore not defined by empty stomach alone. It is defined by hepatic support taking over when intestinal support has gone quiet.

Glycogenolysis as the First Emergency Bridge

When the body stops eating, it does not immediately begin living on newly manufactured glucose. It reaches first for what the liver already stored in advance. That first response is glycogenolysis, the breakdown of hepatic glycogen into usable glucose output. Calling it the first emergency bridge is accurate because it is a short, fast-access system designed to keep blood glucose from falling abruptly while deeper fasting metabolism comes online. The intestine has gone quiet. Portal delivery of meal-derived carbohydrate has collapsed. But the brain, red blood cells, renal medulla, and other glucose-dependent tissues are still drawing from the circulation. The liver solves that gap by opening its glycogen reserve. This is the first handoff from fed physiology to fasting physiology.

The reserve exists because the liver built it during the fed state. Glucose arriving after a meal was taken up by hepatocytes, phosphorylated, and linked into glycogen, a branched storage polymer compact enough to hold large amounts of carbohydrate without the osmotic burden that free glucose would impose inside the cell. Once feeding stops, insulin falls and glucagon rises, and that hormonal shift tells the hepatocyte to reverse direction. Instead of trapping incoming glucose and storing it, the cell begins dismantling glycogen through glycogen phosphorylase activity, releasing glucose units as glucose-1-phosphate and then converting them toward glucose-6-phosphate. The liver has a crucial advantage here. It expresses glucose-6-phosphatase, the enzyme that can remove the phosphate and generate free glucose for release into blood. That final step is what makes hepatic glycogen a circulatory reserve rather than merely an intracellular convenience.

This is why glycogenolysis is an emergency bridge rather than a full solution. It is rapid, efficient, and already built before the fast begins, but it is finite. The liver cannot support the body indefinitely on glycogen because glycogen stores are limited by design. It is a short-horizon buffer, not a long-duration fuel warehouse. What it does exceptionally well is buy time. It prevents the bloodstream from experiencing the immediate disappearance of dietary carbohydrate as an immediate crisis. It keeps blood glucose from collapsing in the early fasting period while the body shifts substrate use, increases fat mobilization, and ramps up gluconeogenesis, the synthesis of new glucose from noncarbohydrate substrates such as lactate, glycerol, and certain amino acids. Glycogenolysis is therefore not the final answer to fasting. It is the opening protection against the metabolic shock of no longer being fed.

The speed of the system is part of its value. Glycogen is highly branched, which means it has many terminal ends where glucose units can be removed quickly. That branching architecture is not incidental chemistry. It is functional design. The liver built glycogen in a form that could be mobilized fast because blood chemistry sometimes needs fast support. A fall in blood glucose cannot wait for a completely new metabolic program to be assembled from scratch. The reserve has to be structurally ready. Glycogen gives the liver that immediate readiness. Stored glucose is converted into a quickly accessible output stream at the exact moment the circulation begins losing intestinal supply.

This early bridge also reveals the liver's larger role in blood chemistry. The liver is not simply storing carbohydrate for itself. It is storing carbohydrate so the bloodstream can remain usable when intake stops. Muscle glycogen is different. Muscle lacks the final enzymatic step needed to release free glucose into the circulation at meaningful scale, so its glycogen is mainly for local work. Hepatic glycogen is different because its destination is systemic stability. The liver is acting on behalf of the body, not merely on behalf of its own cells. That is why glycogenolysis belongs in a chapter about blood chemistry and not just about fuel storage.

When glycogenolysis is weak, delayed, or exhausted too early, the body feels the transition more sharply. The liver has less reserve to buffer the early fast, and the burden shifts more quickly onto gluconeogenesis and other compensatory systems. That does not simply change fuel bookkeeping. It changes how stable the blood remains in the first hours without food. So glycogenolysis deserves to be understood as the first emergency bridge in the literal physiological sense. It is the liver's immediate answer to the silence of the gut, a prebuilt reserve opened just fast enough to keep the circulation livable while the deeper fasting machinery takes over.

Gluconeogenesis as the Deeper Fasting Program

Glycogen can hold the early line, but it cannot carry the whole fast. It is a short reserve, useful because it is already built, limited because it is already finite. Once that reserve begins to fall, the liver has to switch from releasing stored glucose to manufacturing new glucose from noncarbohydrate material. That switch is gluconeogenesis, and it is the deeper fasting program because it does not depend on what was stored yesterday nearly as much as it depends on what the body can recycle and reorganize now. The intestine is silent. Portal delivery of meal-derived sugar has dropped away. The bloodstream still needs glucose. The liver answers by rebuilding it from fragments already moving through the body. Fasting therefore becomes less a story of stored carbohydrate and more a story of metabolic reconstruction.

This reconstruction begins with substrate capture. Lactate returns from peripheral tissues, especially from red blood cells and other sites relying more heavily on glycolysis. Glycerol is released when triglycerides are broken down in adipose tissue. Certain amino acids, especially alanine and other glucogenic amino acids, contribute carbon skeletons after their nitrogen has been removed and handled. These materials are not glucose, but they are close enough in carbon logic to be redirected into glucose synthesis if the liver has enough energy and enzymatic control to do it. That last condition matters. Gluconeogenesis is not a passive reversal of digestion. It is an energetically expensive, highly regulated pathway requiring mitochondrial participation, cytosolic steps, ATP, and directional control over carbon flow. ATP, adenosine triphosphate, is the cell's immediately usable energy currency, and without enough of it the pathway cannot run at meaningful scale.

The liver is uniquely suited for this because it is not trying to survive fasting only for itself. It is trying to preserve blood chemistry for tissues that cannot fully escape glucose dependence. Red blood cells have no mitochondria and therefore cannot oxidize fatty acids directly for their own energy. Parts of the brain still require substantial glucose input even as fasting deepens. Other tissues use glucose selectively or depend on it under certain conditions. The liver reads that demand and converts circulating fragments into new glucose release. Lactate can be turned back toward glucose through the Cori cycle. Glycerol enters through another route and joins the pathway downstream of fat breakdown. Amino acid carbon skeletons enter through several different intermediates depending on their structure. So the fasting liver is not drawing sugar out of storage alone. It is gathering carbon from across the organism and rebuilding a bloodstream fuel from distributed metabolic leftovers.

This is why gluconeogenesis marks a deeper level of fasting physiology than glycogenolysis. Glycogenolysis spends reserve. Gluconeogenesis reorganizes the body around absence. Once the liver is making glucose this way, fasting is no longer just a pause after the last meal. It has become a coordinated state in which adipose tissue, muscle, red blood cells, and liver are all participating in a new division of labor. Adipose tissue releases fatty acids and glycerol. The liver uses fatty acid oxidation to power its own work and converts glycerol and other substrates into glucose. Muscle and other tissues contribute amino acid carbon or lactate depending on conditions. The bloodstream becomes less a recipient of intestinal supply and more a centrally managed product of internal redistribution. The liver is the organ assembling that redistribution into something coherent.

Fat oxidation is critical here because gluconeogenesis costs energy. The liver cannot manufacture glucose for the rest of the body by burning glucose to do it. As fasting deepens, hepatocytes increase fatty acid oxidation in mitochondria, generating the reducing power and ATP needed to support glucose synthesis. This is one of the elegant shifts in fasting metabolism. Fat becomes the liver's main support fuel so glucose can be spared for tissues that need it more urgently. The liver is therefore not simply producing glucose. It is changing its own substrate preference to make glucose production possible. That is why fasting has to be understood as a systems event. One fuel is being used to protect another. One tissue changes its behavior so the blood can carry the right mixture to another.

Hormonal control sharpens the program. Falling insulin removes fed-state suppression. Rising glucagon pushes the liver toward glycogen breakdown first and gluconeogenesis later and more deeply. Cortisol and catecholamines can reinforce substrate mobilization and fasting support depending on timing and stress conditions. But these hormones do not invent gluconeogenesis out of thin air. They act on hepatocytes already built with the enzymes, organelles, transport systems, and mitochondrial machinery required to run it. The liver can respond because its architecture anticipates this state. It stores during abundance so it can bridge scarcity. It bridges scarcity so it can enter manufacture. That sequence is the metabolic grammar of fasting.

So gluconeogenesis is the deeper fasting program because it carries the body beyond reserve into reconstruction. It takes lactate, glycerol, and amino acid carbon skeletons and converts them into new circulating glucose when the meal is gone and glycogen is no longer enough. It is slower and more complex than glycogenolysis, but it is also more durable. The early fast lives on stored glucose. The deeper fast lives on hepatic synthesis. That is the point where the liver stops merely releasing what it had and starts rebuilding what the blood still needs.

Lactate, Glycerol, and Amino Acids as Glucose Substrates

Once glycogen begins to run down, the liver has to stop thinking like a storage organ and start thinking like a reconstruction organ. That is where lactate, glycerol, and amino acids come in. These are not backup sugars waiting quietly in reserve. They are metabolic fragments and intermediates being pulled out of ongoing tissue life and rerouted into glucose synthesis. Gluconeogenesis, the production of new glucose from noncarbohydrate material, depends on them because the bloodstream still has to support glucose-dependent tissues even when the intestine is no longer delivering carbohydrate and hepatic glycogen can no longer carry the full burden. The liver does not create glucose from nothing. It rebuilds it from carbon already moving through the organism.

Lactate is one of the clearest examples because it is produced continuously even outside extreme exercise. Red blood cells, which lack mitochondria, rely on glycolysis and release lactate as an end product. Other tissues can do the same when glycolytic flux is high or oxygen delivery is relatively limited. That lactate enters the bloodstream and returns to the liver, where hepatocytes take it up and convert it back toward pyruvate, then onward through gluconeogenesis into glucose. This recycling loop is called the Cori cycle. What matters physiologically is that lactate is not simply waste the body is trying to get rid of. In the fasting state it is recoverable carbon. Peripheral tissues spend glucose and export lactate. The liver takes that lactate and rebuilds glucose. The body is therefore salvaging part of its own prior carbohydrate use and pushing it back into circulation in a fresh form.

Glycerol enters from a different source and tells a different fasting story. It is released when triglycerides in adipose tissue are broken down, which means when stored fat is mobilized. A triglyceride contains three fatty acids attached to a glycerol backbone. The fatty acids can be oxidized widely as fuel, but the glycerol backbone has a special value for the liver because it can be converted into intermediates that enter gluconeogenesis. In this way adipose tissue is not only supplying fat for energy. It is also supplying a carbon scaffold that can be turned into glucose. This is one of the reasons fasting metabolism is a coordinated system rather than a sequence of isolated organ behaviors. Fat tissue releases both fuel and substrate. The liver uses one part, fatty acids, heavily for its own energy needs and uses the other part, glycerol, as raw material for blood glucose reconstruction. One stored molecule is split into two different physiological jobs.

Amino acids provide another major carbon source, but they come with a cost the body has to manage carefully. Many amino acids can be stripped of their nitrogen group, leaving behind carbon skeletons that enter the gluconeogenic pathway at various points. Alanine is especially important because muscle can release it in meaningful amounts, and the liver can convert it into pyruvate and then toward glucose. Other glucogenic amino acids feed into intermediates such as oxaloacetate or other citric acid cycle components that can be drawn into glucose production. But amino acids are not neutral glucose bricks. Their nitrogen has to be removed and handled, often ending as ammonia that must be converted into urea in the liver before the kidneys can excrete it safely. That means amino acid use for gluconeogenesis is not only a carbon question. It is also a nitrogen disposal problem. The liver has to solve both sides at once. It rebuilds glucose from the carbon and detoxifies the nitrogen burden that process creates.

What makes these three substrates so important is that they reveal the deeper logic of fasting. The liver is not living off a hidden bag of sugar somewhere in the abdomen. It is maintaining blood glucose by pulling carbon from circulating byproducts of glycolysis, from adipose fat breakdown, and from amino acid traffic shaped by whole-body protein turnover and tissue needs. Lactate reflects ongoing glucose use in peripheral tissues. Glycerol reflects fat mobilization. Amino acids reflect the body's willingness to sacrifice part of its protein traffic to preserve circulating glucose. These are not interchangeable sources. Each one carries a different metabolic meaning, a different tissue origin, and a different downstream burden.

So when the liver uses lactate, glycerol, and amino acids as glucose substrates, it is performing a large systems task with local chemistry. It is taking the leftovers of prior fuel use, the scaffold of released fat, and the carbon skeletons of amino acids and converting them into new glucose for the bloodstream. That is the deeper fasting program made visible. The body stops eating, but it does not stop supplying blood sugar. The liver keeps rebuilding it from what the rest of the organism can spare.

Why the Liver Protects Brain Function During Fasting

The brain is one of the main reasons fasting has to be managed so carefully. When food disappears from the gut, many tissues can shift heavily toward fatty acid use, lower their immediate dependence on circulating glucose, or tolerate a temporary change in fuel mix without catastrophic loss of function. The brain has far less freedom. It can adapt partially over time through increased ketone use, but it cannot simply abandon the need for a stable circulating fuel supply the moment a meal is missed. That is why the liver becomes so important as fasting begins. The intestine has stopped delivering carbohydrate. Blood glucose would begin falling unless another organ took over. The liver does exactly that. It protects brain function by turning stored glycogen into circulating glucose first, then by manufacturing new glucose through gluconeogenesis as fasting deepens, and later by producing ketone bodies that help reduce how much glucose the brain still requires. In other words, the liver does not merely survive fasting. It reorganizes the bloodstream so the brain can keep working while food is absent.

The first layer of that protection is speed. Brain tissue cannot wait for a long slow metabolic redesign after the last bite of food. It needs continuity. Hepatic glycogen provides that early continuity. The liver stored glucose as glycogen during the fed state precisely so it could reopen that reserve when intake stops. Once insulin falls and glucagon rises, hepatocytes begin glycogenolysis, breaking glycogen down and releasing free glucose into the circulation. That glucose enters systemic blood and helps preserve the concentration range the brain depends on. This is why hepatic glycogen should be understood as a neurologic safeguard as much as a carbohydrate reserve. The liver is not simply saving sugar for later in a general sense. It is preserving immediate fuel continuity for tissues that cannot afford abrupt interruption, and the brain is at the top of that list.

As fasting continues, the protection becomes deeper and more sophisticated. Glycogen cannot last indefinitely, so the liver shifts toward gluconeogenesis, rebuilding glucose from lactate, glycerol, and glucogenic amino acid carbon skeletons. Lactate returns from glycolytic tissues. Glycerol comes from adipose triglyceride breakdown. Amino acid carbon skeletons come from ongoing protein turnover and tissue metabolism. The liver gathers these substrates and converts them into new glucose so blood sugar does not collapse after the stored reserve thins out. This matters for the brain because neural tissue is not being fed from the gut anymore. It is being fed from hepatic reconstruction. The fasting bloodstream is therefore not a passive remnant of the last meal. It is an actively maintained product of liver work. When the liver is preserving glucose in fasting, it is preserving cognition, wakefulness, coordination, and the electrical stability of a tissue that cannot tolerate fuel disorder for long.

The liver also protects the brain by changing the fuel mix, not only by maintaining glucose. As fatty acid mobilization rises during fasting, hepatocytes increase beta-oxidation and begin generating ketone bodies, especially acetoacetate and beta-hydroxybutyrate. These are water-soluble fuel molecules derived from fat that can circulate to the brain and be used there after adaptation, reducing how much glucose neural tissue must consume. That shift is crucial because it spares glucose at the very moment external carbohydrate has disappeared and glycogen is declining. The liver therefore solves the fasting brain problem from both directions. It keeps producing glucose for the bloodstream, and it creates an alternative brain fuel that lowers the total glucose burden. One pathway preserves supply. The other reduces demand. Together they make prolonged fasting metabolically survivable in a way that pure glucose dependence would not.

There is another layer to this protection that is easy to miss. The liver does not only supply fuel. It also prevents brain injury from metabolic waste during fasting. Amino acid use and broader nitrogen turnover continue, and ammonia remains dangerous to neural tissue if it escapes control. Hepatocytes convert that ammonia into urea so the kidneys can excrete it. So while the liver is rebuilding glucose and generating ketones for the brain, it is also keeping neurotoxic nitrogen from accumulating in the same circulation that feeds the brain. This is an important part of the logic. Brain protection in fasting is not only about sending the right molecules out. It is also about keeping the wrong molecules from building up.

So the liver protects brain function during fasting by acting as the organ that preserves continuity when the gut has gone quiet. It releases stored glucose early, synthesizes new glucose as fasting deepens, generates ketone bodies to reduce neural glucose demand, and keeps ammonia from rising into the same bloodstream it is trying to keep usable. The brain does not endure fasting because it becomes independent. It endures fasting because the liver rebuilds the circulation around its needs.

When the Liver Shifts from Storage Organ to Release Organ

The liver does not stay in one metabolic identity. It changes role according to whether the bloodstream is being fed from the intestine or left to fend for itself. In the fed state, the liver behaves primarily as a receiving and storage organ. Portal blood arrives carrying absorbed glucose, amino acids, fructose, and other post-digestive substrates, and hepatocytes respond by taking part of that incoming load into the cell, storing glucose as glycogen, suppressing endogenous glucose output, and shifting the tissue toward synthesis, packaging, and fed-state handling. But that identity is temporary. The moment intestinal supply begins to fade, the liver starts reading a different set of signals. Portal nutrient delivery falls. Insulin declines. Glucagon rises. The blood is no longer being supported by incoming carbohydrate, yet tissues are still drawing from it continuously. That is when the liver shifts from storage organ to release organ. It stops mainly absorbing the burden of the meal and starts actively protecting the circulation from the absence of the meal.

This shift begins before the person feels dramatic hunger and long before a layperson would imagine anything serious is happening metabolically. The liver is sensitive to the end of input because it stands directly downstream from the intestine. It knows when portal glucose falls. It knows when fed-state hormone levels are receding. It knows when glycogen synthesis is no longer the right answer because the bloodstream has stopped receiving fresh carbohydrate. At that point hepatocytes stop prioritizing storage and begin dismantling what they previously built. Glycogen becomes the first releasable reserve. Glycogenolysis begins. Glucose units are liberated from the glycogen polymer, processed into glucose-6-phosphate, then converted into free glucose that can re-enter the bloodstream. This is the liver's first act as a release organ. It takes stored abundance and turns it into present support.

The change is deeper than glycogen breakdown alone. A storage organ is trying to dampen excess. A release organ is trying to prevent deficit. Those are opposite physiological missions. In storage mode, the liver protects the body from the raw force of absorption. In release mode, it protects the body from the raw force of non-absorption. That means pathways flip direction. Hepatic glucose output, which was suppressed in the fed state, becomes necessary. Glycogen synthase activity gives way to glycogen phosphorylase activity. Gluconeogenesis begins to rise in importance as glycogen reserve starts thinning. Fatty acid oxidation increases because the liver now needs to power glucose production without relying on incoming carbohydrate. The tissue does not merely use up what it had. It reorganizes around a new command: keep blood chemistry livable without help from the gut.

This transition is one of the clearest examples of why the liver is a timing organ, not just a storage organ. Storage by itself is inert. The liver is not inert. It stores with the expectation of later release, and it releases with the expectation that circulation must remain coherent while intake is absent. That is why hepatic glycogen should never be described as a pile of backup sugar sitting in a warehouse. It is a strategically placed short-horizon reserve meant to be opened when intestinal silence begins to threaten blood stability. The same is true later for gluconeogenesis. Once the liver has moved beyond reserve release and into new glucose synthesis, it is no longer just emptying a vault. It is manufacturing continuity.

The importance of this shift becomes easiest to see through the tissues depending on it. The brain cannot tolerate abrupt glucose loss. Red blood cells cannot switch to fatty acids because they lack mitochondria.

Other tissues can partly shift fuel use, but they are still living inside the same blood environment. The liver becomes a release organ because somebody has to step forward when the intestine steps back. It is the organ that changes the meaning of fasting from immediate crisis into managed transition. The person may simply think they have not eaten yet. The liver is already rerouting stored and reconstructed fuel into circulation to keep the brain awake, the blood supplied, and the organism stable.

So the shift from storage organ to release organ is not a dramatic emergency event that only happens in extreme deprivation. It is a normal physiological reversal built into every cycle between feeding and fasting. The liver stores when the meal is arriving. It releases when the meal is gone. That is how it keeps blood chemistry from being governed directly by the clock of eating.

Chapter 11

Amino Acids, Nitrogen, and the Cost of Protein Metabolism

What Happens When Amino Acids Enter Portal Blood

When amino acids enter portal blood, they are no longer food in the ordinary sense. They are dissolved nitrogen-bearing molecules moving toward the liver with several possible futures and one built-in problem. The useful side is obvious. Amino acids are the raw material for protein synthesis, tissue repair, enzyme production, transport proteins, signaling molecules, and many other forms of cellular construction. The dangerous side is less obvious but just as important. Every amino acid carries nitrogen, and nitrogen cannot accumulate casually in the bloodstream without consequences. So when amino acids leave the intestine and enter portal blood, the liver is not simply receiving building blocks. It is receiving building blocks attached to a waste-handling obligation. That is the central truth of protein metabolism. Protein nourishes, but nitrogen has to be managed.

The intestine absorbs amino acids after gastric and pancreatic digestion have broken dietary protein into smaller peptides and free amino acids, and after brush border enzymes on the intestinal surface finish that reduction further. Those amino acids cross the enterocyte, the absorptive cell of the small intestine, and enter the portal circulation on their way to the liver. Once in portal blood they join the same stream carrying glucose, hormones, microbial metabolites, bilirubin-related compounds, and the broader chemical aftermath of the meal. But amino acids are different from glucose in one important respect. Glucose can be stored as glycogen, burned, or redirected without forcing the body to dispose of a large nitrogen load. Amino acids cannot be handled so cleanly. If they are used to build proteins, much of their value stays in the body. If they are oxidized for energy, converted into glucose, or otherwise stripped down for carbon use, their amino group has to be removed. That is where the cost begins.

The liver receives these molecules first because it is the organ best equipped to sort their fate. Some amino acids will be taken up and used directly for hepatic protein synthesis. Albumin, clotting factors, apolipoproteins, transport proteins, and many other liver-made products all depend on this incoming amino acid traffic. Some amino acids will contribute carbon skeletons to glucose production, especially in fasting or when substrate conditions favor gluconeogenesis, the synthesis of glucose from noncarbohydrate material. Some will be oxidized more directly for energy or used to build other nitrogen-containing molecules the body needs. But these fates are not equal. The moment an amino acid stops being used as structure and starts being used as fuel or carbon substrate, its nitrogen becomes a disposal problem. The body cannot burn protein carbon without dealing with protein nitrogen.

That disposal problem is what makes amino acid metabolism more biologically expensive than casual nutrition language usually admits. When an amino group is removed, ammonia or ammonia-generating intermediates appear downstream. Ammonia is small, diffusible, and dangerous, especially to the brain. It cannot simply be allowed to accumulate in blood while the body enjoys the carbon value of amino acids. So the liver has to run a second program alongside amino acid use. It has to trap nitrogen and convert it into urea through the urea cycle, a pathway specifically designed to make nitrogen excretable by the kidneys. That means protein metabolism always carries a double task. The carbon can be useful. The nitrogen must be neutralized. A person can speak casually about protein as fuel, but the liver cannot. The liver has to pay the nitrogen bill.

Portal delivery matters here because it gives the liver first access to this burden before the rest of the circulation sees the full consequence. Amino acids absorbed from the gut do not go wandering through the systemic bloodstream unchecked before anyone decides what to do with them. They reach the liver early, where hepatocytes can extract, sort, and process them while keeping nitrogen disposal tied tightly to carbon use. This is part of why the liver is central to protein metabolism rather than merely involved in it. It sits at the border where absorbed amino acids stop being dietary fragments and become metabolic decisions. Will this amino acid become tissue. Will it become glucose. Will it be oxidized. Will its nitrogen be incorporated into another molecule or stripped and sent toward urea. These are hepatic questions first.

This is also why protein intake can never be evaluated honestly by amino acid presence alone. The body is not only asking how many amino acids arrived. It is asking what fraction can be retained structurally, what fraction will be burned, what nitrogen burden will be generated, how much urea cycle capacity is available, and how efficiently the kidneys can eliminate the final waste. The minute amino acids enter portal blood, they are no longer just nutrition. They are potential structure, potential fuel, and potential toxic burden depending on how the liver handles them. That is the real entry point into protein metabolism. Use and cost arrive together.

Protein Use Versus Protein Breakdown

Protein use and protein breakdown are not opposite moral categories of good and bad metabolism. They are two constant directions of traffic that the body has to balance every hour. Amino acids entering from the diet can be used to build tissue, enzymes, transport proteins, receptors, immune molecules, structural matrix, and countless other protein-based parts of the organism. That is protein use in the constructive sense. But proteins already inside the body are also being broken down continuously. Old proteins are dismantled, damaged proteins are cleared, regulatory proteins are turned over, and amino acids are released back into circulation or into local cellular pools for reuse, oxidation, or conversion. The liver sits in the middle of this moving economy because it receives absorbed amino acids from portal blood, makes many major plasma proteins itself, and helps decide whether amino acid carbon is retained in body structure or diverted into energy and glucose production. So the real contrast is not between protein use and some rare pathological breakdown. The real contrast is between nitrogen being held in organized structure and nitrogen being stripped out because the carbon is being used for something else.

Protein use is the side most people imagine first. Amino acids are incorporated into new proteins through translation inside cells, meaning ribosomes read messenger RNA and link amino acids together in a precise sequence to build a working protein. That process allows muscle repair, enzyme renewal, hormone and receptor synthesis, plasma protein production, immune signaling, connective tissue maintenance, and organ-level upkeep. In the liver this means hepatocytes are constantly turning incoming amino acids into albumin, clotting factors, transport proteins, apolipoproteins, complement proteins, and other secreted products that shape blood chemistry across the whole body. Protein use in this sense is retention. The body is taking amino acid nitrogen and fixing it into useful structure. The carbon and the nitrogen both stay inside organized biological function. That is the high-value outcome of amino acid metabolism.

Protein breakdown is different because the moment a protein is dismantled or an amino acid is diverted away from structure, the body has to make new decisions about what to do with the released parts. Some amino acids can be reused directly to build new proteins. That is internal recycling. But if the carbon skeleton is needed for energy production or for gluconeogenesis, the synthesis of glucose from noncarbohydrate material, then the amino group has to be removed. That process, deamination or transamination depending on the step, creates the nitrogen problem. Carbon can move into fuel pathways. Nitrogen cannot simply float freely in blood without consequence. It tends toward ammonia generation downstream, and ammonia is dangerous enough that the liver has to convert it into urea for renal excretion. So protein breakdown is not only loss of structure. It is the beginning of a disposal burden. Once protein is no longer being retained as protein, nitrogen accounting becomes unavoidable.

This is why protein use and protein breakdown carry very different biological yield. If an amino acid becomes part of albumin, muscle protein, an enzyme, or a transport protein, the body keeps the full value of the molecule in organized form. If that same amino acid is oxidized for energy or turned into glucose, the body keeps only the carbon value and has to pay the nitrogen cost through urea production and renal elimination. One route builds the organism. The other salvages fuel at a price. That price is not hypothetical. Hepatocytes have to run the urea cycle, spend ATP, adenosine triphosphate, to trap nitrogen safely, and then the kidneys have to excrete the resulting urea. The body can do this well, but it is still a cost. Protein used as protein is a different metabolic event from protein used as emergency carbon.

The liver is central to this distinction because it is one of the main organs deciding which side of the line incoming amino acids will fall on.

In the fed state, especially when overall energy is sufficient, more amino acid traffic can be directed toward synthesis and retention. In fasting, low-energy states, stress states, or when amino acid supply exceeds immediate structural need, more amino acid carbon may be diverted toward oxidation or glucose production. Hormones matter here. Insulin favors retention and anabolic use. Glucagon and stress hormones shift the system toward glucose support and fuel provision. But the key point is anatomical and physiological before it is hormonal. Amino acids arrive at the liver first through portal blood, and the liver is the organ that has to decide how much of that traffic becomes body and how much becomes blood fuel plus nitrogen waste.

This is also why protein metabolism should never be reduced to grams eaten alone. The real question is not only how much protein entered the mouth. The real question is how much amino acid traffic the body was able to retain in useful structure versus how much had to be broken down, recycled, oxidized, or converted into glucose. Protein use means nitrogen stays in the body as organized function. Protein breakdown means nitrogen starts moving toward disposal. Both are always happening. Health depends on how well the body keeps the constructive side dominant when it should, and how efficiently the liver pays the nitrogen cost when breakdown becomes necessary.

Deamination and the Production of Ammonia

An amino acid becomes metabolically expensive the moment the body stops using it as structure and starts using it as carbon. That is where deamination enters. The amino acid is useful because it carries both a carbon skeleton and a nitrogen group. If the body wants the carbon for glucose production, energy generation, or conversion into other compounds, it has to separate that nitrogen first. It cannot burn the whole molecule cleanly as-is. Deamination is the process by which that nitrogen is removed, either directly or through linked transfer steps that move the amino group onto another molecule before free ammonia appears downstream. This is not optional cleanup added after protein metabolism. It is built into protein metabolism the moment amino acids stop being retained as protein. The carbon can go one way. The nitrogen has to go another.

In practice, much of this begins with transamination rather than immediate free ammonia release. The amino group from one amino acid is often transferred to a keto acid, commonly alpha-ketoglutarate, producing glutamate. A keto acid is the carbon skeleton form that remains when an amino acid has lost its amino group. This matters because the body does not want free ammonia appearing everywhere in uncontrolled fashion. It consolidates nitrogen traffic first. Glutamate becomes a kind of nitrogen collection point, a molecule that can carry amino groups gathered from multiple amino acids. From there, oxidative deamination can occur, especially in the liver, where glutamate dehydrogenase removes the amino group from glutamate and releases ammonia. That sequence lets the body concentrate nitrogen handling into the organ best equipped to survive the consequence. The liver is not simply receiving amino acids. It is centralizing their nitrogen burden.

Ammonia is the problem hidden inside protein metabolism. It is small, diffusible, and chemically disruptive. In the wrong concentration it interferes with brain function, acid-base balance, and broader metabolic stability. The body cannot allow amino acid carbon to be harvested freely while nitrogen accumulates in parallel. That is why deamination is never a neutral event. Every time the liver strips amino groups to use carbon skeletons for gluconeogenesis, oxidation, or other metabolic purposes, it is generating a toxic burden that has to be neutralized almost immediately. This is one of the reasons protein should never be discussed as if it were simply another fuel class interchangeable with carbohydrate or fat. When glucose is oxidized, there is no nitrogen bill. When fatty acids are oxidized, there is no nitrogen bill. When amino acids are used as fuel, ammonia begins stalking the pathway.

The liver solves this by linking deamination tightly to the urea cycle. Ammonia generated from amino acid metabolism is captured and converted into urea, a far less toxic, water-soluble molecule that the kidneys can excrete in urine. That means deamination is not a standalone reaction. It is the opening move in a detoxification obligation. Hepatocytes remove nitrogen, but they also have to pay for having removed it. ATP, adenosine triphosphate, is spent in the process. Mitochondrial and cytosolic machinery have to stay coordinated. Renal excretion has to be available downstream. So the true physiological cost of using amino acids for carbon is larger than the carbon story alone suggests. The liver gets useful substrate. The body inherits a disposal problem that has to be solved with precision.

This is why deamination marks a threshold in protein metabolism. On one side of that threshold, amino acids are being retained in organized structure, built into enzymes, plasma proteins, muscle proteins, receptors, and tissue matrix. On the other side, the body has decided their carbon is more valuable elsewhere, and the nitrogen has to be stripped and neutralized. That shift is not morally bad and not inherently pathological. It is often necessary. But it is never free. The moment deamination begins, ammonia is being born, and the liver has to keep that birth from becoming poisoning.

Why Nitrogen Disposal Is One of the Liver's Most Life-Preserving Jobs

Nitrogen disposal is one of the liver's most life-preserving jobs because amino acid metabolism creates a kind of waste the body cannot afford to leave loose in circulation. Protein is valuable, but the value is double-edged. Amino acids carry carbon that can be used for structure, glucose production, or energy, and they carry nitrogen that becomes dangerous the moment it is not locked inside organized tissue. When amino acids are broken down, transaminated, or deaminated, that nitrogen begins moving toward ammonia. Ammonia is small, diffusible, and neurologically toxic. It does not need to accumulate in massive quantities before function begins to suffer. The body can tolerate fluctuations in glucose, lipids, and many other metabolites within a range. Ammonia has a much narrower margin. If the liver does not capture and convert it quickly, the bloodstream stops being safe for the brain.

That is why nitrogen disposal is not a secondary cleanup task following real metabolism. It is part of real metabolism. The body cannot use amino acid carbon freely unless it can also neutralize amino acid nitrogen. This is what makes protein metabolism different from carbohydrate and fat metabolism at a fundamental level. Glucose can be stored, oxidized, or converted without generating a nitrogen burden. Fatty acids can be oxidized without forcing the body to detoxify amino groups. Protein always carries that extra obligation. The moment amino acids are diverted away from tissue building and toward fuel use, glucose support, or general carbon extraction, the liver inherits a toxic liability. That liability has to be converted into something the kidneys can safely excrete. The urea cycle is the solution, and the liver is the organ that runs it at organism scale.

The physiological stakes are high because ammonia does not remain politely local. It crosses into systemic circulation and reaches the brain, where it disturbs neural metabolism, osmotic balance, and neurotransmitter handling. That is why failed nitrogen disposal produces some of the most dangerous symptoms in liver failure. Confusion, slowed thinking, sleep reversal, poor attention, altered behavior, asterixis, and hepatic encephalopathy are not vague complications loosely associated with liver disease. They are direct evidence that nitrogen control failed and neurotoxic chemistry escaped hepatic restraint. Asterixis is the flapping tremor seen when the nervous system can no longer maintain steady posture cleanly. Hepatic encephalopathy means brain dysfunction caused by liver failure or severe portal-systemic shunting. These are survival-level events. The liver is not merely cleaning blood for cosmetic biochemical tidiness. It is preventing protein metabolism from poisoning consciousness.

This also explains why the liver sits so centrally in whole-body protein economics. Absorbed amino acids arrive through portal blood. Amino acids released from tissue turnover, stress metabolism, fasting, and gluconeogenesis all contribute to the circulating nitrogen burden in one form or another. The liver stands in the main path of that traffic. It receives amino acid nitrogen early, consolidates it through pathways such as transamination and glutamate handling, and channels the final burden into urea synthesis. Urea is far less toxic than ammonia and water-soluble enough for renal excretion. That conversion is not glamorous, but it is one of the clearest examples of the liver preserving the internal environment by changing the form of a dangerous molecule rather than merely removing it outright. The body survives not because nitrogen disappears, but because the liver turns free nitrogen burden into a transportable excretory product before damage spreads.

The job is especially life-preserving because nitrogen disposal has to continue across multiple metabolic states. In the fed state, dietary amino acids arrive in portal blood and the liver has to sort which will be retained as protein and which will be stripped down with nitrogen sent toward disposal. In fasting, amino acid carbon skeletons may be recruited for gluconeogenesis, which raises the need for nitrogen handling at the same time the body is already under the stress of absent food. In illness, injury, or catabolic states, protein breakdown may rise and push more nitrogen toward hepatic processing. The liver therefore cannot think of ammonia control as an occasional detox problem. It is part of maintaining blood usability under ordinary feeding, fasting, and stress.

This is why a failing liver becomes dangerous so quickly even when many other tissues are still anatomically present. The body can tolerate a surprising amount of structural disorder for a time. It cannot tolerate uncontrolled nitrogen chemistry for long. Nitrogen disposal belongs in the same class as glucose stabilization and bile formation as a core function, but in some ways it is even more obviously tied to immediate survival because its failure reaches the brain so directly. The liver earns its central place in physiology not only by building proteins, handling fuels, and producing bile, but by standing between inevitable protein turnover and systemic ammonia poisoning. That is one of the clearest lines between ordinary metabolism and death.

The Difference Between Using Amino Acids for Structure and Burning Them for Energy

An amino acid can enter the body and end in two very different economies. It can be retained as structure, or it can be dismantled for fuel. Those are not small variations of the same nutritional outcome. They are different biological events with different yields, different costs, and different consequences for the liver. When an amino acid is used for structure, its carbon skeleton and its nitrogen stay inside organized function. The molecule is incorporated into muscle protein, enzymes, transport proteins, receptors, connective tissue, plasma proteins, immune molecules, or some other living architecture the body intends to keep. When an amino acid is burned for energy, the body is no longer treating it as retained tissue capital. It is extracting usable carbon and discarding the nitrogen. That immediately creates a waste problem the liver has to solve. The amino acid may still provide value in both cases, but the value is not the same. One route builds the organism. The other helps power it while forcing the organism to clean up after the burn.

Using amino acids for structure is the higher-yield event in the deepest physiological sense because the whole molecule is being preserved inside ordered biology. The carbon becomes part of protein architecture. The nitrogen remains productively bound inside that architecture. The body keeps both halves. This is what happens when hepatocytes turn incoming amino acids into albumin, clotting factors, apolipoproteins, and other plasma proteins, or when muscle cells incorporate amino acids into contractile proteins, or when connective tissues renew collagen and matrix proteins. The molecule enters, is linked into a longer protein chain, folded, processed, and retained. There is no nitrogen disposal burden attached to that moment because the nitrogen was not liberated. The amino acid did not become waste chemistry. It became tissue. That is what protein is for at its highest biological value.

Burning amino acids for energy is a different calculation. The body is no longer prioritizing retention. It is using the carbon skeleton because fuel is needed, because gluconeogenesis is required, because total amino acid supply exceeds immediate structural demand, or because metabolic conditions are pushing amino acids into oxidation rather than incorporation. To do that, the amino group has to be removed. This usually begins through transamination, where the amino group is transferred onto another molecule such as alpha-ketoglutarate to form glutamate, or through related steps that consolidate nitrogen handling before free ammonia appears. Then the nitrogen burden moves toward ammonia generation, and the liver has to neutralize that burden through the urea cycle. So the moment an amino acid becomes fuel, the body keeps the carbon and has to dispose of the nitrogen. That is a lower-retention outcome by definition. Part of the molecule is being harvested. Part of it is being turned into excretory burden.

This is why amino acids are metabolically more expensive to burn than glucose or fatty acids. Glucose can be oxidized without a nitrogen bill. Fatty acids can be oxidized without a nitrogen bill. Amino acids cannot. The liver has to spend energy to trap the nitrogen safely, convert it into urea, and send it toward renal excretion. ATP, adenosine triphosphate, is required for that process, which means the body is paying an energetic cost just to make amino acid use as fuel survivable. The kidneys then inherit the final elimination burden. So when protein is burned, the metabolic sequence is longer and more costly than casual language suggests. Carbon is salvaged. Nitrogen is detoxified. Energy is spent to make the burning possible. This is not a clean fuel pathway. It is a structurally costly one.

The distinction also matters because it changes how protein should be judged in nutrition. A gram of amino acids retained in tissue is not equivalent to a gram of amino acids oxidized for energy. The chemical entry may look the same on paper, but the biological outcome is very different. In one case the body has built more of itself. In the other it has extracted fuel and generated a disposal burden. That does not mean amino acid oxidation is always wrong. It is often necessary, especially in fasting, low-glycogen states, gluconeogenic support, or when amino acid traffic exceeds the immediate anabolic demand. But it does mean the reader should stop treating protein use as one undifferentiated event. Protein can nourish by becoming structure, or it can serve as emergency carbon while increasing hepatic nitrogen work. Those are not equal forms of value.

The liver sits at the center of this distinction because it is the organ that receives absorbed amino acids first through portal blood and has to decide which route dominates. Will these amino acids become plasma proteins. Will they be retained in the body's protein economy. Will their carbon be recruited for glucose production. Will they be oxidized. If the answer moves toward fuel, the liver also has to handle the resulting nitrogen. So the difference between using amino acids for structure and burning them for energy is not philosophical. It is architectural. One preserves the molecule inside living form. The other splits the molecule, keeps the carbon, and forces the liver to make the nitrogen safe enough to survive.

Why Failed Ammonia Handling Quickly Becomes a Neurologic Problem

Ammonia handling becomes a neurologic problem so quickly because ammonia is one of the few ordinary metabolic wastes that can move from protein metabolism to brain dysfunction with very little anatomical distance. Amino acids are constantly being broken down in the body and by gut microbes. Their nitrogen burden moves toward ammonia. Under normal conditions the liver intercepts that burden through the portal circulation, captures the ammonia, and converts it into urea before the rest of the body has to live with it. Urea is far safer and can be excreted by the kidneys. When that hepatic step fails, the nitrogen burden does not stay in the abdomen and wait politely for a later consequence. It enters systemic blood and begins reaching the brain. That is why failed ammonia handling is not a subtle side issue in liver disease. It is one of the fastest routes by which a metabolic problem becomes a consciousness problem.

The brain is especially vulnerable because ammonia can cross into neural tissue and disrupt the chemistry that keeps neurons and support cells stable. Neurons are the electrically active signaling cells. Astrocytes are one of their major support cells, responsible for buffering the chemical environment around them. Astrocytes try to protect the brain from ammonia by combining it with glutamate to form glutamine through the enzyme glutamine synthetase. At first this is a defensive move. It traps toxic ammonia in a less immediately dangerous form. But the defense creates a second problem. Glutamine accumulates inside the astrocyte, and because glutamine is osmotically active, it draws water with it. The cell begins to swell. When enough astrocytes swell, the brain's microscopic architecture becomes less stable, signaling becomes less clean, and in severe cases cerebral edema, meaning excess water accumulation in brain tissue, can develop. The liver failed to neutralize the nitrogen upstream, so the brain begins trying to neutralize it locally and pays for that effort with swelling.

Ammonia also injures brain function by distorting neurotransmitter handling and energy metabolism. Glutamate is not only a chemical substrate for ammonia trapping. It is also one of the major excitatory neurotransmitters in the brain. When ammonia pressure rises, glutamate handling becomes abnormal. Gamma-aminobutyric acid, usually called GABA, the major inhibitory neurotransmitter, can also become functionally more dominant in the altered chemical setting of hepatic encephalopathy. The result is a brain pushed toward slowed signaling, poor attention, blunted responsiveness, and impaired coordination rather than clean wakeful processing. At the same time ammonia burdens mitochondrial function and disturbs cellular energy balance inside neural tissue. So the problem is not merely that the brain is being exposed to a toxin in a generic sense. The problem is that the toxin is interfering with osmotic control, neurotransmission, and energy production at once.

That is why the symptoms can escalate in a recognizable sequence. Early on the person may become mentally slower, inattentive, irritable, or sleep-disordered, often with reversal of the normal sleep-wake pattern. As the burden rises, speech, concentration, judgment, and motor control degrade further. Asterixis can appear, the flapping lapse in maintained posture that reflects unstable neural control. In more advanced states, confusion deepens into stupor and coma. None of this is mysterious once the physiology is visible. Protein metabolism and gut microbial nitrogen keep generating ammonia. The liver fails to convert enough of it into urea. Ammonia reaches the brain. Astrocytes swell trying to trap it. Neurotransmitter balance shifts. Neural energy handling deteriorates. Thought begins to blur because the blood feeding the brain is no longer chemically safe.

The speed of this transition is what makes ammonia control so life-preserving. The body can tolerate many metabolic imperfections for a time. It cannot tolerate uncontrolled nitrogen chemistry in the circulation for long, especially not when the brain is downstream. This is also why portal-systemic shunting is so dangerous. If blood from the gut bypasses the liver instead of passing through it, ammonia can reach the systemic circulation even if some hepatocyte function remains. The liver does not only need to exist. It needs to stand in the path of the nitrogen burden. When that position is lost, the brain begins paying for protein metabolism directly.

So failed ammonia handling becomes a neurologic problem quickly because ammonia is small, mobile, and chemically disruptive, and because the brain has no good long-term strategy for living in ammonia-rich blood. The liver is supposed to solve that problem before the brain ever sees it. When that solution fails, the nervous system becomes the place where nitrogen toxicity turns visible.

Chapter 12

The Urea Cycle

Step-by-Step Explanation of How Ammonia Becomes Urea

Ammonia does not become urea by a single detox trick. It is moved there through a staged hepatic pathway built to take dangerous nitrogen and lock it into a much safer transport form. That pathway is the urea cycle, and it matters because free ammonia is too reactive and too diffusible to be left circulating after amino acid metabolism. The liver cannot simply hold it, dilute it, or hope the kidneys deal with it directly. It has to convert it. Urea is the conversion product, a small water-soluble molecule that can travel in blood without poisoning the brain and can then be filtered into urine by the kidneys. But the liver does not manufacture urea out of ammonia alone. One nitrogen in urea comes from free ammonia. The second comes from aspartate, an amino acid-related intermediate carrying another nitrogen load. The carbon atom in urea comes from bicarbonate, which is the hydrated carbon dioxide pool used widely in acid-base physiology. So from the beginning, the cycle is not just a disposal tube. It is a structured assembly process. Nitrogen is gathered from more than one source, carbon is inserted, ATP, adenosine triphosphate, is spent, and the final product is built in stages so the bloodstream can survive protein metabolism.

The first major step happens inside the mitochondria of the hepatocyte, and the location matters because this is where ammonia is first trapped in a committed form. Free ammonia combines with bicarbonate under the action of carbamoyl phosphate synthetase I, often abbreviated CPS1, to form carbamoyl phosphate. This is the opening gate of the cycle, and it is energetically expensive because the liver is forcing unstable chemistry into a controlled pathway. CPS1 also requires N-acetylglutamate as an essential activator. That detail matters because it means the liver does not run the cycle at full speed by accident. N-acetylglutamate acts like a signal that amino acid and nitrogen traffic are present and that urea synthesis needs to rise accordingly. Carbamoyl phosphate is therefore the first trapped nitrogen form in the cycle. The ammonia has not become harmless yet, but it has been pulled into a molecule the cell can manage directionally instead of letting it drift through the blood.

The next step is still mitochondrial and introduces the carrier that makes the whole cycle turn. Ornithine, a nonprotein amino acid used mainly as a cycle intermediate rather than as a building block for tissue proteins, combines with carbamoyl phosphate through the enzyme ornithine transcarbamylase, or OTC, to form citrulline. Ornithine is important because the urea cycle is built as a loop rather than a straight disposable chain. The liver uses ornithine, modifies it, moves nitrogen onto it, and then regenerates it later so the cycle can turn again. Once citrulline is formed, it is transported out of the mitochondrion into the cytosol, the fluid compartment outside the mitochondria where the later steps of the pathway continue. Up to this point the cycle has taken one free ammonia-derived nitrogen and attached it to a carbon-containing framework in a controlled way. The nitrogen burden is now inside citrulline rather than loose in circulation.

The third step is where the second nitrogen destined for urea enters the pathway. In the cytosol, citrulline combines with aspartate through the enzyme argininosuccinate synthetase to form argininosuccinate. Aspartate is carrying a nitrogen group that has often arrived through transamination, which is the transfer of an amino group from one molecule to another rather than immediate free ammonia release. This is one reason the urea cycle is so elegant. It does not only dispose of free ammonia directly. It also collects nitrogen that has been consolidated onto amino acid intermediates and brings that burden into the same excretory route. This step also consumes ATP-equivalent energy, again showing that nitrogen disposal is not cheap housekeeping. The liver is investing energy to prevent nitrogen chemistry from becoming lethal. Once argininosuccinate is formed, the molecule now contains the two nitrogens that will ultimately appear in urea.

Argininosuccinate is then cleaved by argininosuccinate lyase into arginine and fumarate. Fumarate is not waste. It can re-enter broader metabolic pathways, linking the urea cycle to the citric acid cycle and other carbon-handling chemistry inside the cell. Arginine, meanwhile, carries the nitrogen burden one step closer to final disposal. In the last true disposal step, the enzyme arginase cleaves arginine into urea and ornithine. Urea is released as the safe excretory product. Ornithine is regenerated and sent back toward the mitochondrion so the cycle can begin again with another load of nitrogen. That regeneration matters because it shows the deeper design of the pathway. The liver is not inventing a new scaffold every time ammonia appears. It is running nitrogen through a reusable carrier system, turning the same cycle repeatedly as long as amino acid metabolism continues to generate burden.

So step by step, the sequence is this. Free ammonia and bicarbonate are trapped into carbamoyl phosphate inside the mitochondrion. Carbamoyl phosphate transfers its nitrogen burden onto ornithine to make citrulline. Citrulline leaves the mitochondrion and meets aspartate in the cytosol, bringing in the second nitrogen that urea will contain. Argininosuccinate forms, then splits into arginine and fumarate. Arginine is finally cleaved into urea and regenerated ornithine. Urea leaves the liver in blood and is excreted by the kidneys. Ornithine returns to turn the cycle again. That is how ammonia becomes urea in real physiological terms. Not by disappearing, but by being assembled into a safer molecule through a costly, compartmentalized, continuously turning hepatic pathway that keeps protein metabolism from becoming brain toxicity.

Mitochondrial and Cytosolic Stages

The urea cycle is split between the mitochondria and the cytosol because ammonia handling is too dangerous and too important to leave in one undifferentiated cellular space. The liver compartmentalizes it. That is the first thing the reader should see. Free ammonia is captured first inside the mitochondrion, where the cell can trap it early and commit it to controlled chemistry before it diffuses farther into the hepatocyte. The later steps then continue in the cytosol, where the nitrogen burden is assembled into its final excretory form and the cycle can connect more easily with the rest of amino acid and carbon metabolism. This split is not a biochemical inconvenience. It is a design choice. The hepatocyte is using compartments to control where toxic nitrogen is first neutralized, where carbon and second-nitrogen inputs are added, and where the final product is released.

The mitochondrial stage is the trap. Ammonia entering this part of the pathway does not remain free for long if the system is working properly. Carbamoyl phosphate synthetase I captures ammonia together with bicarbonate and uses ATP, adenosine triphosphate, to build carbamoyl phosphate. This is the first committed step of the cycle and one of the clearest signs that nitrogen disposal is an energy-requiring survival function rather than passive waste removal. The mitochondrion is the correct place for this opening step because it is already the cell's energy-producing organelle and because early nitrogen capture benefits from being tied closely to mitochondrial metabolism. The next mitochondrial step transfers the carbamoyl group onto ornithine through ornithine transcarbamylase, producing citrulline. At that point the most dangerous part of the nitrogen burden has been forced into a structured intermediate. The mitochondrion has done its main job. It has taken free ammonia and turned it into something that can move forward in the cycle without remaining loose in the cell.

Citrulline then leaves the mitochondrion and enters the cytosol, and that movement marks a real change in the logic of the pathway. The mitochondrial phase is about capture and commitment. The cytosolic phase is about assembly and completion. In the cytosol, citrulline meets aspartate, which contributes the second nitrogen that will ultimately appear in urea. This is important because the liver is not only disposing of free ammonia. It is consolidating nitrogen burden from broader amino acid metabolism. Aspartate brings in nitrogen that has often been transferred there through transamination, which means the cell has already been collecting amino groups from different amino acids rather than letting each one generate free ammonia independently. Citrulline and aspartate are joined by argininosuccinate synthetase to form argininosuccinate, and more ATP-equivalent energy is spent in the process. Again the point is visible. Safe nitrogen disposal costs the liver work.

The cytosolic pathway then moves toward separation and release. Argininosuccinate is cleaved into arginine and fumarate. Fumarate is not discarded as irrelevant debris. It can re-enter broader carbon metabolism, linking the urea cycle to the tricarboxylic acid cycle, also called the citric acid cycle, which is one of the main pathways of oxidative energy metabolism. This connection matters because the urea cycle is not isolated from the rest of hepatic biochemistry. Nitrogen disposal and carbon handling are intertwined. Arginine then becomes the immediate precursor of final disposal. Arginase cleaves it into urea and ornithine. Urea is released for transport in blood and eventual renal excretion. Ornithine is not waste. It is regenerated carrier material. It returns to the mitochondrion so the cycle can begin again. The cytosol completes the build. The mitochondrion receives the carrier back. The loop closes.

This compartmental split also protects the cell by keeping the earliest ammonia-capturing steps near the machinery best equipped to support them while allowing later assembly steps to intersect with cytosolic amino acid traffic and synthetic control. The liver is handling a toxic burden and trying to make it excretable without letting the burden flood the cell in the meantime. Mitochondrial capture reduces that risk. Cytosolic completion lets the cycle integrate with transamination, aspartate supply, arginine handling, and broader metabolic flow. One compartment secures the threat. The other finishes the solution.

So the mitochondrial and cytosolic stages should be understood as two halves of one protective design. The mitochondrion traps free ammonia and commits it to the cycle. The cytosol adds the second nitrogen, completes the assembly of urea, regenerates ornithine, and links nitrogen disposal to wider carbon metabolism. The liver is not merely running a list of enzymes. It is using cellular geography to turn toxic nitrogen into a removable molecule while keeping the process under control at every step.

Why the Cycle Costs Energy

The urea cycle costs energy because the liver is not passively letting nitrogen drift into a safer form. It is forcing dangerous chemistry into order. Ammonia is small, diffusible, and toxic, especially to the brain. The body cannot allow it to remain free in circulation while protein metabolism continues. That means the liver has to capture it, bind it into intermediates, add a second nitrogen from aspartate, and build a final molecule that can travel safely in blood and be excreted by the kidneys. None of that happens downhill. The cell is not watching two unstable nitrogens and a carbon source spontaneously arrange themselves into urea. It is driving the reaction against disorder. That requires energy.

The first energy cost appears immediately because free ammonia has to be trapped before it can escape the liver's control. In the mitochondrial phase, carbamoyl phosphate synthetase I uses ATP, adenosine triphosphate, to combine ammonia with bicarbonate and form carbamoyl phosphate. That step is expensive because it is creating an activated intermediate, a chemically primed molecule that can carry the nitrogen forward into the rest of the cycle. Activation always has a price. The liver is investing energy to convert a toxic free molecule into a committed pathway intermediate. Without that investment, ammonia would remain more mobile than the body can safely tolerate. The first cost of the urea cycle is therefore the cost of locking danger into structure.

More energy is spent later when citrulline is joined to aspartate in the cytosol to form argininosuccinate. This is the step where the second nitrogen that will ultimately appear in urea is brought into the cycle. The joining reaction is not chemically trivial. Two nitrogen-bearing burdens are being assembled into one controlled excretory route, and the cell again has to pay to make that assembly happen. This is one reason protein metabolism is more expensive than casual nutrition language suggests. The body is not only extracting carbon value from amino acids. It is also paying ATP to make their nitrogen survivable. The more nitrogen has to be cleared, the more this matters.

The deeper reason the cycle costs energy is that disposal is never free when the waste is dangerous. The liver is doing precision waste engineering, not letting benign debris wash out in urine by itself. Ammonia must be captured quickly. Nitrogen from different amino acids must be consolidated. Urea must be synthesized in a form that is water-soluble enough to circulate safely and neutral enough not to disturb neural function on the way to the kidney. That transformation requires enzyme control, compartmentalization between mitochondria and cytosol, transporter activity, and activated intermediates. All of those are energy-dependent conditions. The liver is spending metabolic currency to preserve the chemical usability of the blood.

This also explains why amino acids are such a different fuel from glucose or fat. Glucose oxidation does not require the body to spend extra ATP simply to make its waste nitrogen-safe, because there is no nitrogen burden there to begin with. Fatty acids are the same. Amino acids are different because once their carbon is being used for energy or gluconeogenesis, the synthesis of glucose from noncarbohydrate substrates, the nitrogen cannot stay attached and cannot be ignored. The body gets carbon value, but it has to buy safe disposal of the nitrogen. Protein burned as fuel therefore carries a hidden processing tax. The liver has to pay it first.

There is also a systems cost. The liver often powers this work by oxidizing fatty acids and running mitochondrial metabolism hard enough to keep ATP supply available while nitrogen burden is being handled. That means the urea cycle is not only consuming energy directly through its own enzymatic steps. It is leaning on the broader energy economy of the hepatocyte. A liver under hypoxic stress, mitochondrial dysfunction, or severe hepatocellular injury will struggle not only because ammonia is dangerous, but because the machinery required to pay the energy price of disposing it is weaker. The cycle costs energy because safety costs energy. When the liver cannot afford that cost, ammonia control begins to fail.

So the urea cycle should be understood as an expensive but necessary act of chemical containment. ATP is spent because free nitrogen has to be trapped, activated, assembled, and converted into a form the bloodstream and kidneys can live with. The liver is buying time for the brain, safety for the blood, and continuity for protein metabolism. That is why the cycle costs energy. The body is paying to keep amino acid use from becoming self-poisoning.

What Urea Is and Why the Kidneys Can Remove It

Urea is the liver's safer transport form for nitrogen that would otherwise remain trapped in the body as ammonia burden. It is a small water-soluble molecule built from two nitrogens and one carbon-containing unit, assembled through the urea cycle inside hepatocytes after amino acid metabolism has generated nitrogen that can no longer stay productively bound in tissue. That distinction matters. Urea is not a nutrient, not an energy source, and not a random waste fluid already sitting in blood waiting to be filtered out. It is a deliberately manufactured excretory molecule. The liver takes nitrogen that would be dangerous if left as ammonia, packages it into urea, and releases that urea into the bloodstream specifically so another organ, the kidney, can take over the next stage of disposal. Urea is therefore a handoff molecule. It marks the point where hepatic detoxification becomes renal elimination.

What makes urea so useful is not that it is inert in an absolute sense, but that it is vastly safer to circulate than free ammonia. Ammonia is small, diffusible, and neurotoxic. Urea, by contrast, is chemically much less disruptive in the concentrations normally produced by protein metabolism. The liver has converted a high-risk nitrogen burden into a form the bloodstream can carry without immediate poisoning of the brain. That is the real achievement. The body has not made nitrogen disappear. It has changed the form of nitrogen into something compatible with temporary circulation. Urea can travel in plasma as dissolved solute while the kidneys receive it, concentrate it to varying degrees, and move it toward urine. The liver solves the toxicity problem first. The kidneys solve the clearance problem second.

The kidneys can remove urea because urea is water-soluble and small enough to pass through the glomerular filtration barrier, the microscopic capillary filter at the front end of the nephron, which is the kidney's functional unit. Blood enters the glomerulus under pressure, plasma water and many dissolved small molecules pass into the filtrate, and larger proteins and blood cells are normally retained in the circulation. Urea fits the class of molecules that can enter this filtrate. Once there, it moves through the nephron, where some of it may be reabsorbed depending on hydration state, tubular flow, and the kidney's need to manage water concentration, but a substantial portion is ultimately excreted in urine. This is why the liver needs urea to be soluble. Insoluble nitrogen waste would not move through blood cleanly and would not filter through the kidney in this controlled way. Solubility is the transport logic that makes the liver-kidney partnership possible.

That partnership is one of the most important cross-organ agreements in the body. The liver is the organ that can safely convert ammonia into urea. The kidneys are the organs that can efficiently remove urea from the blood. Neither solves the full problem alone. If the liver fails, ammonia rises because the nitrogen was never packaged safely in the first place. If the kidneys fail, urea accumulates because the packaged nitrogen was not removed after the handoff. In the healthy state, though, the sequence is elegant. Amino acid metabolism produces nitrogen burden. The liver traps it in urea. The kidneys filter and excrete the urea. Protein metabolism remains possible because no single tissue is being asked to tolerate uncontrolled ammonia.

Urea also illustrates an important principle about waste in physiology. The body often does not eliminate its most dangerous compounds in their original form. It converts them first into a transport-compatible form that suits the architecture of the next organ in line. That is exactly what is happening here. The liver does not send ammonia to the kidneys and hope for the best. It sends urea, because urea is the version of nitrogen the bloodstream can carry and the kidneys can remove without immediate systemic damage. So when the reader sees urea in blood or urine, the right way to think about it is this: it is the liver's finished package for nitrogen disposal, handed off to the kidneys so protein metabolism does not become self-poisoning.

What Hyperammonemia Is

Hyperammonemia means too much ammonia is circulating in the blood. That sounds simple, but the biology underneath it is severe because ammonia is not a harmless lab number waiting quietly for interpretation. It is a nitrogen waste burden that the body was supposed to capture early, mostly in the liver, and convert into urea before it reached the wider circulation in meaningful quantity. When that containment fails, ammonia remains in a freer, more toxic form than the organism can safely tolerate. The word itself tells the story. Hyper means elevated. Ammonemia means ammonia in the blood. What it really describes is a loss of nitrogen control. Protein metabolism, amino acid breakdown, and gut microbial activity keep generating nitrogen that tends toward ammonia. The liver is supposed to intercept that stream. Hyperammonemia is what you call the condition when the interception is no longer sufficient.

The problem can arise from several directions, but they all converge on the same outcome. The liver may be too damaged to run the urea cycle at adequate scale. Portal blood may bypass the liver through abnormal vascular shunting, meaning ammonia-rich blood from the gut reaches the systemic circulation without enough hepatic first pass. The urea cycle machinery itself may be genetically impaired. In catabolic states, gastrointestinal bleeding, severe infection, high protein breakdown, or other stress conditions, ammonia production can rise high enough that even a compromised liver falls behind more dramatically. Different causes, same chemistry. Too much nitrogen is being converted into ammonia or delivered as ammonia burden, and too little of it is being trapped into urea before it escapes into the blood.

What makes hyperammonemia dangerous is that ammonia is not just present in the bloodstream. It is mobile. It reaches the brain quickly, and the brain handles it badly when levels rise. Astrocytes, the support cells that help maintain the chemical environment around neurons, try to protect the nervous system by combining ammonia with glutamate to form glutamine. That defensive move traps the ammonia temporarily, but it also loads the astrocyte with an osmotically active molecule. Water follows. The cell swells. Once enough of that happens, brain tissue begins losing chemical and structural stability. Neurotransmitter handling shifts. Energy metabolism becomes less clean. Thought slows. Attention degrades. In severe cases cerebral edema, meaning brain swelling, can develop. Hyperammonemia therefore is not merely excess ammonia in an abstract fluid compartment. It is nitrogen toxicity with a direct line to the nervous system.

This is why the symptoms often look neurological before they look chemical. The person may become mentally slower, inattentive, sleep-disordered, irritable, or confused. Coordination worsens. Asterixis can appear, the flapping lapse in maintained posture that reflects unstable motor control. If the burden continues to rise, encephalopathy deepens, and stupor or coma can follow. The liver may be the organ that failed to contain the ammonia, but the brain is often the organ that reveals that failure first. Hyperammonemia is one of the clearest examples in physiology of a waste-handling problem becoming a consciousness problem.

The deeper point is that hyperammonemia means the blood is no longer safe for the kind of protein and nitrogen turnover normal life requires. Amino acid metabolism has continued, gut bacteria have continued producing nitrogenous burden, but hepatic conversion into urea has not kept up. The liver has lost enough of its nitrogen-buffering authority that the bloodstream is now carrying the wrong form of waste. That is what hyperammonemia is in real terms. Not just elevated ammonia, but nitrogen disposal failure made system-wide and made dangerous.

Hepatic Encephalopathy as Failed Nitrogen Control Made Visible

Hepatic encephalopathy is what happens when failed nitrogen control stops being a biochemical problem hiding in the liver and becomes a functional problem unfolding in the brain. It is not best understood as a vague mental complication of liver disease. It is the visible neurologic expression of blood that is no longer being kept safe from gut-derived and metabolism-derived nitrogen burden. Protein turnover continues. Amino acids are deaminated. Gut microbes keep generating ammonia and related nitrogenous compounds. Portal blood keeps bringing that burden toward the liver. Under healthy conditions hepatocytes trap the ammonia, channel it into the urea cycle, and release urea into blood so the kidneys can excrete it. Hepatic encephalopathy begins when that control is no longer sufficient. The liver may be too damaged to run the cycle at adequate scale, or portal blood may be bypassing the liver through shunting, sending ammonia-rich blood into the systemic circulation without enough first pass through hepatocytes. Either way, the result is the same. The bloodstream now carries nitrogen in the wrong form, and the brain is forced to live downstream of that failure.

The word encephalopathy simply means brain dysfunction. In this setting the dysfunction is metabolic before it is structural. The brain is not being destroyed by a large focal lesion. It is being chemically destabilized by blood that contains too much ammonia and a wider burden of incompletely handled nitrogenous substances. Ammonia crosses into neural tissue and astrocytes, the support cells that help maintain the chemical environment around neurons, begin trying to protect the brain by binding ammonia into glutamine. That sounds protective because it is protective at first. But glutamine is osmotically active, which means it pulls water with it. Astrocytes swell. Once enough astrocytes swell, the microscopic architecture of the brain becomes less stable. Signaling becomes less clean. Energy handling becomes less efficient. The chemical environment around neurons becomes harder to defend. So hepatic encephalopathy is not a mystery symptom attached to liver disease. It is nitrogen control failure translated into swollen support cells, altered neurotransmission, and slowed neural function.

This is why the syndrome often begins subtly and then becomes unmistakable. The person may not look profoundly ill at the start. They may simply become slower, less attentive, less organized, more sleep-disordered, or strangely dulled. Sleep-wake reversal is common because the brain's normal timing and arousal systems are chemically less stable. As the burden rises, speech, concentration, judgment, and motor control worsen. Asterixis can appear, the flapping lapse in sustained posture that reveals unstable cortical and motor output. Later the clouding deepens into confusion, disorientation, stupor, and coma. This progression is important because it shows that hepatic encephalopathy is not just "brain fog" and not just "liver disease causing confusion." It is a graded failure of neural performance under the pressure of blood that is no longer being detoxified properly at the nitrogen level.

The gut-liver axis is central to the whole event. Much of the nitrogen burden does not arise only from the body's own protein turnover. It also comes from intestinal bacterial metabolism. The colon is constantly generating ammonia and related compounds from nitrogen-containing substrates, and under ordinary conditions portal flow brings that burden straight to the liver, where it is contained before it can spread systemically. When the liver fails or when portal blood is diverted around it, the brain begins receiving a chemical load it was never meant to see in significant quantity. This is why gastrointestinal bleeding can precipitate encephalopathy so dramatically in a cirrhotic patient. Blood entering the gut lumen becomes a dense protein load for microbial digestion. More nitrogen is liberated. More ammonia is produced. The failing liver cannot keep up. The brain pays the price. The sequence is brutally direct.

What makes hepatic encephalopathy such a useful teaching syndrome is that it exposes the real purpose of hepatic nitrogen control. The liver is not running the urea cycle for elegance or biochemical completeness. It is doing it so thought remains possible. It is converting ammonia into urea so consciousness, attention, motor control, and wakefulness can remain intact while protein metabolism continues. When that conversion fails, the nervous system becomes the place where liver physiology turns visible. The blood still circulates. The heart still beats. The lungs may still oxygenate. But the chemical quality of the circulation has fallen below what the brain can tolerate. Hepatic encephalopathy is failed nitrogen control made visible because the invisible work of ammonia disposal has stopped protecting the one organ that reveals failure fastest.

Chapter 13

Lactate, Pyruvate, and Metabolic Traffic

What Lactate Really Is

Lactate is usually introduced to the public as a waste product, a kind of metabolic exhaust created when the body falls behind and starts running dirty. That framing is too crude to teach real physiology. Lactate is not metabolic trash waiting to be removed. It is a transport form of carbon, a redox solution, and a traffic signal moving between tissues. Redox refers to the handling of electrons inside metabolism, the balance between reduction and oxidation that determines whether pathways can keep moving. When glucose is broken down through glycolysis, pyruvate is produced. If the cell needs to regenerate nicotinamide adenine dinucleotide in its oxidized form, usually written as NAD+, so glycolysis can continue, pyruvate can accept electrons and become lactate. That step is not a sign that metabolism has become useless. It is a way to keep carbon moving and ATP production going under conditions where rapid glycolytic flux outruns the local capacity to oxidize pyruvate immediately inside mitochondria. Lactate is therefore not the opposite of useful metabolism. It is one of the ways metabolism protects continuity under pressure.

This is easiest to see in tissues that rely heavily on glycolysis. Red blood cells have no mitochondria at all, so they turn glucose into lactate continuously. They do not produce lactate because they are failing. They produce lactate because that is how their metabolism is built. Skeletal muscle can increase lactate production when glycolytic demand rises faster than mitochondrial oxidation can match it. The gut, the skin, immune cells, and other tissues also produce lactate under perfectly ordinary physiological conditions. What matters is that lactate is not the end of the story. It leaves one tissue and becomes substrate for another. It circulates in blood. It reaches the liver. It reaches the heart. It reaches oxidative tissues that can convert it back toward pyruvate and use it. This is why the term waste misleads. Waste is something the body is trying to eliminate. Lactate is often something the body is trying to move.

The liver is one of the main reasons that movement matters. Hepatocytes can take up lactate and convert it back into pyruvate through lactate dehydrogenase, the same enzyme system operating in the opposite direction depending on tissue conditions and redox state. From there pyruvate can enter gluconeogenesis, the synthesis of new glucose from noncarbohydrate material. This is the core of the Cori cycle. Peripheral tissues spend glucose and release lactate. The liver receives that lactate, rebuilds glucose from it, and returns glucose to the circulation. The carbon is not lost. It has changed vehicles. Lactate is one of the body's ways of moving carbon from a site of rapid glycolytic use to a site of hepatic reconstruction. That is why the liver belongs in any serious explanation of lactate. Blood lactate is not just evidence of what muscle has done. It is also incoming material for hepatic blood sugar support.

Pyruvate helps make this clearer because lactate and pyruvate are a pair, not isolated molecules with unrelated reputations. Pyruvate sits at a metabolic crossroads. It can move into mitochondria and become acetyl coenzyme A for oxidation. It can be carboxylated into oxaloacetate for gluconeogenesis. It can accept electrons and become lactate. Which direction dominates depends on oxygen availability, mitochondrial throughput, substrate demand, hormonal state, and the redox condition inside the cell. So when lactate rises, the right question is not simply whether the body is producing waste. The right question is what kind of traffic the body is trying to maintain. Is glycolysis running fast. Is redox balance being defended. Is carbon being exported for later recovery by the liver. Is oxygen delivery insufficient relative to demand. Lactate is often the visible consequence of these deeper choices.

This is why lactate should be taught as a traffic molecule rather than a biochemical embarrassment. It lets glycolysis continue when pyruvate cannot be oxidized locally at the required speed. It allows carbon to leave one tissue and be reused elsewhere. It feeds hepatic gluconeogenesis and supports blood glucose during fasting, stress, and high metabolic turnover. It reflects pressure, but pressure is not the same as pathology. Lactate becomes easier to understand the moment the reader stops asking how the body gets rid of it and starts asking where the body is sending it next.

Why Lactate Is Not Simply a Waste Product

Calling lactate a waste product hides the actual logic of metabolism. Waste is something the body is trying to discard because it has no further use for it. Lactate does not fit that definition. The body does not produce lactate merely to get rid of it. It produces lactate to keep glycolysis moving, to preserve redox balance, and to move carbon from one tissue to another in a form the bloodstream can carry. When glycolysis converts glucose into pyruvate, the cell also has to regenerate nicotinamide adenine dinucleotide in its oxidized form, NAD+, if glycolysis is going to continue. One of the fastest ways to do that is to reduce pyruvate into lactate. That step is not a sign that metabolism has broken. It is a way of preventing metabolism from stalling when glycolytic demand is high or when local mitochondrial oxidation is not matching the speed of carbon flow. Lactate is therefore a functional intermediate and a transportable carbon form, not metabolic garbage.

This becomes obvious the moment you ask what happens to lactate after it leaves the producing tissue. Red blood cells produce lactate constantly because they do not have mitochondria at all. They are not failing every second of their existence. They are operating exactly as built. Skeletal muscle can increase lactate production when glycolysis rises sharply. Other tissues do the same under a range of perfectly ordinary conditions. But the lactate does not simply pile up as abandoned residue. It enters the bloodstream and becomes substrate elsewhere. The heart can oxidize lactate. Other oxidative tissues can convert it back toward pyruvate and use it. The liver can take it up and convert it into glucose through gluconeogenesis, the synthesis of new glucose from noncarbohydrate material. That means lactate is often not the end of metabolism in one tissue. It is the beginning of metabolism in another.

The liver makes that especially clear. Lactate arriving in hepatic blood is not treated like trash requiring disposal. It is treated like recoverable carbon. Hepatocytes convert lactate back to pyruvate and then, depending on metabolic state, can route that carbon toward glucose production. This is a major part of the Cori cycle, where peripheral tissues spend glucose and release lactate, and the liver rebuilds glucose from that lactate and returns it to the circulation. A true waste product is something the body is trying to eliminate from the system. Lactate is often being circulated on purpose because it is part of a larger fuel relay. The body is not done with it. It is repositioning it.

The redox side matters just as much. Lactate formation helps regenerate NAD+, and without enough NAD+ glycolysis slows because one of its key reactions can no longer proceed efficiently. In fast-moving metabolic states that regeneration is not optional. It is what allows ATP, adenosine triphosphate, to keep being produced from glycolysis when the cell still needs rapid energy output or cannot oxidize pyruvate at the required rate. So lactate is not merely a leftover carbon molecule. It is part of how the cell keeps electron handling and carbon handling coordinated. It protects flux. It buys time. It allows continued ATP generation under conditions where waiting for full mitochondrial oxidation would not meet immediate demand. A molecule doing that much work should not be dismissed as waste.

This does not mean elevated lactate is meaningless or always benign. It can rise when tissues are under pressure, when oxygen delivery is inadequate relative to demand, when glycolysis is accelerated, when clearance is impaired, or when the liver is too compromised to reclaim and process it properly. But that is different from saying lactate itself is waste. Elevated lactate is often a sign that metabolic traffic is being rerouted under stress. The molecule is telling you something about flux, redox state, perfusion, or hepatic handling. It is a report from metabolism, not a useless residue.

So lactate is not simply a waste product because the body makes it to solve a problem, moves it through blood as useful cargo, and reuses its carbon in other tissues, especially the liver. Waste is something the organism wants gone. Lactate is something the organism often wants moved.

The Cori Cycle and Hepatic Glucose Recovery

The Cori cycle is one of the clearest examples of the liver turning apparent metabolic loss back into usable blood fuel. A peripheral tissue breaks down glucose through glycolysis, produces pyruvate, and then converts much of that pyruvate into lactate when local conditions favor rapid glycolytic flux over immediate mitochondrial oxidation. That lactate leaves the tissue, enters the bloodstream, and travels to the liver. There it is taken up by hepatocytes, converted back toward pyruvate, and pushed through gluconeogenesis, the synthesis of new glucose from noncarbohydrate material. The rebuilt glucose can then re-enter the circulation and support tissues that need it. What looked like spent carbon in one location becomes recovered glucose in another. That is the cycle. It is not a disposal loop. It is a rescue loop.

This matters because glycolytic tissues are often solving an immediate local problem rather than a whole-body one. Red blood cells, which have no mitochondria, produce lactate continuously because they have no oxidative alternative. Working skeletal muscle can produce far more lactate when glycolytic demand outruns what local mitochondria can oxidize at that moment. Other tissues do the same under varying conditions of stress, speed, or oxygen handling. None of those tissues are built to reconstruct blood glucose for the organism as a whole. The liver is. So the Cori cycle is a division of labor. Peripheral tissues get rapid ATP from glycolysis and export lactate. The liver takes the carbon burden back, pays the energetic cost of rebuilding glucose, and returns that glucose to the blood. The body is using one organ's ability to spend energy in order to preserve another tissue's ability to function fast.

The recovery step in the liver is what makes the cycle so important. Lactate arriving in hepatic blood is not discarded as metabolic waste. Hepatocytes convert it back to pyruvate through lactate dehydrogenase, and that pyruvate can then enter gluconeogenesis if the hormonal and energetic setting favors glucose production. This is especially important in fasting, prolonged exercise, stress states, and other conditions where circulating glucose must be defended even though direct intestinal carbohydrate delivery is absent or insufficient. The liver is not inventing glucose out of nothing. It is reclaiming carbon that other tissues have already partly processed and rebuilding it into a form the bloodstream can use again.

There is a cost to this recovery, and that cost tells you something important about metabolic hierarchy. The peripheral tissue got ATP quickly from glycolysis. The liver now has to spend ATP to rebuild glucose from lactate. That means the cycle is not a free lunch. The organism is choosing to preserve blood glucose and support tissue function even at an energetic price. The liver usually pays that price by oxidizing fatty acids and using its own mitochondrial capacity to power gluconeogenesis. This is one reason the liver is such a central metabolic organ. It can absorb energetic cost on behalf of the bloodstream. A muscle cell under pressure gets quick energy and exports lactate. The liver later spends deeper metabolic effort to turn that exported carbon back into glucose. One tissue spends sugar. The liver rebuilds it.

The Cori cycle also clarifies what lactate really is. It is not simply residue from failed metabolism. It is mobile carbon waiting for hepatic recovery. A molecule called waste is usually headed for elimination. Lactate is usually headed for reuse. That is a different biological category. When the liver is functioning well, rising lactate does not automatically mean carbon has been lost to the body. It often means carbon has changed compartments and now requires hepatic work to re-enter the blood as glucose. The cycle is therefore one of the clearest proofs that liver physiology is about traffic control as much as chemistry. The liver receives carbon exported under one set of tissue conditions and reissues it under another.

So the Cori cycle should be understood as a blood-glucose salvage system. Peripheral tissues use glucose and export lactate when their local metabolic conditions demand it. The liver takes that lactate back, converts it into pyruvate, rebuilds glucose through gluconeogenesis, and returns that glucose to the circulation. Hepatic glucose recovery is the reason this carbon is not simply lost after glycolysis. The body spends energy to keep its circulating fuel coherent, and the liver is the organ doing that work.

Pyruvate as a Metabolic Crossroads

Pyruvate matters because it is one of the last molecules in metabolism that still has several futures open at once. It sits at the end of glycolysis, but it is not the end of anything important. It is a decision point. Once glucose has been broken down to pyruvate, the cell has to decide whether that carbon is headed toward oxidation, reduction, glucose recovery, or broader biosynthetic traffic. That is why pyruvate should not be taught as a minor intermediate squeezed between glycolysis and the next chapter. It is the hinge molecule where fast carbohydrate use can become mitochondrial energy production, lactate export, or hepatic glucose rebuilding depending on tissue type, oxygen availability, redox state, and hormonal context. The body keeps returning to pyruvate because pyruvate is one of the places where metabolic direction is chosen.

One of its main futures is mitochondrial entry. Pyruvate can be transported into the mitochondrion and converted by pyruvate dehydrogenase into acetyl coenzyme A, the two-carbon unit that enters the tricarboxylic acid cycle, also called the citric acid cycle. That route commits the carbon toward oxidative energy production. It is the path favored when oxygen delivery and mitochondrial capacity are sufficient and when the cell is trying to extract deeper energy yield from carbohydrate. But pyruvate is not forced down that path. If glycolysis is moving faster than mitochondria can process the incoming carbon, or if the cell needs to regenerate nicotinamide adenine dinucleotide in its oxidized form, NAD+, so glycolysis can continue, pyruvate can instead accept electrons and become lactate. That alternative is not metabolic failure. It is a rerouting decision. The same carbon reaches a different destination because the cell is solving a different problem.

Pyruvate can also move in the opposite strategic direction and become part of glucose recovery. In the liver, and to an important extent in the kidney during prolonged fasting, pyruvate can be carboxylated into oxaloacetate and drawn into gluconeogenesis, the synthesis of new glucose from noncarbohydrate material. This is one reason pyruvate sits at a genuine crossroads rather than a one-way exit ramp. Carbon arriving as pyruvate may be burned. It may be exported as lactate. Or it may be turned back toward glucose and sent into the bloodstream again. The liver depends on this flexibility because fasting, exercise, and stress all require carbon to be rerouted according to whole-body need rather than according to the narrow preference of the cell currently holding it. Pyruvate is one of the key places where the liver can reclaim circulating carbon and turn it back into blood fuel.

The redox side is what makes this crossroads dynamic rather than merely structural. The choice between pyruvate and lactate is partly a choice about electron handling. If the cell needs to regenerate NAD+ quickly to keep glycolysis moving, reducing pyruvate to lactate solves that local redox problem. If mitochondrial oxidation is available and favored, pyruvate can move inward and be oxidized instead. This is why pyruvate cannot be understood only as carbon waiting for the next enzyme. It is also a redox decision point. The cell is asking whether carbon should be pushed deeper into oxidation or temporarily reduced and exported so glycolytic flux can continue. Different tissues answer that question differently depending on their architecture. Red blood cells answer it one way because they have no mitochondria. Working muscle may answer it another way under high glycolytic demand. The liver answers it differently again because it is often receiving lactate or pyruvate as incoming traffic to be reorganized for the bloodstream.

Pyruvate is also a crossroads because it links carbohydrate traffic to amino acid traffic. Through transamination, pyruvate can become alanine, carrying carbon and nitrogen into another transport form that is especially important between muscle and liver. Alanine can then travel to the liver, where the carbon can be used for gluconeogenesis and the nitrogen can be stripped and directed toward the urea cycle. That means pyruvate is not only a carbohydrate intermediate. It is also one of the points where carbohydrate metabolism and nitrogen metabolism begin talking to each other. The liver uses that connection constantly. Carbon and nitrogen rarely travel as separate stories for long.

So pyruvate should be understood as a metabolic crossroads because it sits where several major systems meet. Glycolysis delivers it. Mitochondria can oxidize it. Lactate dehydrogenase can reduce it into lactate. Pyruvate carboxylase can redirect it toward glucose rebuilding. Transamination can move it into amino acid traffic. The fate of pyruvate tells you what the cell is trying to protect at that moment: deep oxidation, rapid glycolysis, blood glucose continuity, or nitrogen-linked carbon transfer. That is why this small molecule matters so much. It is one of the places where metabolic traffic becomes a decision instead of a straight line.

The Liver's Role in Keeping Fuel Forms Interchangeable

The liver keeps the body metabolically flexible by refusing to let fuel stay trapped in only one form. That is one of its deepest jobs. The bloodstream cannot be stabilized if glucose remains only glucose, if amino acid carbon remains only amino acid carbon, if lactate remains only lactate, or if fat-derived carbon cannot be redirected when the fed state ends. The organism survives because the liver can receive one fuel form, strip it down, rebuild it, store it, or export it as something else depending on demand. That is what fuel interchangeability really means.

It is not a vague ability to "use different fuels." It is the biochemical power to move carbon from one molecular vehicle into another so the blood keeps delivering what each tissue needs at the right time. The liver is the main organ doing that traffic conversion at scale.

Glucose is the easiest place to see it. After a meal, hepatocytes take up portal glucose and can immediately burn some, store some as glycogen, or direct some of it toward lipid synthesis if substrate load stays high enough. Later, when external intake stops, that same liver can reverse direction. Glycogen is broken down. Glucose is released. If glycogen begins to thin out, the liver starts rebuilding glucose from lactate, glycerol, and amino acid carbon skeletons through gluconeogenesis. The same organ that stored carbohydrate as reserve can later reconstruct carbohydrate from noncarbohydrate material. That is interchangeability made real. A meal-derived sugar load becomes glycogen. Glycogen becomes circulating glucose. Lactate and glycerol become new glucose. Fuel is not fixed. The liver keeps redrawing it.

The lactate-pyruvate axis shows this even more clearly. Peripheral tissues can convert glucose to lactate when glycolytic demand is high or when their own architecture favors that route, as in red blood cells, which lack mitochondria entirely. That lactate enters the bloodstream and reaches the liver, where hepatocytes convert it back toward pyruvate and then, depending on hormonal state and energetic need, either oxidize it or rebuild it into glucose. The carbon is the same. The form changes according to the problem being solved. In one tissue it is fast glycolytic output. In the liver it becomes substrate recovery. This is why lactate is not waste in the usual sense. It is one of the main ways the body moves usable carbon from a site of rapid use to a site of reconstruction.

Amino acids add another layer because they can serve as structure, but their carbon can also be redirected when the body needs glucose or energy. Once the amino group is removed and the nitrogen is sent toward the urea cycle, the remaining carbon skeleton can enter central metabolic pathways at different points and be converted toward glucose or oxidation. That does not make amino acids ideal fuel in the same way as glucose or fat, because nitrogen disposal makes them metabolically expensive. But it does mean the liver can reclaim their carbon for blood sugar support during fasting or stress. The molecule no longer remains trapped in the category it entered under. Protein-derived carbon becomes glucose carbon when the organism needs that conversion badly enough. That is another form of interchangeability, and the liver is where it is enforced.

Fat enters the same logic through both glycerol and ketone biology. Adipose tissue releases glycerol when triglycerides are broken down, and the liver can turn that glycerol into glucose through gluconeogenesis. At the same time hepatocytes oxidize fatty acids for their own energy and convert part of that fat-derived carbon into ketone bodies when fasting deepens. Ketone bodies are water-soluble fuel molecules that can circulate to tissues such as the brain and muscle, allowing fat-derived energy to travel in a blood-compatible form. So the liver is not only turning noncarbohydrate carbon into glucose. It is also turning fat into a transportable alternative to glucose. One organ is managing conversion in both directions of need. When glucose must be preserved, fat helps support glucose production and ketone generation. When carbohydrate is abundant, excess carbon may be pushed toward lipid synthesis and storage. The liver is constantly translating between abundance and scarcity by changing the form in which energy exists.

This is why fuel flexibility is never just about what a tissue can oxidize locally. It is about whether the blood can keep carrying the right mixture of fuels as feeding, fasting, activity, and stress change. The liver protects that flexibility by sitting in the middle of carbon traffic and refusing to let any one form dominate simply because it arrived first. It can take up fructose and route it into hepatic metabolism. It can store glucose and later release it. It can recover lactate into new glucose. It can strip amino acids down to their carbon skeletons and redirect them while neutralizing the nitrogen burden. It can use fat to power gluconeogenesis and can export ketones when fat-derived carbon needs to reach glucose-sensitive tissues indirectly. That is what makes the liver a master organ of fuel form rather than a mere warehouse.

So the liver's role in keeping fuel forms interchangeable is the reason the body does not live at the mercy of the last meal or the current tissue demand. Carbon can be moved, rebuilt, delayed, released, and reassigned. Glucose can become glycogen. Glycogen can become glucose. Lactate can become glucose. Amino acid carbon can become glucose. Fat can power glucose production or become ketones for export. The liver is the organ that keeps those translations open. Without it, the bloodstream would be trapped in the chemistry of the moment.

Why Metabolic Flexibility Depends Heavily on Hepatic Function

Metabolic flexibility sounds abstract until you force it into the bloodstream. Then the real meaning appears. It is the ability of the body to move cleanly between fed and fasting states, between incoming carbohydrate and stored glycogen, between glucose use and fat oxidation, between lactate release and glucose recovery, between amino acid retention and amino acid carbon salvage, without allowing blood chemistry to become erratic, toxic, or starved. That movement depends heavily on the liver because the liver is the organ that sits at the center of fuel conversion rather than at the edge of fuel consumption. Muscle can burn what it receives. Adipose tissue can store or release fat. The brain can use what arrives and adapt partly to ketones over time. The liver is the organ that decides what form of fuel will arrive in the first place. It receives the first wave after a meal, absorbs part of the excess, stores part of it, rebuilds fuel when intake stops, and changes the mixture of circulating substrates as conditions change. Flexibility therefore is not just a property of mitochondria inside isolated cells. It is a property of hepatic traffic control at the scale of the whole organism.

The fed-to-fasting transition makes this obvious. After a meal the liver takes up portal glucose, suppresses its own glucose output, and stores part of the incoming carbohydrate as glycogen. When food disappears, it reverses direction. Glycogen is opened. Glucose is released. As glycogen thins, lactate, glycerol, and amino acid carbon skeletons are drawn into gluconeogenesis so new glucose can be built. At the same time fatty acid oxidation rises in hepatocytes to power that work, and ketone production begins to increase as fasting deepens. That sequence is metabolic flexibility in real physiological space. The body is not merely switching from sugar to fat because of some motivational slogan about fat adaptation. It is being carried through a controlled transition by an organ that can buffer one fuel form, reconstruct another, and prevent the circulation from crashing during the handoff. If hepatic function is weak, that handoff becomes rough. The bloodstream begins reflecting the gaps between fuel states instead of being protected from them.

The liver also matters because flexibility is not only about access to multiple fuels. It is about access to multiple conversions. Lactate has to become glucose when glucose must be defended. Glycerol has to become glucose when adipose tissue is being mobilized. Amino acid carbon has to be redirected when glycogen is falling and the body still needs circulating sugar. Excess carbohydrate may need to move toward lipid synthesis in the fed state. Cholesterol may need to move toward bile acid synthesis. Fatty acids may need to be oxidized locally while ketones are exported for use elsewhere. These are not small metabolic side paths. They are the reason the organism can survive changing conditions without each tissue being forced to solve every problem alone. Hepatic function is what keeps carbon from getting stranded in the wrong form at the wrong time.

Nitrogen handling makes the dependency even clearer. The body can only be flexible with amino acid use if the liver can safely dispose of the resulting nitrogen burden. An organism that could not run the urea cycle would be far less free to use amino acid carbon for glucose support or energy because ammonia toxicity would quickly become the limiting factor. The same principle applies to lactate recovery, ketone production, glucose output, and bilirubin handling. Flexibility is never free. Every switch in substrate use creates secondary burdens, redox burdens, nitrogen burdens, acid-base burdens, transport burdens, and the liver is one of the main organs absorbing those costs. A body is only as flexible as its central conversion organ is competent.

This is why hepatic dysfunction so often looks like poor adaptability before it looks like total failure. The person may still be alive, still digesting, still storing some fat, still eating meals, but the transitions become unstable. Post-meal glucose handling worsens because the liver no longer buffers the incoming carbohydrate load cleanly. Fasting support worsens because glycogen handling and gluconeogenesis are weaker. Lactate may accumulate more easily because hepatic recovery is impaired. Ketone output may become mistimed or insufficient relative to demand. Amino acid use becomes metabolically more expensive because nitrogen disposal is less secure. The circulation becomes chemically less agile. That is what loss of metabolic flexibility looks like in the real body. Not simply the inability to burn fat, but the inability to move fuel forms, timing, and waste burdens through the blood without disorder.

So metabolic flexibility depends heavily on hepatic function because the liver is the organ that keeps fuel from being trapped in single identities. It stores glucose, releases glucose, rebuilds glucose, oxidizes fat, exports ketones, recovers lactate, redirects amino acid carbon, and neutralizes the waste burdens created by those conversions. Other tissues consume. The liver translates. Without that translation, the body becomes metabolically rigid not because it lacks fuels, but because it loses the organ that knows how to change one into another while keeping the blood livable.

Part IV

The Liver as the Master of Fat Handling

Chapter 14

Dietary Fat Arrives Back at the Liver

Chylomicrons and Chylomicron Remnants

Dietary fat does not arrive at the liver the way glucose and amino acids do. It does not cross the intestine, enter the portal vein in bulk, and present itself immediately to hepatocytes as raw meal chemistry. Long-chain dietary fat takes a more elaborate route because fat is hydrophobic, meaning it does not mix well with water, and blood is a water-based transport medium. The body therefore has to digest fat, emulsify it, absorb it into enterocytes, which are the absorptive cells of the small intestine, rebuild it into transportable particles, send those particles first into lymph rather than straight into portal blood, let peripheral tissues take part of the load, and only then return the reduced particle back to the liver. That returning particle is the chylomicron remnant. This is the real meaning of dietary fat arriving back at the liver. The liver is not the first large blood checkpoint for long-chain dietary fat in the same direct way it is for glucose. It is the receiving organ for fat after the intestine and peripheral tissues have already done part of the handling.

The first particle in that journey is the chylomicron. After bile acids emulsify dietary fat in the intestinal lumen and pancreatic lipase breaks triglycerides into absorbable components, mainly free fatty acids and monoglycerides, those lipid products enter the enterocyte. Inside the enterocyte they are re-esterified, meaning they are rebuilt into triglycerides. Cholesterol is esterified as well, and these lipids are then packaged together with phospholipids and apolipoprotein B-48 into a large lipoprotein particle called a chylomicron. A lipoprotein is a transport particle built to carry hydrophobic lipids through aqueous body fluids. The chylomicron is enormous by lipoprotein standards because it is built to carry a fresh meal-sized fat load. It leaves the intestine through the lymphatic system rather than through the portal vein because it is too large and too lipid-rich for the ordinary immediate portal route. So the body's first answer to dietary fat is not direct hepatic uptake. It is packaging and rerouting.

Once chylomicrons reach the bloodstream through lymphatic drainage, they begin losing triglyceride to peripheral tissues. This happens largely through lipoprotein lipase, an enzyme anchored on capillary surfaces in tissues such as adipose tissue, muscle, and heart. Lipoprotein lipase hydrolyzes the triglycerides inside the chylomicron, releasing free fatty acids that nearby tissues can take up for storage or oxidation. Adipose tissue may re-esterify and store them. Muscle and heart may burn them. What matters for the liver is that the original chylomicron is being progressively stripped of its triglyceride cargo as it circulates. The particle gets smaller, denser, and relatively more enriched in cholesterol and cholesteryl esters as triglyceride is removed. This partially depleted particle is the chylomicron remnant. So the remnant is not a fresh intestinal export. It is the afterimage of a meal after peripheral tissues have already taken their share.

This is why chylomicron remnants matter so much to hepatic fat handling. The liver is receiving not just dietary fat, but dietary fat after distribution. The remnant still carries valuable lipid cargo, especially cholesterol, fat-soluble vitamins, phospholipids, and remaining triglyceride, but it arrives in a form that now signals prior tissue interaction. Apolipoprotein E on the remnant surface helps the liver recognize and take up the particle through hepatic remnant receptors. Once internalized by hepatocytes, the remnant is dismantled and its contents are sorted. Cholesterol may be used for membranes, steroid-related pathways, lipoprotein synthesis, or bile acid production. Fatty acids may be oxidized, re-esterified, stored transiently, or repackaged into very-low-density lipoprotein for re-export. Fat-soluble vitamins may be stored or redistributed. This is the point where the liver re-enters the story decisively. It did not receive the meal fat in its original intestinal form. It receives the leftovers of first delivery and decides what happens next.

That sequence is important because it shows that the liver is not merely a passive endpoint for dietary fat. It is the second major editor of the meal's lipid traffic. The intestine digests and packages. Peripheral tissues extract what they can use immediately. The liver receives the remnant and performs the deeper accounting. What remains worth storing. What should be oxidized. What should be converted into bile acids. What should be repackaged into hepatic lipoproteins and sent back out. What should be held as cholesterol reserve. This is why the liver deserves to be called the master of fat handling. It is not only dealing with endogenous fat made inside the body. It is receiving the partially processed remains of dietary fat after first-pass distribution and bringing those remains back under central control.

The chylomicron and the chylomicron remnant therefore describe two different moments in the life of a meal. The chylomicron is the intestine's first large export particle, built to move newly absorbed fat through lymph and into blood. The chylomicron remnant is what comes back after that fat has already met peripheral tissues. The liver receives the remnant, not the untouched meal, and that difference matters. It means hepatic fat handling begins not at the first bite of fat absorbed, but at the point where the body has already sampled the meal and now needs a central organ to decide the fate of what remains.

How Dietary Fat Moves Before It Returns to the Liver

Dietary fat does not take the straight portal route that glucose and amino acids take. That is the first structural fact the reader needs to hold. Fat enters the mouth as a water-insoluble cargo packed into droplets, membranes, adipose tissue, phospholipids, cholesterol esters, and triglycerides that cannot simply dissolve into the aqueous environment of the gut and then slip cleanly into portal blood. The body has to break that hydrophobic mass into a transportable sequence. In the stomach, mechanical mixing begins to shear fat into smaller droplets, but the major change happens in the small intestine. Bile from the liver and gallbladder is released into the duodenum, and bile acids begin emulsifying the fat, which means breaking larger lipid masses into much smaller suspended droplets so digestive enzymes can reach the surface. Pancreatic lipase then hydrolyzes triglycerides into free fatty acids and monoglycerides. Cholesterol esters and phospholipids are also being acted on by specific enzymes. At this point the meal's fat is no longer intact food fat. It has become a mixture of smaller lipid products floating in the lumen, the hollow interior of the intestine, waiting for the next transport step.

Those products still cannot cross efficiently on their own because the intestinal lumen is watery and lipids do not move well in water. The body solves that by using bile acids to form micelles, small transport assemblies that hold hydrophobic lipid products in a form compatible with the aqueous intestinal environment and deliver them to the brush border of the enterocyte, the absorptive cell lining the small intestine. The micelle is not the final transport particle. It is a ferry across the lumen. Free fatty acids, monoglycerides, cholesterol, and fat-soluble vitamins leave the micelle at the enterocyte surface and enter the cell. Once inside, another major shift occurs. The absorbed lipids are rebuilt. Fatty acids and monoglycerides are re-esterified into triglycerides. Cholesterol may be esterified. Phospholipids are assembled. The enterocyte is not simply passing fat through unchanged. It is reconstructing meal fat into a form the body can export.

That export form is the chylomicron. Inside the enterocyte, triglycerides, cholesterol esters, phospholipids, and fat-soluble vitamins are packaged together with apolipoproteins, especially apolipoprotein B-48, into a large lipoprotein particle built specifically for dietary fat transport. This matters because the bloodstream cannot carry a meal-sized lipid load as free oil. It needs the fat enclosed inside a particle with a phospholipid and apolipoprotein surface compatible with aqueous transport. The chylomicron is the intestine's answer to that problem. But it still does not go directly into the portal vein. It is too large for that route and too specialized for immediate hepatic first pass. Instead it enters the lymphatic system through intestinal lacteals, the small lymphatic vessels inside the villi of the small intestine. From there it moves through lymphatic channels, bypasses the liver's immediate portal checkpoint, and only later reaches the bloodstream through the thoracic duct draining into the venous circulation. This is one of the most important differences between carbohydrate and fat handling. Long-chain dietary fat reaches the circulation first through lymph, not portal blood.

Once chylomicrons enter the bloodstream, they begin losing their triglyceride cargo to peripheral tissues. Capillary-bound lipoprotein lipase hydrolyzes the triglycerides in the chylomicron, releasing free fatty acids that nearby tissues can take up. Adipose tissue may re-esterify and store them. Skeletal muscle and heart may oxidize them for energy.

This means the meal's fat is not returning to the liver untouched. Peripheral tissues are allowed first access to much of the triglyceride load. As this unloading continues, the chylomicron shrinks. It becomes relatively richer in cholesterol and cholesteryl esters and poorer in triglyceride. What remains is the chylomicron remnant, a smaller but still information-rich particle carrying the leftovers of intestinal fat delivery after peripheral tissues have already taken their share. Only then does the liver re-enter the sequence as the central receiving organ.

The remnant is recognized by the liver through apolipoprotein-based receptor interactions, especially involving apolipoprotein E, and taken up by hepatocytes for deeper processing. This is why dietary fat is said to return to the liver rather than simply arrive there first. The liver receives the meal after the intestine has digested and packaged it, after lymph has transported it, and after peripheral tissues have extracted a large portion of its triglyceride content. What comes back is not raw meal fat. It is distributed meal fat after first-pass peripheral use. The liver then decides what happens to the remaining cholesterol, phospholipids, triglyceride, and fat-soluble cargo. Some will be oxidized. Some will be stored. Some will be converted into bile acids. Some will be repackaged into hepatic lipoproteins for re-export. So before dietary fat returns to the liver, it has already passed through a long structural route: emulsification, enzymatic hydrolysis, micellar delivery, enterocyte uptake, re-esterification, chylomicron assembly, lymphatic transport, bloodstream entry, and peripheral triglyceride unloading. The liver does not meet the fat at the beginning of that story. It meets it after the body has already sampled the meal and now needs a central organ to settle the remaining account.

The Liver's Role in Processing Remnants

By the time dietary fat returns to the liver, it is no longer arriving as the original meal. It comes back as a remnant, a particle that has already passed through intestinal packaging, lymphatic transport, bloodstream entry, and peripheral triglyceride unloading. That history matters because the liver is not receiving untouched dietary fat in bulk. It is receiving what remains after adipose tissue, muscle, heart, and other peripheral tissues have already extracted a substantial share of the triglyceride cargo. What comes back is smaller, denser, relatively richer in cholesterol and cholesteryl esters, and marked by the apolipoprotein signals that tell hepatocytes this particle is ready for central processing. The liver's job at this point is not simply to clear debris from circulation. It is to perform the second accounting of the meal's lipid traffic. Peripheral tissues took their first share. The liver now decides the fate of what remains.

Recognition comes first. Chylomicron remnants are taken up by the liver through receptor-mediated pathways, especially involving apolipoprotein E on the remnant surface. That apolipoprotein is one of the main identity tags telling the hepatocyte that this is not just any circulating lipid particle. It is a partially spent dietary transport particle returning for final sorting. The remnant binds to hepatic receptors, is internalized, and then dismantled inside the hepatocyte. Once that happens, the particle stops existing as a transport unit and becomes a set of lipid components that now have to be reassigned. Triglyceride remnants, cholesterol, cholesteryl esters, phospholipids, and fat-soluble vitamins are no longer moving through the blood as a meal-derived package. They are inside the liver, where each class of cargo can be directed toward a different metabolic future.

Cholesterol handling is one of the most important parts of this second accounting. Chylomicron remnants are relatively cholesterol-rich compared with the larger fresh chylomicron that first left the intestine. Once hepatocytes take up that cholesterol, they can use it for membrane maintenance, package it into new lipoproteins, convert some of it into bile acids, secrete some of it into bile, or hold it in storage as cholesteryl ester depending on the broader metabolic state. This is one reason the liver sits at the center of lipid governance rather than merely at the end of lipid clearance. The meal's cholesterol does not simply enter the body and stay where it first landed. It returns to the liver, where it becomes substrate for membrane biology, biliary chemistry, lipoprotein export, or storage. The remnant is therefore not waste. It is unresolved fat traffic brought back for central decision.

The same is true for the remaining triglyceride-derived fatty acids. Once the remnant is taken apart, those fatty acids can be oxidized for hepatic energy, re-esterified into triglycerides, stored transiently, or repackaged into very-low-density lipoprotein, usually abbreviated VLDL, for re-export into circulation. That last point matters because the liver does not only terminate lipid traffic. It also reissues it in a different form. A dietary particle returning as a remnant can therefore become part of a new endogenous transport particle generated by the liver itself. The carbon is the same. The transport logic changes. Intestinal export becomes hepatic redistribution. This is one of the clearest examples of the liver acting as a lipid editor rather than a passive sink.

Fat-soluble vitamins carried in the remnant are also part of this processing role. Vitamins A, D, E, and K travel with dietary fat and often arrive back at the liver inside these particles after first-pass peripheral unloading. The liver can then store, redistribute, or further handle these compounds depending on the vitamin and the broader physiological need. So remnant processing is not simply about clearing triglyceride leftovers and cholesterol burden. It is also part of how the body recovers and centrally manages fat-linked micronutrients after intestinal absorption. The remnant is a composite parcel, and the liver is the organ that opens it.

This is why remnant processing matters so much for blood chemistry. A lipid particle that has completed its first delivery job cannot simply remain circulating indefinitely. It would clutter the bloodstream with partially spent cargo, distort lipid composition, and leave the meal's unresolved fat burden without a final destination. The liver clears that problem by receiving the remnant, dismantling it, and folding its contents into broader hepatic metabolism. Some of that carbon will be burned. Some will be stored. Some will be secreted into bile. Some will be repackaged and sent back out. The circulation is cleaned not by simple deletion, but by absorption into a larger metabolic authority.

So the liver's role in processing remnants is best understood as central lipid adjudication. The intestine built the first transport particle. Peripheral tissues removed much of the triglyceride load. The remnant returns richer in cholesterol and still carrying valuable lipid cargo. The liver recognizes it, internalizes it, dismantles it, and then decides what each component becomes next. That is how dietary fat returns to central control after first distribution. The meal leaves the intestine as chylomicron. It returns to the liver as unfinished business.

Why Fat Arriving at the Liver Is Already Partially Modified by Peripheral Tissues

By the time dietary fat reaches the liver, it is no longer the same fat that left the intestine. That is one of the most important distinctions in lipid physiology. Glucose and amino acids move from the intestine into portal blood and reach the liver first in relatively direct form. Long-chain dietary fat does not. It is absorbed into enterocytes, rebuilt into triglyceride-rich chylomicrons, sent first into lymph, then into the bloodstream, and exposed to peripheral tissues before the liver ever sees the returning particle. That means the liver is not meeting the meal at its original lipid stage. It is meeting the meal after the circulation and peripheral tissues have already acted on it.

The main modifying event is lipoprotein lipase activity in capillary beds of tissues such as adipose tissue, skeletal muscle, and heart. Lipoprotein lipase sits on the vascular surface and hydrolyzes triglycerides carried inside the chylomicron, releasing free fatty acids that nearby tissues can take up. Adipose tissue may re-esterify those fatty acids and store them. Oxidative tissues such as muscle and heart may burn them. This is a major first-pass distribution step. A large fraction of the meal's triglyceride content is therefore removed before the particle ever returns to the liver. The chylomicron shrinks as this happens. It loses triglyceride bulk, changes in density and composition, and becomes relatively enriched in cholesterol, cholesteryl esters, phospholipids, and the apolipoprotein signals that allow the liver to recognize it as a remnant. So the fat arriving at the liver is not fresh dietary transport. It is residual transport after peripheral extraction.

That point matters because the liver is not handling the same physiological question the intestine handled. The intestine's task was absorption and packaging. Peripheral tissues then took their immediate share according to local storage and energy demand. By the time the liver receives the remnant, the question is no longer simply how to distribute meal triglyceride. The question has become what to do with the remaining cholesterol-rich particle and its unresolved lipid cargo. Some of that cargo will be used for hepatic membranes. Some may be converted into bile acids. Some may be stored. Some may be repackaged into very-low-density lipoprotein for re-export. The liver is doing second-stage accounting, not first-stage delivery.

The particle itself proves that peripheral modification has already occurred. A chylomicron remnant is smaller than the original chylomicron because much of its triglyceride load has been hydrolyzed and removed. It is compositionally different because the ratio of triglyceride to cholesterol has shifted. It is functionally different because it is no longer designed for broad first delivery to tissues. It is now designed for hepatic recognition and uptake, largely through apolipoprotein E mediated interactions with hepatic receptors. In other words, the liver is receiving a metabolically edited particle. Peripheral tissues have already written their marks on it.

This is also why the liver's view of dietary fat is different from the public imagination. People often think of fat entering the body and then somehow reaching the liver as the place where it is managed. The real sequence is more distributed. The body lets adipose tissue, muscle, and heart participate first. Only then does the liver receive what remains. That means hepatic fat handling is always partly downstream of peripheral tissue behavior. If adipose tissue is highly storage-active, one kind of remnant profile returns. If muscle is pulling heavily on fatty acids for oxidation, another profile returns. The liver is receiving not only the residue of the meal, but also information about what the periphery already did with that meal.

So fat arriving at the liver is already partially modified by peripheral tissues because long-chain dietary fat is delivered through a two-stage transport logic. First the intestine packages it into chylomicrons and sends it through lymph into blood. Then peripheral tissues unload much of the triglyceride through lipoprotein lipase and take up the released fatty acids for storage or oxidation. What comes back to the liver is the remnant, a reduced, cholesterol-enriched, compositionally altered particle carrying the leftovers of first distribution. The liver is not receiving the meal whole. It is receiving the meal after the body has already started spending it.

The Relationship Between Intestinal Fat Absorption and Hepatic Fat Handling

Intestinal fat absorption and hepatic fat handling are two halves of the same event separated by distance, packaging, and timing. The intestine breaks dietary fat down and gets it across the epithelial barrier. The liver decides what the body is finally going to do with the fraction that returns after first distribution. That distinction matters because fat does not enter the body in a form the bloodstream can simply tolerate on its own. Triglycerides, cholesterol esters, phospholipids, and fat-soluble vitamins arrive in food embedded in hydrophobic structure, meaning structure that resists the watery environment of the gut and blood. The intestine solves the first problem by using bile acids to emulsify fat, pancreatic lipase to hydrolyze triglycerides into absorbable components, and enterocytes to rebuild those components into chylomicrons, large lipoprotein particles designed for transport. But absorption is only the opening move. Once fat has crossed the gut wall, it still needs a second organ to clear remnants, repurpose cholesterol, repackage unused lipid, oxidize part of the carbon, and connect the meal to the larger lipid economy of the body. That organ is the liver.

The quality of intestinal absorption determines what kind of lipid traffic the liver will later receive. If bile delivery is poor, emulsification suffers, micelle formation suffers, and less dietary fat reaches the enterocyte in absorbable form. Micelles are small bile acid-based transport assemblies that ferry lipid digestion products through the watery lumen to the absorptive surface. If pancreatic lipase output is weak, triglycerides are not cleaved efficiently, and the pool of free fatty acids and monoglycerides available for absorption falls. If the intestinal surface itself is damaged, uptake and reassembly inside enterocytes become less effective. In all of those states, the liver receives a different downstream report because the intestine did not hand off the same amount or composition of absorbed fat. So hepatic fat handling begins long before fat reaches hepatocytes. It begins at the level of bile flow, pancreatic enzyme function, micellar transport, enterocyte uptake, and chylomicron assembly in the gut.

Once chylomicrons enter the lymph and then the bloodstream, peripheral tissues begin modifying the meal before the liver sees it. Lipoprotein lipase on capillary surfaces hydrolyzes chylomicron triglycerides and releases free fatty acids to adipose tissue, heart, and skeletal muscle. This means the liver does not receive the original dietary particle. It receives the remnant after first-pass tissue use. That returning remnant is smaller, relatively richer in cholesterol and cholesteryl esters, and marked for hepatic uptake through apolipoprotein signals, especially apolipoprotein E. This is where the relationship between intestinal absorption and hepatic handling becomes especially clear. The intestine determines how much dietary fat successfully enters the body and how it is packaged. Peripheral tissues determine how much of the triglyceride load is removed early. The liver then receives what remains and performs the central accounting. What is oxidized. What is stored. What is turned into bile acids. What is repackaged into very-low-density lipoprotein and sent back out. The meal starts in the intestine, but its unresolved lipid logic ends in the liver.

The liver also depends on the intestine in a deeper feedback sense because bile acids themselves are part of the loop. The liver synthesizes bile acids from cholesterol and secretes them into bile so intestinal fat absorption can occur. Most of those bile acids are later reabsorbed in the ileum and returned to the liver through portal blood in the enterohepatic circulation. So the organ that handles the remnants of absorbed fat is also the organ that supplied one of the main chemical tools required for absorption in the first place. This makes the relationship circular rather than linear. The liver supports intestinal fat absorption by making bile acids. The intestine uses those bile acids to absorb fat and send chylomicron traffic into the body. Peripheral tissues unload part of that traffic. The liver then receives the remnant and decides the fate of what remains. The same organ stands at both the start and the end of the sequence.

This is why disorders of fat handling often cannot be understood honestly by looking only at the liver or only at the intestine. Impaired bile flow will damage absorption first and then change what the liver later receives. Intestinal malabsorption will reduce chylomicron formation and alter hepatic lipid traffic. Overdelivery of absorbable fat from the intestine can increase remnant return and force the liver into heavier repackaging and storage work. A liver already burdened with steatosis, cholestatic dysfunction, or impaired lipoprotein export will then handle those returning remnants differently, which can feed back into broader disturbances in blood lipids and digestion. The two organs are not sharing a topic. They are sharing a single process viewed from different ends.

So the relationship between intestinal fat absorption and hepatic fat handling is best understood as sequential control across compartments. The intestine liberates, emulsifies, absorbs, and packages dietary fat. Peripheral tissues take their first share. The liver receives the edited remainder and integrates it into the body's wider lipid economy. One organ gets fat into the body. The other decides what that absorbed fat will become once first distribution is over.

Why Fat Metabolism Cannot Be Reduced to One Blood Lipid Number

Fat metabolism cannot be reduced to one blood lipid number because fat does not move through the body as one thing, perform one task, or burden one tissue in one way. Dietary fat is digested in the intestine, emulsified by bile acids, hydrolyzed by lipases, rebuilt inside enterocytes, packaged into chylomicrons, delivered first through lymph, unloaded into peripheral tissues, returned to the liver as remnants, repackaged again into hepatic lipoproteins, stored, oxidized, converted into ketones, incorporated into membranes, used for steroid-related chemistry, or secreted into bile. That is already too much traffic for a single number to summarize honestly. A blood lipid measurement can tell you something real, sometimes something important, but it is still a snapshot taken from one compartment after multiple organs have already acted on the fat. It is not the whole pathway. It is one visible trace of a much larger movement.

Even the standard lipid categories are describing different biological objects rather than different readings of the same object. Triglycerides are a storage and transport form of fatty acids. Cholesterol is a structural molecule used in membranes, steroid synthesis, bile acid synthesis, and lipoprotein architecture. Low-density and high-density lipoproteins are not fats themselves but transport particles carrying lipid cargo with different composition and trafficking roles. Chylomicrons move dietary fat. Very-low-density lipoprotein, or VLDL, moves triglyceride exported by the liver. Low-density lipoprotein, or LDL, represents a downstream transport stage after triglyceride unloading from VLDL. High-density lipoprotein, or HDL, participates in reverse cholesterol transport and broader lipid exchange. A single number cannot tell you whether the liver is overexporting triglyceride, whether remnants are being cleared efficiently, whether peripheral tissues are taking up fatty acids properly, whether bile acid synthesis is sufficient, whether hepatic beta-oxidation is strong, or whether membrane needs are being met cleanly. It is reading one surface consequence of a multi-organ transport system.

The liver alone makes that reduction impossible. It can receive chylomicron remnants from dietary fat, synthesize cholesterol, build bile acids from cholesterol, assemble VLDL, oxidize fatty acids for its own energy, generate ketone bodies during fasting, transiently store triglyceride, and redirect lipid carbon depending on feeding state, insulin signaling, and whole-body energy demand. A person can therefore have one blood lipid number sitting at a certain level while the underlying hepatic reality differs dramatically. One liver may be efficiently packaging and exporting triglyceride. Another may be failing to export and accumulating fat internally. One may be converting cholesterol effectively into bile acids. Another may be holding more cholesterol in hepatic pools. One may be producing ketones strongly in fasting. Another may be poor at switching substrate use. The bloodstream may show overlap while the tissue logic underneath it is very different.

Timing complicates it even further. A blood lipid number is measured at one moment, often fasting, but fat handling is highly state-dependent. After a meal, chylomicrons enter circulation and triglyceride traffic changes. In fasting, hepatic VLDL export, adipose tissue lipolysis, fatty acid oxidation, and ketone production all become more relevant. During insulin resistance, the liver may continue producing glucose inappropriately while also overproducing VLDL and accumulating fat internally. During caloric deficit, blood lipid patterns may shift for reasons that do not reflect the same tissue meaning they would have in chronic overfeeding. So a single number is not only incomplete spatially. It is incomplete temporally. It does not tell you which metabolic state generated it.

The same blood lipid value can also come from opposite biological directions. Elevated triglycerides may reflect overproduction of hepatic VLDL, poor clearance of triglyceride-rich particles, excessive post-meal remnant burden, high carbohydrate-driven hepatic lipogenesis, or broader insulin-resistant traffic failure. Cholesterol values can reflect altered synthesis, altered clearance, altered receptor activity, altered bile acid conversion, altered remnant uptake, altered thyroid signaling, or changes in transport particle distribution. This is why serious lipid physiology starts with traffic and compartment, not with moralized numbers. A number can be high for different mechanistic reasons. A number can be normal while hepatic fat handling is still disordered. A number can improve while the underlying transport logic remains poor. The liver does not experience "a cholesterol number." It experiences incoming remnants, intracellular cholesterol pools, biliary demands, membrane demands, lipoprotein export pressures, and oxidation needs.

Fat-soluble vitamins and bile flow push the point even harder. A blood lipid number does not tell you whether bile acids are reaching the intestine effectively enough for fat absorption. It does not tell you whether chylomicron assembly is normal. It does not tell you whether a person is absorbing vitamins A, D, E, and K properly with the meal. It does not tell you whether the liver is converting cholesterol into bile acids cleanly or retaining biliary burden. You can have a number. You still do not have the sequence.

So fat metabolism cannot be reduced to one blood lipid number because fat is not one circulating entity with one meaning. It is a moving system of digestion, packaging, transport, tissue uptake, hepatic reprocessing, storage, oxidation, signaling, membrane construction, and excretion. One blood number may mark part of that system, but it cannot stand in for the whole. The bloodstream gives you traces. The physiology is the route.

Chapter 15

Triglycerides, Cholesterol, and Lipoprotein Packaging

Hepatic Synthesis of Triglycerides

The liver does not only receive triglycerides from the diet. It builds them. That fact matters because it changes the meaning of fat in the bloodstream. A triglyceride is a molecule made of three fatty acids attached to a glycerol backbone, and in the liver it functions less like a final destination and more like a transport and storage form for excess carbon. When hepatocytes synthesize triglycerides, they are often solving a packaging problem. Free fatty acids cannot be allowed to accumulate loosely inside the cell in large amounts because they can disrupt membranes, alter signaling, burden mitochondria, and increase lipotoxic stress, which means direct injury from excessive lipid intermediates rather than from stored neutral fat itself. The hepatocyte therefore esterifies fatty acids into triglycerides, converting a more reactive lipid burden into a more inert storage form that can either be held temporarily in lipid droplets or packaged into lipoproteins for export. Triglyceride synthesis is not automatically pathology. It is one of the liver's ways of making excess carbon and excess fatty acid traffic physically manageable.

The raw materials for hepatic triglyceride synthesis come from several directions at once. Fatty acids can arrive from adipose tissue during fasting or insulin-resistant states, released into blood after lipolysis of stored fat. They can arrive from chylomicron remnant processing after dietary fat has already passed through the intestine and peripheral tissues. They can also be made de novo inside the liver itself through de novo lipogenesis, which means new fat synthesis from nonfat substrates, especially excess carbohydrate. This is where glucose and fructose become especially relevant. If carbohydrate arrives in excess relative to immediate glycogen needs and oxidative demand, the liver can divert that carbon into acetyl coenzyme A and then onward into fatty acid synthesis. The resulting fatty acids can then be esterified with glycerol-derived backbones to form triglycerides. So hepatic triglyceride is not one thing with one origin. It can represent incoming dietary residue, mobilized adipose traffic, or newly manufactured fat built from carbohydrate overload.

The glycerol backbone matters too, because triglyceride synthesis requires more than fatty acids alone. The liver uses glycerol-3-phosphate as the scaffold onto which fatty acids are attached. That glycerol-3-phosphate can be derived from glycolytic intermediates, which means carbohydrate metabolism directly supports triglyceride assembly, or from glycerol entering hepatic metabolism after triglyceride breakdown elsewhere in the body. This is one reason fed-state carbohydrate abundance and hepatic lipogenesis are so tightly linked. The liver is not only receiving carbon that can be turned into fatty acids. It is also receiving glycolytic and glycerol-related substrate that helps build the backbone needed to store those fatty acids as triglyceride. A hepatocyte making triglyceride is therefore integrating multiple streams of carbon at once.

The actual assembly happens through sequential acylation steps, meaning fatty acid chains are attached one by one to the glycerol backbone through enzyme-guided reactions in the endoplasmic reticulum. As each fatty acid is added, the molecule becomes more hydrophobic and more clearly suited for storage or packaging. The end result is a neutral lipid that can be tucked into a lipid droplet or loaded into a lipoprotein particle such as very-low-density lipoprotein, or VLDL, for secretion into blood. This neutralization matters. Free fatty acids are chemically active. Triglycerides are comparatively quiet. The liver often synthesizes triglycerides to protect itself from the more immediate toxicity of free fatty acid excess. That does not mean triglyceride accumulation is harmless. Too much stored triglyceride still marks a traffic problem. But the cell often chooses triglyceride because it is safer than leaving the same carbon in more reactive lipid forms.

This is why hepatic triglyceride synthesis has to be understood in context. In one setting it is part of ordinary fed-state carbon handling and lipoprotein export. In another it is a defensive response to substrate overload, with the liver turning incoming fatty acids or excess carbohydrate into neutral fat because other handling routes are saturated or misaligned. In fasting, hepatic triglyceride synthesis may coexist with fatty acid oxidation and ketone production as the liver sorts incoming adipose-derived fat between immediate energy use and re-esterification. In insulin-resistant states, increased free fatty acid delivery from adipose tissue and increased carbohydrate-driven lipogenesis can push triglyceride synthesis upward at the same time, creating hepatic fat accumulation and rising VLDL export together. The molecule looks the same. The physiological meaning can differ.

The deeper point is that hepatic triglyceride synthesis is part of the liver's larger job as a carbon traffic manager. The liver cannot let every incoming fatty acid remain free, and it cannot oxidize everything immediately. Some lipid burden has to be buffered, neutralized, stored, or exported. Triglyceride synthesis is one of the main ways it does that. It takes excess carbon and excess fatty acid load and turns them into a transportable or storable form the organ can live with for a time. The question is never simply whether the liver makes triglycerides. It always does. The real question is why it is making them, how much it is making, what substrate streams are driving that production, and whether the finished triglycerides are being exported cleanly or accumulating inside the tissue as evidence that the traffic has outrun the exits.

Cholesterol Synthesis and Cholesterol Use

Cholesterol is usually talked about as if it were mainly a blood test problem, a number floating in plasma that has to be judged from a distance. That is far too shallow to teach the liver honestly. Cholesterol is first a structural molecule and a synthetic substrate before it is ever a lab value. The liver synthesizes it because the body needs a stable internal supply that does not depend completely on what happened at the last meal. Hepatocytes build cholesterol from acetyl coenzyme A through a long multi-step pathway centered in the cytosol and endoplasmic reticulum, with 3-hydroxy-3-methylglutaryl coenzyme A reductase, usually called HMG-CoA reductase, acting as one of the major rate-setting control points. That pathway is expensive on purpose. The cell is taking small carbon units and assembling them into a large rigid lipid molecule with highly specific structural properties. A body that had to wait for dietary cholesterol alone would lose too much control over membrane integrity, bile acid supply, steroid substrate availability, and lipoprotein balance. So the liver makes cholesterol not because cholesterol is inherently dangerous and must be restrained, but because cholesterol is necessary and must be governed.

What makes cholesterol valuable is its physical behavior inside membranes. It inserts between phospholipids and changes membrane fluidity, permeability, and domain organization. Too little cholesterol and membranes become less stable and less able to support the correct arrangement of receptors, transporters, and signaling assemblies. Too much in the wrong place and the membrane's behavior shifts in another direction. The body therefore cannot treat cholesterol like inert packing material. It is one of the molecules through which cellular architecture becomes possible. The liver uses cholesterol for its own membranes, packages it into lipoproteins so it can move through blood to other tissues, stores some of it as cholesteryl ester when immediate use is not required, and converts a substantial fraction of it into other necessary molecules. This is why cholesterol synthesis and cholesterol use have to be taught together. The liver does not manufacture cholesterol as an end in itself. It manufactures cholesterol because multiple structural and excretory systems are waiting for it downstream.

One of those downstream systems is bile acid synthesis, and this is one of the clearest ways the liver turns cholesterol from static lipid into active digestive chemistry. Bile acids are made from cholesterol inside hepatocytes and then secreted into bile, where they eventually enter the intestine and emulsify dietary fat. Emulsify means breaking large fat masses into much smaller dispersed droplets so enzymes can work at the surface efficiently. This matters because it reveals one of the liver's main solutions to cholesterol burden. The organ does not only store it or recirculate it. It also destroys part of the cholesterol molecule's original identity by converting it into bile acids, then sends those bile acids into the gut where they become one of the main tools of fat digestion. Cholesterol therefore is not merely being contained. It is being repurposed into a digestive instrument and, through biliary secretion, into a route of elimination as well. That is one of the deepest reasons the liver sits at the center of cholesterol physiology. It controls both supply and one of the major exits.

The lipoprotein side matters just as much because cholesterol cannot travel through plasma as naked hydrophobic cargo. Hydrophobic means poorly compatible with water, and blood is an aqueous medium. The liver solves that transport problem by packaging cholesterol together with phospholipids, triglycerides, and apolipoproteins into lipoprotein particles. Some of this cholesterol comes from de novo hepatic synthesis. Some comes back to the liver in chylomicron remnants after dietary fat has already passed through the intestine and peripheral tissues. Some is taken up from circulating lipoprotein traffic. Once inside the hepatocyte, all of these cholesterol pools enter a common accounting problem. Will the cholesterol be inserted into membranes. Stored as cholesteryl ester. Converted into bile acids. Secreted into bile directly. Loaded into very-low-density lipoprotein, or VLDL, for export. Retained because intracellular demand is high. The answer depends on feeding state, hormone signals, remnant return, biliary needs, and the broader state of hepatic lipid traffic. Cholesterol use is therefore not a single pathway. It is a routing problem inside a central organ.

This is also why cholesterol should never be imagined as a dietary villain whose main importance begins only after absorption. The liver synthesizes large amounts of it because the body cannot outsource membrane biology and bile acid production to meal timing alone. At the same time, cholesterol returning from the diet through remnant uptake is real and significant. So hepatic cholesterol handling always sits at the intersection of endogenous production and exogenous return. The organ is constantly asking whether internal synthesis should rise or fall, whether receptor-mediated uptake should dominate, whether conversion to bile acids should increase, whether storage pools are sufficient, and whether export in lipoproteins should change. Those are not abstract regulatory questions. They determine whether cholesterol stays useful, accumulates in the wrong compartment, or becomes available for conversion and transport in the right one.

The deeper lesson is that cholesterol is one of the liver's great balancing acts between structure and disposal. It is needed for membranes. It is needed as precursor for bile acids and steroid-related biology. It is dangerous only when traffic, storage, export, and conversion lose proportion. A liver that synthesizes cholesterol but cannot convert enough of it into bile acids is facing a different problem from a liver that synthesizes appropriately but is overloaded by remnant return, or from one that overexports lipoprotein cholesterol because other lipid pathways are distorted. The same molecule is sitting in different physiological stories depending on where it is, what form it is in, and what the liver is doing with it next.

So cholesterol synthesis and cholesterol use belong together because the molecule only makes sense inside the full sequence. The liver builds cholesterol from acetyl units because structural continuity and bile acid production require it. It inserts cholesterol into membranes, packages it into lipoproteins, stores some of it as cholesteryl ester, converts some into bile acids, and secretes some through bile. The question is never whether cholesterol exists. The question is where the liver is directing it, what purpose it is serving there, and whether the route still fits the body's real needs.

VLDL, LDL, HDL Explained from Structure and Traffic

VLDL, LDL, and HDL make more sense when they are treated as traffic particles rather than as moral categories of good and bad cholesterol. They are not three free-floating substances with fixed reputations. They are lipoproteins, transport particles built to move hydrophobic cargo through an aqueous bloodstream. Hydrophobic means poorly compatible with water. Triglycerides, cholesteryl esters, and many other lipid molecules cannot simply dissolve into plasma and circulate as naked fat. They need a structural shell. Lipoproteins solve that problem by packaging nonpolar lipid cargo into a core surrounded by a surface layer of phospholipids, free cholesterol, and apolipoproteins, which are the protein components that stabilize the particle and help determine where it goes and how other tissues interact with it. Once the reader sees these particles as architecture moving through blood, the whole subject becomes easier. Structure determines what the particle can carry. Traffic determines what role it is playing at that moment.

VLDL, very-low-density lipoprotein, is the liver's major export particle for endogenously handled triglyceride, meaning triglyceride being packaged and sent out from hepatic metabolism rather than arriving directly from the intestine in a chylomicron. The liver assembles VLDL when hepatocytes are holding triglyceride that needs to be moved into circulation. That triglyceride may come from de novo lipogenesis, which is new fat synthesis from carbohydrate or other substrates, from incoming fatty acids, or from broader hepatic lipid handling. VLDL therefore is not a passive marker floating around to help a lab report happen. It is a structural response to a trafficking problem.

The liver cannot keep unlimited triglyceride stored inside hepatocytes without consequence. It packages that lipid into VLDL and exports it. Structurally, VLDL is large and relatively triglyceride-rich. Traffic-wise, it is an outbound delivery vehicle leaving the liver with energy cargo intended for peripheral tissues.

Once VLDL enters circulation, it does not remain chemically fixed. As it moves through capillary beds, especially in adipose tissue and muscle, lipoprotein lipase hydrolyzes much of its triglyceride cargo, meaning triglycerides are cleaved into fatty acids and glycerol so tissues can take up the fatty acids for storage or oxidation. As that triglyceride load is stripped away, the particle shrinks and becomes denser. It first becomes intermediate-density lipoprotein and then, with further remodeling, becomes LDL, low-density lipoprotein. This is why LDL should not be imagined as a wholly separate creation with no history. It is, in large part, what remains after a triglyceride-export particle has delivered much of its energy cargo. The traffic changed the structure. The structure changed the function.

LDL is therefore better understood as a cholesterol-enriched remnant particle rather than simply bad cholesterol. Its core is relatively richer in cholesteryl ester compared with its VLDL predecessor, and its major traffic role is delivery of cholesterol to peripheral tissues through receptor-mediated uptake. Cholesterol in this context is not decorative excess. It is a structural lipid needed for cell membranes, steroid hormone synthesis, and other biological functions. The problem with LDL is not that cholesterol delivery is inherently pathological. The problem begins when particle number, residence time, oxidative modification, endothelial retention, and broader metabolic conditions make that delivery system more likely to contribute to arterial plaque formation. That is a traffic and tissue-interaction problem, not a one-word essence problem. LDL is a cholesterol-delivery particle moving through a circulatory environment that may or may not handle it well.

HDL, high-density lipoprotein, is built differently and participates in different traffic. It is smaller, denser, and more protein-rich than VLDL or LDL because it carries less core triglyceride and more apolipoprotein relative to its size. HDL is often described as the particle that carries cholesterol away from tissues back toward the liver, and that is broadly true, but the real story is more dynamic. HDL participates in reverse cholesterol transport, meaning it helps collect cholesterol from peripheral cells and shuttle it through the circulation in forms that can eventually be returned to the liver either directly or indirectly. It also serves as a mobile exchange platform, donating and receiving apolipoproteins and lipids during interaction with other lipoproteins. So HDL is not just a garbage truck for excess cholesterol. It is a structurally agile traffic coordinator participating in redistribution, collection, and return.

The density differences between these particles come from composition. Triglyceride is bulky and relatively light. Protein is denser. A large triglyceride-rich particle like VLDL has lower density because more of its volume is occupied by lipid cargo. As triglyceride is removed and the particle shrinks, relative density rises. LDL is denser than VLDL because it carries less triglyceride and proportionally more cholesteryl ester and protein. HDL is denser still because it is comparatively small and protein-rich. This is why the names very-low-density, low-density, and high-density are structural descriptions, not value judgments. They tell you what the particle is built like, which in turn helps explain what kind of traffic it is suited to handle.

The liver sits at the center of this whole system because it is both a packaging organ and a receiving organ. It builds VLDL when triglyceride needs to leave hepatocytes. It helps clear remnant particles. It receives cholesterol traffic returning through HDL-mediated pathways and uses cholesterol for membranes, bile acid synthesis, biliary secretion, or repackaging. So when lipoproteins are taught correctly, they belong inside liver physiology, not at the edge of it. VLDL is outbound triglyceride traffic from the liver. LDL is what remains after much of that cargo has been delivered, a cholesterol-rich delivery particle still moving through the body. HDL is part of the return and redistribution system, helping move cholesterol away from tissues and back toward hepatic handling. Structure tells you what each particle can physically carry. Traffic tells you what job it is doing in that moment. That is the cleanest way to see the system.

Why the Liver Packages Lipids Instead of Simply Storing Them

The liver cannot solve every lipid problem by keeping the fat for itself because the liver is not a warehouse built for indefinite triglyceride accumulation. It is a traffic organ. Fat arrives there from multiple directions at once. Dietary fat returns in altered form after intestinal handling. Fatty acids also arrive from adipose tissue during fasting or metabolic stress. The liver can synthesize new fatty acids and triglycerides from excess carbohydrate and other substrates when the incoming fed-state chemistry pushes in that direction. Cholesterol is being synthesized, modified, secreted into bile, incorporated into membranes, or prepared for transport. If hepatocytes tried to solve that entire burden by permanent intracellular storage, the organ would swell with lipid, distort its own architecture, and begin losing the metabolic and synthetic clarity that makes it useful in the first place. Storage has a place, but lipid traffic in the liver is too large and too dynamic for storage alone to be a viable answer.

Triglyceride is the clearest example. It is a compact energy form, but it is also hydrophobic, which means it does not move comfortably in an aqueous environment like blood. That creates a two-part problem. The liver often needs to clear fatty acids out of its own cytoplasm by esterifying them into triglycerides, because free fatty acids can become chemically disruptive when they accumulate in excess. But once that triglyceride has been assembled, it still has to go somewhere. Keeping too much of it inside the hepatocyte is not neutral. Lipid droplets enlarge, cellular architecture is displaced, organelles are crowded, mitochondrial function becomes easier to impair, oxidative stress rises, and the liver begins drifting toward steatosis, which is pathologic fat accumulation inside hepatocytes. So the liver packages lipids because packaging is one of the main ways to prevent temporary handling from becoming self-injury. The cell converts fatty acids into a safer storage or transport form, then sends part of that load outward rather than allowing the reserve to keep swelling inside the tissue.

That outward movement requires structure. Lipid cannot be exported as naked triglyceride any more than it can circulate as naked cholesterol. Blood is water-based plasma, and hydrophobic cargo needs a transport shell. The liver solves that by assembling lipoproteins, especially VLDL, very-low-density lipoprotein. This particle is a structural solution to a traffic problem. Hepatocytes combine triglycerides, cholesterol, phospholipids, and apolipoproteins into a particle whose outer surface can interact with plasma while its hydrophobic cargo remains protected inside. Apolipoproteins are the protein components that stabilize the particle and help determine how it will be recognized and handled in circulation. Once packaged into VLDL, the triglyceride burden can leave the liver and be delivered to peripheral tissues where fatty acids can be stored or oxidized. Packaging therefore is not a cosmetic detail added after lipid synthesis. It is the exit strategy.

This also explains why simple storage would be metabolically wrong even when the liver has room to hold more fat. The liver is not the final intended destination for all lipid energy arriving after a meal or being mobilized during fasting. Adipose tissue exists as a specialized long-term storage site. Muscle exists as an oxidative site. Other tissues need cholesterol and phospholipids for membranes and signaling-related structure. The liver sits in the middle of this system as a sorter and distributor. If it hoarded lipid instead of packaging and exporting it, it would stop being a traffic controller and become an overloaded depot. The body would lose one of its main coordinators of lipid redistribution. Packaging into lipoproteins lets the liver participate in whole-body fuel and structural lipid management without sacrificing its own working architecture.

The same logic applies to cholesterol. Cholesterol is essential, but its movement has to be governed carefully because it is not freely water-soluble and because it has multiple possible fates. The liver may use cholesterol for membrane needs, convert it into bile acids, secrete it into bile, or incorporate it into lipoproteins for circulation. Simply storing it all would not solve the transport problem or the excretory problem. The liver packages cholesterol because cholesterol often has to move. Some of it has to reach peripheral tissues. Some of it has to be carried back. Some of it has to be routed toward bile. Storage is only one option inside a larger traffic network. Packaging is what makes that network possible.

This is why fatty liver is such an important sign. It shows what happens when lipid arrival, synthesis, or mobilization outpace oxidation and export. The liver keeps packaging lipids because it is trying not to drown in them. When packaging fails, or when the burden becomes too large for packaging and oxidation to keep pace, triglyceride begins to accumulate inside hepatocytes and the organ starts paying for the traffic imbalance itself. So the deeper answer is simple. The liver packages lipids instead of simply storing them because it is built to move chemical burden onward, not to become the final resting place of every fat molecule it touches. Storage buys time. Packaging preserves function.

Apolipoproteins and Transport Logic

Apolipoproteins are the protein components attached to lipoprotein particles, and they matter because fat transport in blood is not solved by lipid cargo alone. The body cannot move triglycerides, cholesteryl esters, and other hydrophobic lipids through plasma as if they were water-soluble nutrients. It has to build particles that can stay suspended, survive circulation, interact with enzymes, dock with receptors, and deliver or receive cargo in the correct tissues. The lipid shell and core solve part of that problem, but without the right proteins on the particle surface the traffic system loses direction. Apolipoproteins give the particle identity. They help stabilize structure, but more importantly they tell the rest of the body what the particle is, where it has been, what it is carrying, and what should happen to it next. In that sense apolipoproteins are not decoration on a fat particle. They are the routing language of lipid transport.

This is easiest to see if lipoproteins are understood as moving packages rather than as static substances. A particle leaving the liver in the form of VLDL, very-low-density lipoprotein, is not only a blob of triglyceride wrapped in phospholipid. It carries apolipoproteins that let it be assembled, secreted, recognized in circulation, and progressively remodeled as it gives up triglyceride to tissues. ApoB-100 is central here. It is the major structural apolipoprotein of VLDL and the particles derived from it, and it gives the particle a durable scaffold while also helping determine receptor interactions later in its life cycle. Without that structural protein, the liver cannot package large lipid cargo into a stable export particle properly. The triglyceride may still exist. The transport logic would be impaired.

Other apolipoproteins act less like scaffolds and more like exchange signals or enzyme controls. ApoC-II is one of the clearest examples because it activates lipoprotein lipase, the enzyme on capillary surfaces that hydrolyzes triglycerides in circulating particles so fatty acids can be taken up by tissue. A triglyceride-rich particle moving through blood is therefore not delivering fuel just because it exists. It needs the right protein instructions on its surface to interact with the right enzyme at the right place. ApoE adds another layer of logic by helping certain remnant particles be recognized and cleared by the liver. So apolipoproteins do not only help lipids leave the liver. They also help determine when triglyceride is unloaded, when a particle is taken back up, and when circulation should stop treating it as a delivery vehicle and start treating it as a remnant.

This is why transport logic matters more than isolated particle names. A lipoprotein is not one thing for its entire life. VLDL leaves the liver rich in triglyceride, then loses triglyceride as lipoprotein lipase acts on it in peripheral tissues. As that cargo is stripped away, the particle shrinks, becomes denser, exchanges apolipoproteins with other particles, and changes its biological role. Intermediate-density particles and then LDL, low-density lipoprotein, emerge from that remodeling process. LDL is therefore not an unrelated molecule floating around for moral judgment. It is what remains after a triglyceride-export process has already happened. Apolipoproteins help govern each step in that transition. They are part of how the body turns one particle into another without losing control of where the lipid is going.

HDL, high-density lipoprotein, makes the same point from the opposite direction. HDL participates in reverse cholesterol transport and in broader exchange among circulating particles, and its major apolipoprotein, ApoA-I, is central to that role. ApoA-I helps HDL act as a collector and coordinator of cholesterol movement rather than as a bulk triglyceride delivery vehicle. HDL is smaller, denser, and more protein-rich because its job is different. It is not hauling large energy loads from the liver to tissue in the way VLDL does. It is participating in redistribution, collection, and return. Again the protein logic and the traffic logic match. A different apolipoprotein pattern helps produce a different particle behavior.

The liver sits at the center of this system because it is one of the main organs assembling these particles and one of the main organs clearing many of their remnants. That means apolipoprotein biology is not peripheral to liver physiology. It is one of the ways hepatocytes turn lipid burden into organized circulation. Triglycerides and cholesterol are chemically real, but they cannot circulate intelligently without being embedded in a transport framework. Apolipoproteins supply that framework's instructions. They tell the bloodstream whether the particle is exporting triglyceride, delivering cholesterol, exchanging surface components, awaiting enzymatic unloading, or ready for receptor-mediated clearance.

So the cleanest way to understand apolipoproteins is this. Lipids provide cargo. The lipoprotein particle provides the vehicle. Apolipoproteins provide the identity, stability, enzyme access, and receptor logic that make the vehicle usable inside a living circulation. Without them, fat transport would lose direction. With them, the liver can package hydrophobic chemistry into particles that do not merely move. They know how to move.

What Happens When Packaging Fails or Overload Outpaces Export

The liver can tolerate a great deal of lipid traffic as long as it can keep that traffic moving. Problems begin when triglyceride is arriving, being synthesized, or being re-esterified inside hepatocytes faster than it can be oxidized or exported in lipoprotein form. That is the real meaning of packaging failure. It does not always mean the liver has lost the biochemical ability to make lipoproteins altogether. More often it means the incoming burden has become too large, the assembly process has become inefficient, or the supporting conditions required for export have been compromised. The hepatocyte keeps receiving fatty acids, keeps converting part of that burden into triglyceride because free fatty acids are more chemically disruptive, but can no longer move enough of the resulting lipid outward through VLDL, very-low-density lipoprotein. At that point the liver begins storing what it had meant to ship. Temporary handling becomes accumulation.

The first visible result is steatosis, the buildup of triglyceride droplets inside hepatocytes. This is not merely fat present in the liver in some benign descriptive sense. It is intracellular traffic backing up. Lipid droplets enlarge, crowding the cytoplasm and displacing organelles that the cell needs for normal work. Mitochondria lose room and often lose performance. The endoplasmic reticulum comes under greater membrane and synthetic stress. Oxidative balance becomes harder to maintain because fatty acid handling, lipotoxic intermediates, and mitochondrial overload start pushing reactive oxygen generation upward. Lipotoxic means injury caused not just by stored neutral fat, but by the more disruptive lipid species and metabolic strain that accompany failed handling. The cell may still be alive and still functioning in part, but it is now doing its job in a more crowded and chemically unstable interior.

This matters because triglyceride storage inside the hepatocyte is partly protective and partly dangerous. It is protective in the sense that esterifying fatty acids into triglyceride can temporarily reduce the burden of more reactive free fatty acids. The cell is trying to stabilize an overload by putting some of it into a less immediately toxic form. But that protective move has limits. If export still cannot keep pace, the growing lipid droplet becomes a sign that the underlying flux problem has not been solved. The hepatocyte is carrying more cargo than it can process cleanly. The fat droplet is therefore not the whole injury. It is the visible residue of a deeper failure in oxidation, packaging, secretion, or overall load management.

Once this backlog persists, the consequences spread beyond fat storage alone. Hepatic insulin signaling often becomes less coherent. Glucose handling becomes rougher. VLDL assembly may continue but in a more distorted pattern, sometimes contributing to elevated circulating triglyceride traffic even while the liver itself remains fat-laden. This apparent contradiction makes sense once the reader sees the system as a traffic problem rather than a simple storage problem. A liver can be exporting too much lipid for the circulation to handle well and still be failing to export enough to clear its own intracellular burden. Overload outpaces export at both levels. The hepatocyte remains congested, and the bloodstream receives excess abnormal traffic anyway.

The inflammatory threshold is crossed when this burden stops being mostly a storage problem and becomes a tissue injury problem. Oxidative stress rises. Mitochondrial function weakens. Endoplasmic reticulum stress increases as protein and lipid handling lose precision. Kupffer cells detect more danger signals. Stellate cells are more likely to activate. Cytokine signaling increases. At that stage the liver is no longer simply fatty. It is reacting to failed lipid management as an injury state. This is how simple steatosis can progress toward steatohepatitis, fibrosis, and broader architectural damage. The transition is not magical. The cell has spent too long trying to contain a burden that never stopped arriving.

Packaging failure can also happen when the structural requirements for lipoprotein assembly are impaired. Export is not just a matter of having triglyceride available. The hepatocyte needs phospholipids, cholesterol, apolipoproteins, intact endoplasmic reticulum and Golgi function, adequate energy, and a membrane transport system capable of assembling and secreting particles correctly. If one of those elements weakens, lipid may be synthesized and trapped without being packaged efficiently. The result looks similar from the outside, hepatocyte fat accumulation, but the inner mechanism may differ. The important principle remains the same. The liver is supposed to process and move lipid, not drown in it.

So when packaging fails or overload outpaces export, the liver begins paying for the traffic imbalance with its own tissue integrity. Fat that should have been routed outward stays behind. Lipid droplets enlarge. Organelles become stressed. Oxidative and inflammatory signaling rise.

Exchange with blood becomes less metabolically precise. The organ that was supposed to coordinate lipid traffic turns into a site of lipid congestion. The liver can store some fat. It cannot remain healthy by becoming the final destination for more lipid than it can ship.

Chapter 16

Fatty Liver

What Steatosis Means at the Cellular Level

Steatosis means that triglyceride has begun accumulating inside hepatocytes in amounts large enough to change the biology of the cell. That definition is more useful than the loose phrase fat in the liver because it forces the reader to ask what kind of fat, in what form, and why it stayed there. The hepatocyte is always dealing with lipid traffic. Fatty acids arrive from the bloodstream, return from adipose tissue, come in through dietary handling, or are newly built from excess carbohydrate and other substrate inside the liver itself. Some of that lipid is oxidized for energy. Some is used to build membranes. Some is converted into signaling molecules. Some is assembled into triglyceride and packaged into lipoproteins for export. So triglyceride inside a hepatocyte is not automatically abnormal. Steatosis begins when lipid arrival and lipid synthesis are outrunning oxidation and export long enough for storage droplets to enlarge and persist. The problem is not that the liver touched fat. The problem is that traffic backed up inside the working cell.

At the cellular level, the hepatocyte often converts incoming fatty acids into triglyceride because free fatty acids are more chemically disruptive than neutral stored fat. Free fatty acids can insert into membranes, disturb mitochondrial function, generate oxidative stress, and create lipotoxic intermediates, which are injurious lipid-related compounds such as diacylglycerols and ceramides that interfere with signaling and damage cell structure. Turning some of that burden into triglyceride is therefore partly defensive. The cell is trying to package reactive lipid into a less immediately dangerous storage form. But that defense has limits. If fatty acid delivery remains high, if de novo lipogenesis, the synthesis of new fat from nonfat substrate, remains active, or if VLDL export and mitochondrial oxidation cannot keep pace, triglyceride droplets enlarge inside the cytoplasm. The droplet is therefore not the first injury. It is the visible sign that the hepatocyte has been forced into prolonged lipid containment.

As these droplets enlarge, the internal geography of the cell begins to change. The hepatocyte is not an empty bag that can swell without consequence. It is a crowded metabolic workstation containing mitochondria, endoplasmic reticulum, Golgi apparatus, lysosomes, glycogen granules, cytoskeletal elements, transport machinery, and a polarized membrane system that has to keep blood-facing and bile-facing tasks distinct. Large lipid droplets displace organelles, distort cytoplasmic organization, and change how the cell handles substrate, redox balance, and secretion. Redox balance refers to the management of electron transfer and oxidative burden inside the cell. Mitochondria may become less efficient. Endoplasmic reticulum stress can rise as lipid handling, membrane turnover, and protein processing become more difficult to coordinate. Bile secretion may become less precise because the cell's energetic and structural order is under pressure. Steatosis therefore is not only stored calories sitting quietly in a liver cell. It is physical crowding of the machinery.

The liver often appears deceptively calm at this stage because triglyceride itself is a relatively neutral storage form compared with the more aggressive lipid species that helped create it. That is one reason simple steatosis can exist before overt inflammation becomes obvious. The cell has partially succeeded in converting a reactive burden into a safer depot. But the success is incomplete. The stored triglyceride indicates that the deeper flux problem is unresolved. Fat is still arriving too fast, being synthesized too readily, or failing to leave efficiently. Mitochondrial oxidation may be insufficient. Lipoprotein assembly and export may be lagging. Insulin signaling inside the hepatocyte may already be changing. So even when the cell is not yet inflamed enough to produce full steatohepatitis, the metabolic ground has shifted. The hepatocyte is living under altered substrate pressure and altered intracellular architecture.

This is why steatosis is best understood as a state of intracellular lipid congestion. Congestion is the right word because the essential event is not simply presence. It is delayed movement. Fatty acids entered a cell that could not fully burn them, fully export them, or fully redirect them into harmless circulation. The hepatocyte answered by storing them as triglyceride droplets, but that answer came at a cost. The larger and more persistent the droplets become, the more the cell's normal working space is compromised. The liver can tolerate some of this for a time, which is why steatosis may be reversible when the pressure falls. But if the burden persists, the droplet stops being a quiet buffer and becomes part of a broader state of mitochondrial strain, oxidative stress, inflammatory signaling, and eventual tissue remodeling.

So what steatosis means at the cellular level is simple in principle even if the biochemistry behind it is dense. The hepatocyte has become a holding site for more lipid than it can process cleanly. It has converted part of that burden into triglyceride to reduce immediate chemical harm, but the price is intracellular crowding and metabolic distortion. The droplet is the cell's attempt to survive a traffic problem. When that traffic problem persists, survival starts looking more and more like the beginning of injury.

How Triglyceride Droplets Accumulate Inside Hepatocytes

Triglyceride droplets accumulate inside hepatocytes when fatty acid traffic into the cell exceeds the cell's ability to burn that fat or export it. That is the core event. The hepatocyte is constantly receiving lipid from several directions at once. Fatty acids arrive from the bloodstream, especially from adipose tissue when stored fat is being released. They arrive indirectly after dietary fat has been digested, absorbed, and redistributed through circulating lipoprotein traffic. They can also be made inside the liver itself through de novo lipogenesis, which is the synthesis of new fatty acids from nonfat substrate, especially excess carbohydrate and fructose. Once these fatty acids are inside the hepatocyte, the cell has limited choices. It can oxidize them in mitochondria for energy. It can use them for membrane phospholipids and other structural lipids. It can package them into lipoproteins for export. Or it can esterify them into triglyceride and store them. Esterify means chemically attaching fatty acids onto a glycerol backbone so the more reactive free fatty acids become a neutral storage molecule. When arrival and synthesis outrun oxidation and export, esterification becomes the cell's main pressure-relief move.

That conversion happens largely around the endoplasmic reticulum, where fatty acids are activated and then attached stepwise to glycerol-derived intermediates until triglyceride is formed. Glycerol-3-phosphate provides the backbone, and enzymes progressively load fatty acids onto it until a fully formed triglyceride exists. This is not trivial bookkeeping chemistry. It is a defensive rearrangement. Free fatty acids and some of their intermediates can damage membranes, disrupt signaling, stress mitochondria, and generate lipotoxic compounds such as diacylglycerols and ceramides. By converting part of that burden into triglyceride, the hepatocyte is trying to lower the immediate chemical threat. Triglyceride is comparatively inert. It is not harmless in the long run when it accumulates excessively, but it is safer than leaving the same lipid load loose in more reactive forms inside the cell.

Once enough triglyceride is made, it begins to collect between the leaflets of the endoplasmic reticulum membrane and then buds outward into the cytoplasm as a lipid droplet. This is the beginning of visible steatosis. The droplet is not a membrane-bound organelle in the same way as a mitochondrion or lysosome. It is a core of neutral lipid coated by a phospholipid monolayer with associated proteins that help stabilize it, grow it, shrink it, or connect it to other parts of lipid metabolism. At first these droplets may be small and numerous. As triglyceride synthesis continues, droplets enlarge, new droplets appear, and existing droplets can fuse. The cell is not storing fat in a tidy separate closet. It is progressively devoting more of its internal space to a backlog it has not been able to clear.

The reason the droplets keep growing is that the imbalance persists. Fatty acid delivery remains high, de novo lipogenesis remains active, mitochondrial oxidation cannot keep pace, or VLDL assembly and export fall short of the burden. VLDL, very-low-density lipoprotein, is the main particle the liver uses to export triglyceride into circulation. If the hepatocyte cannot package enough triglyceride into VLDL, or if the rate of triglyceride formation exceeds even a functioning export system, storage becomes the default overflow compartment. This is why a fatty liver can coexist with high circulating triglycerides. The liver may be exporting too much lipid for the body's good and still not exporting enough to relieve its own intracellular congestion. The problem is not simple absence of export. It is that total traffic has outgrown the available exits.

As the droplets enlarge, they begin changing the cell's internal mechanics. Organelles are displaced. Mitochondria work in a more crowded and stressed environment. Endoplasmic reticulum stress rises because lipid synthesis, membrane handling, and protein trafficking are all taking place in a cell whose interior is being physically remodeled by stored fat. Membrane polarity becomes harder to maintain cleanly. Oxidative stress rises more easily because mitochondrial overload and fatty acid handling generate reactive oxygen species, chemically aggressive oxygen-containing molecules that can injure proteins, lipids, and DNA. So the droplet is not the whole injury. It is the visible residue of a deeper throughput failure. The hepatocyte is storing triglyceride because it could not keep the lipid stream moving.

That is the right way to see droplet accumulation. It is not random fat settling into the liver. It is a structured response to excess lipid flux. Fat arrives too fast, is built too readily, cannot be burned fast enough, and cannot be shipped out fast enough. The hepatocyte converts that burden into triglyceride because neutral storage is safer than chemical chaos. Then the storage itself begins to take over the cell.

Increased Delivery, Increased Synthesis, Reduced Oxidation, Reduced Export

A fatty liver is not created by one mistake. It is created when four streams begin leaning in the same direction. More fat arrives. More fat is built. Less fat is burned. Less fat leaves. Those are the four pressure points that matter. Once the reader sees them clearly, steatosis stops looking like a vague fog of metabolic dysfunction and starts reading like a traffic problem with identifiable causes. The hepatocyte is always handling lipid. Fatty acids come in from blood. New fatty acids can be synthesized from incoming carbohydrate. Some of that burden is oxidized in mitochondria for energy. Some is packaged into very-low-density lipoprotein, VLDL, and exported. Triglyceride droplets begin accumulating when input and internal generation are outrunning the two main exits, oxidation and export. The droplet is the residue of that imbalance.

Increased delivery means the liver is receiving more fatty acid traffic than it can comfortably process. This commonly happens when adipose tissue is releasing large amounts of fat into circulation, especially in insulin-resistant states where lipolysis is not being restrained properly. Lipolysis means breakdown of stored triglyceride into fatty acids and glycerol.

Those fatty acids travel in blood bound to albumin and are taken up by hepatocytes. The liver is then forced to decide quickly whether to oxidize them, re-esterify them into triglyceride, or package them for export. If arrival stays high, re-esterification becomes a defensive move. Free fatty acids are more chemically disruptive than neutral triglyceride, so the cell converts part of the burden into a storage form that is less immediately toxic. But that defense only buys time. High delivery means the liver is being used as a holding site for systemic fat traffic it did not create but now has to contain.

Increased synthesis means the liver is making more fat even while fat is already arriving. This is where excess carbohydrate, especially in a chronically fed insulin-rich environment, becomes part of the lipid story. De novo lipogenesis, the synthesis of new fatty acids from nonfat substrate, rises when hepatocytes are repeatedly exposed to abundant glucose and fructose with a hormonal setting that favors storage and anabolic processing. The cell is no longer only dealing with external fat. It is turning carbon from carbohydrate into fatty acid internally. Fructose is especially liver-centered in its early handling, which makes this pathway even more relevant in the modern diet. The important point is that steatosis is not always fat entering a cell from the outside. Sometimes the hepatocyte is manufacturing the excess burden itself because the incoming nutrient environment keeps telling it to convert surplus carbon into lipid.

Reduced oxidation means the liver is losing one of its major exits. Mitochondria are supposed to take fatty acids and break them down through beta-oxidation, producing acetyl-CoA, reducing equivalents, and ATP, adenosine triphosphate, the cell's immediately usable energy currency. If mitochondrial function is impaired, if oxidative stress is high, if substrate burden is excessive, or if the redox state is already distorted, the liver becomes less able to burn incoming fat cleanly. Fatty acids that might have been oxidized now have to be rerouted. Some are pushed toward triglyceride formation and storage. Some remain in more toxic intermediate forms longer than they should. Reduced oxidation therefore is not only less energy production. It is a narrowing of throughput. The hepatocyte becomes worse at destroying part of the burden it receives, which means more of that burden remains available to accumulate.

Reduced export is the final choke point. The liver does not have to burn every fatty acid it handles. It can also package triglyceride into VLDL and send that lipid burden outward into circulation. But that requires intact lipoprotein assembly, adequate apolipoprotein production, endoplasmic reticulum and Golgi function, membrane integrity, and enough metabolic order to build and secrete the particle correctly. If export falters, triglyceride that was meant to leave stays behind. This can happen because the export machinery itself is impaired or because total burden has become so large that even a functioning export system cannot keep pace. That is why a person can have a fatty liver and elevated circulating triglycerides at the same time. The liver may be exporting too much for the rest of the body's good and still not exporting enough to clear its own backlog.

These four pressures are best understood as converging forces rather than isolated categories. Increased delivery alone can overwhelm a liver with limited oxidative reserve. Increased synthesis can turn a high-carbohydrate environment into a lipid burden even when dietary fat is not the main driver. Reduced oxidation leaves more substrate available for storage. Reduced export traps triglyceride that should have moved out. Steatosis develops when enough of these changes align to keep forcing carbon into droplets faster than it can be burned or shipped away. This is why fatty liver should be read as a dynamic failure of handling rather than as passive fat accumulation. The hepatocyte is being squeezed from all sides. More is coming in. More is being made. Less is being destroyed. Less is leaving. The droplet is what remains after those arithmetic realities become cellular structure.

Why Fatty Liver Is a Traffic Problem, Not Merely a Weight Problem

Fatty liver is too often described as if the central event were body size, as if excess weight somehow settles into the liver by simple overflow. That description is shallow and biologically weak. The liver does not become fatty because it has eyes on the scale. It becomes fatty because lipid traffic through hepatocytes has become disordered. Fat is arriving too fast, being synthesized too aggressively, failing to be oxidized cleanly, failing to be exported efficiently, or some combination of all four. The organ is being asked to process more fatty acid burden than it can move through its normal channels without backlog. That is why steatosis, triglyceride accumulation inside hepatocytes, should be understood first as a flux problem. Weight may correlate with that burden in many people, but the liver is responding to chemical throughput, not to the visual fact of body mass.

The hepatocyte sits at a crossroads of lipid movement. Fatty acids arrive from adipose tissue when stored fat is being released into circulation. They arrive after meals through the broader redistribution of dietary lipid. They are also built inside the liver itself through de novo lipogenesis, the synthesis of new fat from nonfat substrate, especially when carbohydrate and fructose handling keep pushing carbon into fat-making pathways. Once those fatty acids enter the hepatocyte, the cell has only a limited number of possible fates for them. It can oxidize them in mitochondria for energy. It can incorporate them into structural lipids. It can esterify them into triglyceride and export them in VLDL, very-low-density lipoprotein. Or it can esterify them and keep them in intracellular droplets. Fatty liver develops when the incoming and internally generated lipid stream outpaces the cell's capacity to burn or export it. That is a traffic imbalance. The droplet is the traffic jam made visible.

This is why fatty liver can develop in people who are not massively overweight and why not every overweight person has the same hepatic burden. The liver does not care about body size in the abstract. It cares about what is reaching hepatocytes and what hepatocytes are being forced to do with it. A person may have elevated adipose mass but relatively stable hepatic lipid handling. Another may have more severe hepatic fat accumulation because fatty acid release, carbohydrate-driven lipogenesis, insulin signaling, mitochondrial function, lipoprotein assembly, and oxidative stress have combined in a way that floods the liver specifically. The scale is an indirect clue at best. The real event is intracellular lipid management under systemic metabolic conditions.

Fructose helps show this clearly because it is strongly liver-centered in its early handling. Large fructose exposure can drive hepatic substrate pressure in a way that is not captured by body weight alone. The same is true for high rates of adipose lipolysis, which means breakdown of stored fat with release of fatty acids into the blood. If more fatty acids are delivered to the liver than the liver can oxidize or export, triglyceride accumulation begins whether or not a person fits a simplistic visual stereotype. Likewise, if VLDL assembly is strained, if mitochondrial oxidation is weak, or if insulin signaling is disordered enough to keep hepatic lipid synthesis active while glucose and fat traffic remain high, hepatocytes begin storing what they cannot move. The liver is reading biochemical flow, not social categories.

Calling fatty liver a traffic problem also explains why simple storage is only part of the story. Triglyceride droplets form partly because the hepatocyte is trying to defend itself. Free fatty acids and certain lipid intermediates are more chemically disruptive than neutral triglyceride. The cell therefore converts part of the burden into a safer storage form. But this does not solve the underlying traffic disorder. It only contains it temporarily. If the burden continues, droplets enlarge, mitochondria work under greater strain, endoplasmic reticulum stress rises, oxidative stress increases, and inflammatory signaling begins to build. The problem is not that the liver gained weight. The problem is that the liver became a holding site for lipid it was supposed to process and move onward.

This traffic view also makes progression easier to understand. A weight-based story sounds static. A traffic story is dynamic. It explains why the liver can move from simple steatosis to steatohepatitis, fibrosis, and deeper architectural injury. Continued lipid overload means continued conversion of fatty acids into triglyceride, continued organelle crowding, continued redox strain, continued generation of lipotoxic intermediates, and continued inflammatory signaling. Kupffer cells begin reading more danger. Stellate cells become easier to activate. The exchange surface starts paying for a problem that began as excess substrate flux. None of that requires obesity as a standalone explanation. It requires sustained metabolic congestion.

So fatty liver is not best understood as fat passively collecting in an overweight organ. It is better understood as hepatocytes losing control of lipid throughput. Too much arrives. Too much is built. Too little is burned. Too little is exported. The cell responds by storing what it cannot safely process. That storage is the sign. The broken traffic is the cause.

The Difference Between Simple Steatosis and Inflammatory Progression

Simple steatosis means triglyceride has accumulated inside hepatocytes, but the tissue has not yet fully crossed into an injury program dominated by inflammation, cell death, and fibrotic remodeling. That distinction matters because not every fat-loaded liver is behaving the same way. In simple steatosis, the hepatocyte is still primarily solving a traffic problem. Fatty acids are arriving too fast, being synthesized too readily, or failing to be exported or oxidized cleanly, and the cell responds by esterifying part of that burden into triglyceride and storing it in lipid droplets. This is not ideal, but it is partly defensive. Triglyceride is a more neutral storage form than many of the free fatty acids and lipid intermediates that would otherwise remain loose inside the cell. The hepatocyte is trying to reduce immediate chemical harm by packaging the overload into a less reactive state. The liver is congested, but not yet fully inflamed.

That relative calm can be misleading. A steatotic hepatocyte is not normal simply because it is not yet surrounded by obvious inflammatory destruction. Large lipid droplets crowd the cytoplasm, displace organelles, strain mitochondrial function, and increase the burden on the endoplasmic reticulum, which has to keep managing lipid synthesis, membrane turnover, protein folding, and transport in a cell whose internal space is becoming progressively less workable. Oxidative stress may already be rising. Insulin signaling may already be distorted. Bile-facing and blood-facing tasks may already be under metabolic pressure. But in simple steatosis, these shifts remain more weighted toward metabolic congestion than overt tissue attack. The cell is burdened, not yet openly collapsing.

Inflammatory progression begins when that burden stops being mostly a storage problem and starts becoming a damage-signaling problem. The liver crosses into a different state when hepatocytes can no longer contain the lipid load without generating enough collateral injury to alarm the surrounding tissue. Reactive oxygen species rise further. Mitochondria lose efficiency. Endoplasmic reticulum stress intensifies. Lipotoxic intermediates, lipid species more chemically disruptive than neutral triglyceride, accumulate in ways that disturb membranes, signaling pathways, and organelle integrity. Injured hepatocytes begin releasing danger signals and fragments of damaged cellular material. Kupffer cells, the resident macrophages of the liver, detect that shift and respond with inflammatory cytokines. Cytokines are signaling proteins that tell nearby cells the tissue is under threat. At that point the liver is no longer simply holding excess fat. It is reacting to the consequences of failed lipid handling.

This is the beginning of steatohepatitic progression, whether described under older terminology or newer metabolic disease language. The key event is not that there is more fat alone. The key event is that fat burden has become biologically inflammatory. Hepatocytes begin showing ballooning injury, which means they swell and lose their normal structural sharpness as cytoskeletal integrity and intracellular handling deteriorate. Cell death becomes more prominent. Kupffer cell activation increases. Stellate cells in the space of Disse are more likely to activate and begin laying down collagen. Once that happens, the disease is no longer centered only on lipid storage. It is moving into matrix remodeling and architectural injury. The tissue has begun interpreting metabolic overload as chronic wound.

This is why the difference between simple steatosis and inflammatory progression is not merely the amount of visible fat on a slide. It is the state of the hepatocyte and the surrounding tissue response. In simple steatosis, the cell is still using triglyceride storage mainly as a containment strategy. In inflammatory progression, containment is failing. The stored fat is now part of a wider field of oxidative stress, organelle dysfunction, cytokine signaling, hepatocyte injury, and fibrotic activation. One state is overloaded storage. The other is overloaded storage that has started injuring the architecture around it.

The practical importance of this difference is enormous because the liver can sometimes reverse simple steatosis more readily when the lipid traffic burden falls. The exchange surface is still more intact. The inflammatory program is not yet fully entrenched. Stellate activation may be minimal or absent. But once inflammatory progression is established, the disease begins feeding itself. Injured hepatocytes amplify immune signaling. Immune signaling activates stellate cells. Stellate cells distort the exchange architecture. Distorted architecture worsens hepatocyte stress. The problem is no longer fat sitting in a cell. It is fat burden turning into tissue remodeling.

So the cleanest way to see the difference is this. Simple steatosis is a liver trying to contain excess lipid by storing it. Inflammatory progression is a liver no longer containing that burden quietly. The droplets are still there, but now the cell is swelling, the immune system is reading danger, and the scar-building machinery is beginning to wake up.

Why a Fat-Loaded Liver Changes the Organ's Future Vulnerability

A fat-loaded liver is not simply a liver carrying extra stored energy. It is a liver whose working cells have already been forced to reorganize themselves around congestion. That matters because vulnerability is shaped by operating state. A hepatocyte crowded with triglyceride droplets is not the same cell it was before the burden arrived. Its mitochondria are working in a more stressed environment. Its endoplasmic reticulum is handling lipid synthesis, membrane turnover, protein processing, and detoxification inside a cell whose interior has become physically distorted by storage. Its redox balance, the management of electron transfer and oxidative pressure, is under more strain. Its insulin signaling may already be less coherent. Its ability to package and export lipid may already be lagging behind demand. So even before overt inflammation or fibrosis appears, the organ has changed its baseline. It is no longer meeting future stress from a neutral starting point. It is meeting future stress from inside a backlog.

That is why steatosis changes the meaning of the next insult. A healthy liver has reserve. It can absorb a meal, process a drug, handle a bout of inflammation, respond to fasting, and maintain bile secretion while still keeping blood chemistry organized. A fat-loaded liver is already spending part of that reserve to manage its own internal excess. The hepatocyte is already using ATP, adenosine triphosphate, the cell's immediately usable energy currency, to maintain order in a more difficult intracellular environment. It is already diverting fatty acids into triglyceride storage to reduce immediate lipotoxicity. It is already living with altered membrane composition, greater oxidative burden, and less architectural freedom. Add alcohol, a high substrate load, hypoxia, endotoxin from the gut, cholestatic stress, or a drug that requires heavy hepatic metabolism, and the response is no longer occurring in open metabolic space. It is occurring in a cell that was already partially occupied by yesterday's unfinished traffic.

The vulnerability becomes biochemical as well as structural. Triglyceride itself is a relatively neutral storage form, but the conditions that produce steatosis also tend to increase exposure to more dangerous lipid species and to reactive oxygen chemistry. Mitochondria under lipid pressure are more likely to generate oxidative stress. Endoplasmic reticulum stress rises when the cell has to keep synthesizing, folding, packaging, and detoxifying in a distorted lipid environment. Lipotoxic intermediates can interfere with signaling pathways and membrane behavior. This means the fat-loaded hepatocyte is easier to push from storage into injury. The next insult does not need to be enormous. It only needs to arrive in a cell whose defenses are already working harder than they should be. A liver with steatosis is therefore not only storing excess substrate. It is lowering the threshold at which additional burden becomes inflammatory.

This altered threshold affects the whole tissue. Kupffer cells, the resident macrophages of the liver, are more likely to encounter danger signals from stressed hepatocytes. Stellate cells are more likely to receive the cytokine and oxidative cues that begin driving fibrotic activation. Sinusoidal exchange can become less metabolically efficient because swollen, lipid-laden hepatocytes are not handling incoming blood chemistry with the same precision. Bile-facing work can become less clean because polarity and ATP-dependent transport are harder to maintain under organelle strain. So the vulnerability is not only that a fat-loaded hepatocyte is weaker. The vulnerability is that the entire local environment becomes easier to destabilize. What might have remained a transient metabolic challenge in a healthier liver can become sustained inflammatory signaling in a fat-loaded one.

This is why fatty liver matters even before dramatic symptoms or obvious scarring appear. It changes the organ's future by changing the cost of every later demand. The liver can still function, sometimes impressively, but it is doing so from a more precarious state. The next meal, the next toxin, the next infection, the next burst of adipose lipolysis, the next period of hypoxia, the next wave of gut-derived inflammatory traffic, all now strike an organ whose cells are already busy containing excess lipid and protecting themselves from the consequences of that containment. A fat-loaded liver is not just a liver with more fat in it. It is a liver with less room to improvise.

Chapter 17

Ketones and the Fasting Liver

Fatty Acid Delivery in Fasting

The fasting liver does not become idle when food stops arriving. It changes fuel source, traffic source, and biochemical priorities. During the fed state, portal blood delivers carbohydrate, amino acids, and broader post-digestive chemistry from the intestine. During fasting, that stream changes. Intestinal delivery falls, insulin falls, glucagon rises, and one of the main incoming fuel burdens begins arriving from adipose tissue instead. Stored triglyceride in adipocytes is hydrolyzed, meaning broken down, into fatty acids and glycerol. Those fatty acids enter the bloodstream largely bound to albumin, the major plasma transport protein that helps keep hydrophobic molecules mobile in an aqueous circulation. This is the beginning of fatty acid delivery in fasting. The body stops depending on the intestine for immediate fuel and starts mobilizing previously stored lipid so the liver and other tissues can continue supplying energy.

That shift matters because the liver is one of the first organs asked to interpret what fasting means chemically. Falling insulin removes much of the fed-state pressure toward glucose storage and lipid synthesis. Rising glucagon and other counterregulatory signals tell the liver that external fuel supply has dropped and internal reserves now have to be mobilized. Adipose tissue answers by releasing fatty acids into blood. The liver receives that new traffic not as a minor supplement, but as a major fasting substrate stream. Some of those fatty acids are taken up by hepatocytes and directed into mitochondrial beta-oxidation, the stepwise breakdown of fatty acids into acetyl-CoA units that can feed hepatic energy metabolism and ketone production. Other portions of the arriving lipid burden may be re-esterified into triglyceride if delivery outpaces immediate oxidation. So even in fasting, fatty acid arrival is not a single-track event. It is fuel supply entering a cell that still has to decide whether to burn, convert, or temporarily contain what it receives.

The form of delivery is important. Free fatty acids do not circulate freely in the literal sense because their hydrophobic structure makes them poorly suited to travel naked in plasma. Albumin serves as the transport scaffold, carrying these fatty acids through blood to tissues capable of using them. This is one reason fasting chemistry still depends on plasma protein architecture. The fat is mobilized from adipose tissue, but it reaches the liver in a bound transport state that protects the circulation from lipid chaos while still making the substrate available. Once those fatty acids reach the hepatocyte surface, they are transported inward, activated, and routed according to the fasting state. The liver is not only receiving calories. It is receiving the body's signal that stored energy has become the current budget.

Glycerol is released alongside fatty acids during adipose lipolysis, and this matters because fasting fatty acid delivery is part of a larger substrate shift, not a pure fat story. Glycerol returns to the liver and can be used in gluconeogenesis, the synthesis of glucose from noncarbohydrate substrates. So adipose tissue in fasting sends the liver both oxidative fuel and glucose-building material at the same time. Fatty acids help supply hepatic energy and ketone precursor. Glycerol helps support blood glucose. The fasting liver is therefore not merely replacing carbohydrate with fat. It is reorganizing the whole fuel hierarchy so the body can preserve glucose for tissues that still require it while moving more of the general energy burden onto fatty acids and ketones.

The scale of fatty acid delivery in fasting also helps explain why the liver becomes such a central fasting organ. Other tissues oxidize fatty acids too, but the liver sits in a unique coordinating position because it can convert part of this arriving lipid stream into ketone bodies, reshape glucose output, and continue managing the chemistry of blood for the whole organism. It does not just use fat. It interprets fat mobilization as a systems-level condition. When fasting deepens, the stream of adipose-derived fatty acids becomes one of the main drivers of hepatic metabolism. The liver shifts from buffering the chemistry of the meal to buffering the chemistry of the fast.

This is why fatty acid delivery in fasting has to be understood as controlled redistribution rather than simple fat burning. Adipose tissue is not emptying randomly. Hormonal signals are changing the terms of storage and release. Albumin is carrying the liberated fatty acids through blood. The liver is receiving them as incoming fasting fuel, oxidizing a substantial fraction, and preparing to turn part of that carbon load into ketones. The intestine has gone quiet. The adipose tissue has taken over as the major supplier. The liver sits in the middle and makes that transition livable.

Beta-Oxidation Inside Hepatic Mitochondria

Once fatty acids reach the fasting liver, they are not useful simply because they are present. The hepatocyte has to move them into the right compartment, activate them chemically, and then dismantle them in a controlled sequence that can actually support energy production. That sequence is beta-oxidation, and in the liver it occurs primarily inside mitochondria. Mitochondria are the organelles responsible for most aerobic energy production, meaning energy production that depends on oxygen. They are not just passive furnaces burning fat in bulk. They are structured reaction spaces that break fatty acids down two carbons at a time, turning long lipid chains into acetyl-CoA, reducing equivalents, and usable metabolic pressure. Acetyl-CoA is a two-carbon metabolic unit that can enter further oxidative pathways or, in the fasting liver, be diverted toward ketone production. So beta-oxidation is not generic fat burning. It is the controlled disassembly of fatty acids into units the liver can actually use.

The process begins before the fatty acid ever enters the mitochondrial matrix, the inner working compartment where most of the pathway occurs. A fatty acid arriving at the hepatocyte has to be activated first. This means it is linked to coenzyme A, forming fatty acyl-CoA. That activation step costs energy, which is appropriate because the cell is preparing the fatty acid for organized metabolism rather than allowing it to drift through the cytoplasm as chemically disruptive free lipid. Long-chain fatty acyl-CoA molecules then face another problem. They cannot cross the inner mitochondrial membrane freely. The cell solves that with the carnitine shuttle. Carnitine is a carrier system that transfers long-chain fatty acids across the mitochondrial boundary in a regulated way. The fatty acyl group is moved onto carnitine, transported across the membrane, then transferred back onto coenzyme A inside the mitochondrion. This transport step matters because it makes mitochondrial fat oxidation selective and controlled. Entry into beta-oxidation is gated.

Once inside the mitochondrial matrix, the fatty acyl-CoA enters the beta-oxidation spiral. It is called a spiral because the same sequence of reactions repeats over and over, shortening the fatty acid by two carbons each cycle. Each round includes oxidation, hydration, another oxidation, and thiolytic cleavage. The chemical details matter less than the outcome. The chain is progressively shortened, one acetyl-CoA unit at a time, while also generating reduced electron carriers such as NADH and FADH2. These reduced carriers feed electrons into the electron transport chain, the mitochondrial system that uses oxygen to generate ATP, adenosine triphosphate, the cell's immediately usable energy currency. So beta-oxidation does not only produce carbon fragments. It produces reducing power. It is feeding both the carbon economy and the energy economy of the hepatocyte at the same time.

This is why beta-oxidation becomes so important in fasting. The liver is no longer being supplied with abundant dietary glucose through portal blood, but it is still expected to maintain the chemistry of survival. It has to support gluconeogenesis, which is the synthesis of glucose from noncarbohydrate substrates such as lactate, glycerol, and certain amino acids. Gluconeogenesis is energetically expensive. Beta-oxidation supplies much of the ATP and reducing power needed to keep that pathway running. In other words, the fasting liver uses fat partly so it can keep making glucose for tissues that still require it. The arriving fatty acids are not only replacing carbohydrate as fuel. They are financing hepatic glucose output and later ketone production. The liver is burning fat in order to preserve blood chemistry for the rest of the body.

The pathway also helps explain why the liver can become metabolically strained when fatty acid delivery is too high. Beta-oxidation is a powerful solution to fasting fuel demand, but it is not infinite. Mitochondria still need oxygen, intact membranes, adequate cofactors, and a manageable redox state to keep the system running cleanly. If fatty acid delivery outpaces mitochondrial capacity, some of that lipid burden is pushed back toward re-esterification into triglyceride, contributing to steatosis. If mitochondrial function is weak, the liver can become less able to oxidize fatty acids efficiently, which means less energy for gluconeogenesis and more pressure toward intracellular lipid accumulation. Beta-oxidation is therefore one of the key forks between healthy fasting adaptation and hepatic lipid congestion. The liver remains stable when fatty acid delivery and mitochondrial oxidation stay in proportion. It becomes vulnerable when delivery outruns burn capacity.

This pathway also sets up ketogenesis. As beta-oxidation accelerates, acetyl-CoA production can rise to the point where the liver has more two-carbon units than it can fully process through the tricarboxylic acid cycle, especially when oxaloacetate is being diverted toward gluconeogenesis. The tricarboxylic acid cycle, also called the citric acid cycle, is one of the main pathways used to oxidize acetyl-CoA completely for energy. In fasting, some of that oxidative throughput is redirected. The liver begins converting excess acetyl-CoA into ketone bodies instead. That is why beta-oxidation is not just an energy pathway in the fasting liver. It is the upstream source feeding the entire ketone system.

So beta-oxidation inside hepatic mitochondria is best understood as a controlled dismantling of incoming fatty acids to support fasting metabolism. Fatty acids are activated, shuttled into mitochondria, broken down in repeating cycles, and converted into acetyl-CoA, ATP-generating reducing equivalents, and the substrate pressure that later drives ketone production. The liver does not receive fasting fat and simply store it or burn it in a crude way. It channels it through mitochondrial machinery precise enough to keep the fast from becoming systemic fuel failure.

Ketogenesis and Why the Liver Makes Ketones but Does Not Use Them Well Itself

Ketogenesis begins when the fasting liver is receiving more fatty acid carbon than it can simply burn to completion inside its own tricarboxylic acid cycle. Adipose tissue is releasing fatty acids into blood, the liver is taking them up, and hepatic mitochondria are running beta-oxidation hard enough to generate large amounts of acetyl-CoA. Under fed conditions much of that acetyl-CoA can move through the tricarboxylic acid cycle, also called the citric acid cycle, and be fully oxidized. Fasting changes the balance. Oxaloacetate is being pulled toward gluconeogenesis, the synthesis of glucose from noncarbohydrate substrates, because the liver still has to support blood glucose for tissues that depend on it. Once oxaloacetate is being used this way, the liver cannot funnel all of the incoming acetyl-CoA through ordinary oxidative throughput. The carbon load has to go somewhere. Ketogenesis is the answer. The liver begins linking and reshaping acetyl-CoA into ketone bodies, mainly acetoacetate and beta-hydroxybutyrate, with acetone produced as a smaller side product. This is not waste chemistry. It is fuel conversion under fasting pressure.

What the liver is doing here is structurally elegant. It is taking hydrophobic fat-derived carbon and turning it into water-soluble exported fuel. Fatty acids are powerful substrates, but they are awkward for some tissues to rely on directly and they are not distributed through blood in the same clean way as ketone bodies. Ketones solve that problem. Once formed in hepatic mitochondria, they can leave the liver, circulate in plasma without lipoprotein packaging, and reach tissues such as heart, skeletal muscle, and, during prolonged fasting, the brain. Those tissues can then take ketone bodies back up, convert them into acetyl-CoA, and oxidize them for energy. The liver is therefore not merely burning fat for itself. It is reformulating fasting fat into a shared fuel currency the rest of the body can spend.

The important twist is that the liver makes these molecules but does not use them well itself. That is not an accident. It is built into the enzyme map of the hepatocyte. Hepatic mitochondria can generate ketone bodies, but hepatocytes largely lack sufficient activity of succinyl-CoA:3-oxoacid CoA transferase, often called SCOT or thiophorase, the key enzyme needed to convert acetoacetate into acetoacetyl-CoA for efficient ketone utilization. Without that step, the liver can manufacture ketones but cannot re-import them into its own oxidative machinery in the same meaningful way as extrahepatic tissues. This creates a one-way design. The hepatocyte produces ketone bodies and releases them. The heart, muscle, kidney, and brain under fasting adaptation are the tissues meant to use them. The liver is the refinery, not the end user.

That separation of labor matters physiologically. If the liver used ketones freely as fast as it made them, ketogenesis would fail as a distribution system. The whole point is export. The fasting organism needs the liver to keep performing gluconeogenesis, keep managing blood chemistry, and keep converting fat-derived carbon into a fuel other tissues can use so glucose demand falls. The liver itself can meet much of its own energetic needs directly from fatty acid oxidation. It does not need to spend the ketones it is making. In fact, it is more useful to the organism if it does not. The hepatocyte burns incoming fatty acids, uses that oxidation to generate ATP and reducing power for its own work, and sends ketones outward as a separate product. One stream powers the liver. The other feeds the body.

This is why ketogenesis should not be taught as the liver making some mysterious emergency substance when carbohydrate is low. The liver is solving a traffic and allocation problem. Fat is arriving from adipose tissue. Glucose still has to be preserved for select tissues. The tricarboxylic acid cycle cannot absorb all incoming acetyl-CoA because oxaloacetate is being diverted toward glucose production. So the liver converts excess acetyl-CoA into exportable ketone bodies and intentionally leaves their full use to other organs. A liver that made ketones and consumed them itself would be hoarding the solution. The body needs distribution. The liver obliges by making ketones and then staying largely out of their way.

Ketones as Exported Fuel

Ketones are the fasting liver's way of turning incoming fat into a fuel that other tissues can use when direct glucose supply has fallen. That sentence has to be understood structurally, not sloganically. Fatty acids released from adipose tissue during fasting arrive at the liver bound to albumin, enter hepatocytes, and undergo beta-oxidation in mitochondria. Beta-oxidation breaks those long lipid chains down into acetyl-CoA units while also generating reducing equivalents that support ATP production. But the liver is not oxidizing all of that acetyl-CoA to completion inside its own mitochondria. In the fasting state, oxaloacetate is being pulled toward gluconeogenesis, the synthesis of glucose from noncarbohydrate substrates, because the liver still has to support blood glucose for tissues that remain glucose-dependent. That means the tricarboxylic acid cycle cannot always accept the full acetyl-CoA load arriving from fat oxidation. The hepatocyte therefore begins converting excess acetyl-CoA into ketone bodies. This is not waste disposal. It is export.

The main ketone bodies are acetoacetate and beta-hydroxybutyrate, with acetone produced more as a byproduct than as a serious fuel. These molecules are water-soluble enough to travel through blood without needing lipoprotein packaging, which makes them metabolically useful in a way long-chain fatty acids are not. The liver has taken hydrophobic fat, broken it down, and repackaged part of its carbon into a circulating fuel that can cross into other tissues more cleanly. Heart, skeletal muscle, and, during prolonged fasting, the brain can take up ketone bodies and convert them back into acetyl-CoA for energy production. This is why ketones should be understood as exported fuel rather than as a sign that the body is somehow failing. They are the liver's answer to a distribution problem. Stored fat in adipose tissue is abundant, but not every tissue can rely on raw fatty acid traffic in the same way and not every tissue should be forced to live on falling glucose alone. Ketones solve part of that gap.

What makes this especially elegant is that the liver produces ketones for the rest of the body while not using them well itself. Hepatocytes generate the ketone bodies, release them into the bloodstream, and let extrahepatic tissues oxidize them. That division of labor matters. The liver is acting as a converter organ, not just a consumer. It receives fasting fatty acid traffic, processes it, and exports a new fuel form more compatible with the energetic needs of distant tissues. In that sense ketogenesis is an extension of the liver's broader identity. The organ does not merely handle incoming chemistry for itself. It changes the form of incoming chemistry so the rest of the body can survive more coherently.

This export function also helps preserve glucose. The body still needs some glucose during fasting, but if every tissue continued demanding it at fed-state rates, glycogen would disappear faster and gluconeogenic burden would become harsher. Ketones reduce that pressure by giving peripheral tissues another oxidizable substrate. Over time, even the brain begins using ketones to a meaningful degree, which lowers the amount of glucose the liver must keep supplying. That is one of the deeper reasons ketones matter. They do not just provide energy. They change the fuel hierarchy of the fast. The liver is shifting the organism away from exclusive dependence on glucose and toward a mixed system in which fat-derived carbon can circulate in a brain- and muscle-usable form.

So ketones are not simply the biochemical signature of fasting. They are one of its main adaptive tools. The liver takes incoming fatty acid burden, oxidizes it, diverts excess acetyl-CoA into ketone production when full mitochondrial oxidation is no longer the whole answer, and sends those ketones out as soluble distributed fuel. The adipose tissue releases the raw substrate. The liver turns that substrate into a transportable fasting currency. Other tissues spend it.

The Difference Between Normal Fasting Ketone Production and Pathologic Ketoacidosis

Normal fasting ketone production is a controlled adaptation. Ketoacidosis is a control failure. In ordinary fasting, falling insulin and rising counterregulatory signals increase adipose lipolysis, fatty acid delivery to the liver, beta-oxidation, and ketone production, but this happens inside a system that still retains enough insulin effect to restrain runaway ketogenesis and enough whole-body use of ketones to keep pace with much of what the liver is making. Blood ketones rise, but blood pH usually remains normal because production, utilization, renal handling, and buffering stay within a survivable range. Nutritional ketosis or fasting ketosis is therefore a fuel shift, not an acid-base collapse. Diabetic ketoacidosis, by contrast, is marked by severe insulin deficiency together with excess counterregulatory hormone drive, which accelerates lipolysis, floods the liver with fatty acid substrate, drives hepatic ketone production far beyond ordinary adaptive levels, and combines that ketone excess with hyperglycemia, dehydration, and a high-anion-gap metabolic acidosis. The same ketone pathway is present in both states. The difference is scale, restraint, and whether acid production has outrun the body's buffering and clearance capacity.

The liver's role is central in both conditions, but the surrounding metabolic environment is completely different. In fasting, the liver makes ketones because glucose availability has fallen and oxaloacetate is being pulled toward gluconeogenesis, so some acetyl-CoA derived from fatty acid oxidation is diverted into ketone-body synthesis. Peripheral tissues then oxidize those ketones as usable fuel, which helps spare glucose and lowers the burden on hepatic glucose production. In ketoacidosis, that same hepatic machinery is no longer operating under measured fasting control. Severe insulin deficiency removes a major brake on adipose fat release and hepatic ketogenesis, while hyperglycemia and osmotic diuresis strip water and electrolytes from the body. Osmotic diuresis means glucose spilling into urine drags water and salts with it. The liver keeps producing ketones into a circulation that is becoming progressively more concentrated, more acidotic, and less capable of buffering the load. This is why ketoacidosis is not simply "more ketosis." It is ketosis occurring inside a broader metabolic collapse.

The acid-base distinction is the real dividing line. Ketone bodies include acetoacetate and beta-hydroxybutyrate, and when their accumulation becomes pathologic, hydrogen ion burden rises with them and serum bicarbonate is consumed buffering that acid load. In ordinary fasting, ketone concentrations increase, but they usually do not rise far enough to drive the same degree of acidemia because hepatic production remains more tightly limited and tissue use remains proportionate. In diabetic ketoacidosis, ketone production becomes excessive enough that bicarbonate falls and metabolic acidosis becomes clinically important. That difference is why the fasting person may have ketones and still be physiologically adapting, while the person in ketoacidosis is moving toward vomiting, abdominal pain, deep compensatory breathing, neurologic decline, and circulatory instability. The ketones themselves are not the whole story. The story is whether ketone production stayed inside adaptive fuel physiology or crossed into acid accumulation severe enough to disorder the entire internal environment.

So the cleanest way to separate the two is this. Normal fasting ketone production means the liver is exporting an alternate fuel during carbohydrate scarcity while acid-base control is preserved. Pathologic ketoacidosis means the same pathway has escaped normal restraint because insulin signaling is too low, counterregulatory drive is too high, substrate delivery is too intense, and buffering plus clearance can no longer keep blood chemistry stable. One state helps the organism survive the fast. The other is a metabolic emergency.

Why the Liver Is Central to Fuel Switching

Fuel switching means the body is changing not only what fuel is available, but what form of fuel different tissues are being allowed to use at a given time. The liver sits at the center of that transition because it is one of the few organs that can receive incoming substrate, read the hormonal meaning of that substrate, and then change the chemistry of the bloodstream accordingly. In the fed state, it receives portal glucose, amino acids, and fructose directly from the intestine, takes up part of that load, stores glycogen, suppresses its own glucose release, handles amino acid nitrogen, packages lipids, and supports a blood chemistry organized around recent intake. In fasting, that whole pattern has to reverse. Intestinal delivery falls. Adipose tissue begins releasing fatty acids and glycerol. The liver shifts away from storage and toward mobilization, away from buffering a meal and toward sustaining a body that now has to live on reserve. That is why the liver is central. It is the organ that turns a change in substrate supply into a change in circulating fuel form.

This role depends on the liver's unusual metabolic range. A muscle cell can oxidize fuel. An adipocyte can store or release fat. The hepatocyte can do something broader. It can take incoming glucose and store it as glycogen. It can later break glycogen back down. It can build new glucose through gluconeogenesis from lactate, glycerol, and certain amino acids. It can receive fatty acids during fasting, oxidize them in mitochondria, and then convert part of their carbon into ketone bodies when acetyl-CoA supply outruns what can be fully processed through the tricarboxylic acid cycle under fasting conditions. It can also continue packaging and redistributing lipids while performing all of that. This means the liver does not merely switch its own preferred fuel. It manufactures the new fuel pattern the rest of the body will be asked to live on. That is a much larger role.

The hormonal side makes this even clearer. Falling insulin and rising glucagon are not just endocrine background music. They are instructions telling the liver that external fuel delivery has changed and that the bloodstream must now be rewritten. In the fed state, insulin pushes hepatic metabolism toward glycogen synthesis, suppression of glucose output, and accommodation of portal nutrient abundance. In fasting, lower insulin removes that fed-state brake, glucagon shifts the liver toward glycogenolysis and gluconeogenesis, adipose lipolysis increases, and the liver begins receiving a larger fatty acid burden from circulation. The liver then interprets those combined signals by changing what it releases. Less storage. More glucose export. More ketone production. More reliance on fat-derived energy internally. This is why fuel switching is not just a matter of "the body burning fat instead of carbs." The liver is deciding how that shift will be staged, how quickly it will occur, and what molecules will carry it.

The liver is also central because different tissues do not all switch fuels equally or at the same speed. Red blood cells cannot use ketones because they lack mitochondria. The brain continues requiring some glucose even as ketone use rises during prolonged fasting. Muscle can oxidize fatty acids directly, but its choices are not identical to the brain's or the kidney's. The bloodstream therefore needs to carry a coordinated mixture of fuels during transition rather than one universal substrate for every tissue. The liver makes that possible. It keeps glucose available through glycogen breakdown and gluconeogenesis. It converts part of incoming fatty acid carbon into ketones, which are water-soluble enough to circulate as exported fuel. It prevents the fast from becoming a crude collapse in energy availability by changing the form in which stored fuel is delivered to the rest of the organism.

This is why liver dysfunction makes fuel switching harder or more chaotic. A liver that cannot store glycogen properly, cannot release glucose appropriately, cannot oxidize fatty acids efficiently, or cannot generate ketones coherently loses part of its power to smooth the transition between fed and fasting states. The bloodstream then reflects the shift more poorly. Glucose reserve becomes less dependable. Fatty acid traffic may still rise, but the liver is less able to turn that traffic into distributed ketone fuel. The body becomes more vulnerable to the metabolic discontinuity between eating and not eating because one of its main conversion organs is no longer imposing enough order on the transition.

So the liver is central to fuel switching because it is the organ that turns changing substrate availability into a new blood chemistry. It takes the meal and turns it into buffered fed-state stability. It takes fasting fat release and turns it into glucose support and ketone export. It does not merely use fuel. It changes what fuel means for the whole body.

Part V

Bile: The Liver's Outgoing Stream

Chapter 18

What Bile Is Made Of

Bile Acids, Phospholipids, Cholesterol, Bilirubin, Water, Electrolytes

Bile is not a vague digestive juice. It is a structured hepatic secretion built to solve two problems at once. The first is fat handling in the intestinal lumen, the hollow interior of the gut where digestion continues before absorption. The second is disposal of compounds the body cannot simply leave circulating in blood. That dual purpose explains its composition. Bile has to contain amphipathic molecules, compounds with both water-friendly and fat-friendly surfaces, so it can interact with hydrophobic dietary lipid in a watery environment. It also has to contain cargo the liver has decided to push outward, especially bilirubin and cholesterol-derived material. What leaves the hepatocyte in bile is therefore not random fluid. It is a chemically assembled outgoing stream. Bile acids, phospholipids, cholesterol, bilirubin, water, and electrolytes each occupy that stream for a reason. Remove one piece and the function changes. Disturb the ratios and the fluid becomes less stable, less effective, or more dangerous to the tissue carrying it.

Bile acids are the dominant functional detergents of bile. They are synthesized from cholesterol in hepatocytes, then conjugated, meaning chemically linked to amino acids such as glycine or taurine, which makes them more water-compatible and more effective in the intestinal environment. Their shape is the key. One side of the molecule interacts with fat. The other interacts with water. That lets bile acids surround lipid droplets and help break large fat masses into much smaller suspended particles, increasing the surface area available for pancreatic lipase. They also help form micelles, tiny transport assemblies that hold lipid digestion products in a form that can move through the watery intestinal contents toward the absorptive surface. Without bile acids, dietary fat stays physically harder to digest and harder to absorb. The liver is therefore sending chemistry into the intestine that makes the next stage of nutrient extraction possible. Bile acids are also part of cholesterol disposal, because cholesterol is the precursor from which they are made and because they themselves can be lost in stool and later replaced through new hepatic synthesis.

Phospholipids, especially phosphatidylcholine, stabilize this detergent system. Their role is easy to miss if bile acids take all the attention, but the fluid depends on them. Phospholipids sit at interfaces. They help keep cholesterol in solution and protect membranes from the full harshness of bile acid detergent action. Bile acids are useful because they can disrupt lipid structure. That same property would be too injurious if left chemically unbuffered against epithelial surfaces and biliary membranes. Phospholipids soften that edge and help create mixed micellar structures in bile itself, making the secretion more physically stable as it moves through canaliculi, ducts, and the gallbladder. Cholesterol belongs in the same paragraph because it is one of bile's major excreted lipids and one of its major solubility problems. Cholesterol is hydrophobic and poorly water-soluble. Bile acids and phospholipids help keep it suspended in a transportable state. When that relationship fails, cholesterol can precipitate, meaning come out of solution, and that is one of the biochemical preconditions for cholesterol gallstone formation. So cholesterol in bile is not merely cargo. It is also a test of whether bile's internal chemistry is holding together.

Bilirubin gives bile one of its most visible excretory identities. It begins elsewhere as a breakdown product of heme, especially from aging red blood cells, and reaches the liver first in an unconjugated form bound to albumin in blood. Hepatocytes take it up, conjugate it, usually by attaching glucuronic acid, which makes it far more water-compatible, and then secrete it into bile. That outward movement matters because bilirubin is not supposed to remain in circulation. Once it enters bile, it is carried into the intestine, where further microbial metabolism changes its chemistry again and eventually contributes to stool pigment. This is why bile is not only digestive fluid. It is one of the main routes by which the body gets rid of heme-derived waste. If bilirubin cannot enter bile properly, blood levels rise and jaundice appears. The reader should see the sequence clearly. Blood brings the burden in. Hepatocytes conjugate it. Bile carries it out.

Water and electrolytes complete the system by turning that chemical mixture into a flowing secretion rather than a paste of detergent and waste. Water provides volume and mobility. Electrolytes such as sodium, potassium, chloride, and bicarbonate help determine osmotic behavior, pH, and ductal fluid characteristics as bile moves through the biliary tree. Bicarbonate is especially important in the duct system because it contributes to alkalinity and helps shape the final fluid delivered downstream. The watery phase of bile is therefore not biologically empty. It is the transport medium that lets bile acids, phospholipids, cholesterol, and bilirubin actually move. Without adequate water and ionic structure, bile becomes thicker, slower, and more prone to retention or precipitation. The liver does not simply secrete solids into a tube. It builds a flowing chemical solution whose composition has to stay coherent from hepatocyte to canaliculus to duct to gallbladder to intestine.

So what bile is made of tells you what bile is for. Bile acids make fat digestion possible. Phospholipids stabilize the detergent system and help keep cholesterol transportable. Cholesterol enters bile as both cargo and disposal burden. Bilirubin enters as conjugated waste destined for excretion. Water and electrolytes make the whole stream mobile and chemically workable. Bile is therefore best understood as the liver's outgoing fluid for lipid digestion and molecular removal. It is not blood. It is not waste alone. It is the liver's second language, spoken outward into the intestine.

Where Bile Is Formed

Bile is formed first at the level of the hepatocyte. It does not begin in the gallbladder, and it does not begin as a generic fluid already sitting inside large ducts waiting to be released. The gallbladder stores and concentrates bile later. The ducts carry it later. But the actual making of bile begins inside the liver's main working cell. Hepatocytes take up compounds from blood on their sinusoidal side, synthesize bile acids from cholesterol, conjugate bilirubin so it can be excreted, secrete phospholipids and cholesterol into the biliary stream, and then pump this growing mixture across their canalicular membrane into the first microscopic channel of bile flow. That point matters because it keeps the logic straight. Bile is not something the liver merely transports. It is something hepatocytes actively build and secrete.

The first physical site of bile formation is the bile canaliculus. A canaliculus is a tiny channel formed between adjacent hepatocytes. It is not lined at its origin by a separate duct cell. It is created by the apical, bile-facing membranes of neighboring hepatocytes themselves, sealed by tight junctions so bile stays in its own compartment and does not leak backward into the blood-facing side of the tissue. Apical here means the membrane domain specialized for secretion into bile, as opposed to the sinusoidal membrane that faces blood. This is one of the most important structural facts in liver biology. The same hepatocyte that takes glucose, bilirubin, hormones, amino acids, and drug molecules in from blood on one side can secrete bile acids, conjugated bilirubin, phospholipids, and cholesterol out into bile on the other side. Bile is born at that membrane boundary. It begins as directional secretion, not passive drainage.

This means bile formation is already underway before any large bile duct appears. Hepatocytes secrete bile acids into the canalicular space, and because bile acids are osmotically active, they help draw water and solutes into the forming stream. Osmotically active means they influence the movement of water by changing the concentration environment across membranes. Conjugated bilirubin enters the same stream. Phospholipids enter. Cholesterol enters. Water and electrolytes follow as part of the fluid phase. So bile is not formed by mixing ingredients in a late reservoir. It is assembled step by step at the canalicular surface of hepatocytes and then begins moving through a microscopic network built directly into the liver parenchyma, the working tissue of the organ.

From there, the fluid moves outward from canaliculi into progressively larger channels and ductules near the portal region. At that point cholangiocytes, the epithelial cells lining the bile ducts, begin contributing more actively to the final ductal bile composition, especially by modifying water and bicarbonate content. But that comes after bile has already been initiated by the hepatocyte. This sequence matters because it shows that bile formation has two linked stages. The first is hepatocellular secretion, where the core detergent and excretory components are generated and pushed into the canaliculus. The second is ductal modification, where the flowing bile is adjusted as it moves through the biliary tree. If the first step fails, the liver cannot generate proper bile. If the second step fails, the fluid may still exist but become abnormal in volume, alkalinity, or ductal flow characteristics.

The reader should also notice what this means about hepatocyte polarity. Bile can only be formed where it is formed because the hepatocyte has kept its blood-facing and bile-facing work separate. Bilirubin arrives from blood, is taken into the hepatocyte, conjugated internally, and then secreted into the canaliculus. Cholesterol inside the hepatocyte can be converted into bile acids and secreted into the same canalicular stream. The liver is therefore not forming bile in some detached external chamber. It is forming bile at the exact point where intracellular processing meets outward excretion. The canaliculus is the first visible result of that decision.

So where bile is formed is simple in principle even if the chemistry is dense. It is formed first by hepatocytes, at their canalicular membrane, between neighboring cells inside the microscopic liver. The ducts do not create bile from nothing. They receive what hepatocytes have already begun. The gallbladder does not produce bile either. It stores and concentrates what the liver has sent. Bile starts where the hepatocyte turns inward metabolic handling into outward fluid secretion.

Canaliculi and Bile Duct Flow

Bile does not begin in a duct. It begins between hepatocytes. That is the first image the reader needs to hold. The canaliculus is the smallest visible channel of bile flow, a narrow passage formed by the apical, bile-facing membranes of adjacent hepatocytes. Apical here means the membrane domain specialized for secretion into bile rather than exchange with blood. These channels are not lined at their origin by a separate duct cell the way a larger bile duct is. They are carved directly out of hepatocyte polarity. Tight junctions seal their edges so bile stays in its own compartment and does not leak backward into the sinusoidal side of the tissue. This matters because the liver is running two opposite traffic systems through the same parenchyma. Blood is moving through sinusoids toward the central vein. Bile is being secreted into canaliculi and moving away from the hepatocyte field toward the portal region. Canaliculi are the first proof that hepatocytes are not merely changing chemistry inside themselves. They are building an outgoing stream.

That stream begins one transporter at a time. Hepatocytes actively pump bile acids, conjugated bilirubin, phospholipids, cholesterol, and other biliary constituents across the canalicular membrane. Water and electrolytes follow in response to the osmotic pull created by this solute movement, so bile flow is not simply liquid already sitting there waiting to drain. It is secretion-driven flow. The canaliculus fills because hepatocytes keep writing bile into it. Once formed, bile moves through an interlacing canalicular network toward the canals of Hering and then into progressively larger intrahepatic bile ductules lined by cholangiocytes, the epithelial cells of the biliary tree. At that point the fluid is no longer only hepatocyte-made in its final character. Cholangiocytes begin modifying it, especially by adding bicarbonate-rich fluid and reshaping water content. So bile duct flow is a two-stage event. Hepatocytes generate the core biliary stream. Duct cells adjust and convey it.

The direction matters because canalicular flow is not passive drainage down an open pipe. It depends on intact hepatocyte polarity, functioning ATP-dependent transporters, preserved tight junctions, and an open downstream path. ATP, adenosine triphosphate, is the cell's immediately usable energy currency, and bile secretion spends it heavily. If hepatocytes lose energy, lose membrane organization, or lose transporter placement, canalicular secretion slows even before a large duct is blocked. If downstream ducts narrow, inflame, scar, or obstruct, pressure rises backward into the canalicular system and the same stream begins to stall from the opposite end. This is why bile flow has to be understood from canaliculus to duct rather than from duct alone. A person can have cholestasis because the hepatocyte cannot push bile into the canaliculus properly, because the small duct system cannot carry it, or because the larger ducts outside the liver are obstructed. The failure point changes, but the traffic logic stays the same.

This is also why canaliculi and bile ducts belong in the same sentence but not as if they were the same structure. The canaliculus is the birth point of bile flow, built from hepatocyte membranes themselves. The bile duct is the carrying system that receives that secretion and moves it onward toward storage in the gallbladder or delivery into the intestine. One is formed by the working cell. The other is lined by duct epithelium built for transport and modification. The liver works because those structures join cleanly. Bile is secreted into the smallest channels, gathered into larger ducts, adjusted along the way, and then sent outward as a coherent stream. Break the seal, the polarity, the transporter function, or the outflow path, and the stream stops being useful. It becomes retained burden. That is the deeper point. Canaliculi and bile duct flow are not anatomy trivia. They are the architecture that turns hepatic secretion into digestion and excretion.

Why Bile Is Both Digestive Fluid and Waste Route

Bile has a double identity because the liver is solving two different physical problems with one outgoing stream. The first problem is digestion of dietary fat in the intestine. The second is removal of compounds that should not stay in circulation. Those jobs converge because the same chemical properties that make bile useful in the gut also make it useful as a route for exporting certain hydrophobic or conjugated burdens out of the body. Hydrophobic means poorly compatible with water. Dietary lipid is hydrophobic. Cholesterol is hydrophobic. Many compounds the liver wants to move out through the intestine are not well suited to remain dissolved in blood indefinitely or to be filtered cleanly by the kidneys in their original form. So the liver builds a secretion that can both enter the intestinal lumen and carry specific molecular cargo outward. Bile is therefore not half digestion and half excretion by coincidence. It is one fluid designed at the intersection of those two needs.

Its digestive role begins with fat. The intestinal lumen is a watery environment, but dietary fat enters it as large hydrophobic masses that do not mix well with water and therefore do not present much exposed surface to pancreatic lipase. Bile acids solve that physical problem because they are amphipathic molecules, meaning one side of the molecule interacts with water and the other interacts with fat. When delivered into the duodenum, they surround lipid droplets and help disperse them into much smaller particles. That does not chemically digest the fat by itself, but it makes the fat physically accessible to enzymes. Later, bile acids help form micelles, small transport assemblies that keep fatty acids, monoglycerides, cholesterol, and fat-soluble vitamins suspended long enough to move toward the absorptive surface of the intestine. In this sense bile is not optional digestive decoration. It is one of the main fluids that makes lipid absorption possible.

The waste role begins with the same outward path. Once the liver has decided that a compound should leave through the biliary route, secretion into bile allows that compound to exit the liver, pass through the ducts, enter the intestine, and move toward elimination in stool. Bilirubin is the clearest example. It begins as a breakdown product of heme, reaches the liver through blood, is taken up by hepatocytes, conjugated into a more transportable form, and then secreted into bile. Cholesterol also uses this route. Some cholesterol is secreted directly into bile. Some is first converted into bile acids, which both support digestion and create a route by which cholesterol-derived material can leave the body. This is why bile is not just helping with the meal. It is carrying away molecular decisions the liver has already made about what should not remain in the bloodstream.

These two functions are linked by chemistry, not just by location. Bile acids help emulsify fat, but they also make it possible for other hydrophobic molecules to remain in a moving biliary solution rather than precipitating prematurely. Phospholipids help stabilize that mixture and help keep cholesterol suspended. Water and electrolytes create the fluid phase that lets the whole secretion move. So bile does not alternate between being digestive on one day and excretory on another. It is doing both whenever it flows. The same stream that is carrying conjugated bilirubin and cholesterol outward is also preparing the intestinal contents for efficient fat digestion and absorption. That is one reason biliary obstruction creates two classes of failure at once. Digestion worsens because bile acids are not reaching the lumen properly. Excretion worsens because bilirubin and other biliary cargo are no longer leaving efficiently.

This is also why the phrase bile is a waste fluid is incomplete and why the phrase bile is a digestive juice is incomplete. Either one alone cuts the organ in half. A pure waste fluid would not need to be so carefully structured for detergent function in the intestine. A pure digestive fluid would not need to carry bilirubin, cholesterol, and other compounds marked for disposal. The liver uses bile to send forward chemical tools for fat handling and to send outward specific burdens it has already processed. One stream serves the next meal and cleans up the last metabolic cycle at the same time.

So bile is both digestive fluid and waste route because the liver is an organ of outgoing decisions. It does not only absorb from blood and hold. It secretes into a channel that reaches the intestine, and that channel is useful for two reasons. It helps the body extract fat from food. It helps the body remove what blood should not keep carrying.

The Physical Problem Bile Solves in the Intestinal Lumen

The physical problem bile solves is simple to state and hard for the body to ignore. Fat and water do not mix, and the intestinal lumen is a watery space. The lumen is the hollow interior of the small intestine where chyme, the semi-fluid material leaving the stomach, is being chemically processed before absorption. When dietary fat enters that environment, it does not dissolve into the surrounding fluid the way glucose or amino acids can. It tends to pool, coalesce, and form larger hydrophobic masses with relatively little exposed surface area. Hydrophobic means poorly compatible with water. That is a mechanical disaster for digestion, because pancreatic lipase works at the surface of fat droplets, not deep inside a large oil mass. If the surface area stays small, enzyme access stays poor, and much of the dietary lipid remains physically difficult to process. The body therefore needs a way to break one large fat problem into many smaller manageable ones before absorption can even become possible.

Bile supplies that solution through bile acids. These molecules are amphipathic, meaning one side of the molecule interacts with water while the other interacts with fat. That structure allows bile acids to sit at the interface between lipid and the watery intestinal environment. When bile enters the duodenum and mixes with dietary fat, bile acids coat and disperse large fat globules into much smaller droplets. This does not fully digest the fat by itself. It changes the geometry of the problem. Instead of one large poorly accessible lipid mass, the intestine now contains many smaller droplets with vastly greater total surface area. That surface expansion gives pancreatic lipase far more access to triglycerides, which can then be hydrolyzed into fatty acids and monoglycerides. The liver is therefore not sending bile into the intestine as a vague aid to digestion. It is solving a surface-area problem that fat digestion cannot solve alone.

That is only the first half of the physical problem. Even after lipase has acted, the products of fat digestion still face a transport problem. Fatty acids, monoglycerides, cholesterol, and fat-soluble vitamins do not move easily through the watery intestinal contents on their own. They need a way to remain suspended long enough to reach the brush border of the enterocyte, the absorptive cell lining the small intestine. This is where bile acids help again by forming micelles, tiny mixed transport structures that hold these lipid digestion products in a more mobile form. A micelle is not the same thing as an absorbed particle. It is a delivery arrangement inside the lumen. It keeps hydrophobic material from separating out prematurely and helps shuttle that material through the watery unstirred layer near the intestinal surface. So bile does not only help break fat apart. It helps keep the breakdown products moving toward absorption once they have been generated.

This is why bile failure produces such a specific kind of digestive failure. The intestine may still receive dietary fat. Pancreatic enzymes may still be present. The enterocytes may still be capable of absorbing lipids if the lipids reach them properly. But if bile flow is weak or absent, the physical handling of fat breaks down before any of those later steps can work well. Large fat masses remain poorly dispersed. Lipase loses effective access. Micelle formation falls. Cholesterol and fat-soluble vitamins become harder to transport across the luminal water layer. The result is not just less efficient digestion in some vague sense. It is a failure to solve the mechanical incompatibility between hydrophobic food and a water-based intestinal environment.

So the physical problem bile solves in the intestinal lumen is the problem of making fat behave in water long enough for digestion and absorption to proceed. It breaks large lipid masses into smaller accessible droplets and then helps keep the products of that digestion suspended and mobile. Without bile, fat remains chemically present in the gut but physically difficult to use.

Why Water-Insoluble Compounds Need a Route Out

A compound that does not dissolve well in water cannot simply be left in the bloodstream and expected to disappear on its own. Blood is an aqueous transport medium. The kidneys filter water-soluble material out of plasma and into urine with far more ease than they handle large hydrophobic compounds, cholesterol-rich burdens, or molecules that remain bound, precipitate easily, or interact badly with aqueous flow. Hydrophobic means poorly compatible with water. That single property changes everything. A molecule may be chemically real, biologically burdensome, and fully present in the body, yet still have no clean exit through the renal route unless it is first transformed into a more water-compatible form. The liver therefore needs another strategy for compounds that are poorly suited to remain in circulation and poorly suited to urinary elimination. Bile is one of the main answers to that problem.

This is easiest to see with cholesterol. Cholesterol is essential for membranes, steroid hormone synthesis, and bile acid formation, but it is not something the body can simply dissolve freely in plasma and then wash into urine when it becomes excessive. It is hydrophobic, highly structured, and poorly handled by the kidney as a direct waste molecule. The liver solves part of this by incorporating cholesterol into bile, either directly or after converting some of it into bile acids. That does two things at once. It places cholesterol into a moving detergent-rich fluid where it can remain suspended more effectively, and it creates a route into the intestinal tract where that cholesterol-derived burden can leave the body in stool if it is not reabsorbed. The same principle applies to other compounds the liver chooses to secrete into bile. They need an exit compatible with their chemistry.

Bilirubin makes the same point from a different angle. Unconjugated bilirubin is not well suited to circulate freely as a disposable waste. It travels bound to albumin, which keeps it mobile in blood but does not solve the elimination problem. The liver takes it up, conjugates it, meaning chemically modifies it into a more transportable form, and then secretes it into bile rather than relying on it to drift toward renal removal. That route matters because the body is not only dealing with whether a molecule is harmful. It is dealing with whether the molecule can physically leave. A waste product that cannot remain dissolved appropriately, cannot circulate forever, and cannot be filtered well by the kidneys needs another channel. Bile provides one.

This is why excretion is always partly a solubility problem. The body does not remove compounds simply because it dislikes them. It removes them by changing their chemistry or by choosing the route whose fluid environment can actually carry them. Water-soluble compounds can often be sent toward the kidney after enough processing. Water-insoluble compounds may need biliary secretion because the intestine can receive them in a detergent-containing stream and carry them outward with fecal material. The liver stands at the center of that decision. It determines whether a molecule should be conjugated for urine, secreted into bile, stored temporarily, or transformed into something else first. Elimination is therefore not an afterthought added at the end of metabolism. It is built into metabolism from the start.

The need for a route out also explains why bile is not merely a digestive aid. It is a chemical exit stream for compounds that would otherwise linger, accumulate, or create transport instability in blood. If this route is obstructed, retained biliary compounds begin backing up into liver tissue and circulation. Bilirubin rises. Cholesterol handling changes. Bile acids accumulate where they should not. The problem is no longer only digestive. It is excretory. The liver has made its decision about what should leave, but the body has lost the path by which that decision becomes real.

So water-insoluble compounds need a route out because blood is not an all-purpose waste sink and the kidney is not an all-purpose drain. Some molecules are too hydrophobic, too structurally awkward, or too poorly suited to urinary filtration to leave cleanly through water-based excretion alone. The liver solves that by sending them into bile, a detergent-rich outgoing stream that can carry them into the intestine and out of the body. Solubility determines route. Route determines whether burden clears or stays.

Chapter 19

Bile Acids and Fat Digestion

How Bile Acids Emulsify Fat

Fat enters the intestine with a physical disadvantage. It is hydrophobic, which means it does not mix well with water, and the intestinal lumen is a watery space. That creates a mechanical problem before it creates a chemical one. Large fat droplets tend to coalesce, forming bigger lipid masses with relatively little exposed surface area. Pancreatic lipase cannot solve that problem by itself because lipase works at the surface of fat, not deep inside an oil mass. If the surface stays limited, digestion stays limited. The body therefore needs a molecule that can stand at the border between water and fat and force the two into a more workable relationship. That molecule is the bile acid.

Bile acids are built for interface work. They are amphipathic, meaning one side of the molecule is relatively water-friendly and the other side is relatively fat-friendly. That split structure is the whole reason they matter. When bile acids are released into the duodenum and meet dietary lipid, they insert themselves at the surface of the fat droplet. Their hydrophobic face interacts with the lipid. Their hydrophilic face interacts with the surrounding aqueous environment. This lowers surface tension and makes the fat mass easier to break apart into smaller droplets under the mechanical mixing forces already present in the intestine. Emulsification is this physical breaking and dispersal of large fat globules into much smaller suspended droplets. It is not full digestion. It is preparation for digestion.

That distinction matters because bile acids do not cleave triglycerides directly. Pancreatic lipase does that. What bile acids do is make lipase possible at real digestive scale. By dispersing fat into many smaller droplets, they create much more total surface area for lipase to attack. One large droplet has less accessible surface than thousands of smaller droplets containing the same total amount of fat. The liver therefore is not sending bile acids into the gut as a minor chemical supplement. It is solving a geometry problem that enzyme chemistry alone cannot solve. Without emulsification, the fat remains physically resistant to efficient hydrolysis. With emulsification, the enzyme finally has room to work.

There is another layer to this. The products of fat digestion, fatty acids, monoglycerides, cholesterol, and fat-soluble vitamins, are still hydrophobic enough to create transport difficulty in the lumen even after lipase has acted. Bile acids help solve that too by participating in micelle formation. A micelle is a tiny mixed particle in which bile acids organize around lipid digestion products, keeping them suspended in the watery intestinal environment long enough to move toward the absorptive surface of the enterocyte, the main absorptive cell of the small intestine. So bile acids do not only help break fat apart. They help keep the resulting lipid products mobile after digestion has begun. Emulsification opens the problem. Micellar transport carries it forward.

This is why bile acid deficiency or poor bile delivery creates such a characteristic digestive failure. The fat may still be present in the intestine. Lipase may still be present. The absorptive cells may still be there. But the physical handling of the lipid load has collapsed. Large fat droplets remain poorly dispersed. Enzyme access falls. Micelle formation weakens. Fat-soluble nutrients become harder to move across the luminal water layer. The body is left with lipid in the gut that it cannot use efficiently because the first physical barrier, getting fat to behave in water, was never properly solved.

So when the reader asks how bile acids emulsify fat, the answer is structural before it is enzymatic. Bile acids use their amphipathic design to sit at the fat-water boundary, lower surface tension, disperse large lipid masses into smaller droplets, and create the surface conditions required for lipase to work effectively. The liver sends out a detergent system because fat digestion begins as a problem of physics.

Micelles and Why Lipid Absorption Needs Them

Digested fat still has a transport problem even after bile acids have emulsified it and pancreatic lipase has begun breaking triglycerides apart. Emulsification creates smaller droplets and gives lipase access to surface area, but the products of that digestion, mainly fatty acids, monoglycerides, cholesterol, lysophospholipids, and fat-soluble vitamins, are still poorly compatible with the watery environment of the intestinal lumen. Hydrophobic molecules do not move well through that water layer on their own. They tend to aggregate, drift poorly, or remain stranded away from the absorptive surface. The intestine therefore needs another structure after emulsification and after lipase action, a structure that can keep these lipid products suspended and mobile long enough to reach the enterocyte. That structure is the micelle.

A micelle is a tiny mixed transport assembly formed primarily by bile acids surrounding lipid digestion products. The bile acids orient themselves so their water-friendly surfaces face outward toward the intestinal fluid and their fat-friendly surfaces face inward toward the hydrophobic cargo. This lets the micelle carry otherwise poorly water-soluble compounds through a watery lumen without requiring those compounds to dissolve freely. The point is not that the micelle is digesting fat further. The point is that it is making the products of digestion physically transportable. The intestine is full of water. The absorptive membrane is covered by an unstirred water layer and mucus. Lipid digestion products need a shuttle system to cross that last aqueous distance. Micelles provide it.

This is where many simplified explanations break down. People hear that bile helps digest fat and stop there, as if digestion alone solves absorption. It does not. A fatty acid liberated from triglyceride is still a hydrophobic molecule in a watery environment. Cholesterol is still cholesterol. Fat-soluble vitamins are still fat-soluble. Unless those compounds are kept in a transportable state, they do not move efficiently toward the brush border of the enterocyte, the absorptive cell lining the small intestine. The micelle is therefore the second major physical solution bile acids provide. First they help disperse fat so lipase can work. Then they help collect the resulting lipid products and carry them to the absorptive surface. Digestion without micellar transport would leave much of the liberated lipid chemically present but spatially stranded.

The brush border matters here because absorption is a membrane event, not a general soaking process. The lipid cargo has to get close enough to the enterocyte membrane for fatty acids, monoglycerides, cholesterol, and fat-soluble vitamins to leave the micelle and cross into the cell. The micelle itself is not usually absorbed wholesale as a permanent intact particle. It is better understood as a delivery platform. It ferries hydrophobic cargo through the aqueous luminal environment, brings that cargo to the membrane, and then the absorbed lipids move into the enterocyte while bile acids largely remain in the lumen to continue participating in transport until they are later reabsorbed farther down the intestine. This is efficient chemistry. The body is using bile acids repeatedly in the lumen before reclaiming many of them through enterohepatic circulation.

This is why impaired bile flow so quickly becomes a malabsorption problem. If bile acids do not arrive in adequate quantity, micelles form poorly. When micelles form poorly, the products of fat digestion lose mobility in the aqueous lumen. Fatty acids and monoglycerides may still be generated by lipase, but their delivery to the enterocyte weakens. Cholesterol absorption falls. Fat-soluble vitamins such as A, D, E, and K become harder to absorb. The failure is not only enzymatic. It is transport failure inside the lumen. The body has performed part of the digestive sequence but cannot complete the move from liberated lipid to absorbed lipid.

So micelles matter because lipid absorption needs more than cleavage. It needs a way to move hydrophobic digestion products through water to the intestinal surface. Bile acids create that way. They build small transport assemblies that keep lipid cargo suspended, mobile, and deliverable. Without micelles, the gut can break fat apart and still fail to bring much of it into the body.

The Relationship Between Bile Flow and Absorption of Fat-Soluble Vitamins

Fat-soluble vitamins depend on bile because they share the same physical problem as dietary fat. Vitamins A, D, E, and K are hydrophobic molecules. They do not move comfortably through the watery contents of the small intestine on their own, and they are not absorbed efficiently simply because they are present in food. They have to be released from the food matrix, remain associated with the lipid phase during digestion, and then be carried through the aqueous intestinal lumen in a transportable form. That is where bile flow becomes decisive. Bile acids are needed not only to emulsify fat and expand surface area for pancreatic lipase, but also to form micelles, the small transport assemblies that keep lipid digestion products and fat-soluble compounds suspended long enough to reach the absorptive surface. If bile flow is weak, the vitamins may still be in the gut, but their path to the enterocyte becomes physically unstable.

This means fat-soluble vitamin absorption is never just a vitamin story. It is a bile story first, then a lipid digestion story, then an intestinal transport story. Vitamin A in a meal does not enter the body as free vitamin A crossing easily through water. Vitamin D does not float unaided to the intestinal membrane. Vitamin E and vitamin K do not simply diffuse across the lumen because the label said they were present. They travel with the lipid phase. They depend on emulsification to help disperse that phase and on micelles to help move through the watery unstirred layer that sits over the brush border of the enterocyte, the absorptive cell lining the small intestine. Bile flow therefore determines whether these vitamins remain trapped in poorly mobile lipid masses or become deliverable to the membrane where absorption can actually occur.

The relationship is especially clear when bile delivery is impaired. If bile acids do not reach the duodenum in sufficient quantity, fat digestion weakens mechanically and luminal transport weakens immediately after that. Large lipid droplets remain less effectively dispersed. Lipase loses access. The products of digestion, including fat-soluble vitamins carried within that lipid environment, are less effectively incorporated into micelles. Once micelle formation falls, vitamin delivery to the intestinal surface falls with it. The result is not necessarily a total absence of absorption, but a substantial decline in usable uptake. A person may be ingesting the vitamins regularly and still absorb them poorly because the bile-dependent transport step is broken. This is why biliary obstruction, cholestasis, and other disorders of bile formation or flow can produce fat-soluble vitamin deficiency even when intake appears adequate.

Each vitamin then carries its own downstream consequence. Vitamin A deficiency begins affecting epithelial integrity, immune behavior, and visual chemistry. Vitamin D deficiency alters calcium and phosphate handling. Vitamin E deficiency weakens antioxidant protection in lipid-rich tissues and membranes. Vitamin K deficiency disrupts normal clotting because vitamin K is required for proper activation of several coagulation-related proteins. But the point at the liver and intestine level is the same for all four. Their absorption depends on the body's ability to move hydrophobic micronutrients through a watery digestive environment. Bile makes that movement possible. Without it, the vitamins remain nutritionally present but biologically stranded.

So the relationship between bile flow and absorption of fat-soluble vitamins is direct and physical. Bile acids help emulsify dietary fat, generate the micellar transport system, and carry vitamins A, D, E, and K through the intestinal lumen to the absorptive surface. When bile flow is intact, these vitamins can travel with the lipid digestion stream into the body. When bile flow is impaired, the gut may still contain them, but the body loses access to them. The vitamin stays in the meal. The body loses the nutrient.

Why Impaired Bile Delivery Changes the Nutritional Result of a Meal

A meal is not judged by what it contains on the plate. It is judged by what digestion can actually liberate, what the intestine can physically move, and what the body can finally absorb and retain. Impaired bile delivery changes that result because bile is part of the machinery that turns ingested fat into usable nutrition. If bile does not reach the duodenum in adequate quantity, the meal may still contain fat, cholesterol, phospholipids, and fat-soluble vitamins, but the intestine loses one of its main tools for handling them. Large lipid masses remain less effectively dispersed. Pancreatic lipase has less surface to work on. Micelle formation weakens. The products of lipid digestion move less efficiently through the watery intestinal lumen toward the absorptive membrane. The nutritional outcome of the meal therefore changes before absorption is ever measured in blood. The food may still be present. The usable yield drops.

The physical problem comes first. Dietary fat enters a water-based intestinal environment, and fat does not mix comfortably with water. Bile acids normally solve part of that incompatibility by coating lipid droplets and helping break them into smaller suspended particles. That emulsification step expands surface area so pancreatic lipase can hydrolyze triglycerides more effectively. Later, bile acids help form micelles, which are small transport assemblies that keep fatty acids, monoglycerides, cholesterol, and fat-soluble vitamins mobile in the lumen long enough to reach the brush border of the enterocyte, the main absorptive cell of the small intestine. If bile delivery is impaired, that entire sequence weakens. The meal is now being digested under poorer physical conditions. This is why bile failure is not simply a liver or gallbladder problem. It changes the mechanics of what the intestine can actually do with food.

That altered mechanics changes nutrient yield in a specific direction. Fat absorption falls first because fat handling depends most visibly on bile acids. But the problem does not stop at triglyceride calories. Cholesterol absorption changes. Vitamins A, D, E, and K, all fat-soluble vitamins, become harder to absorb because they normally travel with the lipid digestion stream and depend on micellar transport. Certain phospholipids and other lipid-associated compounds also become less available. So the same meal produces a different internal result. Two people can eat the same food, but if one has intact bile delivery and the other does not, they are not biologically eating the same meal. One is extracting a larger fraction of the fat-phase nutrition. The other is carrying more of that chemistry onward toward loss in stool.

This is where the liver's outgoing stream becomes part of nutritional reality. Bile is often spoken about as if it were mainly an excretory fluid, and it is that, but it is also one of the main reasons dietary fat becomes absorbable at all. If bile is obstructed, reduced, or poorly delivered, the body loses access not only to waste clearance through the biliary route but also to a large fraction of the nutrient value embedded in fatty foods. A meal rich in egg yolks, dairy fat, meat fat, or any other lipid-containing food may still test dense on paper, but the intestine cannot extract the same biological return if bile chemistry is missing at the right time and place. The food keeps the calories on the label. The body loses part of the nutrition.

This also explains why impaired bile delivery can create a mismatch between intake and status. A person may be eating enough total fat or enough fat-soluble vitamins and still move toward deficiency if bile-dependent digestion and transport are poor. Over time the consequences spread beyond the intestine. Vitamin K deficiency can destabilize clotting. Vitamin A deficiency can alter epithelial integrity and vision-related chemistry. Vitamin D handling suffers. Vitamin E delivery to tissues weakens. Energy extraction from fat becomes less efficient. Stool becomes more lipid-rich because nutrition that should have crossed into the body stayed in the lumen and left in waste instead. The digestive event was incomplete, and the nutritional result was rewritten.

So impaired bile delivery changes the nutritional result of a meal because nutrition is not defined by ingestion. It is defined by extraction. Bile acids make fat workable in water, make lipase more effective, make micellar transport possible, and make fat-phase nutrients reachable by the intestinal membrane. When bile does not arrive properly, the meal is chemically present but biologically diminished. The plate may look the same. The absorbed result does not.

The Gallbladder as a Storage and Concentration Organ

The gallbladder does not make bile. It receives bile the liver has already made, holds it, and changes its concentration so that a relatively modest organ can deliver a stronger bile pulse at the moment a meal requires it. That is its real place in digestion. The liver secretes bile continuously because hepatocytes do not wait for each meal to begin forming bile acids, conjugated bilirubin, phospholipids, cholesterol, water, and electrolytes. But the intestine does not need bile continuously at the same rate. It needs a more forceful delivery when fat enters the duodenum. The gallbladder solves that timing mismatch. It acts as a reservoir between constant hepatic secretion and meal-triggered intestinal demand. Without that reservoir, bile flow would be more diffuse and less responsive to the digestive event.

Its storage function is anatomical first and chemical second. Bile leaves the liver through the biliary ducts, and when the sphincter at the distal biliary outlet remains relatively closed between meals, some of that bile is diverted backward through the cystic duct into the gallbladder. There it is held rather than immediately emptied into the intestine. This does not mean the gallbladder is a passive sac. Its lining is metabolically active in a specific way. It absorbs water and certain electrolytes out of the stored bile, which reduces volume and increases the concentration of key solutes, especially bile acids, cholesterol, phospholipids, and bilirubin-related constituents. The result is more potent bile in a smaller volume. Concentration matters because digestion is not helped merely by fluid arriving in the duodenum. It is helped by detergent chemistry arriving at sufficient strength to emulsify fat and support micelle formation efficiently.

That concentrating function changes the physical character of the stored bile. As water is removed, the same total amount of bile acids occupies less space, which increases their local concentration. The same is true for cholesterol and bilirubin-containing components, and that is part of why the chemistry has to remain balanced. Bile is not infinitely stable under concentration. It depends on the relationship between bile acids, phospholipids, cholesterol, water, and other solutes remaining within a range that keeps hydrophobic components suspended rather than precipitated. This is one reason gallstones are possible. The gallbladder is doing exactly what it is designed to do, holding and concentrating bile, but if the chemistry becomes unstable during that process, cholesterol or pigment material can begin coming out of solution. So concentration is useful, but it is not chemically neutral. It increases digestive potency while also increasing the importance of bile composition.

The gallbladder therefore changes the digestive force of a meal without changing the fact that the liver remains the source. When fat reaches the duodenum, hormonal signaling, especially through cholecystokinin, tells the gallbladder to contract while the distal biliary sphincter relaxes. Bile is then expelled into the intestine in a more concentrated wave. That wave matters because emulsification and micellar transport depend on enough bile acids arriving when lipid is actually present in the lumen. A slow weak trickle is not the same as a targeted release into a fat-containing meal. The gallbladder sharpens the timing of biliary delivery. It makes the liver's constant work usable in digestive rhythm.

This is why the gallbladder should be understood as a storage and concentration organ rather than as a producer or as a minor accessory pouch. It stores because the liver secretes continuously while digestion happens episodically. It concentrates because the intestine benefits from a stronger detergent pulse than a dilute uninterrupted stream would provide. It stands between bile formation and bile deployment, reshaping not the identity of bile but its timing and strength. The liver writes the fluid. The gallbladder holds it, thickens it, and releases it when the meal demands it.

Cholecystokinin and Meal-Triggered Bile Release

The liver makes bile continuously, but the intestine does not need bile at a constant full force. It needs bile when food, especially fat, arrives in the duodenum. That timing problem is solved hormonally, and cholecystokinin is one of the main signals that does the work. Cholecystokinin, usually abbreviated CCK, is a peptide hormone released primarily from I cells in the upper small intestine when partially digested fat and protein enter the duodenal lumen. The key point is that CCK is not announcing that digestion exists in general. It is announcing that a meal with real digestive demand has reached the segment of intestine where bile and pancreatic secretions now need to arrive. The hormone translates luminal content into coordinated biliary and pancreatic response.

Once released, CCK acts on the gallbladder and the biliary outflow system in a way that turns stored bile into a timed digestive pulse. The gallbladder contracts. The sphincter of Oddi, the muscular valve controlling the final outflow of bile and pancreatic secretions into the duodenum, relaxes. Those two actions have to happen together. Gallbladder contraction without outlet relaxation would create pressure against a closed path. Outlet relaxation without gallbladder contraction would not deliver the same concentrated bile surge. CCK synchronizes both sides of the event. The bile that has been stored and concentrated in the gallbladder is expelled through the cystic duct, into the common bile duct, and into the duodenum at the exact stage of digestion when emulsification and micelle formation are becoming necessary.

This matters because fat digestion is a timing problem as much as a chemistry problem. Dietary fat entering the duodenum arrives as hydrophobic material in a watery environment. Bile acids need to be present there soon enough to emulsify lipid droplets, expand surface area for pancreatic lipase, and later help form micelles that carry lipid digestion products toward the absorptive surface. If bile arrived too early, before the meal, much of its detergent activity would be wasted. If it arrived too late, the fat would sit in the lumen under poor physical conditions for longer than it should. CCK solves that by linking bile release to the actual presence of fat and protein in the upper small intestine. The hormone does not guess. It responds to contact.

CCK also helps coordinate the digestive stream more broadly because bile release and pancreatic enzyme delivery are meant to work together, not as isolated events. Fat in the duodenum needs emulsification from bile acids and hydrolysis from pancreatic lipase. Protein needs pancreatic proteases. The same meal that calls for bile usually calls for pancreatic secretion as well, and CCK participates in both sides of that coordination. This is one reason the upper small intestine functions less like a simple receiving tube and more like a sensory control zone. It detects the chemistry of the arriving meal and recruits the liver-gallbladder-pancreas axis accordingly. Cholecystokinin is one of the main messengers carrying that instruction.

The result is that bile release becomes meal-shaped rather than merely continuous. The liver supplies the fluid continuously in the background. The gallbladder stores and concentrates it between meals. CCK triggers its release when the intestinal lumen proves that the moment has arrived. That is the elegance of the system. Bile is not dumped randomly into the gut. It is held back, concentrated, and then deployed when the nutrient structure of the meal demands it. The intestine sends the signal. The gallbladder answers. The liver's outgoing stream becomes a timed digestive weapon instead of a diffuse trickle.

Chapter 20

Enterohepatic Circulation

Why Bile Acids Are Recycled Instead of Discarded in Large Amounts

Bile acids are recycled because the body cannot afford to treat them as disposable detergent. They are too chemically expensive, too functionally important, and too useful across repeated meals to be thrown away in bulk every time bile enters the intestine. The liver synthesizes bile acids from cholesterol, which already tells you something about cost. This is not a trivial secretory byproduct made from surplus material with no consequence if lost. It is a deliberately built molecule carrying one of the main physical tools for fat digestion. If the body had to replace the entire bile acid pool from scratch after every meal, hepatic cholesterol demand would rise dramatically, synthetic burden would rise, and the efficiency of lipid digestion across the day would be far worse. So the system is built around reuse. The liver sends bile acids out into the intestine to do detergent work, then reclaims most of them and sends them back through the portal circulation for another pass.

That reuse makes sense because bile acids are not consumed in the same way enzymes are not consumed when they catalyze reactions. They participate in emulsification and micelle formation, helping disperse dietary fat and shuttle lipid digestion products through the watery intestinal lumen, but once that work has been done, many of the molecules remain chemically intact enough to be recovered. The intestine therefore treats them less like waste and more like tools that have completed one round of use. Most reabsorption occurs in the terminal ileum, the final segment of the small intestine, where specialized transport systems take bile acids back across the intestinal lining. From there they enter portal blood and return directly to the liver. This loop is called enterohepatic circulation because it links intestine and liver in a repeated circuit. The liver secretes. The intestine uses. The ileum reclaims. The portal vein returns.

This recycling also solves a concentration problem. The body needs a relatively robust bile acid pool available for meal-triggered release, but it does not need or want to synthesize that full pool de novo every time fat enters the duodenum. By recapturing most of what it has already made, the liver can maintain a high-functional pool with only a smaller daily requirement for replacement synthesis. A fraction is still lost in stool, and that loss matters because it is one of the main ways cholesterol-derived material leaves the body permanently. But the normal design is dominated by retention, not waste. The liver makes enough new bile acid to replace the loss, not enough to rebuild the entire circulating pool from nothing each day. That is a much more economical strategy for an organ already carrying enormous synthetic and metabolic workload.

The recycling loop also helps explain why bile acid metabolism belongs to both digestion and blood chemistry. Once reabsorbed from the ileum, bile acids are no longer simply luminal detergents. They are portal cargo again. They return to the liver in blood, are taken back up by hepatocytes, and are resecreted into bile. That means bile acids spend their life alternating between two compartments, the intestine and the portal-hepatic circulation. Their physiology is therefore inseparable from hepatic transport. The liver is not merely making bile acids once and forgetting them. It is receiving them back, recognizing them, and sending them out again. The system is cyclical because fat digestion is cyclical. Meals recur. The bile acid pool is built to recur with them.

This is also why disruption of the ileum can produce a liver and digestion problem at the same time. If bile acids are not reabsorbed properly, more are lost in stool, the returning pool shrinks, and the liver is forced to increase new synthesis to compensate. If loss exceeds replacement capacity, the total bile acid pool falls and bile delivered to future meals becomes less effective for emulsification and micelle formation. Fat digestion worsens. Fat-soluble vitamin absorption worsens. The body then begins paying for a broken recycling loop with reduced nutrient yield. The problem started in recovery, not in original secretion.

So bile acids are recycled instead of discarded in large amounts because they are reusable digestive tools built at real metabolic cost. The body sends them into the intestine to solve the physical problem of fat in water, then reclaims most of them through the ileum and returns them to the liver for another round. Loss is normal in small amounts. Bulk loss would be wasteful. Enterohepatic circulation exists because the liver is not interested in making a brand-new detergent system for every meal when the old one can be recovered, returned, and redeployed.

Ileal Reabsorption and Portal Return

Most bile acids are not meant to finish their life in the stool after one pass through the intestine. They are meant to work in the lumen, help emulsify dietary fat and support micelle formation, then be reclaimed before the small intestine ends. That reclamation happens mainly in the terminal ileum, the final segment of the small intestine. By the time luminal contents reach this region, much of fat digestion has already occurred and the bile acids that helped carry it are still present in the intestinal fluid. The ileum is built to recover them. Its epithelial cells express transport machinery that actively takes bile acids up from the lumen rather than allowing the full detergent pool to be lost downstream into the colon. This is not passive salvage. It is a planned retrieval step built into the digestive sequence from the start.

The most important uptake step occurs at the apical surface of the ileal enterocyte, the absorptive cell lining the intestine, through the apical sodium-dependent bile acid transporter, or ASBT. Apical means the membrane facing the lumen. This transporter uses the sodium gradient across the membrane to pull bile acids into the cell even when simple diffusion would not be enough to recover them efficiently. Once inside the enterocyte, bile acids are not meant to remain there. They are moved across the cell and exported from the basolateral side, the membrane facing blood, into the portal circulation. In other words, the ileal cell acts less like a final absorptive destination and more like a recovery gate. The bile acid has already done its luminal job. Now it is being transferred back into the body's internal transport system so the liver can use it again.

Portal return is what makes this recovery physiologically powerful. Blood draining the intestine enters the portal vein and travels directly to the liver before joining the wider systemic circulation. That means reabsorbed bile acids do not wander the body randomly after ileal uptake. They are sent back to the organ that made them. Hepatocytes then take them up from portal blood with high efficiency and resecrete them into bile. This closes the loop of enterohepatic circulation. The liver sends bile acids into the intestine, the ileum reclaims most of them, the portal vein returns them, and the liver deploys them again. The same molecules can therefore participate in multiple digestive cycles before a fraction is finally lost in stool and replaced by new synthesis from cholesterol.

This return loop is one of the clearest examples of how the liver and intestine function as a single traffic system rather than as isolated organs. The liver does not simply dump bile acids into the gut and hope for the best. It expects them back. The ileum is not simply the tail end of nutrient absorption. It is a recovery organ for one of the body's most important digestive detergents. Portal blood is not carrying nutrients alone. It is carrying reclaimed digestive tools back to the liver for reuse. That is efficient chemistry and efficient anatomy at the same time. The body avoids the cost of synthesizing a full new bile acid pool for every meal by recovering most of what it already made.

The consequences of failure here are predictable and important. If the terminal ileum is diseased, resected, inflamed, or functionally impaired, bile acid reabsorption falls. More bile acids spill into the colon, where they can provoke watery diarrhea by stimulating secretion and motility, and fewer return to the liver through portal blood. The hepatic bile acid pool then begins to shrink unless new synthesis can compensate. If compensation is incomplete, later meals receive a weaker detergent system. Fat digestion worsens. Micelle formation worsens. Absorption of dietary lipid and fat-soluble vitamins falls. The body starts losing nutrient yield because the recovery loop broke. So ileal reabsorption and portal return are not secondary refinements. They are the mechanism that lets bile acids remain a reusable digestive currency instead of a one-time expense.

The Economy of Bile Acid Reuse

The body reuses bile acids because building them is expensive and losing them in bulk would force the liver to keep paying the same synthetic cost over and over again for a tool it has already made. Bile acids are not free secretions. They are synthesized from cholesterol through a multistep hepatic process, conjugated into a more effective transport form, secreted into bile, delivered into the intestine, and then used to solve one of digestion's hardest physical problems, making hydrophobic fat behave in a watery lumen long enough for lipase action and micellar transport to occur. Hydrophobic means poorly compatible with water. A molecule doing that much work, built through that much chemistry, is too valuable to discard casually after one meal. The economy of bile acid reuse begins there. The liver is not interested in manufacturing a brand-new detergent system every time food arrives if the old one can be recovered, returned, and sent out again.

This reuse changes the meaning of bile acids from simple secreted products into a circulating functional pool. The liver maintains a bile acid pool large enough to support repeated digestive cycles, but it does not have to synthesize the whole pool from scratch each day because most of those molecules come back. They are released into the intestine, participate in emulsification and micelle formation, then are reclaimed mainly in the terminal ileum and returned through portal blood to hepatocytes. That return matters because it turns one round of secretion into a loop rather than a one-way loss. The liver replaces only what is lost in stool, not what was merely used. This is efficient at the level of energy, cholesterol handling, and timing. The organ can maintain strong digestive capability across multiple meals without carrying the full synthetic burden of a total pool rebuild every day.

The cholesterol side of this economy is just as important. Cholesterol is the precursor from which bile acids are made, and the body needs cholesterol for membranes, steroid hormone synthesis, and broader structural work. If bile acids were discarded in large amounts after every meal, hepatic demand for fresh cholesterol conversion would rise sharply just to keep fat digestion functioning. Some cholesterol loss through bile acid excretion is normal and useful because it is one of the ways sterol material can leave the body permanently. But the dominant logic is conservation, not waste. The liver does not want to choose between preserving cholesterol-based biology and maintaining digestive detergent supply. Reuse lets it do both. A small fraction is sacrificed. Most of the pool is retained.

This same economy also protects digestive consistency. Meals recur. The need for bile acids recurs with them. If each meal depended on a full new wave of bile acid synthesis before digestion could proceed effectively, the system would be slower, heavier, and less adaptable. Instead, the liver works with a recirculating reserve. Reclaimed bile acids can be rapidly taken up from portal blood and resecreted into bile, which means the organ is operating more like a dispatcher of reusable tools than a factory forced to begin from zero each time. That gives digestion continuity. The intestine receives detergent chemistry from a pool already in motion, not from an emergency manufacturing cycle triggered too late.

This is why disruption of bile acid reuse is so costly even when bile synthesis itself is still possible. Once recovery falls, more bile acids are lost downstream, the circulating pool shrinks, and the liver is forced to spend more cholesterol and more synthetic effort on replacement. If the loss is large enough, replacement cannot fully keep up, and the system begins paying at both ends. Fat digestion weakens because the available bile acid pool is smaller. Hepatic synthetic burden rises because new bile acid production has to increase. The economy breaks. What had been a reusable digestive currency becomes a daily depletion problem.

So the economy of bile acid reuse is not an elegant side note. It is one of the reasons bile physiology is viable at all. The liver makes bile acids at real metabolic cost, sends them into the intestine to solve the physical problem of fat in water, then recovers most of them so the same molecules can do the job again. Reuse lowers synthetic burden, preserves cholesterol economy, stabilizes digestive performance, and keeps bile acid biology from becoming wasteful excess. The body does not throw away its detergent machinery every time it eats. It runs it in cycles.

Why Intestinal Disease Can Become Bile Acid Disease

Bile acid physiology looks like a liver subject until the intestine stops doing its half of the job. Then the whole system shifts. The liver synthesizes bile acids from cholesterol, conjugates them, secretes them into bile, and sends them into the small intestine to emulsify fat and support micelle formation. But that is only the outbound half of the circuit. Most bile acids are supposed to be reclaimed in the terminal ileum and returned through portal blood to the liver for reuse. That recovery step is not optional refinement. It is one of the main reasons the bile acid pool remains large enough to support repeated meals without forcing the liver to rebuild the full detergent system from scratch every day. Once intestinal disease disrupts that recovery, the problem stops being localized to the bowel. The liver begins losing one of its main recycled working fluids. Intestinal disease has now become bile acid disease.

The terminal ileum is the key hinge point because that is where most active bile acid reabsorption occurs. The enterocytes in this region express transport machinery that is built specifically to pull bile acids out of the lumen and move them back into portal blood. If that mucosa is inflamed, resected, ulcerated, fibrosed, functionally damaged, or simply absent, bile acids continue moving downstream instead of being reclaimed. That changes the system immediately. More bile acids spill into the colon, where they are not being used as a carefully regulated detergent for fat digestion anymore. They become irritants to the colonic epithelium, stimulating secretion and motility. The result can be watery urgent diarrhea because the colon is now being exposed to a chemical load that was supposed to be retrieved earlier. But that is only the visible intestinal side of the problem. The quieter hepatic side is that fewer bile acids are coming home.

Once portal return falls, the bile acid pool begins shrinking unless hepatic synthesis can compensate fully. The liver does try to respond by making more, but that compensation has limits. The body normally reuses most bile acids many times. It is not designed to replace massive losses indefinitely without consequence. If ileal disease keeps spilling bile acids into stool, the liver has to spend more cholesterol and more synthetic effort on replacement just to maintain ordinary digestive performance. If that replacement cannot keep pace, the total bile acid pool contracts. Then future meals arrive in an intestine receiving less detergent support than they should. Fat emulsification weakens. Micelle formation weakens. Absorption of fatty acids, cholesterol, and fat-soluble vitamins becomes less efficient. The original problem may have started as ileal inflammation or bowel resection. The downstream result is now a bile acid deficiency state at the level of digestion.

This is why the same intestinal disease can produce two apparently opposite effects at once. In the colon, excess unreabsorbed bile acids can drive diarrhea because too much bile acid has escaped recovery and arrived in the wrong place. In the small intestine during later meals, too little bile acid may be available in the lumen because the circulating bile acid pool has been depleted by repeated loss. Too much downstream. Too little upstream where digestion needs it. That is the logic. The body is not suffering from "bile acids present" or "bile acids absent" in a simple binary sense. It is suffering from misplacement caused by failed recovery. A molecule that was supposed to cycle through liver, bile, small intestine, ileum, and portal return is now leaking out of the loop.

The nutritional consequences can become substantial. A person may eat enough fat and still absorb less of it because the bile acid pool is no longer strong enough to support efficient luminal handling. Vitamins A, D, E, and K may become harder to absorb for the same reason, since they depend on bile acid supported micellar transport through the watery intestinal environment. Cholesterol handling changes. Stool chemistry changes. The meal's lipid phase becomes less usable. At that point the intestine is no longer only diseased as an absorptive surface. It has become the site where the bile acid economy is being broken.

So intestinal disease can become bile acid disease because the intestine is not only a recipient of bile acids. It is the organ that must return them. Damage the recovery site, and bile acids stop behaving like a reusable digestive tool and start behaving like a leaking resource. The colon gets exposed to excess detergent. The liver loses its return stream. The bile acid pool shrinks. Digestion pays the price.

How the Liver Senses and Replaces Bile Acid Losses

The liver does not wait passively to discover that bile acids are being lost. It reads that loss through the return stream. Under normal conditions, most bile acids released into the intestine are reabsorbed in the terminal ileum, returned through portal blood, taken back up by hepatocytes, and resecreted into bile. That returning flux is information. When plenty of bile acids come back, the liver reads that as abundance and restrains new synthesis. When fewer bile acids return, the liver reads scarcity and lifts part of that restraint, allowing bile acid production to rise. The point is not that the liver counts molecules one by one in some conscious way. The point is that bile acid homeostasis is governed by feedback. Return suppresses synthesis. Loss disinhibits synthesis. That is how a recirculating pool behaves when the organ managing it is built around chemical sensing rather than simple depletion.

A major part of that sensing loop begins in the ileum. When bile acids are reabsorbed there, they activate the nuclear receptor FXR, farnesoid X receptor, inside ileal enterocytes. FXR then drives production of FGF19, fibroblast growth factor 19, which enters the portal circulation and travels back to the liver as a hormonal signal reporting that bile acids were successfully recovered from the intestine. In hepatocytes, FGF19 signaling suppresses CYP7A1, cholesterol 7 alpha-hydroxylase, the rate-limiting enzyme in the classic pathway of bile acid synthesis from cholesterol. This is one of the body's main long-loop feedback mechanisms. Strong ileal recovery raises FGF19 and tells the liver to slow fresh bile acid production. Weak ileal recovery lowers that signal and allows synthesis to rise. In other words, the intestine does not merely send bile acids back. It sends a hormonal report about how much came back.

The liver also senses bile acid abundance more directly. Bile acids returning to the liver can activate hepatic FXR, which induces SHP, small heterodimer partner, a transcriptional regulator that also suppresses CYP7A1 expression. So the feedback system is layered. The ileum reports successful recovery through FGF19, and the liver reads returning bile acids directly through its own receptor signaling. Both mechanisms converge on the same practical outcome: when the returning pool is large, de novo synthesis is pushed down. When return falls, that suppression weakens and the liver is permitted to make more bile acids from cholesterol. This is how the organ avoids both wasteful overproduction and uncontrolled depletion.

Replacement therefore is not a dramatic emergency response. It is a graded increase in synthesis when losses rise beyond the usual small fecal leak. The body normally loses only a fraction of the bile acid pool in stool, and the liver replaces that fraction through new synthesis. If ileal recovery drops, more bile acids are lost downstream, portal return falls, FGF19 signaling falls, hepatic suppression of CYP7A1 eases, and bile acid synthesis increases to refill the pool as much as possible. But that compensation has limits. The liver can increase production, but it cannot always fully overcome large or chronic losses caused by ileal disease, ileal resection, or severe interruption of enterohepatic circulation. When loss outpaces replacement, the bile acid pool shrinks and fat digestion begins to suffer. So the sensing system is real and adaptive, but it is not magic.

This is the deeper logic of the system. The liver senses bile acid loss by reading diminished return, both as reduced bile acid flux coming back in portal blood and as reduced ileal FXR-FGF19 signaling. It replaces what it can by increasing synthesis from cholesterol, mainly through lifting suppression of CYP7A1. The intestine therefore is not just where bile acids are used. It is where the main report about bile acid sufficiency is generated. The liver then turns that report into synthesis. Loss in the stool becomes a change in ileal signaling. Ileal signaling becomes a change in hepatic enzyme expression. That is how a digestive detergent pool is kept in balance across repeated meals.

The Liver-Intestine Feedback Loop

The liver and intestine do not relate to each other as one organ making a fluid and another organ merely receiving it. They operate as a loop. The liver synthesizes bile acids from cholesterol, conjugates them, secretes them into bile, and sends them into the intestine. The intestine then uses those bile acids to solve the physical problem of fat digestion in a watery lumen, allowing emulsification, micelle formation, and efficient transport of lipid digestion products toward the absorptive surface. But the loop does not end there. Most bile acids are reclaimed in the terminal ileum, returned through portal blood, taken back up by the liver, and resecreted. This means the intestine is not only a place where bile acids do work. It is also the place where their return is measured. The liver is not only the source of bile acids. It is also the organ waiting to hear how much of its outgoing stream came back. That is the real meaning of the liver-intestine feedback loop. Secretion outward. Use in the lumen. Recovery in the ileum. Return through portal blood. Adjustment of new synthesis in the liver.

What makes this a true feedback system rather than simple recirculation is that the intestine reports bile acid recovery hormonally as well as physically. When bile acids are reabsorbed in the ileum, they activate intracellular sensing pathways in enterocytes, the absorptive cells lining the intestine. One of the main results is production of fibroblast growth factor 19, or FGF19, a hormonal signal that enters portal blood and travels back to the liver. That signal tells hepatocytes that bile acids were successfully recovered downstream, which means the circulating bile acid pool is being preserved and new synthesis can be restrained. At the same time, the liver is reading the direct return of bile acids in portal blood as its own measure of abundance. So the intestine is not only handing molecules back. It is sending a message about the state of the pool. Strong return means slow synthesis. Weak return means release the brake and make more. This is why the loop is elegant. The same molecules that did digestive work in the lumen help generate the report that shapes tomorrow's bile.

That feedback matters because digestion would otherwise become wasteful and unstable. If the liver secreted bile acids without regard for how many were being lost, the body would either overspend cholesterol and synthetic capacity making unnecessary replacement or undersupply future meals with too little detergent support. The intestine prevents that by acting as the measuring station. If bile acid recovery is strong, the liver knows the existing pool is intact and can rely more heavily on reuse. If ileal disease, resection, inflammation, rapid transit, or other disruption reduces recovery, the feedback signal weakens, portal return falls, and the liver responds by increasing bile acid synthesis as much as it can. That increase is not infinite, but it is adaptive. The liver is listening to the intestine and adjusting output based on what the intestine reports back. This is why bile acid physiology cannot be taught honestly as a liver subject alone or an intestinal subject alone. The pool only remains viable because both organs are in constant chemical conversation.

The consequences of breaking this loop are predictable. If the intestine stops reclaiming bile acids properly, more detergent spills into the colon, where it can provoke diarrhea, while less returns to the liver, where the shrinking pool begins to threaten later fat digestion and fat-soluble vitamin absorption. If the liver cannot respond properly to diminished return, replacement falters and the loop weakens further. The system therefore depends on both outbound competence and return competence. The liver sends the tool. The intestine uses it and reports how much survived the trip. The liver then decides how much new bile acid to make. That is the feedback loop in its cleanest form. One organ writes the outgoing stream. The other reads its use and sends the answer back.

Chapter 21

Bilirubin and the Logic of Jaundice

Heme Breakdown and Bilirubin Formation

Bilirubin begins as a disposal problem created by the normal turnover of heme. Heme is the iron-containing ring structure that gives hemoglobin its oxygen-binding function inside red blood cells, and it also appears in other proteins, but the largest day-to-day bilirubin burden comes from the breakdown of aging erythrocytes, red blood cells, after they have reached the end of their usable life. These cells are removed mainly by macrophages in the spleen, liver, and bone marrow. Once the cell is dismantled, globin protein is broken down into amino acids, iron is reclaimed for reuse, and the heme ring is opened enzymatically. That ring opening is the decisive event. It turns a functional oxygen-handling structurc into a wastc prccursor that now has to bc movcd through the body without causing trouble. Bilirubin is therefore not a dietary pigment drifting in from the outside world. It is an endogenous breakdown product, meaning the body makes it itself as part of ordinary internal turnover.

The first major step is conversion of heme into biliverdin by heme oxygenase. This reaction breaks the porphyrin ring, releases iron, and generates carbon monoxide as a byproduct. Biliverdin is then rapidly reduced to bilirubin by biliverdin reductase. At this point the molecule is hydrophobic and poorly water-soluble. That feature matters immediately because it means bilirubin cannot simply dissolve freely in plasma and travel harmlessly to the kidneys for easy urinary disposal. The bloodstream is a water-based transport environment, and unconjugated bilirubin does not fit it comfortably. The body solves that first transport problem by binding bilirubin to albumin in blood. Albumin acts as a carrier, keeping bilirubin mobile enough to reach the liver while limiting its free diffusion. This is the first clue to the logic of jaundice. The problem is never only how much bilirubin exists. The problem is always also what form it is in and whether the body has successfully moved it to the next stage.

That next stage is hepatic uptake and processing, but bilirubin formation itself already explains why the liver has to be involved. Heme breakdown is continuous. Red blood cells are always aging. Macrophages are always dismantling some fraction of them. The body therefore generates a steady bilirubin load whether a person has eaten or not, whether digestion is active or not, whether the individual is healthy or ill. This makes bilirubin a constant housekeeping burden rather than an occasional toxin. The liver stands downstream of that burden because the pigment formed from heme breakdown is not ready for excretion when it first appears in blood. It is albumin-bound, water-insoluble, and chemically incomplete from an elimination standpoint. Bilirubin formation creates the waste. Hepatic conjugation and biliary secretion solve the waste. Jaundice develops when some part of that sequence starts backing up.

The chemistry of formation also explains why bilirubin is tied so tightly to blood-cell turnover. If red cell breakdown increases, more heme is liberated, more biliverdin is formed, and more bilirubin enters circulation before the liver ever has a chance to handle it. In that situation the liver may be normal yet still face a higher incoming burden than usual. This is one reason bilirubin is such a useful molecule for thinking mechanistically. It links hematologic turnover, macrophage processing, plasma transport, hepatic uptake, conjugation, bile secretion, intestinal metabolism, and visible pigmentation all in one sequence. The body is constantly dismantling old oxygen carriers, reclaiming iron, and handing the leftover ring structure forward as bilirubin. That pigment has to keep moving. The moment it stops moving correctly, the yellow logic of jaundice begins.

Unconjugated Versus Conjugated Bilirubin

Bilirubin changes meaning as it changes form. That is why the distinction between unconjugated and conjugated bilirubin matters so much. The molecule begins as a breakdown product of heme, mostly from aging red blood cells. Once macrophages dismantle heme, bilirubin is produced in an unconjugated form. In that state it is hydrophobic, poorly water-soluble, and not ready for direct excretion. It cannot simply dissolve into plasma and drift harmlessly toward the kidneys. The body solves that first transport problem by binding unconjugated bilirubin to albumin in blood. Albumin acts as the carrier that keeps the pigment mobile enough to reach the liver, but this is still an early transport state, not a disposal state. Unconjugated bilirubin is therefore bilirubin before hepatic finishing. It is moving toward the liver, not moving out of the body.

Conjugated bilirubin is what happens after hepatocytes take that incoming pigment up and chemically modify it. Inside the hepatocyte, bilirubin is linked to glucuronic acid, a conjugation step that changes the molecule's solubility and transport behavior. This makes bilirubin far more water-compatible and prepares it for secretion into bile. That shift matters because the liver is not merely holding bilirubin for later. It is changing the chemistry of the molecule so it can enter the biliary stream and leave through the intestine. Unconjugated bilirubin is blood-bound transport on the way in. Conjugated bilirubin is excretory-ready pigment on the way out. The same molecule is present, but the physiological meaning has changed completely.

This difference in solubility is the key to everything that follows. Unconjugated bilirubin is albumin-bound and poorly suited to renal excretion. It stays mainly in the blood compartment unless the liver takes it up. Conjugated bilirubin, by contrast, is much more water-soluble. That means if it accumulates abnormally in blood because bile flow is blocked or hepatocyte secretion is failing, the kidney can filter some of it into urine. That is why conjugated hyperbilirubinemia can darken the urine, while unconjugated bilirubin does not behave the same way. The kidney is not deciding whether bilirubin deserves excretion morally. It is only able to handle what the chemistry of the pigment allows it to carry.

This is also why the two forms point to different kinds of failure. If bilirubin is elevated mainly in unconjugated form, the problem often sits earlier in the pathway. Too much bilirubin may be arriving from increased heme breakdown, or the liver may be failing to take it up or conjugate it efficiently. If bilirubin is elevated mainly in conjugated form, the liver has usually performed the conjugation step, but the route out into bile is impaired or overwhelmed. The pigment has been made excretable, but it has not been excreted. It is backing up into the wrong compartment. So unconjugated bilirubin points more strongly toward upstream load or failed processing. Conjugated bilirubin points more strongly toward secretion failure or obstructed outflow.

The cleanest way to hold the distinction is this. Unconjugated bilirubin is the hydrophobic incoming waste form, carried to the liver on albumin because it is not yet ready to leave the body. Conjugated bilirubin is the liver-processed outgoing form, chemically altered for biliary excretion and, if abnormal reflux into blood occurs, more able to appear in urine. One form belongs to transport toward the liver. The other belongs to excretion through the liver. That is the full logic behind the color, the labs, and the disease patterns that follow.

Canaliculi and Bile Duct Flow

Bile does not begin in a duct. It begins between hepatocytes. That is the first image the reader needs to hold. The canaliculus is the smallest visible channel of bile flow, a narrow passage formed by the apical, bile-facing membranes of adjacent hepatocytes. Apical here means the membrane domain specialized for secretion into bile rather than exchange with blood. These channels are not lined at their origin by a separate duct cell the way a larger bile duct is. They are carved directly out of hepatocyte polarity. Tight junctions seal their edges so bile stays in its own compartment and does not leak backward into the sinusoidal side of the tissue. This matters because the liver is running two opposite traffic systems through the same parenchyma. Blood is moving through sinusoids toward the central vein. Bile is being secreted into canaliculi and moving away from the hepatocyte field toward the portal region. Canaliculi are the first proof that hepatocytes are not merely changing chemistry inside themselves. They are building an outgoing stream.

That stream begins one transporter at a time. Hepatocytes actively pump bile acids, conjugated bilirubin, phospholipids, cholesterol, and other biliary constituents across the canalicular membrane. Water and electrolytes follow in response to the osmotic pull created by this solute movement, so bile flow is not simply liquid already sitting there waiting to drain. It is secretion-driven flow. The canaliculus fills because hepatocytes keep writing bile into it. Once formed, bile moves through an interlacing canalicular network toward the canals of Hering and then into progressively larger intrahepatic bile ductules lined by cholangiocytes, the epithelial cells of the biliary tree. At that point the fluid is no longer only hepatocyte-made in its final character. Cholangiocytes begin modifying it, especially by adding bicarbonate-rich fluid and reshaping water content. So bile duct flow is a two-stage event. Hepatocytes generate the core biliary stream. Duct cells adjust and convey it.

The direction matters because canalicular flow is not passive drainage down an open pipe. It depends on intact hepatocyte polarity, functioning ATP-dependent transporters, preserved tight junctions, and an open downstream path. ATP, adenosine triphosphate, is the cell's immediately usable energy currency, and bile secretion spends it heavily. If hepatocytes lose energy, lose membrane organization, or lose transporter placement, canalicular secretion slows even before a large duct is blocked. If downstream ducts narrow, inflame, scar, or obstruct, pressure rises backward into the canalicular system and the same stream begins to stall from the opposite end. This is why bile flow has to be understood from canaliculus to duct rather than from duct alone. A person can have cholestasis because the hepatocyte cannot push bile into the canaliculus properly, because the small duct system cannot carry it, or because the larger ducts outside the liver are obstructed. The failure point changes, but the traffic logic stays the same.

This is also why canaliculi and bile ducts belong in the same sentence but not as if they were the same structure. The canaliculus is the birth point of bile flow, built from hepatocyte membranes themselves. The bile duct is the carrying system that receives that secretion and moves it onward toward storage in the gallbladder or delivery into the intestine. One is formed by the working cell. The other is lined by duct epithelium built for transport and modification. The liver works because those structures join cleanly. Bile is secreted into the smallest channels, gathered into larger ducts, adjusted along the way, and then sent outward as a coherent stream. Break the seal, the polarity, the transporter function, or the outflow path, and the stream stops being useful. It becomes retained burden. That is the deeper point. Canaliculi and bile duct flow are not anatomy trivia. They are the architecture that turns hepatic secretion into digestion and excretion.

Hepatic uptake, conjugation, and secretion into bile

Excretion is not simply a matter of deciding that a molecule is waste. The body still has to move that molecule through a fluid system that can actually carry it out. That is where solubility becomes decisive. A compound that dissolves easily in water can often remain mobile in plasma, pass through renal filtration more cleanly, and move into urine without requiring an entirely separate transport strategy. A compound that is poorly water-soluble faces a different problem. It may travel awkwardly in blood, require carrier proteins, precipitate more easily, bind tightly to structures that delay clearance, or fail to enter the kidney's excretory pathway in a useful way. The body is therefore not only sorting good from bad or useful from waste. It is sorting what kind of fluid environment a molecule can survive in on its way out.

Bilirubin is one of the clearest examples. When heme is broken down, bilirubin is formed in an unconjugated state that is hydrophobic and poorly compatible with water. It does not circulate as a freely dissolved waste that the kidneys can simply wash away. Instead it travels bound to albumin, which keeps it mobile enough to reach the liver but does not solve the underlying excretion problem. Albumin-bound unconjugated bilirubin is transportable, but it is not yet readily disposable. The liver has to change its chemistry first. Conjugation adds chemical groups, most importantly glucuronic acid, that make bilirubin much more water-compatible and easier to secrete into bile. That transformation is not optional refinement. It is what turns an awkward blood-borne pigment burden into something that can actually leave the body. Solubility determines route, and route determines whether clearance happens at all.

The same logic applies beyond bilirubin. Cholesterol is another molecule whose poor water solubility shapes its fate. It cannot simply circulate naked in plasma or be filtered into urine in meaningful quantity. It needs lipoproteins for blood transport, and when the body wants to remove some of it, the liver often has to secrete it into bile directly or convert it into bile acids first. In other words, the body chooses the biliary route in part because the chemistry of the molecule is not well suited to aqueous renal disposal. Water-insoluble compounds often need a detergent-rich outgoing stream, which bile provides. Water-soluble compounds can often be prepared for urinary loss. The body is always asking the same question. In what fluid can this molecule travel safely and effectively on its way out.

This is why conjugation matters so much in liver physiology. Conjugation is the process of attaching another chemical group to a molecule so its behavior changes. Sometimes the goal is greater water solubility so the compound can remain dissolved in blood and be cleared by the kidneys. Sometimes the goal is compatibility with biliary secretion. In both cases the issue is not moral but physical. The body cannot excrete a molecule simply because it wants to. It has to move that molecule through plasma, bile, urine, or intestinal contents in a form that stays transportable and does not create a new precipitation or retention problem along the way.

So solubility determines excretion because excretion is a transport problem before it is a disposal success. A molecule has to fit the fluid that will carry it out. If it does not, the liver or another organ has to change its chemistry or choose a different route. Waste is not removed by intention. It is removed by compatibility. Hepatic uptake, conjugation, and secretion into bile

What Jaundice Really Means Mechanistically

Jaundice means bilirubin has escaped the sequence that was supposed to move it from blood to bile to intestine. It is not fundamentally a skin-color issue. The yellow color is the visible consequence of a transport and excretion failure that has allowed bilirubin to accumulate in the bloodstream and then deposit into tissues, especially those rich in elastin such as the sclera of the eye. To understand jaundice mechanistically, the reader has to see bilirubin as a moving burden with multiple stages. It begins with heme breakdown, mostly from aging red blood cells. Unconjugated bilirubin is formed, carried through blood bound to albumin, taken up by hepatocytes, conjugated into a more water-compatible form, secreted into bile, moved into the intestine, and then further metabolized and eliminated. Jaundice means that one of those steps has slowed, failed, or been overwhelmed badly enough that bilirubin concentration in blood has risen above what tissue can ignore.

The key is that bilirubin does not behave the same way in every stage. Unconjugated bilirubin is poorly water-soluble and travels bound to albumin. It cannot simply spill into urine in meaningful quantity because its transport state and solubility are wrong for that route. Once the liver conjugates it, usually by attaching glucuronic acid, bilirubin becomes much more suitable for secretion into bile and, if it refluxes abnormally into blood, much more able to appear in urine. This is why jaundice is not one mechanism with one cause. It can begin before the liver, inside the liver, or after the liver. If red blood cell breakdown is excessive, the incoming unconjugated bilirubin load may rise beyond what the liver can process even if hepatocytes are structurally normal. If hepatocytes cannot take bilirubin up, conjugate it, or secrete it properly, bilirubin handling fails inside the liver itself. If bile flow is obstructed, conjugated bilirubin that was already prepared for excretion cannot move forward and begins backing up into blood instead. The yellow color looks similar. The traffic failure is different.

That is why jaundice should be read as a problem of flow, form, and location. Flow asks whether bilirubin is moving from blood into hepatocytes and from hepatocytes into bile. Form asks whether the bilirubin burden is still unconjugated, meaning earlier in the pathway and less water-compatible, or already conjugated, meaning prepared for biliary excretion but now trapped or misdirected. Location asks where the blockage or overload is happening. Too much heme breakdown upstream creates one pattern. Hepatocyte dysfunction creates another. Bile duct obstruction creates another. Once bilirubin rises in plasma, however, the body begins displaying the result at the tissue level. The pigment diffuses into visible structures, the sclera yellows, the skin can yellow, urine may darken if conjugated bilirubin is being filtered by the kidneys, and stool may pale if less bilirubin is reaching the intestine to be metabolized into the pigments that normally color feces. So jaundice is not merely bilirubin present. It is bilirubin present in the wrong compartment because the normal route out has been interrupted.

The cleanest way to think about jaundice is this. The body is always dismantling heme and creating bilirubin. The liver is always supposed to catch that burden, change its chemistry, and push it into bile. The intestine is supposed to receive it and carry the sequence forward toward elimination. Jaundice means that this chain of custody has broken. The pigment remained in circulation long enough, and in high enough concentration, to stain the body visibly. The yellow color is the symptom. The mechanism is failed movement.

Prehepatic, Hepatic, and Posthepatic Patterns of Bilirubin Elevation

Bilirubin elevation is easiest to understand when the reader stops treating jaundice as one condition and starts asking where the traffic problem occurred. Bilirubin has a route. It is formed from heme breakdown, carried in blood to the liver in an unconjugated form, taken up by hepatocytes, conjugated into a more water-compatible form, secreted into bile, and then moved through the biliary tree into the intestine. If the burden rises before it reaches the liver, the pattern is prehepatic. If the handling fails inside hepatocytes, the pattern is hepatic. If bilirubin has already been processed by the liver but cannot get out through the biliary route, the pattern is posthepatic. Those three categories are not arbitrary diagnostic boxes. They are location statements. They tell you where along the bilirubin pathway the movement broke.

Prehepatic elevation means the liver is being overwhelmed by incoming load before hepatocyte handling has even become the primary problem. The classic mechanism is increased red blood cell breakdown or increased heme turnover from some other source. More heme breakdown means more unconjugated bilirubin is formed in macrophages, more bilirubin enters blood bound to albumin, and more of that burden arrives at the liver for processing. The hepatocyte may be structurally normal and functionally capable, but if the inflow is high enough, blood bilirubin rises because production has outrun clearance capacity. The key feature here is that the bilirubin burden is predominantly upstream of hepatic conjugation. The liver is being asked to keep up with a larger supply problem. Since unconjugated bilirubin is poorly water-soluble and albumin-bound, urine bilirubin is usually not the defining feature of this pattern. The failure is not that bilirubin was conjugated and trapped. The failure is that too much arrived before that step could fully absorb the burden.

Hepatic elevation means the problem sits inside the liver cell or within the intrahepatic handling machinery itself. The hepatocyte may fail to take bilirubin up efficiently from blood. It may conjugate it poorly. It may conjugate it successfully but fail to secrete it cleanly into bile. Or multiple steps may be partly impaired at once. This is why hepatic bilirubin disorders often produce mixed patterns rather than a perfectly pure rise in only one form. The reader should picture a congested work cell. Bilirubin is arriving. The hepatocyte is supposed to receive it on the blood-facing side, process it, and dispatch it into bile on the canalicular side. If uptake falters, unconjugated burden remains in blood. If conjugation falters, the liver cannot turn bilirubin into a more transportable excretory form efficiently. If canalicular secretion falters, conjugated bilirubin begins backing up instead of moving forward. Hepatic patterns therefore reflect failure of intracellular transfer and directional secretion. The liver is no longer doing its middle job cleanly.

Posthepatic elevation means the liver has already done much of its work, but the route out has been obstructed. Bilirubin has been taken up, conjugated, and secreted into bile, yet bile flow is blocked or severely impaired somewhere in the extrahepatic or major intrahepatic biliary pathway. This is the pattern of obstruction, stasis, or serious downstream resistance. The problem is not mainly production and not mainly conjugation. The problem is exit. Conjugated bilirubin that was supposed to move into the intestine now accumulates upstream and refluxes back into blood. This is why posthepatic jaundice often comes with dark urine, because conjugated bilirubin is water-compatible enough to be filtered by the kidneys once it appears in blood abnormally, and pale stool, because less bilirubin is reaching the intestine to be converted into the pigments that normally color feces. The body is showing the direction of the blockage. Urine darkens because the blood now carries excretable conjugated pigment. Stool pales because the intestine is no longer receiving the biliary stream it was supposed to get.

These patterns matter because they force the reader to think mechanically instead of visually. Yellow skin alone does not tell you which compartment failed. Bilirubin rose, but the reason it rose depends on whether the burden was generated excessively upstream, mishandled by hepatocytes in the middle, or blocked after hepatic processing on the way out. Prehepatic means too much arriving. Hepatic means the processing cell is failing. Posthepatic means the route out is obstructed. That framework turns jaundice from a color into a map.

Chapter 22

Cholestasis and Obstructed Flow

What Cholestasis Is

Cholestasis means bile is no longer being formed, secreted, or moved in the way the liver and biliary tree are built to require. The word is often treated as if it simply means blocked bile, but the real physiology is broader and more exact. Bile can fail at the level of hepatocyte secretion, where the cell is no longer exporting bile acids, bilirubin, phospholipids, cholesterol, water, and electrolytes into the canalicular stream properly. It can fail within the microscopic duct system, where flow through canaliculi and ductules becomes impaired. Or it can fail farther downstream, where larger ducts outside the liver are narrowed or obstructed. So cholestasis is not one single lesion. It is a state in which bile stops moving normally through an organ that depends on that movement for both digestion and excretion.

That dual dependence is what makes cholestasis so important. Bile is not a secondary fluid the liver can afford to retain casually. It is the outgoing stream that carries bile acids into the intestine for fat emulsification and micelle formation, and at the same time carries bilirubin, cholesterol, and other selected compounds outward for elimination. Once bile flow slows or stops, both functions begin to fail together. The intestine receives less detergent support for fat handling. The liver begins retaining compounds it had already decided should leave. That is why cholestasis is not only a digestive problem and not only an excretory problem. It is both at once. The same stalled stream that reduces fat-soluble vitamin absorption also causes bilirubin and bile acids to accumulate in the wrong compartments.

At the microscopic level, cholestasis is a problem of directional secretion and pressure. Hepatocytes are polarized cells. One surface faces blood. Another faces the bile canaliculus, the tiny channel formed between adjacent hepatocytes where bile begins. That means bile flow depends on ATP-driven transporters, intact canalicular membranes, tight junctions, and an open downstream path. ATP, adenosine triphosphate, is the cell's immediately usable energy currency. If hepatocytes lose the energy or membrane organization required to pump bile constituents into the canaliculus, cholestasis begins even without a stone or duct obstruction. If the canalicular system is intact but pressure rises downstream because ducts are narrowed or blocked, bile can no longer move forward efficiently and begins backing up. So cholestasis can begin from failed secretion, failed channel integrity, or failed drainage. Different starting points. Same final problem. Bile is not exiting on schedule.

That retention is dangerous because bile is chemically useful in one location and chemically injurious in another. Bile acids belong in the intestinal lumen, where their detergent properties help digest fat. They do not belong accumulating inside hepatocytes or bathing biliary epithelium at abnormal concentrations. Conjugated bilirubin belongs moving outward through bile, not refluxing into blood. Cholesterol belongs in a stable biliary solution or being converted into bile acids, not precipitating under stagnant conditions. Once flow stalls, the fluid that was supposed to serve digestion and elimination becomes part of the injury. The liver is not only losing an outgoing stream. It is becoming exposed to its own retained detergent and waste burden.

This is why cholestasis produces such a characteristic pattern of consequences. Bilirubin rises because conjugated pigment can no longer leave cleanly and begins moving back into blood. Urine darkens because conjugated bilirubin is now water-compatible enough to be filtered by the kidneys. Stool may pale because less bilirubin reaches the intestine to be converted into normal fecal pigment. Fat digestion worsens because less bile acid reaches the duodenum. Absorption of vitamins A, D, E, and K can decline because micelle formation weakens. Itching develops in many cholestatic states because bile-related compounds are being retained systemically rather than cleared. These are not disconnected symptoms. They are the visible consequences of one stream failing to move.

So cholestasis is best understood as impaired bile traffic, whether from failed formation, failed secretion, failed microscopic flow, or failed ductal drainage. The liver keeps making the decision that compounds should leave, but the route by which that decision becomes real is narrowing, stalling, or collapsing. Once that happens, the body loses digestive support downstream and accumulates biliary burden upstream. That is the logic of cholestasis. The stream meant to leave no longer leaves.

Intrahepatic Versus Extrahepatic Cholestasis

Intrahepatic and extrahepatic cholestasis are separated by location, but the deeper difference is where the bile stream is failing. Intrahepatic cholestasis means the problem is inside the liver itself, either at the level of hepatocytes, which form and secrete bile, or within the small bile ducts embedded in hepatic tissue. Extrahepatic cholestasis means bile has been formed and sent toward the larger ducts, but an excretory block develops outside the liver in the extrahepatic biliary tree. In both cases bile flow is reduced or stopped. In both cases bilirubin and bile constituents can back up. But the failed compartment is different, and that difference changes how the mechanism should be read.

Intrahepatic cholestasis is a secretion and small-duct problem before it is a plumbing problem. The hepatocyte may fail to move bile acids, bilirubin, phospholipids, cholesterol, water, and electrolytes into the canalicular stream correctly, or the intrahepatic ducts may be inflamed, destroyed, or functionally narrowed. This is the logic behind disorders such as primary biliary cholangitis, where progressive destruction of intrahepatic bile ducts leads to cholestasis and eventual cirrhosis. The route outward is failing inside the organ that makes the fluid. The traffic jam begins in the tissue itself.

Extrahepatic cholestasis is different because it is fundamentally a downstream drainage problem. The liver may still be conjugating bilirubin and secreting bile, but the larger ducts outside the liver cannot carry that flow forward normally. Gallstones in the bile ducts, tumors of the bile ducts or adjacent structures, and other mechanical obstructions can create this pattern. Once the outlet narrows or blocks, conjugated bilirubin and bile acids accumulate upstream, reflux into blood, and the intestine receives less bile than it should. This is why extrahepatic cholestasis often reads like an obstructed exit. The liver made the decision to send compounds out. The duct system could not complete the order.

That location difference is what gives the two patterns their diagnostic logic. Intrahepatic cholestasis suggests a problem of hepatocyte function or small-duct injury within the liver. Extrahepatic cholestasis suggests a mechanical block beyond the liver in the larger biliary channels. Both can produce conjugated hyperbilirubinemia and the broader cholestatic picture, but they do not tell the same structural story. One says the liver cannot generate or move bile properly inside its own architecture. The other says bile was generated, then trapped on the way out.

Why Retained Bile Becomes Toxic to Tissue

Bile is useful because it is chemically aggressive in the right place. That is the beginning of the problem. Bile acids are detergents. They are meant to enter the intestinal lumen and interact with hydrophobic fat, breaking large lipid masses into smaller droplets and helping form micelles that carry lipid digestion products through a watery environment. That same detergent property becomes dangerous when bile does not leave properly and begins accumulating inside hepatocytes, canaliculi, and bile ducts. A molecule built to disrupt lipid organization in the gut is not harmless when it remains trapped against cellular membranes in the liver. Retained bile is therefore not merely stagnant fluid. It is active chemical burden sitting in the wrong compartment.

The toxicity begins at the membrane. Hepatocytes and cholangiocytes, the epithelial cells lining bile ducts, depend on intact lipid membranes to preserve polarity, transporter placement, ion gradients, and compartment separation. Bile acids at excessive intracellular or canalicular concentration can disturb those membranes directly, increasing permeability, impairing transporter function, and making the boundary between bile space and cell interior harder to maintain. Once that starts, the cell loses one of its basic organizational powers, which is keeping the blood-facing world, the cytoplasmic world, and the bile-facing world distinct. A detergent retained in the secretion pathway begins eroding the very surfaces that were supposed to secrete it outward.

Mitochondria also begin paying the cost. A hepatocyte already carries a heavy metabolic workload, and cholestatic retention adds a second burden on top of it. Bile acid accumulation can increase oxidative stress, disturb mitochondrial membranes, impair ATP production, and push the cell toward apoptosis or necrosis depending on the severity and duration of injury. ATP, adenosine triphosphate, is the cell's immediately usable energy currency, and the liver spends it constantly to maintain membrane transport, synthesize proteins, preserve polarity, and drive bile secretion itself. Once retained bile starts lowering that energy reserve, the cell becomes even less able to pump bile outward, which creates a vicious loop. Retention causes injury. Injury impairs secretion. Impaired secretion worsens retention.

This is why cholestatic injury often deepens even after it has already begun. Bile acids accumulate. Membranes become more vulnerable. Oxidative stress rises. ATP falls. Transporters fail. Canalicular flow weakens further. Conjugated bilirubin and cholesterol-related burden remain trapped in the same system. Cholangiocytes become injured as well, and the ductal tree begins to pay for a fluid it was built to carry, not to hold under rising pressure and abnormal concentration. The tissue is no longer simply failing to excrete. It is being exposed to its own outgoing chemistry as a local toxin. Bile was supposed to move. Once it stops moving, its usefulness turns against the liver.

Inflammation follows naturally from that retained burden. Injured hepatocytes release danger signals. Kupffer cells detect tissue stress and amplify cytokine signaling. Stellate cells are more likely to activate, produce collagen, and begin fibrotic remodeling. Cytokines are signaling proteins that tell nearby cells the tissue is under threat. So bile retention does not only injure cells directly. It converts a secretion problem into an inflammatory and architectural problem. The liver begins rebuilding around the retained burden, and that rebuilding usually makes flow worse rather than better. A chemical traffic problem becomes a tissue remodeling problem.

The toxic logic is therefore simple. Bile belongs in motion and in the lumen of the gut. It does not belong stored under pressure inside hepatocytes and ducts at abnormal concentration. Its detergent chemistry, its solute load, and its dependence on intact membrane transport all make it a poor substance to retain. When bile cannot leave, the liver is no longer just missing an excretory route. It is bathing its own tissue in a fluid that was designed to act elsewhere.

Dark Urine, Pale Stool, Itching, Fat Malabsorption

These four findings belong together because they are all different expressions of one failed stream. Bile is supposed to leave the liver, move through the ducts, enter the intestine, carry conjugated bilirubin outward, and deliver bile acids to the lumen for fat digestion. When that stream is obstructed or badly slowed, the chemistry that should have gone into the gut is redistributed into the wrong compartments. Conjugated bilirubin begins backing up into blood instead of moving into bile. Because conjugated bilirubin is water-compatible enough to be filtered by the kidneys, it starts appearing in urine in abnormal quantity. That is why the urine darkens. The kidney is not creating a new pigment problem. It is clearing a pigment that should have left through bile but was pushed into the bloodstream instead. At the same time, the intestine is now receiving less bilirubin. Normally bilirubin reaching the gut is metabolized by intestinal bacteria into downstream pigments, including stercobilin, which gives stool much of its normal brown color. If less bilirubin enters the intestine, less pigment is formed, and the stool becomes pale, clay-colored, or frankly light. Dark urine and pale stool are therefore not random symptoms sitting beside jaundice. They are directional signs. The pigment left the bloodstream through the kidney because it could not leave through bile, and the intestine lost its normal pigment supply because the biliary route failed upstream.

Itching belongs to the same cholestatic logic, though the chemistry is more irritating than visible. When bile flow is impaired, bile acids and other biliary constituents are retained instead of being carried forward into the gut. Those retained molecules do not remain quiet in blood and tissue. They act as circulating pruritogens, meaning compounds that provoke itch, either directly or through signaling pathways that sensitize cutaneous nerve endings. Bile acids themselves are part of that retained burden, though the itch of cholestasis is not always explained by bile acids alone. Other retained biliary signals and mediators are also thought to contribute. The important physiological point is simpler and stronger than any single molecule. A fluid that was built to leave is now circulating abnormally, and the body begins feeling that mistake through the skin. The person is not itching because the skin became the original site of disease. The person is itching because bile chemistry has escaped its proper route and entered systemic sensory biology.

Fat malabsorption completes the picture because bile is not only an excretory stream. It is one of the main digestive tools required for handling dietary lipid. Bile acids are the detergent molecules that emulsify fat in the intestinal lumen and help form micelles, the small transport assemblies that carry fatty acids, monoglycerides, cholesterol, and fat-soluble vitamins through the watery intestinal environment toward the absorptive surface. If bile delivery falls, the meal's lipid phase becomes physically harder to digest and harder to absorb. Large lipid droplets remain less effectively dispersed. Pancreatic lipase loses surface access. Micelle formation weakens. Absorption of fat and vitamins A, D, E, and K falls. The meal may still contain the same fat. The intestine no longer extracts the same usable result. This is why cholestasis changes stool color and nutrient yield at the same time. Less bilirubin reaches the gut, so stool pales. Less bile acid reaches the gut, so fat handling worsens.

So dark urine, pale stool, itching, and fat malabsorption are best read as a single physiological sentence written in four dialects. The liver conjugated bilirubin, but the biliary route failed, so the kidney had to clear some of that pigment instead and the urine darkened. The intestine lost bilirubin input, so stool lost pigment and became pale. Retained biliary compounds circulated abnormally, so the skin began to itch. The lumen received too little bile acid, so lipid digestion and absorption weakened. One blocked stream. Four different tissues forced to display the consequence.

The Chemistry Behind Cholestatic Symptoms

Cholestatic symptoms are not a random cluster of discomforts attached to liver disease. They are the visible and sensory consequences of bile chemistry being trapped, misdirected, or withheld from the place it was supposed to go. Under normal conditions, hepatocytes secrete bile acids, conjugated bilirubin, cholesterol, phospholipids, water, and electrolytes into the canalicular stream, that stream moves through the biliary tree, and the intestine receives it as both digestive support and waste disposal. In cholestasis, that movement slows or fails. The result is two simultaneous chemical distortions. Compounds that should have exited through bile accumulate in liver tissue and blood, while the intestinal lumen is deprived of the detergent-rich fluid it needs for normal fat handling. Symptoms then appear wherever those two distortions become biologically obvious. The blood shows one set of consequences. The intestine shows another. The skin, urine, and stool simply reveal where the chemistry went wrong.

Jaundice and dark urine come from conjugated bilirubin failing to leave by its normal route. Bilirubin begins as a heme breakdown product, arrives at the liver bound to albumin in an unconjugated state, is taken up by hepatocytes, and then conjugated into a more water-compatible form so it can be secreted into bile. In cholestasis, that conjugated bilirubin is made, but forward flow into the intestine is impaired. Pressure builds backward. Conjugated pigment refluxes into blood instead of continuing into bile. Once it is in blood, the kidney can filter it because the conjugated form is sufficiently water-soluble to enter urine. That is why the urine darkens. The pigment has been rerouted into a fluid that was never meant to be its primary exit. At the same time, less bilirubin reaches the intestine, which means less is converted by gut bacteria into the downstream pigments that normally color stool. Pale stool is therefore not just a digestive curiosity. It is proof that biliary pigment never made it into the bowel in adequate amount.

Itching comes from retained biliary chemistry acting systemically rather than luminally. Bile acids are useful in the intestine because they behave like detergents and help manage the fat-water interface. In blood and tissue, that same class of molecules becomes irritant chemistry. Cholestatic pruritus, itching caused by cholestasis, is not explained by a simple skin rash or superficial inflammation. It reflects circulating pruritogenic burden, meaning molecules retained during cholestasis that can activate itch pathways. Bile acids are part of that burden, though not necessarily the only molecules involved. Other retained biliary constituents and signaling mediators likely contribute as well. The important principle is that the body is now carrying in blood what it meant to confine to bile. The person begins feeling bile retention through the skin because the bloodstream has become the wrong transport medium for compounds that were supposed to move outward into the intestine.

Fat malabsorption follows because the intestine has lost one of its main chemical tools. Bile acids are needed to emulsify dietary fat and to build micelles, the small transport assemblies that keep fatty acids, monoglycerides, cholesterol, and fat-soluble vitamins mobile in the watery intestinal lumen. Without adequate bile delivery, large lipid droplets remain less effectively dispersed, pancreatic lipase loses surface access, and the products of fat digestion move less efficiently toward the absorptive membrane. The chemistry of the meal changes. The same plate of food now yields less usable fat, less cholesterol uptake, and poorer absorption of vitamins A, D, E, and K. This can lead to steatorrhea, fat-rich stool, because lipid that should have been digested and absorbed stays in the intestinal contents and leaves the body instead. A cholestatic person is therefore not only retaining the wrong compounds internally. They are failing to extract the right compounds from the meal.

This is why cholestatic symptoms should be read as one integrated chemical pattern. Conjugated bilirubin backs into blood and darkens urine. Too little bilirubin reaches the gut and stool pales. Retained biliary molecules circulate and itching begins. Too little bile acid reaches the lumen and fat absorption falls. The symptoms differ because the affected compartments differ, but the logic is unified. Bile was supposed to move. Once it stops moving, the body has to display that failure everywhere the misplaced chemistry now lands.

Why a Blocked Outflow Path Changes Digestion and Tissue Injury at the Same Time

A blocked outflow path changes digestion and tissue injury at the same time because bile is doing two jobs with one stream. It is the liver's outgoing digestive fluid and the liver's outgoing waste route. When bile moves normally, bile acids reach the duodenum and help emulsify dietary fat, build micelles, and support absorption of lipid and fat-soluble vitamins. At the same time, conjugated bilirubin, cholesterol, and other selected compounds are carried outward through that same biliary flow toward the intestine. Once the outflow path is blocked, both functions fail together. The intestine is deprived of the detergent chemistry it needs for the meal, and the liver begins retaining the very compounds it had already decided should leave. A single obstruction therefore creates two kinds of damage at once. The meal becomes less usable. The tissue becomes more burdened.

The digestive side is the easier part to picture. Fat enters a watery intestinal lumen and needs bile acids to behave properly in that environment. Without adequate bile delivery, large lipid droplets remain less effectively dispersed, pancreatic lipase loses surface access, and micelle formation weakens. Fatty acids, monoglycerides, cholesterol, and fat-soluble vitamins move less efficiently toward the absorptive membrane. The meal may still contain the same fat and the same vitamins, but the intestine cannot extract the same result. The person is now eating under impaired physical conditions of digestion. That is why cholestatic states can produce fat malabsorption, pale stool, and deficiency of vitamins A, D, E, and K. The bile stream did not merely fail as an excretory channel. It failed as a digestive tool.

The tissue injury side begins because bile is chemically safe only when it is moving through the compartments designed to carry it. Bile acids are detergents. Conjugated bilirubin is meant to be in transit, not retained. Cholesterol is meant to remain suspended within a stable biliary solution or be converted into bile acids, not accumulate behind a rising pressure gradient. When the ductal path is blocked, these compounds begin backing up into canaliculi, hepatocytes, and bile ducts. The liver is no longer only missing an exit. It is now being exposed to retained biliary chemistry at abnormal concentration. Membranes begin suffering because detergents that belong in the intestinal lumen are now pressing against cells inside the liver. Transporters fail more easily. ATP-dependent secretion becomes harder to maintain. ATP, adenosine triphosphate, is the cell's immediately usable energy currency. Once bile retention starts impairing energetic and membrane integrity, the hepatocyte becomes less able to clear the retained burden, which makes the obstruction even more injurious.

That is why obstruction is not a passive backup like water behind a dam. It is chemically active retention. Bile acids begin contributing to oxidative stress and membrane injury. Cholangiocytes, the epithelial cells lining the bile ducts, become exposed to a stagnant concentrated fluid they were meant to carry, not hold under pressure. Hepatocytes begin refluxing conjugated bilirubin into blood because forward movement into bile is no longer possible. Kupffer cells read tissue stress. Stellate cells receive inflammatory cues and are more likely to activate. Fibrotic remodeling becomes easier to start because the tissue is now interpreting retained bile as chronic injury rather than ordinary traffic. So the blocked path does not merely stop a fluid. It transforms a useful outgoing stream into a local toxin.

This is why the symptoms point in two directions at once. Dark urine and jaundice reflect retained conjugated bilirubin being pushed back into blood and then filtered by the kidney. Pale stool reflects loss of bilirubin reaching the intestine. Itching reflects circulating biliary compounds that should have been cleared. Fat malabsorption reflects bile acids failing to reach the meal. These are not separate diseases occurring side by side. They are one blocked stream being felt in different tissues. The liver is injured because bile stayed. Digestion is impaired because bile did not arrive.

So a blocked outflow path changes digestion and tissue injury at the same time because bile was always serving both domains at once. The liver sends it outward to help digest fat and to remove compounds that should not remain in the body. Obstruct the path, and the intestine loses a digestive fluid while the liver retains a damaging burden. One stream fails. Two systems pay.

Part VI

The Liver as the Body's Protein Factory

Chapter 23

Albumin

What Albumin Is

Albumin is the liver's major plasma protein, the most abundant single protein dissolved in the liquid portion of blood, and one of the clearest examples of how the liver does not merely clean blood but helps build it. Hepatocytes synthesize albumin continuously and release it into the circulation, where it becomes part of the physical structure of plasma itself. That point matters because albumin is often described too casually as a carrier protein, which is true but incomplete. Albumin is also one of the main determinants of oncotic pressure, the protein-generated pulling force that helps keep water inside the vascular compartment instead of letting it drift too freely into surrounding tissue. In other words, albumin is not an optional blood accessory. It is part of what makes blood behave like blood rather than like diluted fluid leaking everywhere it can.

Structurally, albumin is a soluble protein built to stay in circulation and interact with many different small molecules at once. It is synthesized in hepatocytes on the rough endoplasmic reticulum, processed through the Golgi apparatus, and then secreted into the blood-facing side of the cell. Once in plasma, it serves as a molecular shuttle for compounds that do not travel cleanly on their own or that benefit from buffered transport. Fatty acids bind to albumin during fasting as they move from adipose tissue toward liver and muscle. Bilirubin in its unconjugated form travels bound to albumin because it is poorly water-soluble and cannot circulate freely in a safe, useful way. Many drugs also bind albumin to varying degrees, which changes how much of the drug remains free in plasma and how rapidly it distributes into tissues or is cleared. So albumin is both structural and logistical. It holds fluid where it belongs and carries molecules that would otherwise move less predictably.

Its role in oncotic pressure is where the physiology becomes visible. Blood plasma contains dissolved proteins, and albumin contributes a large share of the osmotic force generated by those proteins because it is present at high concentration and remains largely within the vascular space. Water moves according to solute forces, and albumin helps pull water inward toward the capillary lumen rather than allowing it to escape unchecked into the interstitial space, the fluid-filled compartment around cells. This does not mean albumin acts alone, because hydrostatic pressure, capillary permeability, sodium handling, lymphatic drainage, and vascular integrity all matter. But albumin is one of the main reasons plasma can retain water as a circulating fluid rather than surrendering it too easily to tissue spaces. When albumin falls significantly, fluid distribution changes. Edema becomes easier to generate. Ascites, fluid accumulation in the abdominal cavity, becomes easier to sustain. The body has not merely lost a protein. It has lost part of the force keeping vascular fluid where it belongs.

This is why albumin has to be understood as a synthetic product of the liver rather than as a general blood fact. Hepatocytes are constantly writing part of the physical character of plasma by making this protein. The liver is not just filtering what the blood already contains. It is manufacturing one of the molecules that determines how blood carries fluid and cargo through the body. A decline in albumin therefore is not only a lab abnormality. It is evidence that hepatic synthetic output has changed enough to alter vascular behavior. Blood chemistry and blood physics begin shifting together.

Albumin also helps reveal the broader logic of liver function. The same organ that takes up bilirubin from blood also makes the protein that carries unconjugated bilirubin before hepatic uptake. The same organ that receives fatty acids from adipose tissue also makes the protein that transports many of those fatty acids through plasma. The same organ that metabolizes drugs also makes one of the main proteins affecting how much of those drugs remain free in circulation. Albumin therefore sits at the junction of transport, fluid balance, and hepatic synthesis. It is one of the clearest proofs that the liver is not merely downstream from digestion and metabolism. It is upstream from the very composition of the blood that performs those tasks.

So what albumin is, at its core, is a liver-made plasma protein that stabilizes circulation. It keeps water in the vascular space more effectively, carries a wide range of molecules through blood, and reflects the liver's synthetic power every hour of the day. When albumin is present in the right amount, blood holds itself together better. When albumin falls, the body begins learning how dependent plasma was on hepatic protein manufacture all along.

Why It Matters for Fluid Balance

Albumin matters for fluid balance because blood is not just liquid moving through tubes. It is a compartment that has to hold water against the outward push of vascular pressure. Capillaries are constantly exposed to hydrostatic force, the physical pressure of blood pushing fluid outward toward the surrounding interstitial space, the fluid-filled environment around cells. If nothing pulled water back toward the vascular side, a much larger fraction of plasma water would leave the circulation and remain in the tissues. Albumin supplies a major part of that inward pull. Because it is abundant in plasma and largely retained within the vascular compartment, it generates oncotic pressure, the osmotic force created by proteins that attracts water toward the bloodstream. This is why albumin is not merely a transport protein floating in blood. It helps decide where body water stays.

The importance of that becomes obvious when albumin falls. The capillary wall is not a solid pipe. It allows continuous exchange of water and solutes, and fluid distribution across that wall depends on the balance between outward hydrostatic pressure and inward oncotic pressure. Albumin is one of the major reasons that balance does not tip too far toward tissue flooding. When albumin concentration drops, the vascular side loses part of its ability to retain water. Fluid leaves the capillary more easily and is less effectively drawn back. The result is edema, which is excess fluid in the interstitial space, or ascites, which is fluid accumulation in the abdominal cavity. The body has not necessarily gained new water overall. It has misplaced it. That is a more precise way to think about liver-related fluid retention. The problem is not only excess volume. It is loss of the protein force that helps keep existing volume in the right compartment.

This is especially important in liver disease because the liver is the organ that makes albumin. If hepatocyte synthetic function declines, albumin production falls, and plasma begins losing one of its main structural proteins. That shift does not stay inside the laboratory value. It changes the physical behavior of circulation itself. Water becomes easier to push out of the vascular space. Peripheral edema becomes more likely. Ascites becomes easier to sustain, especially when portal hypertension is also present and raising hydrostatic pressure on the venous side. This is why low albumin and portal hypertension are such a dangerous combination. One increases the pressure pushing fluid outward. The other weakens the force pulling fluid inward. The circulation begins losing its grip on its own water.

Albumin also matters because fluid balance is not a cosmetic detail. Once fluid leaves the bloodstream too easily, organ perfusion can become less stable even when the body looks swollen. A person can be edematous and still have an underfilled effective circulation, meaning the vascular system is not holding and distributing fluid with the same efficiency. That can trigger renal sodium and water retention through neurohormonal compensation, which worsens the swelling without solving the original compartment problem. The body reads poor effective circulatory filling and tries to defend it, but if albumin remains low, much of the retained fluid is still vulnerable to leaking into tissue spaces rather than restoring clean vascular stability. The result is a self-reinforcing pattern of fluid misplacement.

So albumin matters for fluid balance because it helps make plasma a water-holding compartment rather than just a watery stream under pressure. It provides a major share of the inward protein force that opposes capillary fluid loss, stabilizes vascular volume, and limits the tendency of water to collect in tissues and body cavities. When albumin falls, the body does not simply lose a blood protein. It loses part of the physical discipline that keeps fluid where circulation needs it most.

Oncotic Pressure Explained Clearly

Oncotic pressure is the inward pulling force created by proteins dissolved in plasma, especially albumin, that helps keep water inside the bloodstream instead of letting too much of it drift out into surrounding tissue. The easiest way to understand it is to stop thinking of blood vessels as sealed pipes and start thinking of capillaries as exchange surfaces. Water is always being pushed and pulled across those walls. Blood pressure inside the capillary creates an outward force, called hydrostatic pressure, that tends to push fluid into the interstitial space, the fluid-filled space around cells. Plasma proteins create the opposing force. Because large proteins such as albumin are concentrated within the vascular compartment and do not cross the capillary wall freely under normal conditions, they attract water back toward the bloodstream. That inward protein-driven pull is oncotic pressure.

The reason albumin matters so much is not that it is the only protein in plasma, but that it is present in high concentration and remains largely where the circulation needs it. Water moves toward solute, and albumin is one of the main solutes the body has positioned inside blood to keep vascular water from escaping too easily. So every moment the capillary is balancing two pressures. Hydrostatic pressure pushes fluid out. Oncotic pressure pulls fluid in. Normal circulation depends on neither side winning completely. Some fluid leaves the capillary to nourish tissues, and some is drawn back or later returned through lymphatic drainage. Oncotic pressure is one of the main reasons that outward leak does not become uncontrolled flooding.

This is why low albumin changes the body so visibly. If plasma protein concentration falls, especially through loss of albumin, the inward pull weakens. The capillary wall has not necessarily changed, and blood pressure may still be pushing fluid outward as usual, but now there is less force bringing that fluid back toward the vascular side. The result is edema, excess fluid in the tissues, or ascites, fluid collecting in the abdominal cavity. The body has not simply gained water. It has lost control over where that water stays. That is the clearest practical meaning of oncotic pressure. It is one of the forces that keeps circulating fluid circulating.

The concept becomes even more important in liver disease because the liver makes albumin. When hepatic synthesis falls, plasma oncotic pressure can fall with it. Then the circulation loses part of its inward holding force at the same time other pressures, especially portal and venous pressures, may be rising. Fluid is pushed out harder and pulled back less effectively. Once that happens, the person can look swollen and still have poor effective vascular filling, because water has shifted into the wrong compartment. Oncotic pressure therefore is not an abstract lab concept. It is one of the reasons blood can remain a stable fluid compartment rather than dissolving outward into tissue.

Drug Binding and Molecule Transport

Albumin matters for drug binding because blood is not just carrying molecules in a neutral dissolved state. Many compounds do not move through plasma as fully free agents. They spend part of their time bound to proteins, and albumin is one of the main proteins providing that transport surface. This is especially important for hydrophobic drugs, weak acids, and other small molecules that would otherwise distribute, precipitate, or clear differently if left entirely unbound in aqueous plasma. Binding does not mean the drug disappears. It means the drug enters a reversible transport relationship. Some fraction remains free and immediately available to cross membranes, interact with receptors, be metabolized, or be excreted. Another fraction remains protein-bound, which limits its immediate tissue access and helps stabilize its movement through circulation. Albumin is therefore not only carrying nutrients and endogenous wastes. It is helping shape the pharmacologic behavior of many compounds the body encounters.

This binding changes the meaning of concentration. A lab value may tell you how much total drug is present in plasma, but the biologically active fraction is often the unbound fraction, the portion not attached to albumin or another carrier protein. The free fraction is generally the part that can leave the bloodstream easily, cross into tissues, reach target receptors, be taken up by hepatocytes, or pass through the kidney for filtration. The bound fraction serves more like a circulating reserve. It is still in the body, but it is not all acting at once. That matters because albumin can slow rapid swings in free concentration by buffering the molecule in transit. A drug that is partly albumin-bound may stay in circulation longer, distribute more gradually, and release active free compound over time rather than hitting tissues all at once. The bloodstream is not merely delivering the drug. It is modulating its presentation.

This same logic applies to endogenous molecules as well. Unconjugated bilirubin travels bound to albumin because it is poorly water-soluble and cannot circulate safely in a free state. Free fatty acids released during fasting also use albumin as a transport scaffold because hydrophobic lipid cannot simply dissolve into plasma and travel cleanly on its own. Some hormones and signaling molecules depend on carrier proteins for similar reasons, though other proteins besides albumin often dominate those systems. The broader point is that albumin helps turn plasma into a workable transport medium for molecules that would otherwise move badly in water or exert their effects too abruptly if left entirely free. It is one of the reasons blood can carry chemically diverse cargo without dissolving into transport chaos.

Binding also matters because transport is not passive. A molecule bound to albumin is generally less able to cross membranes until it dissociates. That means albumin influences where compounds can go, how quickly they can get there, and how available they are for metabolism and excretion. The liver, which makes albumin, is therefore also helping shape the downstream handling of many of the same compounds it later metabolizes. A drug bound tightly to albumin may reach hepatocytes or renal filtration surfaces more gradually because only the free portion is immediately accessible. A drop in albumin can therefore change drug behavior without changing the total amount of drug present. The same dose now produces a higher free fraction, which can increase tissue exposure, pharmacologic effect, or toxicity. This is one reason low albumin matters clinically. The body has not only lost a protein. It has lost part of the buffering system that controls how fast certain molecules become biologically available.

So drug binding and molecule transport are part of albumin's deeper role in circulation. Albumin is not just carrying cargo from one place to another like a passive cart. It is shaping the timing, solubility, and compartment access of that cargo. It keeps poorly water-compatible molecules mobile, buffers the free concentration of many drugs, and helps plasma function as a controlled transport environment rather than a simple fluid container.

What Happens When Albumin Production Falls

When albumin production falls, the blood begins losing one of the proteins that helps make it a stable vascular compartment rather than a watery fluid under pressure. The decline may look like a number on a lab panel at first, but the real event is physical. Albumin contributes a large share of plasma oncotic pressure, the inward protein-driven force that helps keep water inside the circulation. Capillary hydrostatic pressure is always pushing fluid outward toward the interstitial space, the fluid-filled compartment around cells. Albumin helps oppose that outward push. Once hepatic synthesis falls and plasma albumin drops far enough, the bloodstream loses some of its ability to retain water. Fluid leaves the vascular space more easily, and less of it is drawn back effectively. The result is not simply low protein in the abstract. The result is fluid misplacement.

That misplacement becomes visible as edema and ascites. Edema means excess fluid in tissues. Ascites means fluid accumulating in the abdominal cavity. In liver disease, this often becomes especially pronounced because low albumin does not act alone. Portal hypertension may already be raising hydrostatic pressure on the venous side, which pushes even more fluid outward. The vascular compartment is now being squeezed from both directions. More pressure is forcing fluid out, and less oncotic force is pulling it back in. The circulation becomes less able to hold its own volume. A person can therefore become swollen and still have a poorly filled effective circulation, meaning the bloodstream is not carrying and distributing fluid with normal efficiency even though fluid is accumulating elsewhere. The body reads that underfilling and responds by retaining sodium and water through neurohormonal systems, which often worsens the edema without fixing the underlying compartment failure.

Albumin decline also changes how molecules move through blood. Many drugs, bilirubin, free fatty acids, and other compounds spend part of their time bound to albumin rather than moving entirely in a free state. Once albumin falls, the free fraction of some drugs can rise, which means the same total concentration may produce greater immediate tissue exposure or altered clearance. Unconjugated bilirubin transport becomes less buffered. Fatty acid handling in plasma becomes less orderly. The bloodstream has not merely lost a water-holding protein. It has lost one of its main transport scaffolds. This is one reason low albumin changes the behavior of the whole internal environment rather than only fluid distribution. Transport chemistry becomes less stable at the same time that vascular fluid balance becomes less stable.

A fall in albumin also tells you something about the liver itself. Albumin is synthesized continuously by hepatocytes. It does not vanish from plasma because one meal was missed or because digestion was poor for a day. When production falls, it often means hepatic synthetic capacity has declined enough to change the composition of plasma over time. That makes low albumin one of the clearest signs that liver disease is no longer only about inflammation, fat accumulation, or altered enzymes. It is now affecting the organ's ability to manufacture core structural proteins for the circulation. The liver is not just failing to clear waste. It is failing to build part of the fluid the body depends on.

So when albumin production falls, the consequences spread in two directions at once. Water leaves the bloodstream more easily, and molecular transport becomes less buffered. The tissues swell, the abdomen may fill, drug behavior may shift, and the circulation becomes less physically coherent. A low albumin level is therefore not a minor lab abnormality attached to liver disease. It is evidence that the liver has begun losing one of its most important whole-body construction jobs.

Why Edema and Ascites Often Reflect Synthetic Decline

Edema and ascites often reflect synthetic decline because the liver is one of the organs that helps make plasma capable of holding water where it belongs. Albumin is central to that job. Hepatocytes synthesize albumin continuously and release it into blood, where it contributes a large share of plasma oncotic pressure, the inward protein-driven force that helps keep fluid inside the vascular compartment. Capillary hydrostatic pressure is always pushing water outward into surrounding tissue. Albumin helps oppose that push. When hepatic synthetic function declines and albumin production falls, the circulation loses part of its inward holding force. Water begins leaving the bloodstream more easily and is less effectively pulled back. That is the beginning of edema in tissue and ascites in the abdominal cavity. The problem is not only extra fluid. It is weakened fluid discipline.

This is why edema and ascites are often less about total body water in the abstract and more about compartment failure. The body may still contain the same water, or later retain even more of it, but once albumin falls, more of that water shifts out of the vascular space and into places where it is less useful. Interstitial tissues swell. The peritoneal cavity begins collecting fluid. Circulation becomes less effectively filled even while the body looks fluid-overloaded. That paradox matters. A person with low albumin and advanced liver disease can be visibly swollen and still have poor effective vascular volume because so much fluid has been displaced out of the bloodstream. The body reads that underfilling through the kidneys and neurohormonal systems and responds by retaining sodium and water, which often worsens the edema and ascites rather than correcting the original distribution problem.

Ascites in particular makes the link to synthetic decline especially clear because the abdomen becomes a favored site for fluid accumulation when two forces converge. One is low oncotic pressure from reduced albumin synthesis. The other is elevated hydrostatic pressure from portal hypertension. Portal hypertension pushes fluid outward from the splanchnic circulation, the blood supply of the abdominal digestive organs, while low albumin reduces the plasma force that would normally help retain fluid intravascularly. More pressure pushing out. Less protein pulling in. The abdominal cavity becomes the reservoir for that failed balance. This is why ascites in liver disease so often signals more than portal resistance alone. It usually means the protein-building side of hepatic function has also begun to weaken.

Edema follows the same logic in peripheral tissues. The capillary bed is always balancing outward pressure against inward oncotic pull. If albumin falls, the bloodstream loses one of its main reasons to keep water contained. Fluid begins drifting into the interstitial space more easily, especially in dependent tissues where hydrostatic forces already favor outward movement. Legs swell. Soft tissues take on excess fluid. The visible symptom is swelling, but the real event is synthetic decline expressing itself through altered vascular physics. The liver has not merely failed a lab value. It has failed part of the force system that normally keeps plasma water inside circulation.

So edema and ascites often reflect synthetic decline because a failing liver is no longer making enough of the protein architecture that helps blood behave as a contained fluid compartment. The body may keep retaining water in response, but that only magnifies the misplacement. Once albumin falls, fluid starts leaving the bloodstream more easily. Once portal pressure rises with it, the abdomen becomes an even easier place for that fluid to collect. The swelling is the symptom. The deeper mechanism is that the liver is no longer building plasma strongly enough to hold the circulation together.

Chapter 24

Clotting Factors and Coagulation

Which Clotting Proteins the Liver Makes

The liver does not merely influence coagulation from a distance. It manufactures most of the protein machinery that allows blood to clot in the first place. That is why bleeding risk in liver disease is never just a vascular problem or a platelet problem. It is often a synthesis problem. Hepatocytes build fibrinogen, which later becomes the fibrin mesh that stabilizes a clot. They build prothrombin, also called factor II, which is converted into thrombin, the enzyme that drives some of the most important amplification steps in the clotting sequence. They also synthesize factors V, VII, IX, X, XI, and XII, along with other proteins that help the coagulation system assemble, propagate, and stabilize itself in flowing blood. When the liver is working well, these proteins are being produced continuously and released into plasma so the bloodstream remains ready to respond when a vessel is injured. When hepatic synthesis falls, the blood may still be present in full volume, but it is no longer carrying the same clot-building equipment.

Some of these proteins deserve special attention because they depend not only on hepatic synthesis but also on correct vitamin K handling. Prothrombin and factors VII, IX, and X are vitamin K-dependent clotting factors. That means they require vitamin K-mediated modification after synthesis so they can bind calcium properly and anchor to phospholipid surfaces during clot formation. This is one reason bile flow and coagulation are tied together. If bile delivery to the intestine falls, fat absorption worsens, vitamin K absorption can decline, and the liver may be left synthesizing the protein backbone while still failing to produce fully functional clotting factors at normal efficiency. The reader should see the architecture clearly. The liver makes the proteins. The intestine, with bile support, helps secure one of the vitamins required to finish some of them correctly. A biliary problem can therefore become a coagulation problem even before hepatocyte synthesis has completely collapsed.

It is also important to understand that the liver makes most, but not every, major hemostatic protein in exactly the same way. Von Willebrand factor, which helps platelets adhere to damaged vessel surfaces, is produced mainly by endothelial cells and megakaryocytes rather than by hepatocytes. Factor VIII does not follow the same simple hepatocyte-centered rule either and is produced largely by endothelial cells, including important contribution from liver sinusoidal endothelial cells. So when we say the liver makes the clotting proteins, the precise version is that it makes most of the circulating coagulation machinery, especially the major soluble factors that drive thrombin generation and fibrin formation, but the full hemostatic system also includes proteins with important extraheptic sources. That nuance matters because it prevents the reader from flattening coagulation into one organ alone while still recognizing that the liver is the dominant protein factory behind the cascade.

This is why hepatic synthetic decline changes coagulation so powerfully. A failing liver is not only losing detoxification capacity or bile flow. It is losing the ability to keep the blood stocked with the proteins required to build and stabilize clot. The person may still have platelets, vessels, and blood volume, but the plasma no longer carries the same concentration or quality of coagulation factors. Bleeding risk rises, clot architecture weakens, and laboratory measures such as prothrombin time begin reflecting that synthetic loss. The body does not clot because blood exists. It clots because the liver has spent every prior day manufacturing the proteins that make clotting possible.

Fibrinogen, Prothrombin, and the Cascade in Plain Language

Blood clotting can sound abstract until the reader stops imagining it as a mystical cascade and starts seeing it as a controlled construction project in flowing plasma. A vessel wall is damaged. Platelets arrive and begin forming a temporary plug. That plug helps, but it is not strong enough on its own to resist ongoing blood flow. It needs reinforcement. The coagulation system supplies that reinforcement by converting soluble proteins already circulating in plasma into an insoluble stabilizing mesh at the site of injury. Two of the most important proteins in that sequence are prothrombin and fibrinogen, and both depend heavily on the liver. The liver makes the raw materials that allow a soft platelet plug to become a more durable clot.

Prothrombin is an inactive precursor protein made by hepatocytes and released into blood. Inactive here matters. The bloodstream cannot be full of fully active clotting enzymes wandering freely, or blood would clot in the wrong place. So the body circulates many of these factors in a quiet form until injury provides the right surface and the right sequence of activation. When the coagulation system is triggered, prothrombin is converted into thrombin, an active enzyme with enormous importance. Thrombin is one of the main turning points in clot formation. It does not build the clot by itself, but it turns a soluble protein environment into a clot-building environment. If the reader wants one transition to remember, it is this one: prothrombin is the quiet precursor, thrombin is the active cutter that begins hardening the response.

Fibrinogen is the other major player. It is a soluble plasma protein, also made by the liver, circulating quietly in blood until thrombin acts on it. Once thrombin is generated, it cleaves fibrinogen and converts it into fibrin monomers, which then polymerize, meaning they link together into long strands. Those strands form the structural mesh that stabilizes the developing clot. Platelets alone make a temporary seal. Fibrin turns that seal into something reinforced. The clot becomes less like loose cellular debris collecting at an injury and more like a woven protein net holding the damaged area together against pressure and flow. This is why fibrinogen matters so much. It is the major soluble precursor from which the final structural scaffold of the clot is built.

The cascade is the sequence that makes this conversion happen only where and when it is needed. In plain language, one clotting factor activates the next, and the steps amplify as they go. Small early triggers become larger local enzyme activity, and that amplification converges on generation of thrombin. Thrombin then acts on fibrinogen and also strengthens the process further by accelerating upstream and downstream steps. That is why clotting is called a cascade. It is not one switch flipping once. It is a staged activation sequence in which inactive proteins become active enzymes, those enzymes activate more proteins, and the system builds enough force at the injury site to convert flowing plasma into a localized clot. The blood is carrying the ingredients in advance. Injury tells the system where to assemble them.

This also explains why the liver matters so much to hemostasis, the body's control of bleeding. If hepatocytes are not making enough prothrombin, fibrinogen, and the other soluble clotting factors, the bloodstream may still contain platelets and may still try to respond to vessel injury, but the protein machinery needed to generate thrombin and build fibrin is weaker. The result is not only "thinner blood" in some vague sense. The result is impaired clot construction. The platelet plug forms less securely. The fibrin mesh forms less effectively. The clot is less stable. That is the cleanest plain-language way to see it. The liver makes the quiet proteins that blood carries in reserve. Injury activates them. Thrombin turns fibrinogen into fibrin. The clot becomes real.

Anticoagulant Proteins Also Made by the Liver

The liver does not build only the proteins that make blood clot. It also builds some of the proteins that keep clotting from spreading too far or starting in the wrong place. That matters because coagulation is not supposed to behave like a one-way emergency switch. It is supposed to be local, timed, and self-limited. Hepatocytes therefore synthesize several major natural anticoagulants, especially antithrombin, protein C, and protein S. Antithrombin restrains activated clotting enzymes, especially thrombin and factor Xa. Protein C, once activated, helps shut down factors Va and VIIIa, and protein S acts as an important cofactor helping activated protein C do that work effectively. These proteins are not optional refinements added after the clot is built. They are part of the same plasma architecture that makes clot formation controllable rather than explosive. The liver is therefore writing both halves of the system at once: the proteins that build the clot and the proteins that stop the clot from becoming a circulating threat.

This is why liver disease does not produce a simple story of "poor clotting" and nothing else. When hepatic synthetic function declines, procoagulant proteins can fall, but anticoagulant proteins can fall too because they are also liver-made. In other words, the blood may lose parts of both the accelerator and the braking system. That is one reason coagulation in liver disease is unstable rather than neatly predictable. A person may bleed more easily in one context and still remain vulnerable to thrombosis in another because the liver is no longer maintaining the balanced protein environment that normal hemostasis requires. The old shorthand that liver disease automatically means a straightforward bleeding state misses this architecture. The liver is not making one clotting system. It is making a regulated clotting field.

So the deeper point is simple. The liver does not merely supply the raw materials to seal a wound. It also supplies part of the restraint that keeps clotting from overrunning the vessel. Antithrombin, protein C, and protein S are the clearest examples of that restraint, and their hepatic origin is one of the reasons a failing liver changes coagulation so profoundly. The body loses not only the proteins that help build a clot. It loses some of the proteins that tell the clot where to stop.

Why Liver Disease Creates Unstable Coagulation Rather Than a Simple Bleeding Disorder

Liver disease does not create a clean one-direction coagulopathy because the liver makes both sides of the hemostatic system. Hepatocytes synthesize most of the procoagulant factors needed to generate thrombin and fibrin, but the liver also synthesizes major anticoagulant proteins such as antithrombin, protein C, and protein S. When hepatic synthetic function declines, the blood is not simply losing the machinery that builds a clot. It is also losing part of the machinery that restrains clotting once it begins. That is why modern liver thinking no longer treats cirrhosis or severe hepatic dysfunction as a straightforward bleeding state. The coagulation system becomes rebalanced, but that balance is fragile and easily tipped by infection, renal failure, bleeding, dehydration, endothelial injury, portal hypertension, or invasive procedures. The result is instability, not a simple slide toward hemorrhage alone.

The old bleeding-only model survives because some parts of the picture do point in that direction. Prothrombin time and international normalized ratio can rise as hepatic synthetic capacity worsens, platelets may fall from hypersplenism and reduced thrombopoietin signaling, vitamin K-dependent factor activity can be impaired, and advanced liver disease can also disturb fibrinolysis, the system that dissolves clots. All of that can increase bleeding risk in the right clinical setting. But those abnormalities do not tell the whole story because anticoagulant proteins fall too, and standard clotting tests such as PT and INR do not capture the full balance between procoagulant and anticoagulant forces in cirrhosis. This is one reason INR alone is a poor predictor of procedural bleeding risk in cirrhotic patients. The blood may look "thin" on paper while still retaining or even shifting toward thrombotic potential under the wrong conditions.

That is the real reason liver disease creates unstable coagulation. The system loses reserve on both sides at once. The injured liver is no longer maintaining a broad, buffered hemostatic field. It is producing less of the proteins that build clot, less of the proteins that brake clot, and doing so in a patient whose platelets, endothelium, portal circulation, and fibrinolytic environment may already be abnormal. So the person with liver disease may bleed from varices, mucosa, or procedures in one moment and still remain vulnerable to portal vein thrombosis or other thrombotic events in another. The blood is not simply unable to clot. It is clotting under degraded supervision.

Vitamin K and Bile-Linked Coagulation Problems

Vitamin K becomes a coagulation problem when bile flow becomes a digestion problem. The connection is physical before it is hematologic. Vitamin K is a fat-soluble vitamin, which means it does not move efficiently through the watery intestinal lumen on its own. It depends on the same bile acid-driven lipid handling that dietary fat depends on. Bile acids are needed to emulsify fat and support micelle formation, and micelles are the small transport assemblies that carry hydrophobic lipid products and fat-soluble vitamins toward the absorptive surface of the intestine. When bile delivery to the gut falls, vitamin K absorption falls with it. The vitamin may still be present in the meal, but the intestine loses part of the machinery required to bring it into the body. Cholestasis therefore can create a coagulation defect without first destroying every hepatocyte, because the liver may still be synthesizing clotting proteins while the intestine is no longer delivering one of the vitamins required to make several of those proteins fully functional.

The liver-made factors most clearly affected are the vitamin K-dependent clotting proteins: factor II, which is prothrombin, and factors VII, IX, and X. These proteins need vitamin K-dependent modification after synthesis so they can bind calcium properly and assemble on phospholipid surfaces during clot formation. Without enough vitamin K, the protein backbone may still be produced, but the clotting factors are not finished correctly for normal function. This is why a cholestatic person can begin showing prolonged coagulation times even when the immediate issue is bile flow rather than catastrophic hepatocyte loss. The liver wrote the proteins. The intestine failed to deliver the fat-soluble cofactor needed to complete them.

This is also why vitamin K helps distinguish some cholestatic coagulation problems from more severe synthetic failure. In chronic cholestasis, fat malabsorption can produce vitamin K deficiency and prolong the prothrombin time or international normalized ratio. If vitamin K is replaced and the clotting abnormality improves, that suggests bile-linked malabsorption was playing an important role. If correction does not occur, then impaired hepatic synthetic capacity becomes a stronger concern. The underlying logic is straightforward. A bile problem can create a coagulation problem by starving the body of vitamin K. A failing liver can create a coagulation problem by no longer making the factors well. Sometimes both are present at once.

So vitamin K and bile-linked coagulation belong in the same sentence because bile is part of the absorption pathway for the vitamin, and the vitamin is part of the finishing chemistry for several major clotting factors. Obstruct bile flow, and the meal's fat phase becomes less usable. Vitamin K uptake falls. The liver may still synthesize the factor proteins, but their functional maturation suffers. The result is a coagulopathy born not only from liver injury, but from failed delivery of a fat-soluble cofactor that normal clotting depends on.

PT and INR as Rough Windows Into Hepatic Synthetic Function

Prothrombin time, or PT, measures how long plasma takes to clot after tissue factor and other reagents are added, and the international normalized ratio, or INR, is the standardized way of expressing that result across laboratories. In liver physiology, these tests matter because the liver synthesizes most of the soluble clotting proteins required for the extrinsic and common coagulation pathways, especially factors II, V, VII, and X and fibrinogen-related support within the broader system. When hepatic synthetic capacity falls enough, PT tends to lengthen and INR rises. That is why these tests are often read as windows into liver synthetic function. They are not measuring hepatocytes directly. They are measuring the downstream clotting behavior of plasma that no longer contains the same full complement of liver-made coagulation proteins.

The reason they are only rough windows is that they reflect one important slice of hepatic output, not the entire synthetic landscape and not the whole coagulation balance. PT and INR become abnormal mainly when the liver is failing badly enough for clotting factor production to fall meaningfully, which is why they may remain normal in mild hepatocellular dysfunction and can even stay normal in compensated cirrhosis for a time. When they do rise in chronic liver disease, that usually signals progression toward more advanced synthetic failure. In acute liver injury, a rising PT or INR is especially ominous because some of the affected factors, particularly factor VII, have short half-lives, so the test can worsen quickly when synthesis is collapsing. This makes PT and INR useful, but it also means they are weighted toward more serious dysfunction rather than subtle early change.

They are also rough because they do not tell the whole truth about bleeding risk. The liver makes anticoagulant proteins as well as procoagulant ones, and routine PT and INR do not capture that broader rebalanced but unstable hemostatic state in cirrhosis. This is why INR can rise in liver disease while still failing to predict procedural or spontaneous bleeding reliably, except perhaps at more extreme elevations. The test sees part of the lost clotting machinery, but it does not see the full architecture of coagulation, anticoagulation, endothelial contribution, platelet behavior, and fibrinolysis moving together in the same patient. So PT and INR are valuable as markers of hepatic synthetic decline, but they are poor as simple stand-alone measures of "how likely this person is to bleed."

The cleanest way to think about them is this. PT and INR are plasma behavior tests that become abnormal when the liver is no longer maintaining enough of the clotting-factor output required for normal assay clot formation. That makes them useful as rough windows into synthetic function and disease severity, which is one reason INR is included in major liver prognostic frameworks such as MELD. But they remain rough windows because they do not capture all liver-made proteins, they may stay normal until dysfunction is substantial, and they do not summarize the full coagulation reality of liver disease. They show that the liver is falling behind on one crucial synthetic job. They do not show everything the failing liver has already changed.

Chapter 25

Acute-Phase Proteins and Inflammatory Signaling

What Acute-Phase Proteins Are

Acute-phase proteins are liver-made plasma proteins whose production shifts when the body enters an inflammatory state. They are not general background proteins drifting in blood at fixed concentration regardless of what the tissue environment is doing. They are part of the liver's response to injury, infection, stress, necrosis, and broader immune activation. When inflammatory cytokines rise, especially interleukin-6 and related signals, hepatocytes change what they prioritize. Some proteins are produced in larger amounts because the body now needs a different plasma environment. Others fall because resources are being redirected and because the functions most useful in quiet physiology are not the same functions most useful during systemic defense or tissue injury. Acute-phase proteins therefore are best understood as a reconfiguration of blood chemistry under inflammatory command. The liver is not merely receiving inflammation as a message from somewhere else. It is helping rewrite the bloodstream so the body can act on that message.

This matters because inflammation is not only a cellular event inside injured tissue. It is also a circulating event. Once infection, trauma, necrosis, or immune stimulation begins, the body needs proteins in plasma that can bind microbial structures, activate complement, modify iron availability, alter coagulation tendency, clear damaged material, and change how immune cells behave. Hepatocytes provide much of that shift. They increase synthesis of proteins such as C-reactive protein, fibrinogen, serum amyloid A, hepcidin, and several complement components. Each of these proteins serves a different purpose, but together they create a new plasma state that is less about ordinary transport and more about defense, containment, and resource control. Acute-phase proteins are therefore not just biomarkers of inflammation. They are part of the mechanism of inflammation.

The phrase acute-phase can be misleading if the reader hears it as meaning only brief illness. The point is not merely that these proteins rise in acute conditions. The point is that they belong to a phase of physiology in which the body has shifted out of ordinary baseline maintenance and into an injury-response mode. That mode can be transient, but it can also become chronic if inflammatory signaling does not resolve. The liver then continues writing inflammatory bias into the bloodstream day after day. So acute-phase proteins should not be imagined as rare emergency molecules appearing only during dramatic infection. They are part of how plasma changes whenever inflammatory signaling reaches the liver strongly enough to alter its synthetic priorities.

This is one of the clearest examples of the liver acting as a system-level coordinator rather than as a mere detox organ. A damaged tissue in one part of the body can signal through cytokines, and the liver responds by changing blood protein composition everywhere. That means the liver is helping convert local injury into systemic physiology. The blood becomes different because the liver has decided it now needs to be different. A fibrinogen rise changes clotting potential. Hepcidin changes iron traffic. C-reactive protein changes how microbial and tissue debris are recognized and handled. Complement proteins alter immune attack and clearance. The liver is writing inflammatory intent into plasma.

So what acute-phase proteins are, at the simplest level, are liver-made proteins whose production is upshifted or downshifted when inflammatory cytokines tell hepatocytes that the body has entered a defense state. They are not passive indicators. They are part of the body's attempt to reorganize circulation, immunity, and resource handling around injury.

C-Reactive Protein, Fibrinogen, Complement Proteins

These proteins matter because they show what the liver does when inflammation stops being local and starts becoming circulatory. A damaged tissue, an infection, a necrotic lesion, or a strong cytokine signal does not remain confined to the original site. Once hepatocytes receive that inflammatory message, they begin rewriting plasma composition so blood itself becomes part of the response. C-reactive protein, fibrinogen, and complement proteins are three of the clearest examples. They are not random lab markers that happen to rise during illness. They are liver-made tools that change how the bloodstream recognizes danger, handles clotting, and participates in immune defense.

C-reactive protein, often abbreviated CRP, is one of the best known acute-phase proteins because it rises quickly when inflammatory signaling reaches the liver, especially through interleukin-6. Its name comes from its ability to bind the C-polysaccharide of pneumococcus, but its biological role is broader than that old naming history suggests. CRP binds to damaged cells, microbial surfaces, and certain exposed phospholipids that appear when tissue is injured or membranes are disturbed. That binding helps mark targets for clearance and can also help activate complement, the plasma protein system that amplifies immune recognition and attack. So CRP is not simply announcing that inflammation exists. It is participating in the labeling of what the body now treats as abnormal. The liver increases CRP production because inflammatory blood needs a better way to identify damaged surfaces and microbial burden moving through circulation.

Fibrinogen belongs in the same section because inflammation is not only about recognition. It is also about containment. Fibrinogen is a soluble plasma protein made by the liver that later becomes fibrin when thrombin cleaves it during coagulation. In ordinary hemostasis that matters for stopping bleeding, but during inflammation its rise also reflects the body's tendency to shift blood toward a more reactive, more sealant-ready state. A tissue under threat often needs barrier reinforcement, local clot support, and stronger capacity to wall off injury. Higher fibrinogen contributes to that environment. It also changes blood rheology, meaning how blood flows physically, because plasma with more fibrinogen behaves differently from plasma with less. This is why fibrinogen is both a clotting protein and an acute-phase protein. The liver is using the same molecule to support hemostasis and to reshape inflammatory blood into a more containment-oriented fluid.

Complement proteins push the point even further. They are a family of plasma proteins, many of them synthesized largely by the liver, that circulate in inactive form until triggered. Once activated, complement proteins participate in opsonization, the coating of microbes or debris so phagocytic cells can recognize and engulf them more easily, chemotaxis, the recruitment of immune cells toward sites of injury or infection, and direct membrane attack against certain targets. In simple terms, complement turns plasma into an immune-active medium. The bloodstream is no longer only carrying cells and nutrients. It is carrying a distributed defense network that can identify targets, amplify danger signals, and help destroy or clear what does not belong. The liver is central to this because it supplies much of the raw complement system before activation ever begins.

Taken together, CRP, fibrinogen, and complement proteins show the same deeper truth. The liver does not only metabolize inflammation after the fact. It helps manufacture the protein environment in which inflammation becomes systemic. CRP marks danger. Fibrinogen supports containment and clot structure. Complement proteins amplify recognition, recruitment, and attack. When these proteins rise, the blood becomes a different kind of fluid. It is no longer optimized only for transport and exchange. It is optimized for defense, containment, and clearance. That is what acute-phase signaling means at the plasma level. The liver receives the inflammatory message and writes it into circulation.

Why the Liver Changes Plasma Composition During Inflammation

The liver changes plasma composition during inflammation because blood has to become a different kind of fluid when the body moves from ordinary maintenance into defense and damage control. In the baseline state, plasma is optimized mainly for transport, nutrient distribution, waste movement, fluid balance, and stable interaction with tissues under normal exchange conditions. That version of blood is not enough when cells are dying, microbes are present, barrier surfaces are breached, or immune signaling is rising. The body then needs a circulation that can identify damaged material, bind microbial structures, alter iron handling, support clot reinforcement, recruit immune cells, and help contain spread. Hepatocytes are one of the main cell populations that make this transition possible. They receive cytokine signals from the injured or infected body and respond by changing what proteins they release into plasma. The blood becomes different because the situation has become different.

This matters because inflammation is not solved locally. A damaged lung, an infected gallbladder, an inflamed joint, a necrotic tissue bed, or an intestinal breach can begin in one place, but the organism has to respond through a shared fluid compartment. Plasma is that compartment. If the liver kept producing the exact same protein mix during severe inflammation that it produced during quiet physiology, the bloodstream would be poorly equipped for the demands now placed on it. So hepatocytes increase production of acute-phase proteins such as C-reactive protein, fibrinogen, serum amyloid A, complement components, and hepcidin, while some other proteins fall in relative priority. This is not random. It is a deliberate redistribution of synthetic effort. The liver is reallocating protein production toward what the body needs most under inflammatory conditions.

The change is functional at every level. C-reactive protein helps mark damaged cells and microbial structures for clearance. Complement proteins increase the plasma's ability to recognize, amplify, and attack targets or tag them for phagocytosis. Fibrinogen supports clot formation and changes the blood toward a more containment-oriented state, which matters when the body is trying to limit spread of injury or infection. Hepcidin alters iron traffic, restricting iron availability in ways that can disadvantage microbes while also changing systemic iron handling. These are not merely markers rising on a lab panel. They are instructions written into the fluid itself. The liver is turning plasma into an active participant in defense.

This also explains why inflammation changes more than immunity in the narrow sense. Once plasma composition shifts, coagulation behavior shifts, iron distribution shifts, vascular interactions shift, and the way immune cells encounter circulating proteins shifts. The bloodstream becomes more adhesive to injury logic and less devoted to ordinary transport neutrality. That is useful in the short term because the body needs a more armed circulation. But if the inflammatory state persists, the same hepatic response can become part of chronic disease. The liver keeps writing inflammatory priorities into blood day after day, and the circulation remains biased toward defense, containment, and altered resource handling even when that state starts damaging the host itself.

So the liver changes plasma composition during inflammation because the body cannot defend itself with baseline blood alone. Local injury sends the alarm. Cytokines carry it systemically. The liver rewrites the protein environment of plasma so circulation itself becomes part of the response. The blood is no longer just transporting what the body contains. It is being retooled for what the body is trying to survive.

Hepatic Response to Cytokine Signaling

The liver does not respond to inflammation by waiting for damage to spread into its own tissue first. It responds because cytokines arriving in blood tell hepatocytes that the state of the organism has changed and that plasma now needs to be rewritten. Cytokines are signaling proteins released by immune cells, injured tissues, endothelial cells, and other stressed cellular populations. They do not carry calories, oxygen, or waste. They carry instruction. When hepatocytes receive cytokine signals, especially interleukin-6, but also interleukin-1, tumor necrosis factor, and related inflammatory messengers, they shift from ordinary baseline synthesis into an acute-phase program. That shift is one of the clearest examples of the liver acting as a whole-body coordinator rather than as a passive downstream organ. The liver is reading the inflammatory state of the body and then changing the composition of blood in response.

This change begins at the cell surface and moves inward through transcriptional control. Cytokines bind their receptors on hepatocytes and activate intracellular signaling pathways that alter gene expression, meaning the pattern of which proteins the cell is now being instructed to make more of, make less of, or prioritize differently. A healthy liver making albumin, transport proteins, coagulation proteins, bile components, and metabolic enzymes under ordinary conditions can therefore be pushed into a different synthetic hierarchy when inflammatory signaling rises. C-reactive protein increases. Fibrinogen increases. Serum amyloid A increases. Complement proteins increase. Hepcidin increases. Some other proteins, including albumin, may fall in relative priority. The hepatocyte is not simply becoming inflamed. It is reallocating its synthetic budget toward a blood chemistry more suited for defense, containment, and resource control.

That reallocation matters because plasma must do different work in an inflamed body. C-reactive protein helps bind damaged cells and microbial structures. Complement proteins increase the circulation's ability to tag, recruit, and attack. Fibrinogen pushes blood toward a more sealant-ready state, supporting clot formation and tissue containment. Hepcidin changes iron movement so less iron is available in plasma and from storage release, which can limit microbial access to a nutrient many pathogens rely on. This means cytokine signaling is not only making the liver louder. It is making plasma more defensive. The bloodstream stops being optimized only for transport and becomes increasingly optimized for surveillance, containment, and altered nutrient allocation under threat.

The hepatic response to cytokines also explains why inflammation can spread systemically without every organ being infected or injured directly. A local abscess, a damaged intestine, an inflamed joint, or a necrotic tissue bed can release cytokines into blood, and the liver then converts that local alarm into systemic protein change. The body does not need bacteria in the bloodstream for the blood itself to become inflammatory. It only needs the liver to hear the cytokine signal and act on it. This is why acute-phase proteins are often detectable at levels that seem disproportionate to the visible site of injury. The liver has amplified the message by changing the plasma environment of the entire organism.

There is a cost to this, which is why chronic cytokine signaling becomes dangerous. A short inflammatory program may be adaptive. The body needs a stronger defense fluid for a limited period. But when cytokine exposure persists, the liver keeps producing an altered plasma state day after day. Albumin may remain suppressed. Fibrinogen may stay high. Hepcidin may continue trapping iron inappropriately. Complement and other inflammatory proteins may keep circulating at elevated levels. The liver is still doing what the signal told it to do, but the signal has stopped being useful and started becoming part of chronic disease. That is one reason long-standing inflammation changes energy, coagulation, iron biology, vascular behavior, and tissue repair so profoundly. The liver is still writing the message into blood.

So the hepatic response to cytokine signaling is best understood as an instruction-driven change in plasma manufacture. Cytokines tell hepatocytes that the body has entered an inflammatory state. Hepatocytes answer by changing what proteins they make and release. The blood becomes a different fluid because the liver has been told it needs to be.

The Liver as a Participant in Immunity, Not a Passive Bystander

The liver is not sitting downstream from the immune system waiting to be told what happened elsewhere. It is part of the immune system's operating surface. That has to be understood structurally first. Portal blood flows into the liver directly from the intestine carrying not only nutrients, but bacterial fragments, microbial metabolites, endotoxin, food antigens, damaged-cell debris, and inflammatory signals generated at the gut barrier. No organ receiving that kind of traffic can remain immunologically neutral. The liver has to decide constantly what should be cleared quietly, what should be tolerated, what should be signaled onward, and what should provoke a stronger defensive response. That is not passive filtering. That is immune judgment embedded into tissue architecture.

Kupffer cells make this easiest to see. These resident macrophages sit in the sinusoids, the low-pressure exchange channels where portal blood slows and becomes readable. They do not merely engulf debris after the important work is done. Their sampling is part of the important work. They clear microbial fragments before those fragments reach the wider circulation, interpret the difference between ordinary gut-derived background traffic and more dangerous inflammatory burden, and release cytokines when the tissue needs to shift from quiet clearance into active defense. Cytokines are signaling proteins that tell neighboring cells and distant organs that the inflammatory state has changed. So the liver is not only receiving immune signals from the body. It is generating them.

Hepatocytes themselves are part of this immune participation. They are usually described as metabolic cells, which they are, but they also respond to inflammatory signaling by changing what they synthesize and secrete into plasma. Acute-phase proteins such as C-reactive protein, fibrinogen, complement components, and hepcidin are part of that response. Once those proteins enter circulation, the blood becomes a different kind of fluid. It is better equipped for recognition of damaged surfaces, microbial tagging, altered iron handling, and containment of tissue injury. That means the liver is not standing aside while immune cells do the real work. It is rewriting the protein environment in which immune work becomes systemic.

The sinusoidal endothelial cells participate too. They are not passive lining cells coating the inside of a vascular channel. They regulate permeability, shape what kinds of cellular and molecular interactions occur at the sinusoidal surface, and contribute to the liver's balance between tolerance and activation. That balance matters enormously because portal blood is never fully sterile and never fully quiet. The liver cannot mount maximal inflammatory aggression against every meal and every microbial fragment that crosses the gut barrier in low amount. If it did, digestion itself would become a daily inflammatory crisis. So hepatic immunity is built around discrimination. It has to clear without panicking. It has to tolerate without becoming blind. That is an active task, not a passive absence of response.

This is why the liver is one of the body's great immune border organs. It sits at the interface between the outside world, represented chemically by the intestine, and the internal circulation that must remain coherent enough for every organ to survive. It screens what just entered. It dampens what should not become systemic overreaction. It amplifies what truly signals danger. It manufactures proteins that reshape inflammatory blood. It clears debris, helps direct iron away from pathogens, and coordinates with innate immune cells already positioned in the tissue. None of that is bystander behavior. It is participatory immune work written directly into the organ's daily function.

The distinction becomes even sharper in chronic disease. A liver repeatedly exposed to excess endotoxin, ongoing gut barrier dysfunction, alcohol, steatotic stress, viral injury, or cholestatic burden does not merely suffer collateral inflammation. It helps perpetuate the inflammatory field through Kupffer cell activation, cytokine release, acute-phase synthesis, and stellate cell recruitment into fibrotic remodeling. The organ is not just being damaged by immunity. It is contributing to how that immune environment evolves. That is why liver disease so often carries systemic inflammatory consequences. The organ involved is one that normally helps decide how much local danger becomes whole-body chemistry.

So the liver is a participant in immunity because it does not only metabolize after the immune system speaks. It helps determine what gets counted as danger, what gets cleared, what gets tolerated, and what gets written into plasma as a systemic response. The gut sends the first report. The liver reads it, edits it, and often answers back.

How Infection, Injury, and Inflammation Shift Liver Output

The liver does not keep making the same blood proteins when the body enters a state of threat. Infection, tissue injury, and systemic inflammation change the instructions arriving at hepatocytes, and hepatocytes respond by changing what they release into plasma. That shift is one of the clearest examples of the liver acting as a systems organ rather than as a local metabolic machine. In ordinary conditions, hepatic output is weighted toward transport proteins, albumin, coagulation balance, bile production, nutrient handling, and the broader maintenance of blood chemistry under baseline physiology. Once inflammatory cytokines rise, especially interleukin-6 along with related signals from injured tissue and activated immune cells, that baseline changes. The liver starts reallocating its synthetic effort. Some proteins rise sharply because the body now needs a more defensive circulation. Other proteins fall in relative priority because the bloodstream is being repurposed for containment, recognition, and resource control rather than for quiet stability alone.

This is why inflammation changes plasma composition so fast. Hepatocytes increase production of acute-phase proteins such as C-reactive protein, fibrinogen, complement components, serum amyloid A, and hepcidin. Each one changes blood behavior in a different direction. C-reactive protein helps mark damaged cells and microbial surfaces for clearance. Fibrinogen pushes plasma toward a more clot-capable, containment-oriented state. Complement proteins make the bloodstream more immunologically active, improving tagging, amplification, and attack against targets that should not remain unchallenged. Hepcidin alters iron traffic by restricting iron release and absorption, which can limit microbial access to iron while also changing the body's own iron availability. This is not the liver merely reacting to inflammation after the fact. It is the liver manufacturing the protein environment that lets inflammation become systemic physiology.

At the same time, some ordinary hepatic outputs are de-emphasized. Albumin commonly falls during sustained inflammatory signaling because the liver is redirecting synthetic capacity away from baseline transport and oncotic maintenance and toward acute-phase priorities. That does not mean albumin instantly disappears, but it does mean the blood begins shifting away from a pure transport-and-balance profile and toward a defense-and-containment profile. The consequences can spread widely if this state persists. Plasma becomes more inflammatory, more prothrombotic in some respects, more restrictive in iron handling, and less devoted to maintaining some of the quieter stabilizing functions that dominate in health. The liver is telling the bloodstream that the organism is no longer in ordinary conditions.

The trigger for this shift can come from many places. A bacterial infection can send endotoxin and cytokine signals into circulation. Tissue necrosis can release damage-associated molecular patterns that tell the liver cells are dying somewhere in the body. Gut barrier disruption can increase portal delivery of microbial fragments and inflammatory cues. Even sterile injury without infection can provoke the same hepatic response if the tissue stress is large enough. This is why the liver is so central to whole-body illness. A local wound or infection becomes a change in plasma behavior because the liver hears the alarm and rewrites what the blood contains.

The cost is that prolonged inflammatory signaling turns an adaptive hepatic response into part of the disease burden. A short-term rise in acute-phase output can help the body survive infection or injury. A chronic rise means the liver keeps producing inflammatory plasma, keeps suppressing some baseline outputs, and keeps biasing circulation toward defense long after that bias has stopped being cleanly protective. That is one reason chronic inflammation changes coagulation, iron biology, fluid balance, and nutritional state so profoundly. The liver is still doing what the body asked it to do. The problem is that the body never stopped asking.

So infection, injury, and inflammation shift liver output by changing the instructions reaching hepatocytes. The organ reduces some baseline priorities, raises acute-phase synthesis, and turns plasma into a more defensive fluid. Blood is no longer just carrying oxygen, nutrients, and waste. It is being reauthored by the liver to help the organism survive threat.

Part VII

Endocrine and Mineral Authority

Chapter 26

Iron Control and Hepcidin

Iron Absorption and the Need for Tight Control

Iron has to be controlled tightly because it is both indispensable and dangerous in the same molecule. The body needs it for oxygen transport, mitochondrial energy production, and a wide range of enzymatic reactions built around electron transfer. Hemoglobin depends on it. Myoglobin depends on it. Many oxidative enzymes depend on it. But iron is also chemically reactive. In the wrong amount, in the wrong compartment, or in the wrong binding state, it can help drive oxidative injury by catalyzing reactions that generate damaging free radicals. That means iron cannot be treated like a harmless mineral drifting in and out according to appetite. It has to be absorbed carefully, transported carefully, stored carefully, and withheld carefully. The body has no large active pathway for dumping excess iron on demand. That single fact changes the whole logic. Because iron loss is limited and relatively slow, control has to happen mainly at the level of entry and release rather than at the level of easy disposal.

Absorption begins in the small intestine, especially the duodenum, where dietary iron is taken up by enterocytes, the absorptive cells lining the gut. But the intestine is not allowed to behave as though every iron-containing meal deserves full entry into the bloodstream. Iron in food arrives in different forms, and those forms matter. Heme iron from animal tissue is absorbed differently from non-heme iron found in many other foods. Once iron reaches the enterocyte, the question is not only whether it entered the cell. The larger question is whether it will be allowed to leave that cell and enter the circulation. That exit step is crucial. An enterocyte can hold iron temporarily, but for the body to gain systemic iron, the mineral has to be exported across the basolateral membrane into blood through ferroportin, the main known cellular iron exporter. From there iron is bound by transferrin in plasma and carried toward marrow, liver, and other tissues. So intestinal iron absorption is not a single membrane crossing. It is a staged handoff, and the body regulates the final release with unusual seriousness.

That seriousness exists because iron is not just another nutrient. Too little iron weakens erythropoiesis, the production of red blood cells, reduces oxygen-carrying capacity, and compromises oxidative metabolism. Too much iron creates a different kind of injury. It accumulates in tissues, especially the liver, heart, pancreas, and endocrine organs, where it can drive oxidation, protein injury, fibrosis, and functional decline. A body that absorbs iron recklessly would therefore trade one problem for another. It might protect itself from deficiency briefly while quietly building a storage toxicity that has no fast route out. This is why iron physiology is built around guarded traffic rather than free access. The intestine receives the mineral. The body still decides whether the mineral gets to stay.

The liver sits at the center of that decision because it is one of the main organs reading total-body iron status and writing the hormonal instructions that determine how much iron enters circulation from the gut and from storage sites. That hormone is hepcidin, and once the reader understands it, iron control becomes much more intelligible. The body does not control iron primarily by flushing excess out in urine or bile. It controls iron by restricting the release of iron into blood when stores are sufficient or when inflammation makes free iron more dangerous. Entry from the intestine and release from macrophages and storage cells are therefore the main gates. Tight control is necessary because iron is too useful to waste and too reactive to leave ungoverned. The body cannot afford either extreme. It cannot starve the marrow of iron. It cannot let the circulation become a free metal stream.

Hepcidin as the Liver-Made Hormone That Governs Iron Traffic

Hepcidin is the liver-made hormone that decides whether iron is allowed to move. That is the cleanest way to understand it. The body does not control iron mainly by throwing excess away, because it has no large rapid excretory route for iron comparable to the way it can dump water-soluble waste through the kidney. Iron control therefore has to happen at the gates of entry and release. The intestine can take iron up from food, macrophages can recover iron from aging red blood cells, and storage sites such as hepatocytes can hold substantial iron reserve. But none of those pools matters to the rest of the body unless iron is allowed to leave the cell and enter plasma. Hepcidin governs that step. It is made by hepatocytes and released into blood as a systemic instruction telling iron-exporting cells whether they are permitted to hand iron forward.

Its target is ferroportin, the main known iron exporter on the surface of enterocytes, macrophages, and certain storage cells. Ferroportin is the door through which iron leaves the cell. Enterocytes use it to move dietary iron into the bloodstream. Macrophages use it to release recycled iron after they break down old red blood cells. Hepatocytes and other storage cells use it when reserve iron needs to be mobilized. Hepcidin binds to ferroportin and causes it to be internalized and degraded. That means the door is physically removed from the cell surface. Once ferroportin is gone, iron can still exist inside the cell, but it cannot leave efficiently. This is why hepcidin is not a vague regulator. It is a traffic stopper. High hepcidin traps iron in cells and lowers iron delivery into plasma. Low hepcidin leaves ferroportin available and allows more iron to enter or re-enter the circulation.

This arrangement gives the liver enormous authority over iron biology. If body iron stores are high, hepatocytes increase hepcidin production. More hepcidin means less ferroportin activity, which means less intestinal iron enters blood and less stored or recycled iron is released from macrophages and hepatic stores. The body protects itself from iron overload by slowing iron traffic at the level of export. If iron is low or erythropoietic demand rises, meaning the bone marrow needs more iron to build red blood cells, hepcidin production falls. Less hepcidin means more ferroportin survives on the cell surface, so more dietary iron can enter blood and more recycled iron can leave macrophages for reuse. The liver is therefore not watching iron passively. It is deciding whether iron remains trapped in local compartments or becomes available for systemic use.

Inflammation changes this logic in a way that reveals how closely iron control and immunity are tied together. During inflammatory states, cytokine signaling, especially through interleukin-6, can increase hepcidin production even if total-body iron is not excessive. The result is iron sequestration. Iron is trapped in enterocytes, macrophages, and storage compartments, and less reaches plasma. From one angle this is defensive. Many microbes need iron, and lowering freely available iron in circulation can be part of host defense. From another angle it is costly, because the marrow may now have less usable iron for erythropoiesis even while total body iron is still present. This is one reason chronic inflammation can produce anemia of inflammation or anemia of chronic disease. The iron is not necessarily absent. It is being withheld. Hepcidin is the hormone enforcing that withholding.

This makes hepcidin one of the clearest examples of the liver acting as an endocrine authority rather than only a metabolic processor. The intestine absorbs the iron. Macrophages recycle it. The marrow spends it. But the liver writes the hormone that decides whether those movements are allowed. Hepcidin turns iron physiology from a simple nutrient story into a traffic-control story. Iron in the body is not the same as iron available to the body. Hepcidin determines the difference.

Ferroportin Explained Simply

Ferroportin is the doorway iron uses to leave a cell. That is the simplest and most useful way to understand it. Iron may be present inside an enterocyte after absorption from food, inside a macrophage after breakdown of an old red blood cell, or inside a storage cell such as a hepatocyte, but none of that iron helps the rest of the body unless it can get back into the bloodstream. Ferroportin is the exporter that makes that possible. It sits on the cell membrane and moves iron outward from the inside of the cell to the outside, where plasma proteins can take over and carry the iron forward. If ferroportin is open and present, iron can circulate. If ferroportin is removed or suppressed, iron becomes trapped where it sits.

This is why ferroportin matters so much in three places at once. In the intestine, enterocytes absorb iron from the lumen, but systemic iron gain depends on whether ferroportin lets that iron exit the cell and enter blood. In macrophages, which recycle iron from aging red blood cells, ferroportin determines whether recovered iron is released for reuse or held back inside the recycling cell. In storage cells, ferroportin helps decide whether reserve iron remains stored or is mobilized into plasma. The body therefore does not control iron mainly by deciding whether a cell can touch it. It controls iron by deciding whether the cell can release it. Ferroportin is the release point.

That is why hepcidin, the liver-made iron hormone, has so much power. Hepcidin does not need to change every iron-handling reaction inside the body. It only needs to target ferroportin. When hepcidin binds ferroportin, the exporter is pulled off the cell surface and degraded. The door disappears. Iron remains inside the cell, but it is no longer moving efficiently into blood. In the intestine that means less dietary iron enters the circulation. In macrophages that means less recycled iron becomes available to the marrow. In storage cells that means less reserve iron is mobilized. One hormone. One transporter. Whole-body iron traffic changes.

So ferroportin should be thought of as the final common gate of iron release. Iron in a cell is not the same as iron available to the body. Ferroportin is what turns intracellular iron into circulating iron. Without it, the mineral is present but stranded.

Why Inflammation Changes Iron Handling

Inflammation changes iron handling because iron is useful to the host and useful to microbes at the same time. That makes free access to iron dangerous during infection and costly during chronic immune activation. The body cannot treat iron like a neutral mineral that simply enters from food, circulates, and gets handed out wherever demand is high. Iron supports hemoglobin synthesis, mitochondrial respiration, and many enzymatic reactions built around electron transfer, but bacteria and other pathogens also need iron for growth and metabolism. Once inflammation begins, the body starts reading circulating iron not only as a nutrient resource, but as a liability that may need to be withheld. The liver is central to that decision. It responds to inflammatory cytokines by changing the hormonal instructions that govern iron traffic, and the result is that iron begins staying inside cells instead of moving freely through plasma.

The key hormone is hepcidin. During inflammatory signaling, especially when interleukin-6 rises, hepatocytes increase hepcidin production. Hepcidin then binds to ferroportin, the iron exporter on enterocytes, macrophages, and certain storage cells, and causes that exporter to be internalized and degraded. Once ferroportin disappears from the cell surface, iron can still be present inside the cell, but it can no longer leave efficiently. That means dietary iron absorbed by enterocytes is more likely to stay trapped until the enterocyte is shed. Iron recovered by macrophages from aging red blood cells is more likely to remain inside the macrophage instead of being released back into plasma for reuse. Iron stored in hepatocytes or other reserve sites is less available for mobilization. Inflammation therefore changes iron handling by closing the export gates rather than by removing iron from the body outright.

This is a defensive move, at least in the short term. Lower iron availability in plasma can make the internal environment less favorable for invading organisms that depend on accessible iron. The body is not trying to eliminate iron absolutely. It is trying to hide it. That is the real logic. Iron shifts from a circulating resource to a sequestered resource. But the same mechanism that can help with host defense also creates a cost for the host if the signal persists. The bone marrow may still need iron to make red blood cells, yet much of the body's iron is now locked inside macrophages and storage compartments instead of circulating on transferrin where erythropoiesis can use it efficiently. This is why chronic inflammation can produce anemia even when total body iron is not truly depleted. The iron is present. The traffic is blocked.

That is the difference between iron deficiency and inflammatory iron restriction. In true deficiency, the body lacks enough iron overall. In inflammatory states, the body may have iron, but the liver has hormonally declared that iron should not move freely. Hepcidin rises, ferroportin falls, plasma iron availability drops, and erythropoiesis begins operating under restricted supply. The marrow experiences scarcity even though the reticuloendothelial system and liver may still be holding substantial iron stores. This is one of the clearest examples of the liver acting as endocrine authority over mineral traffic. It is not mining iron from tissues itself. It is issuing the order that determines whether iron will circulate.

So inflammation changes iron handling because the body stops treating iron as a simple nutrient and starts treating it as contested chemistry. The liver hears the inflammatory signal, raises hepcidin, removes ferroportin from the main export surfaces, and shifts iron out of plasma and into storage and recycling cells. In the short term that can be protective. In the long term it can starve red blood cell production while iron remains trapped in reserve.

Iron Overload and Iron Restriction as Two Different Failure States

Iron overload and iron restriction are opposite failures of the same traffic system. In iron overload, too much iron is allowed to enter or remain available, and the metal begins accumulating in tissues, especially the liver, heart, and endocrine organs, where excess ferrous iron can drive oxidative damage through free radical chemistry. In iron restriction, the body may have iron on hand, sometimes even normal or increased stores, but too little of that iron is being allowed into the circulating pool where marrow and other tissues can actually use it. The first state is excess metal burden. The second is trapped availability. Both are harmful, but they injure the body for opposite reasons.

Iron overload happens because iron balance is controlled mainly at absorption and release, not by a large regulated excretory pathway. The body loses iron only in limited, relatively unregulated ways such as shedding cells, sweat, menstruation, and enterocyte turnover, which means chronic excess entry or excess retention has nowhere fast to go. Iron then accumulates, and because excess ferrous iron can catalyze formation of reactive oxygen species, tissues begin paying for that storage with oxidative injury to lipids, proteins, and nucleic acids. This is why iron overload is not just "high iron." It is excess iron becoming tissue chemistry.

Iron restriction is different. Here the body may not be lacking total iron so much as refusing to release it. In inflammatory states, cytokines increase hepatic hepcidin, and hepcidin binds ferroportin, the iron export protein, causing its degradation. Once ferroportin is removed from enterocytes, macrophages, and hepatocytes, iron becomes trapped inside those cells instead of moving into plasma. The marrow then sees a relative iron shortage even while macrophages and storage cells may still be iron-rich. That is the logic of anemia of inflammation or chronic disease. The iron is present, but the traffic is blocked. Ferritin can remain elevated because stored iron is not the same thing as usable circulating iron.

So these are two different failure states of mineral authority. Iron overload means the gates of entry and release have been too permissive for too long, and tissue storage becomes toxic. Iron restriction means the liver has closed those gates too tightly, often under inflammatory instruction, and functional iron availability collapses despite preserved or elevated stores. One state burdens organs with excess metal. The other starves blood production and oxidative work by withholding that same metal from circulation.

The Liver as the Organ That Decides Whether Iron Moves or Stays Trapped

Iron may be present in the body, but presence is not the same as permission to circulate. That distinction is where the liver becomes decisive. The intestine can absorb iron from food. Macrophages can recover iron from aging red blood cells. Hepatocytes and other storage sites can hold reserve iron. But none of those pools tells you whether iron is actually available to the marrow, to mitochondria, or to the wider circulation at a given moment. The real question is whether iron is being allowed to leave the cell. The liver answers that question hormonally. By producing hepcidin, it governs the fate of ferroportin, the iron exporter on enterocytes, macrophages, and storage cells. When hepcidin rises, ferroportin is removed, and iron stays where it sits. When hepcidin falls, ferroportin remains on the surface, and iron is allowed to move into plasma. This is why the liver should be understood not merely as an iron storage organ, but as the organ that decides whether iron traffic is open or closed.

That authority matters because the body has no generous rapid excretory route for iron. It cannot afford sloppy control. Too much free movement risks overload and tissue injury. Too much withholding risks functional deficiency even when total iron is still present. The liver therefore operates as the central adjudicator between these extremes. If body iron stores are adequate or high, hepcidin rises and the liver effectively orders iron to remain trapped in absorptive, recycling, and storage cells. If iron is scarce or erythropoietic demand is high, hepcidin falls, ferroportin survives, and iron is released into circulation where transferrin can carry it to the tissues that need it. The marrow does not negotiate directly with the intestine. The liver stands in the middle and writes the hormone that determines whether absorbed or recycled iron gets to move.

This also explains why inflammation changes iron biology so profoundly. During inflammatory signaling, the liver raises hepcidin even when total body iron is not excessive. The result is sequestration. Enterocytes absorb less usable iron into blood. Macrophages hold on to recycled iron. Storage sites become less willing to release reserve. Plasma iron falls. The marrow begins experiencing scarcity while tissue iron may still be present in substantial quantity. In that state the body is not lacking iron in the simple sense. It is being denied iron by its own traffic authority. The liver has judged movement too dangerous and chosen retention instead.

So the liver decides whether iron moves or stays trapped because it controls the hormone that governs the export step. Iron in a cell is not yet available iron. Only once the liver allows ferroportin to remain active does that iron become part of circulating physiology. The mineral may be in the body. The liver determines whether the body gets to use it.

Chapter 27

Vitamin D, Thyroid Hormone, and Hormone Conversion

The Liver's Role in 25-Hydroxylation of Vitamin D

Vitamin D does not become physiologically useful in the form it first enters the body. That is the first correction the reader needs to make. Whether vitamin D is synthesized in skin after ultraviolet light exposure or absorbed from food, it begins as a precursor, not as the finished hormone-like molecule that blood tests and downstream tissues are really tracking. The liver is the first major conversion site in that sequence. It takes incoming vitamin D and performs 25-hydroxylation, a chemical step in which a hydroxyl group is added at the 25 position of the molecule. This produces 25-hydroxyvitamin D, also called calcidiol, the major circulating storage and transport form measured in blood. Without that hepatic step, vitamin D remains upstream of the form the body actually uses to assess supply and to feed the next stage of activation.

That conversion matters because vitamin D has to be made compatible with whole-body traffic before it can become a regulated endocrine signal. The liver is not simply tagging the molecule with an extra oxygen and moving on. It is changing the fate of the compound. Raw vitamin D entering from skin or intestine does not yet function as the main circulating reserve. Once hepatocytes 25-hydroxylate it, the molecule becomes much more useful as a stable blood-borne intermediate. It can then circulate bound largely to vitamin D binding protein and albumin, remain available over time, and serve as the substrate delivered onward to the kidney and other tissues for further activation. The liver therefore is not the final site of vitamin D activation, but it is the organ that turns vitamin D into a form the endocrine system can work with.

This is why the common blood test for vitamin D status measures 25-hydroxyvitamin D rather than the original parent compound. The body is not asking how much raw vitamin D recently entered. It is asking how much has successfully passed through hepatic conversion and now exists in the major circulating pool. That makes the liver central to interpretation. A person can have sunlight exposure or dietary intake, but unless the liver performs this conversion step, the molecule has not yet entered the main transport form that reflects real systemic availability. The test is therefore not only a nutrient measure. It is also, indirectly, a readout of hepatic participation in hormone precursor handling.

The chemistry also fits the broader logic of the liver as a conversion organ. Vitamin D is a fat-soluble secosteroid, which means it belongs to the broader family of lipid-like hormone-related molecules that often require transport proteins, controlled activation steps, and tissue-specific modification before they become fully functional. The liver is already built to handle this class of problem. It modifies steroid-related compounds, packages lipids, makes binding proteins, and converts raw incoming molecules into transportable intermediates. 25-hydroxylation fits that pattern. The liver receives a precursor and turns it into a circulating form suited for regulated endocrine use rather than immediate uncontrolled action.

This also helps explain why liver disease can alter vitamin D status in ways that are deeper than poor intake alone. If hepatocyte function declines enough, 25-hydroxylation may become less effective, vitamin D binding protein production may be altered, bile-related fat absorption may already be impaired upstream at the intestinal level, and the whole system begins losing efficiency before the kidney ever sees the substrate it is supposed to activate further. The final hormonal problem may show up as poor calcium balance, altered bone metabolism, or low measured 25-hydroxyvitamin D, but the sequence often began with failure of hepatic conversion or hepatic support for transport. The liver therefore belongs at the front of the vitamin D story, not as a footnote between sunlight and bone.

So the liver's role in 25-hydroxylation of vitamin D is to take a precursor and turn it into the main circulating intermediate that the rest of the endocrine system can actually use. Skin or diet provides the raw molecule. The liver performs the first major activation step. The bloodstream carries that converted form onward. Without the liver, vitamin D remains earlier in the sequence than the body needs it to be.

Why the Circulating Vitamin D Test Measures a Liver-Made Intermediate

The circulating vitamin D test measures a liver-made intermediate because the body does not track raw vitamin D as the most useful indicator of status. Vitamin D entering from skin or diet is still too early in the pathway. It is a precursor, not yet the main circulating reserve form that reflects what the body has successfully absorbed or synthesized and then processed into endocrine traffic. The liver changes that. Hepatocytes take incoming vitamin D and perform 25-hydroxylation, adding a hydroxyl group that turns the parent compound into 25-hydroxyvitamin D, also called calcidiol. That is the form that accumulates in blood in meaningful quantity and persists long enough to act as the body's main storage and transport intermediate. So when the lab measures vitamin D status, it is not asking how much parent molecule briefly entered the system. It is asking how much has already passed through the liver and entered the major circulating pool.

This matters because parent vitamin D is too dependent on recent exposure to serve as a stable index on its own. Sun exposure can vary day to day. Dietary intake can vary meal to meal. Absorption can vary with bile flow and intestinal fat handling. A blood test that focused only on the incoming parent compound would be reading recent arrival more than actual body availability. The liver-made 25-hydroxy form is more useful because it smooths that variability. It represents vitamin D that has already been taken up, processed, and placed into the bloodstream in a form suitable for wider distribution. It is therefore a better reflection of body stores and substrate availability for further activation. The test is measuring the part of the pathway that has already survived the first major conversion step.

It is also important that 25-hydroxyvitamin D is not the final fully active hormonal form. The kidney performs the next major activation step, producing 1,25-dihydroxyvitamin D, the more tightly regulated form that acts as the major hormonal signal in calcium and phosphate physiology.

But that final form is not the best general status marker because its concentration is much lower, its regulation is tighter, and it can remain normal or even rise in some deficiency states due to compensatory control. The liver-made 25-hydroxy intermediate is different. It is abundant enough, stable enough, and representative enough of the body's reserve to function as the main clinical window into vitamin D status. That is why the standard circulating test is centered there. It measures the substrate pool the kidney and other target systems are being given to work with, not merely the moment-to-moment hormonal endpoint.

So the circulating vitamin D test measures a liver-made intermediate because the liver is where vitamin D first becomes a durable endocrine resource rather than a newly arrived precursor. The body wants to know what has been successfully converted into the main transport and reserve form, not just what recently entered through skin or gut. The blood test is therefore reading hepatic participation in vitamin D biology as much as it is reading vitamin D itself.

Thyroid Hormone Transport Proteins

Thyroid hormone does not travel through blood mainly as free naked signal. It travels bound. That matters because thyroxine, T4, and triiodothyronine, T3, are lipophilic molecules, chemically comfortable in membranes and poorly suited to drift unescorted through an aqueous plasma environment. The bloodstream therefore needs carrier proteins that can hold these hormones in circulation, buffer sudden swings in free concentration, and deliver a stable reservoir from which tissues can draw what they need. The liver is central to that system because it synthesizes several of the main plasma proteins that carry thyroid hormone, especially thyroxine-binding globulin, transthyretin, and albumin. These proteins do not merely move hormone from one place to another. They shape how much hormone remains free, how long it stays in blood, and how stable the endocrine signal feels to the body over time.

Thyroxine-binding globulin, usually abbreviated TBG, is the highest-affinity major carrier. It binds T4 especially tightly and carries much of the circulating hormone despite being present at far lower concentration than albumin. High affinity means it holds hormone strongly, which is useful because the body does not want wild fluctuations in free thyroid hormone with every shift in secretion or tissue demand. Transthyretin, formerly called prealbumin, binds thyroid hormone with lower affinity than TBG but still participates meaningfully in transport, especially for T4. Albumin binds thyroid hormone more loosely, but because albumin is so abundant in plasma, it still contributes substantially to the overall transport pool. So the blood is not carrying thyroid hormone through one protein alone. It is carrying it through a layered transport system in which different proteins provide different combinations of strength, abundance, and exchange behavior.

This is why the distinction between total hormone and free hormone matters so much. Most circulating T4 and T3 are protein-bound at any given time. Only a small fraction remains free, and that free fraction is generally the biologically active portion most immediately available to enter cells and engage receptors after intracellular conversion and handling. The transport proteins therefore act like a buffering reservoir. They keep a large amount of hormone in circulation without letting all of it act at once. If free hormone falls, some bound hormone can dissociate and replenish the active pool. If hormone secretion rises abruptly, much of that extra hormone can be buffered by binding rather than instantly flooding tissues in full active proportion. The bloodstream is not simply carrying thyroid hormone. It is smoothing thyroid hormone.

The liver matters here because changes in hepatic protein synthesis can change the transport landscape even when thyroid gland output has not changed to the same degree. If TBG rises, total circulating thyroid hormone may rise because more binding capacity is present, even while free hormone remains relatively stable. If binding proteins fall, total hormone may fall without a proportional fall in free hormone. This is one reason interpretation of thyroid labs requires structural thinking. A blood test measuring total T4 is partly measuring thyroid output and partly measuring the protein environment the liver helped build. Liver disease, malnutrition, inflammatory states, hormone states such as estrogen excess, and other shifts in plasma protein production can all change that environment. The gland may be normal. The transport system may not be.

So thyroid hormone transport proteins reveal the liver's endocrine role in a very precise way. The liver is not making T4 and T3, but it is making much of the plasma architecture that determines how those hormones travel, how stable their circulating pool remains, and how free hormone is buffered against sudden change. The thyroid writes the signal. The liver helps shape the bloodstream that carries it.

T4 to T3 Conversion and Hepatic Participation

The thyroid gland does not release its entire message in final form. Much of what it sends into blood is thyroxine, T4, a prohormone that still needs peripheral conversion before the body receives the fuller metabolic force of triiodothyronine, T3. That conversion matters because T3 is the more potent signaling molecule at the tissue level. The liver participates in this process by serving as one of the major high-blood-flow organs where deiodination occurs, especially through type 1 iodothyronine deiodinase, often abbreviated D1. Deiodination means removal of an iodine atom from the hormone molecule, and the direction of that removal matters. Outer-ring deiodination converts T4 into active T3. Inner-ring deiodination converts T4 into reverse T3, or rT3, an inactive form. The liver therefore is not just carrying thyroid hormone through plasma. It is helping decide what form of thyroid signal the bloodstream and tissues will actually receive.

This fits the liver's broader identity as a conversion organ. Hepatocytes already modify steroid-related compounds, generate circulating intermediates, and reshape hormone traffic before it reaches the rest of the body in finished physiological form. Thyroid hormone follows that same logic. T4 arrives in blood bound largely to transport proteins, and the liver participates in turning part of that circulating pool into T3 through D1-mediated outer-ring deiodination. That does not mean the liver is the only site of conversion. Other tissues contribute, and local tissue conversion is important throughout the body. But the liver remains one of the major peripheral participants in generating circulating T3, especially because of its blood flow and its expression of D1. The thyroid writes much of the precursor. The liver helps sharpen the signal.

The mechanism also explains why liver participation in thyroid physiology is more than passive transport. A hormone precursor moving through hepatocytes can be activated or redirected depending on which deiodinase pathways dominate. If outer-ring deiodination predominates, more active T3 is generated. If inner-ring deiodination is favored elsewhere in the system, more inactive reverse T3 appears instead. This is one reason serious illness, calorie deprivation, and some drug exposures can shift thyroid hormone patterns without the thyroid gland itself being the primary source of failure. Peripheral conversion changes.

Serum T3 may fall. Reverse T3 may rise. The liver is one of the tissues inside that peripheral conversation, not an observer standing outside it.

So T4 to T3 conversion and hepatic participation should be understood as part of the body's endocrine finishing work. The thyroid secretes a large amount of T4. The liver helps convert part of that circulating precursor into more active T3 through deiodination. In doing so, it participates directly in determining how much thyroid signal remains stored as precursor, how much becomes active hormone, and how much endocrine force the rest of the body is actually exposed to.

Estrogen, Cortisol, and Other Hormone Transformations

The liver does not only transport hormones through blood. It changes them. That is the first principle. Many circulating hormones arrive at the liver after they have already signaled in target tissues, and the liver's job is to alter their chemical form so they can be inactivated, redistributed, extended in circulation in altered form, or prepared for biliary or urinary excretion. Steroid hormones are a major part of this workload. In the liver they undergo phase I metabolism, which includes oxidation, hydroxylation, and reduction, and phase II metabolism, which includes conjugation reactions such as glucuronidation and sulfation that make the molecules more water-compatible and easier to clear. The liver is therefore not a passive highway for endocrine traffic. It is one of the main organs deciding how long a hormone remains active, what metabolites it becomes, and what route it takes on the way out.

Estrogen shows this clearly. Hepatic metabolism can hydroxylate estrogens through cytochrome P450 pathways, creating downstream estrogen metabolites with different biological implications, and the liver also conjugates estrogens through glucuronidation and sulfation, both of which are major ways of reducing free active estrogen exposure and promoting excretion. Sulfation can also create circulating sulfate conjugates that alter the transport and half-life of estrogen-related molecules rather than simply erasing them instantly. The point is that estrogen does not simply rise and fall according to ovarian output alone. The liver participates in determining how much remains active, how much is converted into other metabolites, and how much is pushed toward elimination. This is one reason altered liver function can change the endocrine environment without the endocrine gland itself being the original source of failure.

Cortisol follows the same broader logic even though the details differ. Cortisol is not meant to remain in circulation indefinitely in one fixed form. The liver participates in glucocorticoid metabolism through reduction, oxidation, and conjugation pathways that help terminate or reshape steroid activity and prepare metabolites for clearance. In plain physiological terms, the liver helps determine how much cortisol remains as active circulating signal and how much has already been moved toward metabolic disposal. So when liver function changes, cortisol biology can change with it, not necessarily because adrenal output changed first, but because one of the major organs responsible for steroid transformation is no longer processing the hormone with the same efficiency.

The same principle extends beyond estrogen and cortisol. Steroid hormones in general do not simply vanish after receptor binding. They are oxidized, reduced, hydroxylated, sulfated, glucuronidated, and otherwise chemically rewritten so their activity, solubility, and route of elimination change. The liver is one of the main sites where this happens. That is why endocrine physiology cannot be taught honestly as gland secretion alone. The gland writes the first message. The liver helps decide how long that message remains readable, what altered versions of it circulate, and when it is finally removed.

Hormonal Symptoms That Can Appear When Hepatic Processing Falls Behind

When hepatic hormone processing falls behind, the clearest symptoms often come from changed steroid balance rather than from one single hormone simply rising or falling in isolation. The liver normally helps transform, conjugate, and clear estrogens, cortisol-derived metabolites, and many other circulating steroid signals while also making the transport proteins that buffer endocrine traffic in plasma. Once that handling weakens, the hormonal field becomes less disciplined. In men with cirrhosis, the visible pattern can shift toward feminization and hypogonadal change: gynecomastia, spider angiomas, palmar erythema, loss of axillary or chest hair, female-pattern pubic hair, testicular atrophy, erectile dysfunction, infertility, and loss of libido are all recognized findings in chronic liver disease. This is not because the liver suddenly becomes a sex gland. It is because the organ that normally reshapes and clears steroid traffic is no longer keeping endocrine ratios, transport, and disposal under the same control.

The thyroid side looks different. The liver is one of the major organs participating in peripheral T4 to T3 conversion through type 1 deiodinase, and in illness or advanced liver dysfunction that conversion can fall, with lower T3 and higher reverse T3 becoming more common. Mechanistically, the body is no longer sharpening as much circulating prohormone into active hormone, and the bloodstream begins carrying a duller thyroid signal. Clinically this can contribute to a low-T3 pattern of fatigue, slowed metabolic tone, weakness, and reduced physiologic drive, even when the thyroid gland itself is not the primary site of failure. The important distinction is that these symptoms do not always read like clean textbook primary hypothyroidism. They often appear as part of a broader endocrine flattening in which hormone transport, conversion, and clearance have all become less precise because the liver's processing work is lagging behind the body's signal traffic.

Chapter 28

IGF-1, Thrombopoietin, and Systemic Signaling

Growth Hormone and Hepatic IGF-1 Production

Growth hormone does not do all of its work directly at the tissues it is famous for affecting. A major part of its body-wide effect is routed through the liver. The pituitary releases growth hormone into blood, but one of the main organs that hears that signal and turns it into a longer-range anabolic message is the liver. Hepatocytes respond by producing insulin-like growth factor 1, or IGF-1, a peptide signal that carries much of the growth-promoting and protein-building logic associated with growth hormone into the wider circulation. This matters because growth hormone on its own is not the whole story. The hormone released by the pituitary is the first instruction. Hepatic IGF-1 production is one of the main ways that instruction becomes a distributed systemic program. The liver is not just another target tissue receiving growth hormone. It is one of the organs that converts endocrine command into endocrine output.

That conversion makes physiological sense because the liver sits at the intersection of nutrient handling, protein synthesis, and blood chemistry control. Growth is not merely a matter of telling cells to divide or tissues to enlarge. Growth requires amino acid availability, glucose management, lipid handling, and a blood-borne signal that can report whether the organism is in a condition where anabolic investment is even possible. IGF-1 becomes part of that report. When the liver responds appropriately to growth hormone, it releases IGF-1 into blood, and that circulating molecule supports growth plate activity, protein synthesis, tissue maintenance, and broader anabolic tone in multiple organs. The pituitary therefore does not simply command growth in a direct one-step line. It speaks to the liver, and the liver helps write the hormonal language the rest of the body receives.

This is also why IGF-1 reflects more than pituitary output alone. A person can have growth hormone present and still show altered IGF-1 production if hepatic function, nutritional state, inflammatory burden, or metabolic context are poor. The liver is not a neutral relay that always translates the signal with equal force. It is a synthetic organ reading growth hormone in the context of substrate availability and internal stress. Severe liver disease, malnutrition, and inflammatory states can all lower IGF-1 production because the organ is no longer positioned to support a full anabolic message. In that setting the problem is not only that the pituitary spoke weakly. It is that the liver did not or could not answer with the same normal endocrine output. This is one reason growth biology has to be taught through the liver as well as through the pituitary. The bloodstream is carrying a liver-made growth signal, not just a pituitary one.

The distinction between growth hormone and IGF-1 also helps the reader see a larger principle about liver physiology. The liver is not only a processor of nutrients and waste. It is an endocrine amplifier and interpreter. A gland upstream releases a hormone. The liver receives that hormone and produces another signal with different stability, reach, and tissue consequences. That is exactly the kind of work a central metabolic organ is well suited to do. Hepatocytes are already synthesizing major plasma proteins, transport proteins, acute-phase proteins, clotting proteins, and hormone-related intermediates. IGF-1 belongs naturally in that company. It is one more example of the liver taking a message, integrating it with the state of the body, and then releasing a signal that changes the behavior of distant tissues.

So growth hormone and hepatic IGF-1 production should be understood as a two-step endocrine sequence. The pituitary sends the initiating signal. The liver receives it and generates a major circulating mediator of growth and anabolic activity. Growth hormone announces the program. The liver helps make the program systemic.

Why Liver Function Influences Growth and Tissue Maintenance Signals

Growth and tissue maintenance do not depend on hormones alone. They depend on whether the body can turn hormonal instruction into a usable circulating program, and the liver is one of the main organs that performs that translation. A gland can release a signal, but that signal still has to move through blood, meet transport proteins, encounter the right metabolic context, and often be converted into another form before distant tissues receive a coherent message. The liver sits in the middle of that sequence. It makes insulin-like growth factor 1, IGF-1, in response to growth hormone. It makes albumin and other plasma proteins that shape how molecules travel. It participates in thyroid hormone conversion. It reads inflammatory signals and can shut down or redirect anabolic priorities when the organism is under threat. This means the liver is not a passive background organ during growth or repair. It helps decide whether the blood will carry a growth-promoting message, a maintenance message, or a restraint message.

IGF-1 is the clearest example. Growth hormone from the pituitary does not simply strike bone and muscle and force them to grow by direct command alone. A large part of its body-wide effect depends on hepatic response. Hepatocytes receive the growth hormone signal and produce IGF-1, which then circulates as one of the main anabolic signals supporting protein synthesis, tissue growth, and maintenance of lean structure. If liver function is poor, that translation weakens. The pituitary may still be speaking, but the bloodstream carries less of the downstream message. The same principle extends beyond childhood growth. Adult tissues are constantly repairing proteins, maintaining connective structures, renewing cellular components, and preserving muscle and organ integrity. Those are not one-time developmental events. They are ongoing maintenance costs, and the liver helps determine whether the endocrine environment supports them.

This is also why liver dysfunction often produces a physiology that looks catabolic, meaning biased toward breakdown rather than building, even when calorie intake alone does not explain the pattern. A damaged liver has less synthetic reserve, less ability to generate IGF-1, less ability to maintain stable plasma protein architecture, and often less ability to convert or clear hormones in the correct proportions. Inflammation compounds this further. Cytokines shift hepatic output toward acute-phase proteins and away from some anabolic priorities. Thyroid signaling may flatten as peripheral conversion becomes less favorable. Steroid handling changes. Nutrient traffic becomes less orderly. The result is that the bloodstream starts carrying weaker signals for growth and repair at the same time it carries stronger signals for defense and conservation. The body begins acting as though survival is the priority and long-horizon tissue investment has become too expensive.

So liver function influences growth and tissue maintenance signals because the liver is one of the organs that turns endocrine instruction into circulating biological reality. It does not merely receive hormones. It amplifies some, converts some, restrains some, and builds the plasma environment in which those signals travel. When the liver is functioning well, growth and repair messages move through blood with more coherence. When hepatic function falls behind, the organism loses part of its anabolic language. The tissues then do not merely lack nutrients. They lose the bloodstream's permission to build.

Thrombopoietin and Platelet Production

Platelets do not appear in blood simply because bone marrow feels like making them. Their production is driven by a liver-centered signal. That signal is thrombopoietin, usually abbreviated TPO, a hormone made predominantly by the liver that tells the marrow how strongly it should support megakaryocyte growth and platelet release. Megakaryocytes are the large marrow cells that eventually shed platelets into the circulation. This matters because platelet biology is often taught as if it belongs mainly to the bone marrow and the spleen, with the liver entering the story only later when portal hypertension enlarges the spleen and platelet counts fall. That is incomplete. The marrow builds the platelets, but the liver supplies much of the hormonal instruction that keeps that construction line running. If the liver is not producing thrombopoietin appropriately, the marrow is not receiving the same signal to expand megakaryocyte output, and platelet production weakens before the platelets ever face sequestration or destruction elsewhere.

Thrombopoietin works by binding its receptor on megakaryocyte precursors and mature megakaryocytes, pushing the lineage toward growth, maturation, and platelet release. In plain terms, it tells the marrow that circulating platelet support needs to be maintained. This is one more example of the liver acting as an endocrine authority rather than as a passive metabolic organ. Hepatocytes are already making albumin, clotting proteins, acute-phase proteins, and hormone-related transport molecules. Thrombopoietin belongs in that same pattern. The liver is helping determine whether the blood will carry enough platelets to maintain vascular repair and hemostatic readiness. Platelets are not simply fragments floating around for clotting. They are continuously consumed in small amounts, continuously aged out, and continuously needing replacement. The bloodstream stays platelet-competent because the marrow is being hormonally supported to keep the line moving.

This also explains why liver disease so often carries thrombocytopenia, a low platelet count, as one of its blood signatures. The usual teaching emphasizes splenic sequestration from portal hypertension, and that is real. An enlarged spleen can trap and hold more platelets, reducing the circulating count. But that is not the whole mechanism. A failing liver may also produce less thrombopoietin, which means the marrow is receiving less stimulus to generate new platelets in the first place. One problem lowers output. Another increases pooling and removal from the circulating compartment. The platelet count then falls from both directions at once. This is why thrombocytopenia in liver disease is not just a spleen story. It is also a synthetic and endocrine story. The liver is no longer writing the same platelet-production signal into blood.

So thrombopoietin and platelet production belong inside liver physiology because the liver is one of the organs deciding whether the marrow is told to keep building. The marrow executes the work. The liver helps issue the command. When hepatic function is intact, platelet renewal is better supported. When hepatic output declines, the blood may begin losing not only clotting proteins and albumin, but also part of the hormonal instruction that keeps platelet numbers up.

Angiotensinogen and Blood Pressure Relevance

Angiotensinogen is one of the clearest examples of the liver writing a systemic control signal into the blood long before the rest of the body seems to be involved. It is a protein made mainly by hepatocytes and released into plasma in an inactive precursor form. By itself it does not constrict a vessel, raise blood pressure, or move sodium. What it does is provide the raw substrate for the renin-angiotensin system, the hormonal cascade that helps defend circulating volume and vascular pressure when perfusion begins to fall. This matters because blood pressure is not controlled only by the heart, the kidneys, or the vessels in isolation. It is also controlled by whether the bloodstream contains the precursor proteins required for those organs to generate a pressure-saving response. The liver supplies one of the most important of those precursors.

The sequence begins when renin, an enzyme released from the kidney, cleaves angiotensinogen into angiotensin I. Angiotensin I is then converted, largely by angiotensin-converting enzyme, into angiotensin II, the active peptide that narrows blood vessels, stimulates aldosterone release, supports sodium retention, and helps preserve blood pressure when the body senses low effective circulating volume. So the liver is not the organ pulling the final lever, but it is the organ stocking the line with the molecule that lets the lever work. Without hepatic angiotensinogen in plasma, the kidney can release renin and still have far less substrate to act on. That is the deeper relevance. The liver is not only metabolizing the bloodstream. It is provisioning one of the endocrine systems that determines whether circulation can defend itself under stress.

This is why angiotensinogen belongs in a chapter on endocrine and mineral authority rather than in a narrow blood pressure chapter alone. The liver makes a precursor that becomes part of a major pressure-regulating axis. The kidney reads perfusion and releases renin. The vascular system responds to angiotensin II. The adrenal gland responds with aldosterone. The body then shifts sodium and water handling, vascular tone, and effective circulatory filling. But the sequence begins with the fact that the bloodstream already contained a liver-made substrate waiting to be cut. That is a very liver kind of role. The organ places a quiet precursor into circulation, and that precursor becomes physiologically important only when another tissue calls on it under the right conditions.

This also helps explain why blood pressure problems are often really volume-signaling problems in disguise. The renin-angiotensin system is not just about squeezing arteries tighter. It is about defending effective circulation when the body believes perfusion is threatened. Angiotensinogen matters because it is part of how that threat response becomes possible. A liver making plasma proteins is therefore also helping determine how strongly the body can activate one of its main blood pressure rescue systems. The vessels may tighten downstream. The kidney may trigger the cascade. But the liver helped load the bloodstream with the molecular potential for that response before the crisis ever began.

The Liver as Endocrine Tissue in Plain Sight

The liver is usually introduced as a metabolic organ, a detox organ, a bile organ, or a storage organ, and all of that is true. But those descriptions become incomplete the moment you watch what the liver actually releases into blood. An endocrine tissue is a tissue that makes signals or signal-shaping molecules and sends them into the circulation to change the behavior of distant organs. By that standard, the liver is endocrine tissue in plain sight. It makes hepcidin and decides whether iron is allowed to move or stay trapped. It makes insulin-like growth factor 1, IGF-1, in response to growth hormone and helps determine whether the body stays in a growth-supporting or growth-restrained state. It makes thrombopoietin and influences whether the bone marrow is told to keep producing platelets. It makes angiotensinogen and stocks the bloodstream with a precursor required for one of the body's main blood pressure defense systems. It converts vitamin D into its main circulating intermediate. It helps convert thyroid hormone. It makes transport proteins that determine how hormones travel, how much stays free, and how much remains buffered in reserve. The liver does not just respond to endocrine signals. It writes endocrine consequences into the blood.

What makes this easy to miss is that the liver often acts upstream of the visible symptom. A person may think about low platelets as a marrow issue, altered thyroid tone as a gland issue, poor growth signaling as a pituitary issue, iron restriction as a marrow or dietary issue, or blood pressure regulation as a kidney and vessel issue. But the liver is often shaping the bloodstream those organs are working inside. It is making the precursor, the carrier, the intermediate, or the controlling hormone that decides whether the downstream tissue receives a strong message, a weak message, or no usable message at all. That is classic endocrine logic. The difference is that the liver does not always behave like a gland with one named hormone and one obvious target. It behaves like a broader endocrine authority. It changes mineral traffic, hormone availability, growth tone, clotting readiness, plasma volume behavior, and inflammatory protein composition by altering what blood contains before the target tissue ever gets its turn.

This is also why liver disease produces symptoms that seem to belong to other systems. The person may present with edema, low platelets, altered iron handling, feminizing steroid patterns, low IGF-1, weaker thyroid conversion, poor vitamin D status, or unstable blood pressure biology, and the shallow reading is to treat each of those as a separate endocrine or hematologic problem. The deeper reading is that the liver is no longer manufacturing the signals, precursors, binding proteins, and conversion products that normally keep those systems coherent. The gland downstream may still exist. The receptor may still exist. The target tissue may still be present and waiting. But the bloodstream no longer carries the same instruction set. A failing liver does not merely stop clearing toxins. It stops writing part of the body's endocrine language.

So the liver should be understood as endocrine tissue in plain sight because it governs distant physiology through circulating molecules every day, even when no one is calling it a gland. It is making commands, permissions, restraints, and transport frameworks and releasing them into plasma where other organs are forced to live with the result. The liver does not only metabolize what the body already is. It helps decide what the body will be allowed to become.

Why an Organ Known for Metabolism Is Also a Hormone Organ

The distinction between metabolism and hormones only looks clean when physiology is taught in pieces. In the body, the two are inseparable. Metabolism decides what substrates are available, what can be stored, what can be converted, what can be released, and what burden must be cleared. Hormones decide when those choices should shift, which tissues should change behavior, and whether the body is in a fed state, a fasting state, a growth state, an inflammatory state, a reproductive state, or a conservation state. An organ sitting at the center of nutrient traffic, blood chemistry, protein synthesis, steroid transformation, and circulating substrate control cannot remain merely metabolic. The liver handles too much of the body's moving chemistry not to become endocrine. It does not just respond to signals coming from elsewhere. It changes those signals, builds the transport environment they travel in, and releases new signals of its own into blood.

This is the deeper reason the liver so often disappears inside endocrine stories while quietly governing them. The pituitary may release growth hormone, but the liver answers with insulin-like growth factor 1, which carries much of the anabolic message into the wider circulation. The intestine may absorb iron, but the liver writes hepcidin, the hormone that decides whether that iron is allowed to enter plasma or remain trapped in cells. The marrow may produce platelets, but the liver makes thrombopoietin, the signal that helps tell the marrow to keep that production line moving. The thyroid may release T4, but the liver helps determine how much of that prohormone is converted into more active T3 and how the hormone is transported through plasma. Vitamin D may come from skin or diet, but the liver performs the first major hydroxylation step that turns it into the main circulating intermediate. These are not side notes attached to metabolism. They are endocrine decisions made by a metabolic organ.

The liver is able to do this precisely because metabolism is the raw material of hormone authority. A hormone organ that does not understand nutrient state, inflammatory state, mineral state, transport capacity, and excretory burden can only send simplified messages. The liver sees far more than that. It receives portal blood from the intestine, reads fasting fat delivery from adipose tissue, tracks inflammatory cytokines, monitors iron load, transforms steroid hormones, manufactures plasma proteins, and continuously rewrites what kind of blood the rest of the body will live in. That gives it unusual leverage. It can restrain iron movement when inflammation makes free iron dangerous. It can lower albumin and raise acute-phase proteins when defense takes priority over ordinary transport. It can alter hormone availability by changing transport proteins, by converting precursors, or by metabolizing active compounds into more excretable forms. In other words, the liver does not merely supply energy chemistry. It supplies endocrine context.

This is why liver disease so often produces symptoms that look hormonal even when no endocrine gland is structurally primary. Platelets fall because thrombopoietin signaling weakens. Iron becomes unavailable because hepcidin stays high. Estrogenic effects appear because steroid transformation and clearance change. Thyroid-related energy tone can flatten because peripheral conversion shifts. Growth and repair signals decline because IGF-1 production falls. The failing liver has not suddenly turned into a broken thyroid, gonad, pituitary, or marrow. It has stopped performing the metabolic conversions, protein synthesis, transport shaping, and endocrine signaling that those systems were quietly depending on. A body can still have glands and still lose hormonal coherence if the liver is no longer writing the bloodstream correctly.

So an organ known for metabolism is also a hormone organ because metabolism at this level is not just fuel burning. It is control over what enters blood, what form it enters in, what remains active, what gets transformed, what gets withheld, and what message the circulation carries next. The liver is endocrine not because it resembles a classical gland in shape, but because it governs distant tissue behavior through molecules released into blood. It metabolizes, and in doing so, it instructs.

Chapter 29

Vitamin and Mineral Storage

Glycogen as Quick-Access Carbohydrate Reserve

Glycogen is the liver's nearest glucose reserve, the form in which the organ holds carbohydrate close enough to circulation to matter within hours rather than weeks. That distinction matters because the body does not store every energy form for the same purpose. Adipose tissue is the long-horizon calorie depot. Glycogen is different. It is rapid-access carbohydrate held for short-interval blood sugar support. The liver builds it when glucose is abundant after a meal, packing thousands of glucose units into a dense branched polymer that can sit inside hepatocytes without creating the osmotic chaos that free glucose would create if it were stored in large quantity as individual molecules. Osmotic chaos is the right phrase here because glucose is highly osmotically active. Leave too much of it free inside a cell and water movement becomes disruptive. Glycogen solves that by turning dispersed sugar into compact reserve. The liver is not just storing carbohydrate. It is restructuring carbohydrate into a form the cell can hold safely and release quickly.

That release speed is the point. Blood glucose has to remain usable between meals, overnight, and during the first phase of fasting even when the intestine has gone quiet. The brain, red blood cells, renal medulla, and other glucose-requiring tissues cannot wait for the full slower machinery of long fasting adaptation to take over. The liver answers this by breaking glycogen back down through glycogenolysis, releasing glucose-6-phosphate from the branched polymer and then converting it to free glucose that can leave the hepatocyte and re-enter the bloodstream. This is one of the liver's great timing functions. It takes the concentrated surge of a meal, stores part of it in a compact intracellular reserve, and then meters part of that reserve back into blood as the meal recedes. Glycogen therefore is not only storage. It is delayed release built into tissue.

This is why glycogen belongs inside a chapter on storage even though it behaves differently from vitamin and mineral storage. Vitamins and minerals are often held as reserves against depletion. Glycogen is held against interruption. Its scale is smaller and its turnover faster, but the logic is similar. The liver does not trust the body's needs to line up perfectly with the timing of intake. It stores what will be needed soon but not continuously supplied. That is what makes glycogen such an important liver reserve rather than a generic carbohydrate deposit. It is stored where the bloodstream can be protected by it.

The branched structure also matters mechanistically. Glycogen is not one long simple chain. It is highly branched, which creates many terminal ends where enzymes can add or remove glucose units quickly. That architecture gives the liver a major practical advantage. It can build glycogen rapidly in the fed state and mobilize it rapidly in the early fasting state. The reserve is not only chemically compact. It is kinetically accessible. The body does not need a storage form that is merely dense. It needs one that can be opened fast when blood glucose begins to fall. Glycogen provides both.

So glycogen is best understood as the liver's quick-access carbohydrate reserve because it exists to bridge the gap between intake and need. The meal arrives in bursts. Blood glucose cannot live in bursts. The liver stores part of the surge as glycogen, then opens that reserve when external supply falls. It is short-horizon energy, structurally packed for safe holding and rapid release, placed exactly where blood chemistry can be defended by it first.

Vitamin A Storage in Stellate Cells

Vitamin A is not stored in the body as a vague nutritional surplus. It is held in a specific tissue, in a specific cell type, in a specific chemical form, because the molecule is too important and too biologically active to be left circulating freely without control. The liver is the main reserve organ, and within the liver the key storage cells are the hepatic stellate cells sitting in the space of Disse, the narrow compartment between the sinusoidal blood channel and the hepatocyte surface. These cells do not dominate liver physiology the way hepatocytes do, but they occupy a strategically perfect location for reserve management. Nutrients arriving from portal blood are being processed nearby, hepatocytes are handling transport and conversion nearby, and stellate cells sit just outside the main exchange lane where they can receive retinoid traffic and hold it in reserve without carrying the full metabolic burden of the hepatocyte. This is not random storage. It is protected storage placed inside the organ that already governs nutrient traffic.

The stored material is not simply loose vitamin A. It is largely held as retinyl esters inside lipid droplets. That detail matters because storage is a chemistry problem, not just a location problem. Retinol, the transport form of vitamin A alcohol, is useful in circulation when bound to the right carrier proteins, but inside tissue a reserve has to be dense, stable, and less likely to create uncontrolled signaling. Esterification solves that. The retinol molecule is linked to a fatty acid and packed into lipid droplets, giving the stellate cell a chemically quieter, more compact reserve form. This lets the liver hold substantial vitamin A without forcing the molecule to remain active in plasma or free in the cytoplasm where it could distort membranes, gene signaling, or oxidative balance. The body is not storing a vitamin label. It is storing a controlled retinoid reserve in a form meant for delayed use.

That reserve matters because vitamin A is not a minor accessory nutrient. It is involved in epithelial integrity, immune regulation, visual chemistry, cellular differentiation, and gene expression through retinoic acid signaling. Differentiation means the process by which cells maintain or acquire specialized identity rather than remaining generic. A molecule participating in that level of tissue instruction cannot be handled casually. Too little compromises barrier tissues, vision, and immune competence. Too much becomes toxic because the same signaling power that makes vitamin A useful also makes excess exposure disruptive. Stellate cell storage gives the body a buffer between intake and use. It lets intermittent dietary supply become a more stable physiological reserve. The meal may be irregular. Tissue demand may change. The liver smooths that mismatch by holding a retinoid pool the body can draw on later.

The placement of stellate cells in the space of Disse also reveals something important about how the liver thinks about storage. The liver does not only hold bulk fuel like glycogen. It also holds high-authority molecules whose effect reaches far beyond calories. Stellate cells are close enough to the sinusoidal exchange field to receive retinoids entering from blood, but they are not themselves the main blood-processing cells. That division of labor is elegant. Hepatocytes dominate throughput. Stellate cells dominate reserve, at least for vitamin A. One cell type handles the moving stream. Another quietly holds a strategic micronutrient in waiting. This is one reason liver storage should never be taught as if it were all the same. Glycogen, iron, copper, and vitamin A are not being stored for the same reasons or in the same compartments. The liver uses different cells and different chemistries for different kinds of reserve.

This storage role also explains why stellate cells matter so much later in liver disease. A healthy stellate cell storing retinyl esters is a quiet reserve cell. Under chronic injury, it can lose those lipid droplets, lose that storage identity, and transform into a collagen-producing fibrogenic cell. The reserve cell becomes part of the scar-building machinery. That shift is one of the great tragedies of chronic liver injury because the same cell that once helped preserve nutrient reserve and tissue stability begins helping bury the exchange architecture in matrix. So vitamin A storage in stellate cells is not a side note in micronutrient physiology. It is one of the liver's clearest examples of quiet stratègic reserve, and one of the clearest examples of how a healthy support cell can become something far more dangerous when tissue order breaks.

Vitamin B12 Reserve and Why Deficiency Can Develop Slowly

Vitamin B12 deficiency often develops slowly because the body does not treat B12 like a nutrient that must be spent the week it is eaten. It treats it like a strategic reserve. B12, also called cobalamin, is required for DNA synthesis, red blood cell production, and key methylation and mitochondrial reactions, but the body stores a substantial amount of it, and the liver is the main storage organ. That changes the timeline completely. A person can stop absorbing B12 today and still carry enough hepatic reserve to keep blood counts and neurologic function apparently stable for months or even years before the deficit becomes obvious. The deficiency clock does not start with symptoms. It starts when input stops. Symptoms appear only after storage, recycling, and tissue demand have spent enough of the reserve to expose the shortfall.

The reason the reserve is so durable begins with how B12 is handled after absorption. B12 is absorbed in the terminal ileum after binding intrinsic factor, the gastric protein that protects and escorts it through the intestine. Once absorbed, it does not simply float through blood in a loose free state. It binds transport proteins, enters tissues in a controlled way, and a large fraction is ultimately stored in the liver. Hepatocytes can retain significant amounts of cobalamin, which means the liver is acting as a micronutrient warehouse for a molecule the body cannot synthesize on its own but also cannot afford to run out of quickly. That reserve matters because DNA synthesis and nervous system maintenance do not tolerate abrupt failure well. The body smooths the risk by storing B12 in advance.

Deficiency also develops slowly because B12 is partly recycled. The liver secretes some B12 into bile, but much of that biliary B12 can be reabsorbed again downstream if the absorption machinery is intact. That creates a kind of internal conservation loop. Intake matters, but the body is not relying only on each new meal to keep B12 status alive. It is reusing part of what it already holds. This is one reason a person can eat poorly for a time and still look biochemically stable, or lose gastric acid or intrinsic factor and not develop dramatic deficiency immediately. The reserve is still being drawn down gradually, not emptied in a single pass. The liver is holding the store, and the intestine is helping salvage part of the traffic.

The delay becomes dangerous because the slowness can hide the mechanism. When B12 finally falls far enough, the consequences appear in tissues that depend on rapid cell division and precise methylation chemistry. The marrow begins producing large abnormal precursor cells because DNA synthesis is impaired. This is why megaloblastic anemia develops. The nervous system begins suffering because myelin maintenance and methylation-dependent pathways are no longer being supported cleanly. Methylmalonic acid can rise because methylmalonyl-CoA mutase, a B12-dependent mitochondrial enzyme, is no longer working efficiently. Homocysteine can rise because methionine synthase, another B12-dependent enzyme, is falling behind. By the time these changes are visible, the person is often no longer showing a new deficiency. They are showing exhaustion of a long-protected reserve.

So vitamin B12 deficiency develops slowly because the liver stores it heavily, the body conserves it through biliary recycling, and tissue demand can be buffered for a long time before the reserve becomes critically low. That delay is protective at first. It becomes deceptive later. A person can be absorbing poorly for years before the blood finally admits what the liver has been quietly covering.

Iron and Copper Storage

Iron and copper are not stored in the liver because the liver happens to have extra room. They are stored there because both metals are useful enough to keep and dangerous enough to contain. Iron is required for hemoglobin, mitochondrial respiration, and a wide range of electron-transfer reactions, but free iron also drives oxidative injury when it sits in the wrong chemical state. Copper is required for enzymes involved in energy production, antioxidant defense, connective tissue formation, and iron metabolism, but unbound copper is chemically reactive and can damage proteins, lipids, and DNA. The liver solves the same problem for both metals in slightly different ways. It receives them, binds them, packages them, and decides whether they will be held, released, or pushed toward excretion. Storage is therefore not passive accumulation. It is controlled sequestration, meaning the metal is kept in a chemically restrained form so the body can use it later without letting it injure tissue in the meantime.

Iron storage in the liver occurs mainly in hepatocytes and in reticuloendothelial cells such as Kupffer cells, depending on how the iron arrived and what the body is doing with it. Some iron enters the liver after intestinal absorption and transferrin-mediated delivery in blood. Some arrives through macrophage recycling after old red blood cells are dismantled and their iron is handed back into circulation or retained locally. Once iron reaches the liver, it is not left loose in cytoplasm. It is stored primarily in ferritin, a protein shell built to hold iron in a safer, more chemically restrained form. Ferritin is the first-line storage state, compact enough for reserve and protective enough to keep reactive iron from catalyzing unnecessary oxidative chemistry. When iron burden rises beyond ordinary ferritin handling, more iron can accumulate as hemosiderin, a less readily mobilized aggregate form that often reflects more chronic or excessive storage. This is one of the key differences between normal reserve and overload. Ferritin is controlled storage. Hemosiderin is storage beginning to harden into burden. The liver is central here because it is both a major iron reservoir and one of the organs writing hepcidin, the hormone that decides whether iron stays trapped in cells or is allowed back into blood through ferroportin.

Copper storage follows a related but not identical logic. Copper absorbed from the intestine reaches the liver through portal blood, and the hepatocyte becomes the main cell deciding whether that copper will be incorporated into cuproenzymes, exported into circulation in a controlled transport form, or secreted into bile for elimination. Inside the hepatocyte, copper is not supposed to remain free. It is bound to chaperone proteins and storage proteins such as metallothionein, which help buffer the metal and keep it from behaving like loose catalytic shrapnel in the cytoplasm. This matters because copper is needed, but it is not gentle. A body that allowed free copper to accumulate would pay in oxidative and mitochondrial injury very quickly. The liver therefore stores copper as part of a traffic system, not just as a reserve system. Some copper is held. Some is used. Some is loaded onto ceruloplasmin, the major copper-carrying protein released into blood. Some is secreted into bile. That last point is crucial. Unlike iron, which the body is poor at excreting in large controlled amounts, copper has a major exit route through bile, which makes hepatic biliary function central to copper balance.

This is why iron and copper disorders feel different even though both metals accumulate in the liver. Iron overload often reflects too much entry or too much release into circulation over time in a system with weak excretory escape. Copper overload often reflects failed hepatocellular handling or failed biliary export, as in Wilson disease, where copper begins accumulating in the liver because the normal route into bile is impaired. The chemistry of storage then becomes the chemistry of injury. Ferritin and metallothionein can buffer only so much. Beyond that point, the reserve organ becomes a target organ. Oxidative stress rises. Mitochondria suffer. Fibrosis becomes easier to trigger. The important principle is the same for both metals. The liver stores them because the body needs a central authority over dangerous useful elements. It is not enough to absorb them. They have to be bound, counted, and either held or moved under strict control.

So iron and copper storage in the liver should be understood as mineral governance rather than simple warehousing. Iron is held mainly in ferritin and, under heavier burden, hemosiderin, while the liver also writes the hormonal instructions that decide whether iron can circulate. Copper is buffered within hepatocytes, incorporated into useful transport and enzyme systems, and cleared largely through bile when disposal is needed. Both metals prove the same larger truth. The liver does not only store what the body may need later. It stores what the body cannot afford to leave chemically loose.

Why the Liver Can Buffer Deficiency for a Long Time Before Symptoms Appear

The liver can buffer deficiency for a long time because it is not only a processing organ. It is a reserve organ. That difference matters more than most people realize. A deficiency state does not begin the day intake falls. It begins the day intake falls below need, but symptoms only appear when the body has already spent enough reserve that tissue function can no longer be defended quietly. The liver is one of the main reasons that delay exists. It stores glycogen for short-horizon glucose support. It stores vitamin A in stellate cells as retinyl esters. It holds substantial vitamin B12 reserve. It stores iron in ferritin and, under heavier burden, hemosiderin. It buffers copper through intracellular binding systems. In each case the principle is the same. The body does not want every meal interruption or every temporary absorption failure to become an immediate systemic crisis. The liver absorbs the first impact by serving as a depot between intake and symptom.

That buffering works because storage changes time. A nutrient or substrate entering the body today does not have to be spent today if the liver can capture it in a controlled form and release it later. Glycogen is the clearest short-term example. The meal ends, but the bloodstream still needs glucose, so the liver draws from a polymer built earlier and delays the appearance of deficiency. Vitamin A and B12 show the same principle over a much longer horizon. These are not just nutrients floating in plasma waiting to disappear at the speed of daily use. They are held in reserve, in specific cells, in specific chemical forms, so the tissues can continue functioning long after intake has become irregular or absorption has begun to fail. A person may therefore be living on yesterday's hepatic decisions for weeks, months, or even years before the deficit becomes obvious in blood or tissue.

This is why deficiency symptoms are often late findings rather than early ones. The liver is quietly paying the bill while the upstream problem is already present. Poor intake, impaired bile flow, loss of intrinsic factor, ileal disease, chronic inflammation, or altered mineral handling may already be reducing what enters or what circulates, but the organ keeps releasing stored material and stabilizing the bloodstream for as long as it can. That delay is protective, but it is also deceptive. It can make a deficiency look sudden when in reality the deficiency process has been unfolding silently for a long time. The body did not fall apart overnight. The liver simply ran out of reserve yesterday after spending months masking the shortage.

The deeper point is that symptoms appear not when a nutrient first becomes scarce, but when buffering fails. The liver can cover interruption for a while because it stores, binds, releases, and prioritizes. Once the reserve thins enough, tissues finally begin revealing what the liver had been preventing the whole time. The symptom is late. The depletion began much earlier.

Storage as Hidden Reserve and Hidden Vulnerability

Storage looks protective because most of the time it is protective. The liver stores precisely because the body cannot afford to live at the speed of immediate intake. Meals are intermittent, absorption is imperfect, and physiological demand changes from hour to hour and season to season. A reserve organ smooths those discontinuities. Glycogen lets the bloodstream survive the hours after a meal ends. Vitamin A held in stellate cells allows epithelial and immune function to continue even when intake is uneven. Vitamin B12 stored in hepatic reserve can postpone overt deficiency for a long time after absorption has already begun to fail. Iron and copper can be bound, contained, and mobilized when needed rather than left circulating freely in chemically dangerous forms. In all of these cases, storage is hidden strength. The person appears stable because the liver is covering the gap between what the body needs and what the environment is currently supplying.

But the same reserve creates hidden vulnerability because a buffered system can be declining for a long time without looking unstable. Storage delays the symptom, and that delay can be mistaken for safety. A person can have falling intake, impaired absorption, chronic blood loss, inflammatory iron sequestration, cholestatic fat-soluble vitamin malabsorption, or progressive liver injury and still appear functionally intact for a while because the reserve is being drawn down quietly. The blood remains more normal than it should. The tissues remain more stable than they would be without hepatic storage. The deficiency or transport problem is already real, but the liver is spending its reserve to hide it. This is why storage can make disease feel sudden when it finally appears. The crisis is visible late, but the depletion or transport failure began much earlier.

Storage also becomes vulnerability when the thing being held is dangerous in excess. Iron is the clearest example. Stored properly, it is strategic reserve. Stored excessively, it becomes oxidative burden.

Copper follows the same logic. Trapped safely, it is useful mineral reserve. Trapped excessively or exported poorly, it becomes tissue poison. Even triglyceride in the hepatocyte begins this way. The cell stores it partly as a defensive move, converting more disruptive fatty acid burden into a less immediately toxic form. But if the traffic problem persists, the droplet stops being quiet protection and starts becoming evidence that the organ is carrying more than it can safely manage. Storage begins as adaptation. Prolong it under the wrong conditions and it becomes part of the injury.

That is the deeper truth. Storage is never just a warehouse function. It is a timing strategy with a cost. It protects the organism from short-term interruption, but it can also conceal decline, delay recognition, and convert burden into a quieter form that eventually becomes harder to reverse. The liver's reserves are one of the reasons the body survives instability so well. They are also one of the reasons pathology can deepen in silence.

Part VIII

The Gut-Liver Axis

Chapter 30

Everything Absorbed Is Not Equally Safe

Why Portal Blood Carries More Than Nutrients

Portal blood is often imagined as if it were a delivery stream filled with finished nutrition on its way to the rest of the body. That picture is wrong in the way that matters. The intestine does not send the liver a purified nutrient solution. It sends the chemical consequences of digestion. Some of that is useful substrate, glucose, amino acids, minerals, water-soluble vitamins, short-chain fatty acids, and many other absorbed compounds the body needs. But the same blood also carries microbial metabolites, bacterial fragments, endotoxin, food-derived chemicals, drug molecules, alcohol, hormone fragments, bile acid return, and a broad range of compounds that are inside the body only because they crossed the gut barrier, not because they have already been judged safe or useful. Portal blood is therefore not simply nutrient-rich blood. It is chemically mixed blood, and the liver sits where it does because somebody has to read that mixed report before systemic circulation is forced to live with it.

This is the core logic of the gut-liver axis. The intestine is not only a site of absorption. It is also a border zone between the organism and a lumen full of food particles, digestive products, microbial life, microbial debris, and foreign molecules. Crossing the intestinal wall is not the same thing as being fully admitted into the controlled chemistry of the body. Once a compound crosses into portal blood, it has entered internal transport, but it has not yet escaped hepatic review. That is why the portal vein matters so much. It does not carry purified fuel directly to the heart. It carries the raw aftermath of digestion directly to the liver. The liver then decides what is extracted, what is stored, what is transformed, what is detoxified, what is secreted into bile, and what will be allowed onward into the wider circulation.

That mixed burden is not a design flaw. It is the inevitable cost of absorbing from a living intestine. The gut lumen is not sterile. The epithelial barrier is selective but not absolute. Digestion itself generates chemical complexity rather than simplicity. Proteins are broken down into amino acids and peptides, but bacterial fermentation products appear alongside them. Carbohydrates yield glucose and other sugars, but microbial metabolites may rise with them. Lipids are digested and absorbed, but bile acids return and lipopolysaccharide may also be present in low amounts. Drugs taken by mouth enter through the same absorptive path the meal uses. Alcohol uses it too. Plant compounds, additives, and countless other exogenous molecules enter the same portal route. This is why absorbed does not mean harmless. It only means crossed. The liver exists downstream because crossing is not the end of the story.

The immune side of this is just as important as the metabolic side. Kupffer cells in the liver sinusoids sample portal blood for bacterial fragments and inflammatory signals because the liver knows the gut is sending more than fuel. Hepatocytes and sinusoidal endothelial cells also participate in reading this traffic. The liver has to tolerate ordinary low-level microbial evidence without turning every meal into an inflammatory crisis, yet it also has to clear enough of that burden to keep the rest of the circulation from being exposed too freely. So portal blood should be understood as a filtered threat stream as well as a nutrient stream. Useful compounds arrive in it. Potentially disruptive compounds arrive in it. The liver is the first organ built to tell the difference.

This is why the phrase everything absorbed is not equally safe belongs at the beginning of the gut-liver axis. Intestinal absorption is not a moral stamp of approval. It is an entry event. What follows depends on form, concentration, microbial context, hepatic transformation, biliary excretion, and systemic handling. Portal blood carries what the gut has allowed through, not what the whole organism has already agreed to keep. The liver stands in the middle because the body cannot afford to let every absorbed molecule speak for itself.

Microbial Fragments, Endotoxin, Alcohol, Drugs, Food Compounds

Portal blood carries more than nourishment because the intestine is not a clean nutrient pipeline. It is a living border full of microbes, digestive byproducts, partially processed food chemistry, and foreign compounds taken in by mouth. Once material crosses the intestinal epithelium, it enters the portal circulation whether it is beneficial, burdensome, neutral, or actively disruptive. That means the liver receives a mixed chemical report every minute. Microbial fragments come with the territory because the intestinal lumen contains dense microbial life, and the barrier separating that world from blood is selective, not absolute. Small amounts of bacterial structural material, lipopolysaccharide from gram-negative organisms, peptidoglycan fragments, bacterial DNA, and microbial metabolites can all reach portal blood. Lipopolysaccharide, often called endotoxin, matters especially because it is not nutritionally relevant cargo. It is an immune stimulus. The liver has to clear or neutralize enough of this burden to keep the rest of the body from reading ordinary gut traffic as systemic invasion.

Alcohol belongs in the same sentence because it enters through the same absorptive route but imposes a completely different type of burden. Once absorbed, ethanol moves rapidly into portal blood and reaches hepatocytes before systemic dilution has fully softened the concentration profile. The liver then has to oxidize it, converting ethanol first into acetaldehyde and then further into acetate. Acetaldehyde is especially reactive and injurious, which means alcohol is not simply extra caloric substrate. It is a redox burden and a chemical stressor arriving directly from the gut. The liver is therefore not just filtering alcohol away from the body. It is paying for its metabolism with altered NADH to NAD+ balance, oxidative stress, and downstream effects on fat handling, inflammation, and mitochondrial function. The fact that alcohol uses the portal route is one reason its effects are so strongly liver-centered even before the rest of the body feels them fully.

Drugs taken by mouth follow the same broad route and reveal the same principle. The intestine absorbs them. Portal blood delivers them. The liver meets them first. This is why first-pass metabolism exists. Many oral drugs are partly transformed by hepatocytes before the parent compound ever reaches the wider circulation in full. That transformation may activate the drug, inactivate it, slow it, speed it, or produce metabolites that are easier to excrete. The liver is therefore functioning as a biochemical customs station for pharmacology as well as for nutrition. The same is true for food compounds more broadly. Polyphenols, plant defense chemicals, additives, preservatives, sweeteners, fermentation products, and countless other molecules absorbed from the gut enter the same portal stream. Some are relatively inert. Some require conjugation. Some interact with transporters and enzymes. Some contribute to the liver's total xenobiotic burden, xenobiotic meaning a compound foreign to the body's own structural and signaling design. The portal vein does not discriminate morally. It delivers what crossed.

This is why the gut-liver axis has to be taught as a mixed-input system rather than as a nutrient-delivery system alone. Microbial fragments provoke immune reading. Endotoxin provokes defensive signaling. Alcohol imposes redox and toxic-metabolite stress. Drugs demand enzymatic handling and often biotransformation. Food compounds may nourish, burden, or simply occupy processing capacity depending on their chemistry. The liver receives all of this before the general circulation does because the body cannot afford to let a meal, a drink, a medication, and the microbial reality of the intestine enter the bloodstream without review. Portal blood is not food in transit. It is the chemical aftermath of oral life, and the liver is the first organ forced to decide what part of that aftermath the rest of the body will be allowed to keep.

The Intestinal Barrier as the Liver's Upstream Partner

The intestinal barrier is not a wall in the crude sense. It is a living sorting surface. That distinction matters because the gut is being asked to do two opposite things at once. It has to let nutrients cross into the body, and it has to keep microbes, toxins, digestive debris, and excessive inflammatory traffic from crossing with them. The liver inherits the consequences of how well that sorting surface performs. Portal blood does not arrive at the liver as an independent stream with its own private logic. It arrives as the output of the intestinal barrier. If the barrier selects well, the liver receives a cleaner post-digestive report. If the barrier leaks, loosens, inflames, or becomes chemically disordered, the liver receives a more dangerous one. That is why the intestinal barrier should be understood as the liver's upstream partner rather than as a separate digestive detail. The liver processes what the barrier permits.

That barrier is built in layers, and each layer matters. Mucus sits above the epithelium and helps keep large amounts of luminal microbial content from pressing directly against absorptive cells. The epithelial cells themselves, especially enterocytes, form the immediate cellular lining. Between them sit tight junctions, protein complexes that regulate how much can move between cells rather than through them. Beneath that surface lies immune tissue, secretory immunoglobulin A, antimicrobial peptides, and a resident immune network trained to tolerate food and commensal microbes while still responding to danger. None of this is static. The barrier is continuously sampling, digesting, transporting, excluding, and repairing itself. It is not a brick wall. It is a living customs border. Nutrients cross through controlled transport systems and membrane events. Microbial fragments and toxins are supposed to be limited, neutralized, or kept luminal. The liver then receives whatever this layered interface failed to exclude or successfully selected for transfer.

This is why intestinal permeability cannot be reduced to a slogan about leaky gut without losing the real physiology. The question is not whether the barrier is open or closed in some absolute sense. The question is what is crossing, in what quantity, in what chemical form, and under what immune conditions. Glucose crossing through enterocyte transporters after starch digestion is not the same event as lipopolysaccharide crossing in higher-than-normal burden from gram-negative bacteria. Peptides and amino acids moving through nutrient transport systems are not the same as bacterial DNA fragments or inflammatory microbial metabolites reaching portal blood. The liver reads these differences immediately because it sits directly downstream. Kupffer cells sample the incoming stream. Hepatocytes metabolize, conjugate, and redirect what they can. Sinusoidal endothelial cells participate in the immune reading of the same blood. The intestinal barrier and the liver are therefore performing one extended act of selection in two locations. The barrier decides what enters portal blood. The liver decides what exits portal blood into the rest of the body.

When the barrier is functioning well, this partnership is efficient. Nutrients cross in controlled quantity. Microbial burden remains low enough for quiet clearance rather than inflammatory escalation. Bile acids complete their intestinal work and return through enterohepatic circulation in an orderly way. Food compounds, drug molecules, and microbial metabolites may still reach the liver, but they do so at levels the organ was built to handle. Portal blood remains chemically mixed, but not chaotic. The liver can then focus on buffering glucose, handling amino acids, packaging lipids, conjugating bilirubin, transforming hormones, and clearing low-level immune burden without turning the entire circulation into a field of alarm. The barrier upstream has reduced the liver's work to a solvable problem.

When the barrier deteriorates, the character of portal blood changes before the rest of the body sees any obvious symptom. Tight junction regulation can become less disciplined. Mucus quality can change. Epithelial cells can become inflamed or injured. Luminal bacteria can move closer to the epithelial surface, and more microbial products can cross into portal blood. Now the liver receives a different message. More endotoxin. More inflammatory metabolites. More immune stimulation. More burden for Kupffer cells. More cytokine signaling. More oxidative stress inside hepatocytes. More pressure on bile flow, lipid handling, and acute-phase protein synthesis. A barrier defect in the intestine therefore becomes a liver problem quickly, not metaphorically but mechanically. The liver is not hearing about gut dysfunction from a distance. It is bathing in its downstream chemical result.

This is why the intestinal barrier should be thought of as the liver's upstream partner in maintaining blood chemistry. The intestine is not simply digesting food. It is editing the first version of what the liver will have to process. The liver is not simply detoxifying whatever shows up. It is working on the draft the barrier hands it. A strong barrier makes the liver's job smaller, cleaner, and more selective. A weak barrier turns the liver into a frontline immune and detoxification organ more often than it should have to be. The partnership is constant whether the person knows it or not. Every absorbed meal, every microbial fragment held back or allowed through, every bile acid returned, every toxin excluded or admitted becomes part of the conversation between gut and liver.

Why Liver Physiology Begins in the Gut

Liver physiology begins in the gut because the liver is not primarily reacting to isolated molecules already sorted into neat categories. It is reacting to the chemical aftermath of digestion. The intestine is where food is broken down, where bile acids meet fat, where pancreatic enzymes reduce larger structures into absorbable fragments, where microbes ferment, transform, and shed their own compounds, and where the body first decides what will be allowed across the epithelial border into internal transport. By the time blood reaches the liver through the portal vein, the decisive upstream events have already happened. Carbohydrate has become glucose and other sugars. Protein has become amino acids and peptides. Lipid has been emulsified, hydrolyzed, repackaged, and partly reassembled. Bacterial fragments, endotoxin, microbial metabolites, drug molecules, alcohol, and food-derived chemicals may have crossed alongside nutrients. The liver is not the first digestive organ in this sequence. It is the first major internal organ forced to interpret the gut's output.

That is why the phrase liver physiology begins in the gut is not rhetorical. It is anatomical and chemical. The liver sits where portal blood arrives before the general circulation sees it because the body cannot afford to let intestinal absorption become systemic chemistry without review. The gut does not send the liver pure nourishment. It sends a mixed report. Some of that report is valuable substrate that must be buffered, stored, converted, or released in controlled fashion. Some of it is potentially inflammatory. Some of it is chemically foreign. Some of it is incompletely processed burden that now requires hepatic transformation, conjugation, sequestration, or biliary disposal. The liver receives glucose surges that need damping, amino acid traffic that needs sorting, fructose that needs early hepatic handling, microbial signals that need immune interpretation, and compounds that may need detoxification before they are allowed further into blood. In that sense the liver is downstream from digestion in the deepest possible way. It inherits the consequences of what the gut has absorbed.

The gut also determines how hard the liver has to work. A well-functioning intestinal barrier, disciplined digestion, and orderly bile-dependent absorption send the liver a stream that is still chemically mixed but more physiologically manageable. A compromised barrier, disordered microbiota, poor bile delivery, rapid transit, inflammatory mucosal injury, or heavy toxin exposure changes portal blood before the liver ever gets a chance to respond. More endotoxin may cross. More microbial fragments may reach the sinusoids. Nutrient absorption may become less orderly. Bile acid recovery may weaken. The liver then begins its work from a more hostile starting point. Kupffer cells have more immune burden to clear. Hepatocytes face more inflammatory and xenobiotic pressure. Acute-phase signaling becomes easier to trigger. Stellate cells become easier to activate. This is why liver disease and gut dysfunction so often travel together. The liver is processing the river the gut sends it.

Even the classic functions people associate with the liver are gut-conditioned from the start. Glycogen storage begins as a response to intestinal glucose delivery. Ketogenesis later depends on the contrast created when intestinal delivery stops and fasting fat traffic takes over. Bile secretion matters because the intestine needs it for fat handling and because bile acids are later reclaimed from the ileum and returned to the liver. Bilirubin disposal ends in the intestine because bile is the route out. Acute-phase protein synthesis can be driven by inflammatory signals that began at the intestinal barrier. Iron regulation through hepcidin becomes relevant partly because the intestine is the gate of iron entry. So many hepatic functions are really second-stage responses to first-stage intestinal events. The liver is not an isolated metabolic factory generating its own problems to solve. It is the organ that receives what digestion, absorption, microbial ecology, and mucosal barrier function have already made real.

So liver physiology begins in the gut because the gut writes the first draft of internal chemistry. The liver receives that draft, edits it, rejects parts of it, stores parts of it, converts parts of it, and sends the revised version onward as blood the rest of the body can live on. If the reader wants to understand why the liver stores glycogen, clears endotoxin, makes bile, regulates iron, buffers blood sugar, or becomes inflamed under modern metabolic burden, the starting point is upstream. The liver begins where the intestine ends.

How Intestinal Permeability Changes Hepatic Burden

Intestinal permeability changes hepatic burden because the liver does not receive an abstract summary of gut function. It receives whatever the barrier allowed through. The intestinal wall is supposed to act as a selective interface, letting digested nutrients cross while holding back excessive microbial fragments, endotoxin, poorly processed luminal debris, and inflammatory traffic. Once that selectivity weakens, portal blood changes before the rest of the body has had any chance to respond. More bacterial structural material, more lipopolysaccharide, more food-derived antigenic burden, more microbial metabolites, and more inflammatory signaling reach the liver directly. The portal vein does not dilute this into a neutral stream first. It delivers it straight into the sinusoidal bed, where Kupffer cells, hepatocytes, and sinusoidal endothelial cells are forced to read and handle a blood chemistry that has become more immunologically loaded than it was designed to receive in ordinary digestion. The liver's burden changes not because the organ itself chose a new job, but because the gut handed it a dirtier first draft.

The first consequence is immune work. Kupffer cells, the resident macrophages positioned in the sinusoids, are built to clear low-level microbial evidence without turning every meal into systemic alarm. That quiet clearance depends on burden remaining within a tolerable range. When permeability rises, more endotoxin and bacterial fragments arrive, and the liver has to escalate from quiet screening toward louder inflammatory behavior. Cytokine release increases. Sinusoidal endothelial behavior changes. Hepatocytes are no longer handling only glucose, amino acids, bilirubin, and ordinary post-digestive chemistry. They are now metabolizing inside an inflammatory field. This matters because the liver is a high-flow exchange organ. The same tissue surface used for nutrient buffering and hormone handling is now being asked to clear and interpret a larger immunologic load. A more permeable intestine therefore does not merely "stress the liver" in a vague sense. It changes the composition of portal blood in a way that converts hepatic physiology from metabolic governance toward metabolic defense.

That immune shift quickly becomes a metabolic shift. Inflammatory cytokines alter hepatocyte priorities. Acute-phase protein synthesis rises. Albumin synthesis may fall in relative priority. Glucose handling becomes less clean. Lipid handling becomes more vulnerable to congestion and steatosis. Mitochondrial stress rises because the hepatocyte is now trying to maintain bile secretion, blood chemistry control, protein synthesis, and detoxification while also absorbing the cost of inflammatory signaling and oxidative pressure. Oxidative pressure matters here because endotoxin-driven immune activation and hepatocyte stress increase reactive oxygen species, and those reactive molecules damage membranes, proteins, and mitochondrial function if the burden persists. The liver is therefore paying twice. It is clearing what crossed the barrier, and it is also paying the downstream metabolic cost of having to clear it.

The biliary side is affected too. Cholestatic tendencies become easier to provoke in tissue already living under inflammatory load, because bile secretion depends on intact hepatocyte polarity, ATP-dependent transport, and stable canalicular membranes. ATP, adenosine triphosphate, is the cell's immediately usable energy currency, and the more it is spent on inflammation, oxidative defense, and injury response, the less margin the hepatocyte has for clean directional secretion. At the same time stellate cells in the space of Disse are more likely to receive the kind of cytokine and oxidative signals that push them out of quiet retinoid storage and into fibrogenic activation. Once that happens, extracellular matrix begins accumulating in the exchange space, sinusoidal function becomes less efficient, and the original permeability problem starts translating into architectural liver burden. The gut leaked. The liver first cleared the burden, then began remodeling under it.

This is why intestinal permeability has to be understood as a load amplifier. It does not invent new hepatic machinery. It forces existing hepatic machinery to work under dirtier, more inflammatory, and more metabolically expensive conditions. Nutrient traffic is still arriving. Hormone fragments are still arriving. Microbial signals are still arriving. The difference is that the balance has changed. The liver is now spending more of its capacity on immune containment and less on quiet high-efficiency regulation. Over time that shift makes steatosis, cholestasis, fibrotic signaling, and acute-phase dominance easier to sustain. The intestine did not merely become more open. It changed what kind of work the liver has to perform every minute after that.

Why the Liver Is the Organ That Absorbs Digestive Mistakes

The liver is the organ that absorbs digestive mistakes because it sits exactly where the body has decided mistakes must be intercepted. The intestine is not a perfect sorting machine. It is a living border carrying out digestion, absorption, microbial negotiation, and barrier control all at once. Meals arrive in irregular size and composition. Sugars can be absorbed too quickly. Fructose can arrive in a load the rest of the body was never meant to see directly. Microbial fragments and endotoxin can cross in higher amounts than they should. Alcohol, drugs, food additives, and plant compounds enter through the same portal route as nutrients. Bile acid recovery can be disordered. Partially processed chemical burden can move forward alongside useful substrate. The body cannot stop all of this at the level of the gut without also stopping life-sustaining absorption. So it places the liver immediately downstream and gives it the first right of refusal. The intestine absorbs. The liver pays for what should not have been passed forward unreviewed.

That is why so many different physiological errors converge there. A glucose surge too steep for systemic blood is buffered first by hepatocytes. Excess amino acid traffic and nitrogen burden are handled there. Fructose is taken up there. Alcohol is oxidized there. Drugs taken by mouth undergo first-pass transformation there. Endotoxin and microbial fragments are sampled there by Kupffer cells. Bile acids returning from the intestine are received there. Bilirubin is taken from blood and pushed toward bile there. The liver is not simply one more organ affected by digestive events. It is the organ built to absorb their consequences before those consequences become the chemistry of the entire organism. This is why portal blood is so important. It does not carry a finished nutritional gift. It carries the aftermath of digestion, and the liver stands where aftermath becomes decision.

The phrase digestive mistakes should be taken broadly. It includes not only obvious pathology such as increased permeability, maldigestion, microbial overgrowth, inflammatory barrier damage, poor bile delivery, and toxin exposure, but also ordinary biochemical mismatches between what the gut lets through and what the bloodstream can safely tolerate all at once. The liver smooths those mismatches. It stores part of the glucose surge as glycogen. It converts some burdens into more excretable forms. It packages lipids when they cannot remain free. It holds iron when free circulation would be dangerous. It raises acute-phase proteins when portal blood begins carrying more immune burden than quiet tolerance can absorb. It makes bile to solve the next meal's fat problem even while disposing of what the last metabolic cycle has already marked for removal. In this sense the liver is not merely cleaning up digestive errors. It is continuously preventing normal digestive imperfection from becoming systemic disorder.

This is also why liver disease so often follows gut burden. A healthier gut hands the liver a stream that is chemically mixed but manageable. A damaged gut hands it a stream richer in endotoxin, microbial fragments, inflammatory signals, unstable nutrient delivery, and luminal chemistry that should have stayed on the other side of the barrier. The liver can absorb that burden for a while. It can buffer, clear, conjugate, store, neutralize, and secrete. But the more often it has to absorb digestive mistakes, the more it begins to pay in steatosis, cholestasis, inflammatory signaling, fibrotic activation, and synthetic strain. The liver is protective precisely because it takes the hit first. That protection is real. So is the cost.

Chapter 31

Immune Tolerance and Immune Defense

The Liver's Challenge: Constant Exposure Without Constant Overreaction

The liver lives under an impossible-looking assignment. It receives portal blood from the intestine every hour of the day, and that blood is never a sterile nutrient stream. It carries food-derived compounds, microbial metabolites, bacterial fragments, endotoxin, bile acid return, damaged-cell debris, and the low-grade inflammatory evidence that naturally comes from maintaining a microbial world on the other side of a thin epithelial barrier. If the liver responded to every one of those arrivals as if it were bloodstream sepsis, digestion itself would become a daily inflammatory injury. If it ignored them completely, the rest of the body would be exposed too freely to microbial burden and chemically disruptive traffic that should have been intercepted upstream. So the liver has to do two opposite things at once. It has to tolerate constant exposure, and it has to remain capable of real defense. That is the challenge. Not whether it can react. Whether it can react selectively enough to survive its own position.

This is why hepatic immunity is built around discrimination rather than maximal aggression. Kupffer cells sit in the sinusoids as resident macrophages, directly in the path of portal blood, and they are expected to clear bacterial fragments, endotoxin, and debris without turning every meal into a cytokine storm. Liver sinusoidal endothelial cells help shape that environment further, controlling how blood meets tissue and how immune contact is staged at the exchange surface. Hepatocytes also participate by responding to inflammatory signals, altering protein synthesis, and helping decide whether the bloodstream will remain in a baseline transport state or shift into acute-phase defense. The liver therefore is not merely a filter catching particles. It is an immune interpretation organ. It is constantly asking the same question in chemical form. Is this ordinary gut traffic that should be absorbed into quiet clearance, or is this the beginning of injury, invasion, or overload that needs to be escalated?

Tolerance in this setting does not mean weakness. It means controlled non-escalation in the face of ordinary burden. That burden is not imaginary. Small amounts of microbial material really do reach portal blood. Bile acids really do return from the intestine. Food antigens and metabolic byproducts really do arrive after every meal. The liver has to prevent that ordinary exposure from becoming systemic noise. It does so by clearing quietly, containing locally, and limiting unnecessary amplification. A well-functioning hepatic immune environment allows the body to digest, absorb, and live with a colonized intestine without paying for that arrangement in constant fever, vascular inflammation, or tissue-level alarm. In other words, immune tolerance is one of the conditions that makes ordinary eating biologically possible.

Defense begins when the incoming traffic changes character or scale. More endotoxin, more bacterial fragments, more danger signals from injured hepatocytes, more inflammatory metabolites from a compromised gut barrier, or more cholestatic or steatotic stress inside the liver itself can push the tissue out of quiet clearance and into active inflammatory response. Kupffer cells release cytokines. Hepatocytes alter acute-phase protein synthesis. Complement production rises. Stellate cells become easier to activate if the signal persists. What had been tolerance becomes mobilization. This shift is necessary when the burden is genuinely dangerous, but it is also costly. A liver living in chronic escalation begins to damage itself in the process of defending the body. The same immune machinery that once kept portal exposure survivable now helps drive fibrosis, altered blood chemistry, and broader systemic inflammation.

So the liver's challenge is not simply that it is exposed a lot. Many barrier-associated tissues are exposed. The liver's challenge is that it sits at the point where absorbed chemistry first becomes internal blood and where microbial evidence from the gut first becomes the problem of a central organ. It has to hold the line without hardening into permanent suspicion. It has to clear what should not circulate without declaring war on the ordinary chemistry of eating. That is why immune tolerance and immune defense belong in the same chapter and in the same sentence. In the liver, they are not opposites. They are the paired skills required to survive constant portal exposure without letting either chaos or overreaction take over.

Antigen Presentation in the Liver

Antigen presentation in the liver means that incoming proteins, peptides, lipids, microbial fragments, and cell-derived debris are not only cleared there. They are displayed there. An antigen is a molecular fragment that immune cells can recognize as something worth reading, whether that fragment came from a microbe, a dying host cell, a food-derived compound, or a transformed cell. Presentation means that a liver cell takes up that material, processes it, loads pieces of it onto major histocompatibility complex molecules, and then shows those pieces to T cells. This is one of the reasons hepatic immunity is so unusual. The liver is not just filtering portal blood mechanically. It is turning what arrives in portal blood into an immune conversation. Kupffer cells and liver sinusoidal endothelial cells are central to that work, and the liver's overall antigen-presenting environment is strongly shaped toward tolerance rather than constant inflammatory escalation.

What makes the liver different is that it uses more than one kind of presenter. Kupffer cells act as macrophage antigen-presenting cells inside the sinusoids. Liver sinusoidal endothelial cells also take up blood-borne antigen and present it, despite not being classical professional dendritic cells. Dendritic cells are present in the liver as well, but the liver also relies heavily on these non-classical or non-professional antigen-presenting cells, which gives the organ a very different immune tone from lymph nodes or an acutely inflamed tissue. Even hepatocytes themselves can participate. They express antigen-presentation machinery, and their polarity toward the sinusoidal side places that machinery where passing lymphocytes can encounter it through the fenestrated sinusoidal interface. So the liver is not showing antigen through one dedicated immune cell alone. It is presenting incoming material through a layered tissue network built directly into blood flow and parenchyma.

This matters because antigen presentation in the liver often produces a different outcome from antigen presentation in a conventional inflammatory organ. The liver is well known for a bias toward immune tolerance. Reviews of hepatic antigen presentation describe antigen-specific T-cell responses in the liver as tending toward tolerance, and presentation by liver cell types often results in incomplete activation of CD8-positive T cells, with initial proliferation followed by deletional or dysfunctional outcomes rather than durable full effector immunity. That is not an accident. The liver sits downstream from the gut and is exposed constantly to harmless but immunologically readable material. If every antigen presentation event there produced maximal cytotoxic activation, normal digestion and microbial coexistence would become immunologic self-destruction. Hepatic antigen presentation therefore has to preserve the ability to defend while keeping a strong built-in leaning toward restraint.

Hepatocyte presentation makes this logic especially visible. Hepatocytes are not classical immune cells, but they can present antigen to CD8-positive T cells in vivo, and the published literature describes this interaction as contributing to tolerance and even removal of virus-specific T cells from the repertoire in some settings rather than straightforward protective amplification. In plain terms, the liver can show antigen in a way that teaches the immune system to stand down, exhaust, or prune itself instead of always escalating. That makes sense in an organ that sees endless portal exposure, but it also helps explain why chronic liver infections and liver tolerance are such difficult immunologic subjects. The same architecture that prevents overreaction can also blunt effective clearance when the antigen truly matters.

So antigen presentation in the liver should be understood as immune reading built into a high-flow metabolic organ. Kupffer cells, liver sinusoidal endothelial cells, dendritic cells, and even hepatocytes can all participate in showing antigens to T cells. The result is not usually the same kind of sharp inflammatory decision made in a lymph node responding to an obvious invader. The liver presents under conditions of constant exposure, which means presentation there is inseparable from tolerance. The organ is not merely seeing what came from the gut. It is deciding how much of that incoming molecular evidence should become immunity and how much should be neutralized by silence.

Kupffer Cell Surveillance

Kupffer cell surveillance begins with location. These cells are not wandering immune responders that happen to enter the liver when something goes wrong. They are resident macrophages already positioned inside the sinusoidal blood channels where portal blood first slows and becomes readable. That placement tells you what their job is before any microscope stain does. Portal blood arriving from the intestine is never only a nutrient stream. It carries microbial fragments, endotoxin, damaged-cell debris, food-derived antigens, bile acid return, and a broad range of chemical material that crossed the gut barrier but has not yet been judged by the wider organism. Kupffer cells sit in that stream because the liver needs a local population of immune cells that can inspect, engulf, interpret, and respond before the rest of the circulation is forced to absorb the burden. Surveillance is not an added task. It is their reason for being there.

What they are watching for is not just living bacteria in the dramatic sense. They are constantly sampling low-level microbial evidence from the gut, especially lipopolysaccharide, peptidoglycan fragments, bacterial DNA, and other structural residues that appear in portal blood even under ordinary conditions. They also encounter apoptotic debris, aged blood-cell remnants, and tissue-derived danger signals. This is why Kupffer cell surveillance is fundamentally different from the immune behavior of a macrophage sitting deep in a relatively sterile tissue. The liver's macrophage has to read a stream full of ambiguous material. Some of it is routine. Some of it is harmful. Some of it is only harmful if the amount is high enough or if the surrounding tissue is already stressed. Surveillance in this setting means discrimination under continuous load.

That discrimination is what keeps digestion from becoming chronic systemic alarm. If Kupffer cells interpreted every meal-associated trickle of microbial material as open invasion, the liver would become an organ of constant inflammatory escalation. Instead, much of their work is quiet. They phagocytose, meaning they engulf and digest particulate material, without always broadcasting a crisis signal. They clear debris, sample endotoxin, and help prevent ordinary portal exposure from turning into whole-body inflammation. This is immune restraint, but it is not weakness. It is active, selective tolerance built into tissue that has no choice but to live downstream from the gut. Surveillance means noticing constantly without reacting maximally every time.

When the burden rises or changes character, the tone shifts. Higher endotoxin load, barrier failure in the intestine, cholestatic stress, steatotic injury, viral infection, alcohol-related injury, or significant hepatocyte damage all alter what Kupffer cells are seeing. Now surveillance becomes signaling. They release cytokines such as tumor necrosis factor and interleukins, recruit other immune responses, and begin reshaping the local environment. Hepatocytes change synthetic priorities. Acute-phase proteins rise. Sinusoidal endothelial behavior changes. Stellate cells receive stronger inflammatory cues and may begin moving toward fibrogenic activation. In this way Kupffer cell surveillance is not simply observation. It is the first switch in a larger hepatic response system. The liver's blood chemistry, inflammatory tone, and long-term architecture can all begin changing because what Kupffer cells are reading in the sinusoid has changed.

This is why Kupffer cells have to be understood as both protective and potentially destructive. In health they are part of the reason portal blood is survivable. They intercept microbial fragments and debris before that burden reaches the wider circulation too freely. In chronic stress they can become one of the main engines by which gut-derived burden, lipid injury, alcohol, or cholestatic retention is translated into cytokine dominance, oxidative stress, and eventual fibrosis. The same surveillance system that prevents daily immune chaos can become part of chronic tissue remodeling if the incoming signal never quiets down.

So Kupffer cell surveillance is best understood as constant immune reading at the gateway between gut and blood. These cells sit in the portal stream, clear what should not circulate, tolerate what should not provoke alarm, and signal when the burden becomes dangerous enough that quiet clearance is no longer sufficient. They do not merely patrol the liver. They determine how much of the gut's chemical reality the rest of the body is allowed to experience.

Tolerance Versus Inflammatory Activation

The liver cannot survive by choosing only one immune posture. If it were built only for tolerance, portal blood from the intestine would deliver microbial fragments, endotoxin, food antigens, damaged-cell debris, and other chemically noisy material into a tissue that never knew when to defend itself. If it were built only for inflammatory activation, the ordinary act of eating would become a daily immunologic injury. The liver sits downstream from a colonized gut, not from a sterile nutrient tube. That means it has to read constant exposure without collapsing into either blindness or paranoia. Tolerance and inflammatory activation are therefore not two separate chapters in hepatic immunity. They are the two edges of the same decision. Every incoming portal stream forces the liver to ask whether what arrived is routine burden that should be absorbed into quiet clearance or danger severe enough to justify changing the chemistry of blood and the behavior of tissue.

Tolerance is the quieter state, but it is not passive. It is an actively maintained decision not to escalate in response to ordinary portal traffic. Kupffer cells clear microbial fragments and debris without turning every small exposure into a cytokine storm. Sinusoidal endothelial cells shape how antigens and inflammatory signals are presented to circulating immune cells. Hepatocytes continue buffering glucose, handling amino acids, conjugating bilirubin, making bile, and synthesizing plasma proteins while living under the low-grade microbial reality of the gut. This works because the liver is built to treat a certain amount of portal contamination as normal physiology rather than as immediate crisis. Small quantities of endotoxin, food-derived antigens, and microbial metabolites are not interpreted as proof that the body is under invasion. They are interpreted as the ordinary consequence of absorbing from a living intestine. Tolerance is what keeps normal digestion from becoming systemic inflammation.

Inflammatory activation begins when the character, amount, or context of the incoming burden changes enough that quiet clearance is no longer safe. More endotoxin. More bacterial fragments. More barrier failure in the gut. More hepatocyte injury from fat, alcohol, or cholestatic stress. More dying cells releasing danger signals into the local environment. Under those conditions Kupffer cells shift from scavenging to signaling. Cytokines rise. Acute-phase protein synthesis is reprogrammed. Complement and fibrinogen production increase. Endothelial behavior changes. Stellate cells begin receiving the kind of cues that can push them toward fibrotic activation. What had been a tolerant high-flow exchange organ starts behaving like an inflamed defensive organ because the portal stream has ceased to look like ordinary digestive aftermath and started to look like real threat. Inflammation is not the liver forgetting tolerance. It is the liver deciding tolerance has become too expensive.

The problem is that chronic liver disease often lives in the unstable space between these two states. The liver continues receiving constant portal exposure, so it cannot abandon tolerance completely without self-destruction. But persistent gut-derived inflammatory burden, steatosis, cholestasis, viral injury, or alcohol-related stress keep pushing it toward activation. The tissue then enters a damaged compromise. It is no longer quietly tolerant, but it is not solving the upstream burden either. Kupffer cells remain activated. Hepatocytes keep shifting protein synthesis toward inflammatory priorities. Stellate cells begin laying down collagen. The organ starts remodeling under a pressure it was designed to buffer, not to endure indefinitely. This is how the liver can become simultaneously immunologically restrained and chronically inflamed. It is trying to live with a burden that no longer deserves simple tolerance but cannot be answered with full destructive aggression every hour of the day.

So tolerance versus inflammatory activation is not a moral distinction between weak immunity and strong immunity. It is a problem of fit. Tolerance fits ordinary portal exposure. Activation fits real threat. The liver has to keep moving between those states without mistaking one for the other. When it succeeds, digestion remains compatible with systemic stability. When it fails, the same organ that protected the body from the gut begins turning portal burden into chronic inflammation and tissue injury.

Why the Liver Filters Bacterial Traffic from the Intestine

The liver filters bacterial traffic from the intestine because the gut is not a sterile nutrient source. It is a colonized chamber packed with microbes, microbial fragments, fermentation products, and immune-provoking molecules sitting only one epithelial barrier away from the bloodstream. Every day that barrier allows nutrients to cross, but it never creates a perfectly clean stream. Small amounts of bacterial structural material, especially lipopolysaccharide from gram-negative organisms, along with peptidoglycan fragments, bacterial DNA, and other microbial residues, enter portal blood as part of ordinary life downstream from a living intestine. If that material were allowed to pass unfiltered into the systemic circulation in the same way glucose or oxygen pass, the whole body would be forced to live under a much higher inflammatory burden. The liver sits between gut and general blood flow because someone has to decide how much of that microbial reality is tolerated quietly, how much is cleared, and how much truly deserves escalation.

This filtering begins as blood-handling architecture, not as an abstract immune principle. Portal blood flows into the liver sinusoids, where the stream slows and becomes readable. Kupffer cells sit directly in that flow as resident macrophages, already positioned to engulf bacterial fragments, endotoxin, and debris before they move farther downstream. Liver sinusoidal endothelial cells and hepatocytes participate in the same broader filtering environment by shaping what crosses, what is signaled, and what is metabolically altered. This means the liver is not filtering bacterial traffic the way a coffee filter strains particles from water. It is filtering biologically. It is sampling, engulfing, interpreting, neutralizing, and deciding whether a given burden belongs to the category of ordinary portal noise or to the category of actual threat. The organ is turning a chemically dirty stream into a more tolerable one before the rest of the body has to deal with it.

That filtering is necessary because bacterial traffic from the intestine is not automatically proof of infection. A colonized gut always generates some low-level microbial evidence in portal blood. If the liver treated every fragment as a reason for systemic alarm, digestion itself would become a chronic inflammatory event. The body would overreact to its own normal arrangement. So the liver has to clear a great deal of bacterial traffic without overreacting to it. This is one of the deepest reasons hepatic immunity is built around selective tolerance. The organ is not trying to sterilize portal blood absolutely. It is trying to reduce microbial burden to a level that can be lived with while preserving the capacity to respond when that burden becomes excessive, more inflammatory, or structurally dangerous.

When that filtering role is overwhelmed or impaired, the consequences spread quickly beyond the liver. More endotoxin and bacterial fragments reach the systemic circulation. Cytokine signaling rises. Acute-phase protein synthesis is pushed harder. Hepatocytes begin working under greater inflammatory pressure. Kupffer cells shift from quiet scavenging toward louder immune activation. Stellate cells become easier to activate, and fibrosis becomes easier to sustain if the burden persists. This is why the liver filters bacterial traffic from the intestine not only to protect itself, but to protect the organism from the immunologic cost of living with a colonized gut. The intestine absorbs from a microbial world. The liver makes that arrangement survivable.

When This Balance Breaks and Inflammation Becomes Chronic

The balance breaks when the liver stops receiving a manageable portal burden and starts receiving a stream that demands defense more often than tolerance can safely absorb. Under healthy conditions, low-level microbial fragments, endotoxin, food antigens, and ordinary digestive debris are cleared quietly enough that the organ can remain a high-flow metabolic and synthetic tissue rather than hardening into a chronic immune organ. But once the incoming burden becomes heavier, more frequent, or more chemically aggressive, the same machinery that once preserved calm begins staying switched on. Increased intestinal permeability, chronic endotoxin delivery, steatotic hepatocyte stress, alcohol-related injury, cholestatic retention, viral burden, and repeated cell death all push the liver in the same direction. Kupffer cells stop behaving mainly like quiet scavengers and begin behaving like persistent inflammatory amplifiers. Cytokines stop arriving as brief instructions and begin acting like background chemistry. The organ no longer returns fully to baseline between insults because the insults never fully stop.

That is the turning point. In a healthy liver, inflammation is an event. In a chronically burdened liver, inflammation becomes the environment. Hepatocytes keep trying to maintain bile secretion, glucose control, protein synthesis, and detoxification, but they are now doing this work inside a tissue field increasingly saturated with inflammatory signaling, oxidative stress, and immune cell activation. Acute-phase protein production stays elevated. Albumin and other quieter baseline priorities may fall in relative emphasis. Sinusoidal endothelial behavior changes. Microvascular flow becomes less physiologically open. The liver is still trying to perform its ordinary jobs, but it is performing them in an atmosphere that increasingly punishes ordinary function. The exchange organ becomes harder, louder, and metabolically less graceful.

Once that state persists, stellate cells become one of the major reasons inflammation stops being just a signaling problem and starts becoming a structural one. These cells normally sit in the space of Disse storing vitamin A and helping preserve the exchange environment. Under chronic inflammatory pressure they lose that reserve identity, activate, and begin laying down collagen and extracellular matrix. The quiet transfer space between blood and hepatocyte thickens. Sinusoids lose their permissive character. Portal resistance rises. Hepatocytes become less well supplied and less able to meet the moving blood chemistry they are supposed to govern. Chronic inflammation therefore does not just make the liver angry. It makes the liver physically less able to be a liver.

This is why chronic liver inflammation is so dangerous. The original burden may have been microbial traffic from the gut, fat congestion inside hepatocytes, retained bile acids, alcohol metabolism, or viral injury, but once the inflammatory program is sustained, the tissue begins helping perpetuate its own damage. More injury produces more cytokines. More cytokines activate more stellate cells. More matrix distorts more sinusoids. Worse sinusoidal exchange creates more hepatocyte stress. The balance that once existed between tolerance and defense has now broken in a specific direction. The liver is no longer tolerating enough to stay quiet, and no longer resolving enough to return to calm. It begins living in self-reinforcing immune activation.

So inflammation becomes chronic when the liver is forced to defend against a burden that has become too constant to clear quietly and too persistent to resolve. The organ that once protected the body from the gut now starts carrying the chemistry of unresolved alarm in its own tissue. What was supposed to be temporary defense becomes the new baseline, and the price is paid in fibrosis, synthetic decline, impaired bile flow, distorted blood handling, and slow architectural collapse.

Chapter 32

Alcohol, Drugs, and Xenobiotics

What Xenobiotics Are

Xenobiotics are compounds the body did not make for its own structure, fuel system, or signaling design. The word simply means foreign to life inside the organism. That does not automatically mean poisonous. Some xenobiotics are medicines. Some are food additives, preservatives, sweeteners, colorants, flavoring chemicals, plant defense compounds, alcohols, solvents, environmental contaminants, smoke-derived molecules, or industrial residues. Some arrive by mouth. Some enter through air, skin, or water. The important point is not whether the molecule is legal, natural, synthetic, prescribed, or accidental. The important point is that the body has to decide what to do with a chemical that is not one of its ordinary internal building blocks. Once that chemical enters blood, the liver becomes one of the main organs forced to make that decision.

This matters because the bloodstream cannot carry every foreign compound indefinitely and still remain a stable internal environment. Some xenobiotics are lipid-soluble and would linger in membranes or adipose tissue if they were not transformed. Some are reactive enough to damage proteins, lipids, or DNA if allowed to remain unchanged. Some interfere with hormone systems, neural signaling, mitochondrial function, or redox balance. Some are relatively harmless until the liver itself converts them into a more reactive intermediate. This is why xenobiotic handling is not just "detox" in the loose popular sense. It is chemical triage. The liver has to determine whether the compound should be oxidized, reduced, hydrolyzed, conjugated, secreted into bile, returned to blood in altered form for renal excretion, or stored temporarily because immediate export is not yet possible. The body is not simply getting rid of something. It is trying to make a foreign molecule more compatible with survival.

The portal route makes this especially important for orally consumed xenobiotics. A drug, a drink, a plant compound, a food additive, or an industrial contaminant taken in by mouth may cross the intestinal barrier and enter portal blood before the rest of the circulation sees much of it. That means the liver receives a concentrated first pass at a compound the organism may never have evolved, expected, or wanted. Hepatocytes then expose it to enzyme systems, especially cytochrome P450 enzymes in the smooth endoplasmic reticulum, that begin changing its chemistry. Sometimes this makes the molecule less active and more water-compatible. Sometimes it activates a prodrug into its useful therapeutic form. Sometimes it creates a reactive intermediate more dangerous than the parent compound. Xenobiotics therefore should not be imagined as substances the liver simply erases. They are substances the liver has to engage chemically, and that engagement can itself become part of the burden.

This is one reason the category matters so much. Nutrients are supposed to enter physiology and become part of tissue, energy production, or signaling under ordinary rules. Xenobiotics enter physiology without belonging to those ordinary rules. They have to be screened, altered, contained, or expelled because the body did not build them and did not build itself around them. The liver is central because it is the high-flow organ sitting between absorption and systemic distribution, with the enzyme machinery, transport systems, and biliary access required to turn foreign chemistry into something the organism can either tolerate or remove. A xenobiotic is therefore not defined mainly by where it came from. It is defined by what the body has to do once it arrives.

Why the Liver Is the Main Organ for Foreign Compound Handling

The liver is the main organ for foreign compound handling because it is the first major internal organ downstream from the gut and because it is built for chemical transformation at scale. Those two facts together matter more than either one alone. A foreign molecule taken in by mouth does not usually enter the general circulation first. It enters the portal circulation and reaches the liver before the rest of the body sees its full concentration. That gives the liver first possession of a compound the organism did not make and may not be able to tolerate in its incoming form. At the same time, hepatocytes are packed with the organelles, enzymes, transporters, and biliary access needed to change, package, or export that compound. The liver therefore is not just in the right place. It is in the right place with the right machinery.

That machinery is what makes the liver a chemical-processing organ rather than a simple filter. Foreign compounds, xenobiotics, are not all handled the same way. Some need oxidation or reduction to expose a chemical handle for later conjugation. Some need hydrolysis. Some are made less active by transformation. Some are activated into a more functional or more toxic form by the first metabolic step. The cytochrome P450 system in the smooth endoplasmic reticulum is especially important here because it performs much of this first-stage modification. Conjugation pathways then attach groups such as glucuronic acid, sulfate, or glutathione, making the compound more water-compatible, less reactive, or more fit for biliary or urinary excretion. This is why the liver is not mainly trapping foreign compounds physically. It is rewriting their chemistry so the body can survive having absorbed them.

The organ's structure also makes this possible at high throughput. Portal blood slows in the sinusoids, the endothelium is fenestrated, and hepatocytes sit with a large microvillous blood-facing surface across the space of Disse. This lets dissolved compounds in plasma reach hepatic transporters and enzymes efficiently. The same hepatocyte that is buffering glucose, synthesizing albumin, producing bile acids, and converting ammonia into urea is also sampling drugs, alcohol, additives, plant compounds, and environmental chemicals arriving from the gut. Few organs operate at that level of mixed responsibility. The liver does, which is exactly why foreign compound handling became one of its defining jobs. It is already the organ of post-absorptive judgment. Xenobiotics are simply one more class of incoming material that must be judged before systemic circulation is allowed to live with it.

The biliary route strengthens this role even further. Some foreign compounds or their metabolites are not well suited to renal excretion alone, especially if they remain hydrophobic or are being secreted in larger conjugated forms. The liver can move these into bile and send them into the intestine, creating an exit route that the kidney cannot provide as cleanly. So the liver is not only transforming foreign molecules. It is often deciding which fluid system will carry them out. Blood to kidney. Bile to intestine. Temporary storage. Further modification. The organ sits at the intersection of transformation and elimination, which is exactly where foreign chemistry becomes a physiological problem.

This is why oral drugs, alcohol, and food-borne chemicals are so strongly liver-centered. The intestine absorbs them. Portal blood delivers them. The liver gets first review and often first injury. A compound can be harmless in low dose and burdensome in high dose. It can be relatively quiet until cytochrome P450 activation makes it reactive. It can be detoxified, delayed, conjugated, or retained depending on what the hepatocyte has the capacity to do at that moment. The liver carries that burden because the rest of the organism is safer if the first concentrated encounter with foreign chemistry happens in the one organ built to metabolize it.

So the liver is the main organ for foreign compound handling because it combines strategic position with chemical capability. It sees orally absorbed compounds first, exposes them to transformation systems designed for exactly this class of problem, and then routes them toward tolerance, temporary retention, urinary clearance, or biliary excretion. The body cannot prevent every foreign molecule from entering. It survives because the liver stands where entry becomes judgment.

First-Pass Metabolism

First-pass metabolism is what happens when a compound absorbed from the intestine reaches the liver before it reaches the rest of the body in full. The phrase matters because oral exposure is not the same as systemic exposure. A drug swallowed by mouth, alcohol, a plant compound, an additive, or some other xenobiotic may cross the intestinal barrier and enter portal blood, but that only means it has entered the route to the liver. It has not yet become the final concentration the brain, muscle, kidney, endocrine tissues, and wider circulation will actually experience. The liver stands in the middle of that route and gets first review. It can extract part of the incoming compound, transform it, conjugate it, inactivate it, activate it, delay it, or direct it toward bile before the bloodstream beyond the liver ever sees the full original burden. That is first-pass metabolism. The intestine absorbs. The portal vein delivers. The liver edits.

This is why swallowing a compound is never equivalent to injecting it directly into the bloodstream. Portal delivery changes the terms of exposure. The liver can remove a substantial fraction of some drugs on the first pass, which means the amount reaching systemic circulation in active form may be much lower than the amount swallowed. In other cases, the liver converts a relatively inactive parent compound into an active form, which means the first pass is not merely subtractive. It can be generative. Some compounds become more water-compatible and easier to excrete after hepatic transformation. Some become less active and safer. Some become more reactive and potentially more dangerous because the first transformation produces an unstable intermediate that still has to be neutralized by a second protective step. So first-pass metabolism is not simply the liver "detoxing" what came in. It is the liver deciding what chemical form the rest of the body will actually receive.

The machinery for this sits inside hepatocytes. Compounds arriving from portal blood cross the sinusoidal interface and meet transporters, enzymes, and intracellular conversion systems, especially the cytochrome P450 family in the smooth endoplasmic reticulum. Phase I reactions such as oxidation, reduction, and hydrolysis begin altering the molecule. Phase II reactions such as glucuronidation, sulfation, and glutathione conjugation may then make it more water-compatible, less reactive, or more fit for biliary or urinary excretion. The liver is therefore not acting as a simple sponge for foreign molecules. It is reauthoring them. A swallowed compound enters one chemical life in the gut and may leave the liver in another.

This matters because first-pass metabolism protects the body and burdens the liver at the same time. It protects the body by reducing how much unchanged foreign chemistry reaches the wider circulation. It burdens the liver because hepatocytes have to do the work of transformation, pay the redox cost, generate the transport machinery, and sometimes survive reactive intermediates created during the process. Alcohol shows this clearly. It enters through the gut, reaches the liver early, and is oxidized there in a way that changes redox balance and generates acetaldehyde, a more reactive intermediate. Many oral drugs follow the same principle in less obvious ways. The person swallows one compound. The liver receives it first and decides whether the body will meet less of it, more of its metabolite, or a chemically altered mixture of both.

So first-pass metabolism is the liver's first right of refusal on orally absorbed chemistry. It is the reason a molecule entering the body by mouth does not automatically become the same molecule reaching systemic blood. The gut grants entry. The liver determines form.

Phase I and Phase II Biotransformation Explained Step by Step

Biotransformation begins when a compound enters the body in a form the body cannot simply leave alone. That compound may be a drug, alcohol, a plant alkaloid, a food additive, a steroid hormone metabolite, an environmental chemical, or some other xenobiotic, meaning a molecule foreign to the body's own structural design. It may also be an endogenous compound that now needs clearance rather than continued activity. The liver receives much of this burden first, especially after oral absorption, because portal blood carries it directly to hepatocytes. Once the compound reaches the hepatocyte, the problem is not only whether the molecule is harmful. The problem is whether the molecule is chemically fit to remain in circulation, to enter bile, or to be filtered by the kidney. Many are not. They may be too lipid-soluble, too inert, too membrane-compatible, or too chemically awkward to leave the body efficiently in their incoming form. Biotransformation is the liver's way of changing that. It is a stepwise rewriting of the molecule so its activity, solubility, reactivity, and route of elimination become more manageable.

Phase I is the first rewrite. This phase usually introduces or exposes a reactive chemical handle on the molecule. The main reactions are oxidation, reduction, and hydrolysis. Oxidation often adds oxygen or removes electrons. Reduction does the opposite. Hydrolysis uses water to split part of the molecule apart. Much of Phase I work is carried out by cytochrome P450 enzymes in the smooth endoplasmic reticulum. These enzymes do not exist to "detox everything" in one clean move. They exist to alter structure. A molecule that was relatively stable and lipophilic, meaning comfortable in fat and membranes but poorly suited to water-based excretion, may become more polar after oxidation. Polar means it now carries a chemical feature that interacts more readily with water or with the enzymes of the next step. Sometimes that makes the molecule less active. Sometimes it activates a prodrug into its useful form. Sometimes it creates a reactive intermediate more dangerous than the parent compound. That last point matters. Phase I is not always protective by itself. It often creates a chemically exposed version of the molecule that now must be handled quickly and correctly or the hepatocyte itself begins paying the price.

That is why Phase I should be understood as exposure chemistry, not completion chemistry. The liver is taking a molecule that was too hidden, too fat-soluble, or too chemically self-contained and making it more available for further processing. A hydroxyl group may be added. An ester bond may be cleaved. A ring structure may be altered. The molecule becomes more recognizable to the cell's later conjugation systems. But it may also become more electrophilic, meaning more eager to react with proteins, lipids, or DNA. Acetaldehyde after alcohol oxidation is a classic example of a metabolite more dangerous than the parent compound that produced it. So the first phase opens the problem. It does not always solve it. The hepatocyte has exposed the molecule so it can now be neutralized, packaged, or directed out more effectively, but the transition itself can be the most dangerous moment in the sequence.

Phase II is the second rewrite, and this is where the liver usually makes the molecule easier to carry out of the body. In Phase II, the hepatocyte attaches another chemical group to the compound or to the metabolite produced in Phase I. This is called conjugation. Common conjugates include glucuronic acid in glucuronidation, sulfate in sulfation, glutathione in glutathione conjugation, amino acids such as glycine, or methyl groups in methylation. These attachments usually make the molecule more water-compatible, less reactive, or more recognizable to export systems. Glucuronidation is especially important because it can turn poorly soluble compounds into conjugated forms that move more easily into bile or urine. Glutathione conjugation is crucial when the molecule is highly reactive, because glutathione can bind dangerous electrophilic intermediates and prevent them from attacking cellular proteins and membranes. Phase II is therefore packaging chemistry. The liver is no longer merely exposing the molecule. It is attaching something to it that changes its fate.

The sequence, step by step, is easier to see when pictured as traffic control. A foreign compound arrives in the hepatocyte from portal blood. If it is too lipid-soluble or too chemically closed off, Phase I modifies it, often by oxidation through cytochrome P450 enzymes, creating a more polar or more reactive version. That new version is then handed to Phase II systems, which attach a conjugate such as glucuronic acid, sulfate, or glutathione. Now the molecule is more water-compatible, less free to lodge in membranes, and better suited for export. At that point transporters can move it into bile for intestinal elimination or back into blood for renal clearance. Some molecules skip much of Phase I and move directly into conjugation. Others depend heavily on both stages. The important point is that biotransformation is not random. The liver is changing structure first, then changing transport destiny.

This is also why liver injury and drug toxicity are so often tied to biotransformation rather than to the parent compound alone. The hepatocyte is the cell performing the chemistry, and if Phase I generates too many reactive intermediates or Phase II lacks the resources, enzyme capacity, or glutathione needed to neutralize them, the cell becomes the first tissue exposed to the danger. The liver does not simply erase foreign compounds. It enters into chemical combat with them. Phase I sharpens the molecule into something more exposed. Phase II binds, buffers, and prepares it for exit. When the sequence works, the foreign compound leaves in a form the body can tolerate. When it fails, the molecule or its metabolites begin injuring the very cell that tried to save the organism from it.

Cytochrome P450 Systems in Plain Language

Cytochrome P450 systems are one of the liver's main chemical editing tools. They are not a single enzyme and they are not a magic detox machine. They are a large family of enzymes, most heavily concentrated in the smooth endoplasmic reticulum of hepatocytes, that take incoming molecules and begin changing their structure so the body can deal with them differently. Many compounds arriving from portal blood are too lipid-soluble to be excreted easily, too chemically stable to leave the body unchanged, or too biologically active to remain in circulation indefinitely. Drugs, alcohol-related metabolites, steroid hormones, plant compounds, food chemicals, and countless xenobiotics fall into this category. Xenobiotic simply means foreign to the body's own ordinary structural chemistry. The cytochrome P450 system takes these molecules and performs the first major act of biotransformation, usually by oxidation, which means inserting oxygen or otherwise changing electron structure in a way that exposes the molecule to the next stage of handling.

The easiest way to picture this is not as a garbage disposal, but as a cutting and reshaping bench. A molecule enters the hepatocyte in a form that fits badly with water-based excretion. The P450 enzyme grabs it, modifies part of its surface, and makes it chemically less hidden. Sometimes a hydroxyl group is added. Sometimes a bond becomes easier to break. Sometimes the molecule becomes more polar, meaning more compatible with aqueous transport. This often helps because a more polar molecule is easier to conjugate later and easier to move into bile or urine. But this first step is not always gentle. The P450 system can create an intermediate that is more reactive than the original compound. That is one of the reasons liver chemistry is dangerous work. The hepatocyte is not always neutralizing the molecule in one clean motion. It is often making the molecule chemically exposed enough that Phase II systems can finally bind it, buffer it, and send it out.

This is why the word detoxification is too soft if it is left unexplained. Sometimes cytochrome P450 activity reduces toxicity. Sometimes it activates a prodrug into its useful therapeutic form. Sometimes it creates the very intermediate that injures the liver if the next protective step cannot keep up. Acetaldehyde after ethanol oxidation is a familiar example of how a metabolite can be harsher than the parent compound. Many drugs behave the same way in broader principle. The liver receives a molecule that is stable enough to survive absorption, then uses P450 chemistry to alter it, but that alteration may temporarily increase reactivity before glucuronidation, sulfation, or glutathione conjugation finish the job. So the P450 system should be understood as exposure chemistry. It opens the molecule up. It does not guarantee safety by itself.

These enzymes also matter because they are adaptable. The liver can increase or decrease certain P450 activities depending on what the body is exposed to, which means one drug, food compound, or chemical can change how another is handled. Some compounds induce these enzymes, meaning the liver builds more of the relevant machinery and begins clearing certain molecules faster. Others inhibit them, meaning metabolism slows and the same dose of a drug may stay active longer or rise to a more dangerous concentration. This is one reason foreign compound handling is never just about the compound in isolation. The hepatocyte is responding to a chemical crowd, not a single molecule standing alone. A drug entering a liver already handling alcohol, inflammatory stress, nutrient deficiency, or multiple medications is entering a different metabolic landscape than the same drug entering a quiet liver with full reserve.

So cytochrome P450 systems in plain language are the liver's first-line chemical editors. They take foreign or excess endogenous molecules that cannot simply be left alone, oxidize or otherwise alter them, and prepare them for the next stage of transformation. They help the body clear hormones, handle drugs, process plant compounds, and survive oral exposure to chemistry the organism did not design for itself. But they do this by making molecules more chemically exposed, not by making them instantly harmless. The liver is performing controlled molecular surgery. When the sequence works, the compound leaves in a more manageable form. When it does not, the cell doing the editing becomes the first place the damage appears.

Why "Detox" Is Too Vague to Explain What the Liver Is Actually Doing

"Detox" is too vague because it collapses a long sequence of distinct chemical jobs into one soft reassuring word. The liver does not look at a bad molecule and simply remove it by force of virtue. It receives compounds of radically different structure, source, and behavior, and each one creates a different problem. Some are hydrophobic and persist in membranes. Some are already reactive and need to be neutralized quickly. Some are relatively inert until a liver enzyme converts them into a more dangerous intermediate. Some are endogenous compounds such as bilirubin, steroid hormones, ammonia-derived nitrogen, and bile acids that are not foreign at all, yet still need transformation or controlled exit. Others are true xenobiotics, drugs, alcohol, additives, plant defense chemicals, smoke-derived compounds, industrial residues. The liver is not performing one act called detoxification. It is sorting, oxidizing, reducing, hydrolyzing, conjugating, packaging, storing, secreting, and excreting. Each verb describes a different kind of work. "Detox" hides the architecture.

The first reason the word fails is that it does not tell you what the real problem with the molecule is. Is the compound too lipid-soluble to leave in urine. Is it chemically reactive and capable of injuring proteins or DNA. Is it harmless until cytochrome P450 enzymes expose a reactive handle. Does it need to be pushed into bile rather than blood because water-based excretion will not carry it cleanly. Does it need to be converted from ammonia into urea because nitrogen in its current form is neurotoxic. Does bilirubin need conjugation because in its unconjugated form it cannot be excreted properly. The body is always solving specific chemical incompatibilities. "Detox" sounds like purification. The liver is doing something more exact. It is changing molecular form so transport, retention, and excretion become physically possible.

The second reason the word fails is that the liver's work is often not subtractive in the simple sense. Sometimes the liver makes a compound less active. Sometimes it makes a prodrug active. Sometimes it converts ethanol into acetaldehyde, which is more dangerous than ethanol itself. Sometimes it adds glucuronic acid or sulfate to a molecule and makes it more water-compatible. Sometimes it binds a reactive intermediate to glutathione to keep that intermediate from attacking the hepatocyte itself. Sometimes it packages cholesterol into bile or lipoproteins instead of destroying it. Sometimes it stores iron or copper because immediate release would be more dangerous than temporary retention. None of that is well described by the fantasy of the liver as a chemical drain. It is a conversion organ, not just an elimination organ.

The third reason "detox" is too vague is that it hides cost. Hepatic handling is expensive. It uses ATP. It uses reductive capacity. It uses glutathione. It uses membrane transporters. It exposes hepatocytes to intermediate compounds that can injure mitochondria, endoplasmic reticulum, and membranes if throughput is too high or protective systems are too weak. A person hears "detox" and imagines a waste-removal service. The hepatocyte experiences redox shifts, enzyme induction, oxidative stress, and the need to decide whether the molecule will go back into blood, into bile, into temporary storage, or into a more reactive state that still requires another step before safety is restored. The liver is doing chemistry under load, not washing toxins down a drain.

The cleaner way to speak is to ask what the liver is doing to this specific class of molecule. It may be performing Phase I biotransformation through cytochrome P450 oxidation. It may be performing Phase II conjugation through glucuronidation, sulfation, or glutathione binding. It may be converting ammonia to urea. It may be conjugating bilirubin for biliary excretion. It may be packaging lipids into VLDL. It may be secreting bile acids and cholesterol into bile. It may be transforming estrogen, cortisol, and thyroid-related compounds into more transportable or less active forms. It may be withholding iron through hepcidin. These are real mechanisms. "Detox" is what remains after those mechanisms are blurred beyond usefulness.

So the problem with the word is not that it is completely false. The liver does make harmful chemistry less compatible with the body and more compatible with exit. The problem is that the word is too small for the work. It hides the sequence, the cost, the compartments, the enzymes, the transport decisions, and the fact that some of the danger is created during the handling itself. The liver is not detoxing in the abstract. It is rewriting molecules so blood can survive them and the body can get rid of them.

Chapter 33

Oxidative Stress, Reactive Intermediates, and Hepatic Burden

Why Transforming Compounds Can Generate Danger

The liver does not neutralize every compound by making it disappear gently. Often it has to make the compound more dangerous before it can make it excretable. That is the first thing the reader needs to see. Many foreign molecules arrive in a form that is too chemically stable, too lipid-soluble, or too structurally hidden to be cleared cleanly. If the liver left them untouched, they would linger in membranes, adipose tissue, or circulation in ways the body could not manage well. So hepatocytes begin by changing the molecule. Cytochrome P450 enzymes oxidize it. Other systems reduce it or hydrolyze it. A new chemical handle appears. A hydroxyl group may be added. A bond may be weakened. The molecule becomes more exposed to the next step. That exposure is useful because it prepares the compound for conjugation and elimination. It is also dangerous because the newly exposed form is often more reactive than the parent compound. The liver is not simply removing a toxin. It is opening a stable molecule into a chemically active intermediate and then racing to contain what it just created.

This is where oxidative stress begins to make sense. Oxidation in this setting means electron movement, and electron movement inside a living cell is never morally neutral. It has to be controlled tightly. Cytochrome P450 systems use oxygen and reducing equivalents to alter incoming compounds, but in doing so they can leak partially reduced oxygen species, chemically aggressive molecules such as superoxide and hydrogen peroxide that are capable of injuring lipids, proteins, and DNA if they accumulate beyond the cell's defensive capacity. These are called reactive oxygen species. At the same time, the metabolite created from the compound may itself become electrophilic, meaning electron-hungry and eager to bind to cellular proteins, membranes, or nucleic acids in ways that distort function. So transforming a compound can generate two classes of danger at once. The enzyme system can produce oxidative burden as part of the transformation process, and the transformed molecule can become more reactive than the molecule that arrived.

This is why Phase I biotransformation is such a precarious moment. The parent compound may have been relatively inert in the short term. The intermediate is now chemically unfinished and potentially more harmful. If Phase II conjugation follows quickly, the hepatocyte can often regain control. Glucuronidation, sulfation, glutathione conjugation, and related pathways attach groups that make the compound more water-compatible, less reactive, or easier to export into bile or urine. But if the incoming burden is too high, if glutathione stores are depleted, if ATP is low, if mitochondrial function is already impaired, or if inflammation and nutrient deficiency have weakened the cell's reserve, then the gap between Phase I exposure and Phase II containment becomes the place where injury happens. Glutathione matters especially because it is one of the cell's major sacrificial defenses against reactive intermediates. It can bind and neutralize electrophilic compounds before they attack more important structures. Once glutathione runs low, the intermediate begins looking for another target, and that target is often the hepatocyte itself.

Mitochondria pay heavily for this burden because they are already carrying the organ's energetic load. Hepatocytes need ATP to maintain membrane polarity, run transporters, secrete bile, synthesize proteins, and sustain the whole architecture of blood-facing and bile-facing traffic. Reactive intermediates and reactive oxygen species injure mitochondrial membranes, distort electron transport, and worsen redox balance, which is the controlled handling of electron flow inside the cell. Once redox balance deteriorates, ATP production weakens, and the hepatocyte becomes even less able to complete safe biotransformation. The cell enters a trap. It needs energy and reducing power to neutralize the burden, but the burden is damaging the very organelles that supply that energy and reducing power. This is one reason hepatic injury from drugs, alcohol, and other xenobiotics can spread so quickly from molecular chemistry into organ dysfunction. The transformation machinery and the energy machinery are physically close and physiologically entangled.

The endoplasmic reticulum is burdened too. Much of the cytochrome P450 system sits in the smooth endoplasmic reticulum, and when foreign compound load remains high, membrane chemistry, calcium handling, lipid composition, and protein processing all begin working under harsher conditions. The hepatocyte is not only metabolizing a foreign molecule. It is reorganizing parts of its own interior around the stress of doing so. This is why hepatic burden is a better phrase than simple toxicity. The danger is not only the external compound. The danger is the total cost of transforming that compound inside a cell already busy with glycogen handling, bile production, clotting protein synthesis, hormone conversion, and blood chemistry control. A quiet liver can absorb some of that cost. A steatotic liver, inflamed liver, cholestatic liver, or nutrient-depleted liver absorbs it less well.

So transforming compounds can generate danger because metabolism is not a magical washing process. It is controlled chemical violence. The liver takes molecules that cannot be left alone and subjects them to reactions that often make them more exposed, more oxidized, and temporarily more hazardous in order to make final clearance possible. When the sequence is completed cleanly, the body survives the chemistry and the compound leaves. When the sequence stalls in the middle, the reactive intermediate stays behind and the hepatocyte becomes the first casualty of the work it was trying to perform.

Reactive Oxygen Species and Oxidative Stress

Reactive oxygen species are oxygen-containing molecules that are chemically eager to react. That eagerness is what makes them useful in small, controlled amounts and dangerous when they accumulate beyond the cell's ability to contain them. The main examples include superoxide, hydrogen peroxide, and the hydroxyl radical. Superoxide is produced when electrons leak onto oxygen prematurely. Hydrogen peroxide is less reactive than the hydroxyl radical but still dangerous if it is not cleared or compartmentalized. The hydroxyl radical is especially destructive because it reacts almost immediately with whatever is nearby. The liver produces reactive oxygen species continuously because the liver is an organ of oxidation. Mitochondria generate them during energy production. Cytochrome P450 enzymes generate them while transforming drugs, alcohol, and other xenobiotics. Peroxisomes generate them during fatty acid handling. Kupffer cells generate them as part of immune defense. So reactive oxygen species are not rare invaders entering a peaceful cell from outside. They are partly the cost of doing liver work at all.

That is why oxidative stress is not the same thing as merely having reactive oxygen species present. The hepatocyte always has some. Oxidative stress begins when production outpaces control. Control means antioxidant systems, glutathione, catalase, superoxide dismutase, peroxiredoxins, membrane repair, reducing equivalents such as NADPH, and the structural integrity of the organelles that are supposed to keep reactive chemistry compartmentalized. As long as these defenses match the burden, reactive oxygen species can even serve useful roles in signaling, microbial killing, and adaptation. But once the burden rises too high, or the defensive systems fall behind, the chemistry changes character. The reactive oxygen species are no longer functioning as controlled intermediates inside a bounded system. They begin attacking the system itself.

The main targets are predictable. Lipids in membranes are vulnerable because reactive oxygen species can initiate lipid peroxidation, a chain reaction in which unsaturated fatty acids inside membranes are oxidized and the membrane begins losing its normal fluidity, barrier function, and signaling behavior. Proteins are vulnerable because oxidation can alter amino acid side chains, distort enzyme shape, damage transporters, and weaken structural proteins. DNA is vulnerable because oxidative injury can damage bases and strand integrity. Mitochondria are especially vulnerable because they are both a major source of reactive oxygen species and one of the first structures injured by them. Once mitochondrial membranes and electron transport are damaged, ATP production falls. ATP, adenosine triphosphate, is the cell's immediately usable energy currency. That creates a trap. A stressed hepatocyte needs energy to restore order, but oxidative stress is injuring the very organelles that provide that energy.

This is why oxidative stress spreads beyond chemistry into physiology so quickly. A membrane-damaged hepatocyte secretes bile less cleanly. A mitochondrially weakened hepatocyte oxidizes fat less efficiently and becomes easier to push toward steatosis. A protein-damaged hepatocyte handles drugs, hormones, and nutrient traffic less precisely. Oxidative injury also activates inflammatory pathways, because damaged lipids, proteins, and mitochondrial components begin acting like danger signals. Kupffer cells respond. Cytokines rise. Stellate cells receive stronger fibrogenic cues. The liver is no longer just dealing with reactive molecules. It is now dealing with the tissue consequences of those molecules, inflammation, cell death, impaired secretion, and remodeling of the exchange architecture.

So reactive oxygen species and oxidative stress should be understood as a problem of pace and containment. The liver always generates reactive oxygen chemistry because the liver always runs oxidation-heavy work. Trouble begins when that chemistry escapes the systems meant to keep it brief, local, and survivable. Reactive oxygen species are part of the price of metabolism. Oxidative stress is what happens when the bill comes due and the hepatocyte cannot pay it cleanly.

Glutathione in Liver Defense

Glutathione is one of the liver's main sacrificial defense molecules, and it only makes sense when the reader sees how dangerous normal hepatic work already is. The hepatocyte is constantly oxidizing fatty acids, transforming drugs, processing alcohol, handling hormones, running mitochondrial respiration, and exposing foreign compounds to cytochrome P450 chemistry that can generate reactive intermediates. Reactive intermediates are unstable forms of a molecule produced during metabolism that are more eager to bind proteins, lipids, or DNA than the parent compound was. A cell doing this much chemistry needs something small, abundant, and chemically ready to intercept danger before that danger reaches more important structures. Glutathione is that molecule. It is a tripeptide made from glutamate, cysteine, and glycine, but the important fact is not its size. The important fact is that it carries a reactive sulfur group on cysteine, and that sulfur group can donate reducing power or bind hostile compounds before they attack the hepatocyte itself.

This gives glutathione two major jobs. The first is redox defense. Redox refers to the controlled handling of electrons inside the cell. Reactive oxygen species such as hydrogen peroxide are constantly generated during mitochondrial work, peroxisomal oxidation, immune signaling, and xenobiotic metabolism. Hydrogen peroxide is dangerous not because it looks dramatic on a page, but because if it is not cleared, it can help generate far more destructive chemistry. Glutathione helps neutralize this burden through enzymes such as glutathione peroxidase, which use reduced glutathione to convert peroxides into less dangerous products. In that act, glutathione itself is oxidized and has to be regenerated. That regeneration depends on glutathione reductase and NADPH, nicotinamide adenine dinucleotide phosphate in its reduced form, which supplies the electrons needed to recharge the system. This is why glutathione should not be imagined as a one-time antioxidant pill floating through the liver. It is part of a cycling intracellular defense economy. The hepatocyte spends it, oxidizes it, and rebuilds it continuously because oxidative chemistry in the liver is continuous.

The second major job is conjugation. Some molecules are too reactive to be managed by redox buffering alone. They need to be chemically trapped. Glutathione can bind electrophilic intermediates, compounds that are electron-hungry and therefore prone to attack cellular proteins, membranes, and nucleic acids. When glutathione conjugates one of these intermediates, the molecule becomes less free to damage the hepatocyte and more suited for onward processing and excretion. This is one of the reasons glutathione sits so close to the center of hepatic survival. The liver is often forced to make dangerous intermediates in order to make final clearance possible. Glutathione stands in the gap between exposure and catastrophe. It catches what Phase I metabolism has sharpened before that sharpened chemistry cuts into the cell.

This is also why glutathione depletion is so consequential. A hepatocyte with low glutathione is not merely missing a helpful antioxidant. It is losing one of its main permissions to keep doing liver work. Reactive oxygen species accumulate more easily. Lipid peroxidation rises. Protein thiols are damaged. Mitochondria lose integrity. Membrane transport becomes less precise. Bile secretion weakens because ATP-dependent polarity and transporter function are harder to preserve in an oxidatively stressed cell. The same oxidative burden that once would have been contained locally now starts spreading into inflammation, necrosis, apoptosis, and fibrotic signaling. The liver becomes less able to handle alcohol, less able to tolerate high xenobiotic burden, less able to survive fat congestion, and less able to complete the very biotransformation work that made glutathione necessary in the first place. The trap is obvious. The cell needs glutathione to survive transformation, but transformation is one of the things that consumes glutathione.

This is why glutathione belongs near the center of any honest account of liver defense. The liver does not survive because it never generates dangerous chemistry. It survives because it has built a buffered system for generating dangerous chemistry without letting that chemistry escape into the tissue unchecked. Glutathione is one of the main molecules that makes that possible. It reduces peroxides, binds reactive intermediates, protects proteins and membranes, and helps preserve mitochondrial function while the hepatocyte carries out oxidation-heavy work for the rest of the organism. When glutathione is abundant and regenerating well, the liver can absorb chemical insult without losing itself immediately. When glutathione falls behind, the same workload that once looked like normal hepatic function starts becoming self-injury.

When Metabolism Creates Compounds More Damaging Than the Starting Material

People often imagine liver metabolism as a one-way safety machine, as though a foreign compound arrives dangerous and leaves gentler simply because the liver touched it. That is not how much of hepatic chemistry works. The liver often has to make a molecule more chemically exposed before it can make it more excretable. A stable lipid-soluble compound can linger in membranes, circulate too long, or resist elimination if it is left unchanged. So hepatocytes use Phase I enzymes, especially cytochrome P450 systems, to oxidize, reduce, or otherwise alter the molecule. This first step may add oxygen, expose a reactive site, loosen a bond, or change the electron structure enough that the compound becomes a better target for Phase II conjugation. The problem is that this halfway state can be more dangerous than the original material. The parent compound may have been chemically quiet. The metabolite may now be electrophilic, meaning it is electron-hungry and inclined to attack proteins, lipids, and DNA. The liver is not always neutralizing danger in one motion. It is often opening a stable molecule into a more aggressive one so the next protective step can finally grab it.

That is why hepatic metabolism can become self-injury when throughput outruns containment. The classic logic is simple even when the chemistry is not. A compound arrives in a form too hidden to leave. Phase I metabolism exposes it. Now the liver needs glutathione, glucuronidation, sulfation, or another conjugation pathway to bind, buffer, or export what it just created. If that second step happens quickly, the reactive intermediate is only a passing threat. If glutathione is depleted, if ATP is low, if the cell is already inflamed, steatotic, cholestatic, or oxidatively stressed, the intermediate remains free long enough to attack the hepatocyte itself. ATP, adenosine triphosphate, is the cell's immediately usable energy currency, and a metabolically strained liver has less of it available for repair, transport, and orderly completion of biotransformation. The danger therefore is not only the chemical identity of the starting molecule. It is the gap between exposure and containment. The liver creates the sharper form first, then has to survive long enough to finish the job.

Alcohol shows the logic cleanly. Ethanol itself is disruptive, but one of the main reasons alcohol injures the liver is that its metabolism produces acetaldehyde, a more reactive compound that binds proteins, disturbs membranes, alters mitochondrial function, and drives inflammatory signaling. Acetaminophen shows the same principle in another register. Most of it is handled safely, but a fraction is converted into N-acetyl-p-benzoquinone imine, or NAPQI, a highly reactive intermediate that is normally neutralized by glutathione. When glutathione becomes insufficient, NAPQI begins binding hepatocellular proteins instead. The same logic extends beyond these famous examples. Many xenobiotics, meaning foreign compounds the body did not make for its own structure or signaling, are not dangerous mainly because of what they are on entry. They become dangerous because the liver must change them in order to make eventual clearance possible.

This is why hepatic burden is not a simple matter of how toxic the incoming world is. It is also a matter of how much dangerous transitional chemistry the liver is being forced to generate inside itself. The hepatocyte is a cell that survives by doing risky work in a controlled way. It oxidizes. It reduces. It conjugates. It traps nitrogen. It secretes bile acids. It handles hormones, drugs, alcohol, and food compounds while still maintaining polarity, protein synthesis, and blood chemistry control. When metabolism creates a compound more damaging than the starting material, the liver is temporarily carrying a sharper chemical weapon inside the very cell that must disarm it. If the sequence completes, the body survives the transaction and the compound leaves in a safer form. If the sequence stalls, the intermediate remains, and the liver becomes the first tissue injured by the chemistry it created in order to protect everything else.

The Difference Between Successful Transformation and Collateral Injury

Successful transformation means the liver changed a compound without letting the intermediate chemistry escape into the tissue. Collateral injury means the liver began that same transformation but could not contain the cost of doing it. The distinction is not whether metabolism happened. In both cases metabolism happened. The difference is whether the sequence reached completion under controlled conditions or stalled in the dangerous middle. A foreign compound, a drug, alcohol, a hormone metabolite, a plant chemical, or some other xenobiotic arrives in the hepatocyte in a form that may be too lipid-soluble, too stable, or too poorly configured for direct elimination. The liver begins Phase I work, often through cytochrome P450 enzymes, exposing the molecule by oxidation or other structural change. That first step is useful because it makes later conjugation possible. It is also the most precarious step because the compound often becomes more chemically aggressive before it becomes safer. If the hepatocyte has enough glutathione, enough ATP, enough reductive capacity, and intact Phase II machinery, the intermediate is captured, conjugated, and exported. That is successful transformation. If not, the exposed intermediate stays free long enough to attack the cell that made it. That is collateral injury.

The liver succeeds when the timing holds. The incoming compound is altered. The altered compound is immediately bound, neutralized, or made more water-compatible. Transporters then move it into bile or back into blood for renal excretion. The hepatocyte spends energy and antioxidant reserve to protect itself through the whole sequence, but the burden remains inside the cell's handling capacity. Membranes stay intact. Protein thiols are preserved. Mitochondria keep producing ATP. Endoplasmic reticulum function remains organized enough to keep synthesis and detoxification coordinated. The molecule leaves in a form the body can tolerate better than the form that entered. In this state the liver has done what it is built to do. It received foreign chemistry, rewrote it, and protected the organism from broader exposure without paying too much local damage.

Collateral injury begins when one of those protective layers is too thin for the burden being imposed. The compound may arrive in too high a dose. The hepatocyte may already be steatotic, cholestatic, inflamed, nutrient-depleted, or hypoxic. Glutathione may already be low. Mitochondria may already be struggling. Phase I still exposes the compound, but Phase II cannot keep up. Now the reactive intermediate begins binding cellular proteins, attacking membrane lipids, distorting enzymes, and pushing oxidative stress upward. Reactive oxygen species rise from the same organelles already under pressure. Calcium handling destabilizes. Mitochondrial membranes become more vulnerable. ATP falls. Once ATP falls, the hepatocyte becomes even less capable of finishing the transformation cleanly, because the cell needs energy to run transporters, regenerate defenses, preserve polarity, and maintain the architecture that keeps bile, blood, and cytoplasm in their proper compartments. The work of protection becomes part of the damage.

This is why successful transformation and collateral injury are separated less by category than by reserve. The chemistry may begin the same way in both cases. What changes is whether the cell had enough margin to carry the reaction through. A quiet well-supplied hepatocyte can often survive chemistry that injures a crowded oxidatively stressed hepatocyte badly. The parent compound may be identical. The outcome differs because the cell performing the transformation was different. This is one reason liver disease amplifies vulnerability to later toxic injury. A burden that was once transformable becomes injurious when reserve is thinner.

So the difference between successful transformation and collateral injury is the difference between a controlled chemical handoff and a failed one. In the successful version, the liver exposes the compound, captures the exposed form, and sends it out before the cell itself becomes the target. In the injured version, the liver opens the molecule but cannot finish the job before the intermediate begins cutting into membranes, proteins, and mitochondria. The same metabolism that protects the body becomes the source of tissue damage when the cell can no longer survive the chemistry it is being asked to perform.

Why Chemical Handling Always Carries a Cost When the Load Stays High

Chemical handling always carries a cost because the liver does not neutralize burden by waving it away. It has to spend itself against it. Every compound that enters in a form the body cannot safely keep has to be taken up, exposed to enzymes, moved across membranes, oxidized, reduced, conjugated, buffered, packaged, and exported. That work consumes ATP, reducing equivalents, glutathione, amino acids, membrane turnover, transporter activity, and mitochondrial reserve. ATP, adenosine triphosphate, is the cell's immediately usable energy currency. Glutathione is one of the liver's main sacrificial defense molecules against reactive oxygen species and electrophilic intermediates. Reducing equivalents such as NADPH provide the electron power needed to regenerate protective systems after they have been spent. None of this is free. A small burden can be absorbed because reserve is larger than demand. A high persistent burden changes the arithmetic. The hepatocyte is no longer performing occasional defensive chemistry inside an otherwise stable cell. It is living in defense mode.

The cost begins with transformation itself. Phase I enzymes such as cytochrome P450 systems often have to make a compound more chemically exposed before later steps can make it more excretable. That exposure generates reactive intermediates and oxidative byproducts, which means the hepatocyte is not only changing the molecule. It is also generating local danger in the process of changing it. If the next protective step comes fast enough, glutathione conjugation, glucuronidation, sulfation, or another buffering move, the intermediate remains temporary. If the load stays high, those buffering systems are forced to cycle faster and faster. Glutathione has to be spent and regenerated repeatedly. Mitochondria have to keep supplying energy while also absorbing the oxidative spillover generated by the very chemistry they are helping finance. The endoplasmic reticulum has to keep hosting enzyme activity and membrane traffic under conditions that become progressively less quiet. High load therefore means the liver is not only handling more compounds. It is manufacturing more transitional danger inside itself every hour those compounds continue arriving.

That is why the burden spreads beyond one enzyme family into the architecture of the whole cell. Mitochondria become less efficient because oxidative stress injures membranes and electron transport. Endoplasmic reticulum stress rises because membrane composition, protein handling, and xenobiotic metabolism are all being pushed at once. Bile secretion becomes harder to maintain because transporter systems and hepatocyte polarity depend on ATP and intact membranes. Fat oxidation becomes less clean, which means steatosis becomes easier to sustain. Protein synthesis begins competing with defense chemistry for the same metabolic reserve. The hepatocyte can keep many of these plates spinning for a while, but the point is that it is now paying with core housekeeping functions, not with some isolated detox budget sitting off to the side. A liver under chronic chemical load is spending the same reserve it needs for blood chemistry control, bile formation, protein synthesis, and tissue survival.

The other reason the cost always rises when the load stays high is that high load changes signaling, not just throughput. Persistent reactive oxygen species, retained intermediates, and membrane injury are read by the cell as damage, not merely as work. Kupffer cells begin seeing more danger signals. Cytokines rise. Stellate cells receive stronger fibrogenic cues. Acute-phase protein priorities can begin displacing quieter synthetic priorities. What began as chemical handling becomes inflammatory handling, then structural handling. The liver is no longer only processing compounds. It is remodeling itself around the cost of processing compounds. This is why a chronically burdened liver drifts toward fibrosis, steatosis, cholestatic fragility, and synthetic decline even when the original insult was "just" chemical exposure. The chemistry does not stay chemical. It becomes tissue behavior.

So chemical handling always carries a cost when the load stays high because the liver has to pay in energy, antioxidant reserve, membrane integrity, mitochondrial performance, and inflammatory calm every time it transforms a compound. A low burden can be finished cleanly and forgotten. A high persistent burden turns transformation into attrition. The molecule may still leave. The liver does not leave unchanged.

Part IX

How the Liver Repairs Itself and How It Loses That Ability

Chapter 34

Liver Regeneration

Why the Liver Has a Unique Regenerative Reputation

The liver has a unique regenerative reputation because it is one of the few major organs that can lose a substantial amount of working mass and still rebuild enough functional tissue to restore throughput. That reputation is real, but it is often misunderstood. The liver does not regenerate because it is magical, immortal, or exempt from biological cost. It regenerates because its architecture and cell biology are built around surviving fluctuating load. Hepatocytes are not terminally passive cells with no reserve for re-entry into growth. Under the right conditions they can leave a relatively quiescent state, re-enter the cell cycle, divide, and restore lost parenchyma, the working tissue of the organ. This matters because the liver is not a decorative structure that can afford to remain partially disabled after injury. It is the main organ for blood chemistry control, bile formation, plasma protein synthesis, nutrient conversion, and xenobiotic handling. If it could not rebuild after partial loss, ordinary survival after injury would be much harder than the body can tolerate.

The key is that regeneration in the liver is functional before it is dramatic. The organ does not need to recreate every microscopic contour immediately. It needs to restore enough hepatocyte mass, vascular connection, and biliary continuity that blood can once again move through an organ capable of doing liver work. That is why the liver's regenerative reputation is tied so closely to its workload. Portal blood still arrives. Ammonia still needs handling. Bilirubin still needs conjugation and biliary secretion. Albumin and clotting factors still need to be synthesized. The regenerative drive is not just wound closure. It is restoration of a metabolic machine the rest of the body cannot easily replace. A damaged skin surface can scar and still cover the body. A damaged liver has to return to chemistry.

This is also why liver regeneration is not the same thing as simple healing. Healing can mean collagen deposition, scar formation, and structural sealing. Regeneration means restoration of parenchymal cell mass and working capacity. The liver is unusual because it can do a remarkable amount of true parenchymal recovery when the injury is partial, the extracellular scaffold is sufficiently preserved, blood flow is still supportable, and the inflammatory burden does not keep destroying the tissue as fast as it tries to rebuild. Hepatocytes can proliferate. Nonparenchymal cells can help coordinate the response. Growth signals rise. Quiescent cells are recruited back into active work. The organ begins replacing lost functional units rather than merely walling off the wound.

That reputation, though, only belongs to the liver under the right conditions. The same organ famous for regeneration can lose that ability when injury becomes chronic, when the exchange architecture is buried under fibrosis, when bile retention remains toxic, when steatotic stress and oxidative burden stay high, or when inflammatory signaling never turns off. In that environment the liver is still trying to repair itself, but the repair logic becomes distorted. Stellate cells lay down collagen. Sinusoids lose openness. Hepatocytes divide inside an increasingly scarred field. Nodules may form, but nodular growth is not the same thing as clean regeneration. The organ can become full of regenerative intent and still move farther away from real functional recovery because the architecture being rebuilt is no longer the architecture of a healthy high-flow exchange organ.

So the liver's regenerative reputation is deserved, but it should be understood correctly. The organ is unusually capable of rebuilding working mass because hepatocytes can re-enter growth and because survival requires restoration of chemistry, not just closure of injury. But regeneration is not infinite, and it is not independent of structure. The liver can recover impressively when the tissue framework still supports true rebuilding. It loses that gift when chronic injury forces repair to happen inside a field that has already forgotten how a liver is supposed to be built.

What Regeneration Is and What It Is Not

Regeneration is the restoration of working liver tissue, not merely the survival of injured liver tissue. That distinction matters because the word is often used too loosely. A liver that shrinks its necrotic zone, walls off injury, lays down collagen, and keeps the person alive has not necessarily regenerated in the true physiological sense. Regeneration means that hepatocytes and the supporting tissue environment have restored enough parenchyma, the functional working tissue of the organ, that blood chemistry control, bile formation, protein synthesis, and metabolic throughput can return toward normal. The liver's reputation comes from this capacity to replace lost working mass, not from a vague ability to endure damage heroically. A scar can preserve structure. Regeneration restores function.

What it is, at the cellular level, is controlled re-entry into growth by cells that are usually quiescent, meaning not dividing actively under ordinary conditions. Hepatocytes can leave that resting state when enough tissue has been lost and when the surrounding environment still supports real rebuilding. Growth factors rise. Cytokine patterns change. The extracellular scaffold, the structural framework outside the cells, helps guide where new tissue should organize itself. Sinusoidal flow, bile drainage, and matrix integrity all matter because the liver is not just replacing cells in open space. It is rebuilding a high-flow exchange organ that has to receive blood on one side, form bile on the other, and keep those two traffic systems distinct. Real regeneration means the architecture of function is being restored, not simply that cell number is increasing.

What regeneration is not is scar formation. Fibrosis is repair in the broad sense, but it is not regeneration. Collagen can stabilize an injured area, reduce the risk of structural collapse, and wall off damage, but collagen does not conjugate bilirubin, buffer post-meal glucose, synthesize albumin, or make bile acids. Stellate cells turning into matrix-producing cells may help the tissue survive acute disruption, but if that response dominates, the liver is moving away from regeneration even while appearing busy with repair. This is why a chronically injured liver can look active on histology and still be losing function. The tissue is responding. It is not necessarily rebuilding a normal liver.

Regeneration is also not the same thing as nodular growth. In chronic liver disease, hepatocytes may proliferate inside a scarred field and produce regenerative nodules, but that term can mislead if it sounds automatically healthy. A nodule of hepatocytes growing inside fibrotic distortion is not equivalent to restoration of normal lobular architecture, open sinusoids, orderly bile flow, and clean portal-to-central blood gradients. Cells may be dividing. The organ may still be losing itself. True regeneration requires not only hepatocyte proliferation, but recovery of the structural conditions that let those hepatocytes behave like liver tissue rather than like trapped islands of survival inside scar.

So regeneration should be defined narrowly and honestly. It is the restoration of functional hepatic mass inside an architecture still capable of supporting real liver work. It is not scar. It is not compensation alone. It is not nodules by themselves. It is not the body merely holding on. The liver deserves its regenerative reputation because it can truly rebuild under the right conditions. The tragedy of chronic liver disease is that repair can continue long after true regeneration has become harder and harder to achieve.

Hepatocyte Re-Entry Into the Cell Cycle

A hepatocyte in the healthy adult liver is usually not dividing. That does not mean it is inert. It means it is working. These cells spend their lives handling portal blood, making bile, building plasma proteins, storing glycogen, converting ammonia into urea, transforming hormones, and processing foreign compounds. Under ordinary conditions they sit in a quiescent state, often called G0, which means outside the active cell cycle. The cell cycle is the ordered sequence by which a cell prepares its DNA, duplicates it, and divides into two daughter cells. The liver's regenerative power begins with the fact that hepatocytes are quiescent, not permanently finished. They are resting from division, not disqualified from it. When enough tissue is lost, or enough functional mass disappears, the surviving hepatocytes can be called back into that cycle. This is one of the liver's great differences from more rigid organs. The working cell can become a replacement cell without ceasing to be a liver cell in principle.

That re-entry does not happen by simple mechanical loss of tissue alone. It happens because the chemistry around the surviving hepatocyte changes. Cytokines and growth factors begin rewriting the local environment. Cytokines are signaling proteins released by immune and tissue cells. Growth factors are signals that push cells toward survival, growth, and division. The injured liver first primes hepatocytes, which means it makes them competent to hear the regenerative command. A quiescent hepatocyte does not leap straight from ordinary metabolic work into DNA replication. It has to be pushed out of the resting state. Signals such as tumor necrosis factor and interleukin-6 help create that primed condition, while growth factors such as hepatocyte growth factor and epidermal growth factor help drive the cell forward into G1, the first active growth phase of the cycle. From there the cell may cross the restriction point, commit to DNA synthesis in S phase, and later move through mitosis, the physical division step. The important point is that the liver does not regenerate because hepatocytes are always dividing. It regenerates because they can be convinced to divide when the organ's lost mass makes that biologically necessary.

This re-entry is tightly tied to structure. A hepatocyte does not divide in a vacuum. It divides inside a tissue that still needs sinusoidal blood flow, polarity, membrane transport, bile canaliculi, extracellular matrix guidance, and enough mitochondrial energy to survive the work of growth. DNA replication is expensive. Protein synthesis rises. Membrane production rises. The cell has to keep being a hepatocyte while also becoming two hepatocytes. That is why regeneration works best when the underlying scaffold is still largely intact. If the extracellular matrix is too distorted, if sinusoids are capillarized, if fibrosis has thickened the exchange space, if bile retention is toxic, or if chronic inflammation keeps injuring the cells as fast as they try to re-enter the cycle, the command to divide becomes less likely to restore real liver architecture. Cells may proliferate, but proliferation alone is not enough. The tissue has to remember how to be a liver while it is growing.

This is also why hepatocyte re-entry into the cell cycle is not the same thing as uncontrolled growth. Regeneration is supposed to stop. Once enough functional mass has been restored, inhibitory signals rise, growth factor dominance falls, and the liver returns toward quiescence. A normal regenerative response is therefore temporary, targeted, and burden-driven. The organ senses loss, recruits surviving hepatocytes back into division, restores working mass, and then shuts the program down. When that shutdown fails, or when regeneration is forced to happen repeatedly inside chronic injury, the process becomes less clean. Cells divide in a scarred field. Nodules form. Architecture drifts. The liver may still be trying to replace what was lost, but the re-entry program is no longer happening inside a quiet stable organ. So hepatocyte re-entry into the cell cycle is one of the foundations of liver regeneration, but it only becomes true restoration when the signals that start growth, guide growth, and stop growth remain under control.

Priming, Proliferation, and Termination Phases

Liver regeneration does not happen as one continuous blur of growth. It happens in phases, and each phase solves a different problem. The first problem is that a hepatocyte doing ordinary liver work is usually not dividing. It is sitting in a quiescent state, outside the active cell cycle, busy with bile formation, glucose handling, plasma protein synthesis, detoxification, and the rest of hepatic throughput. That cell cannot jump directly from quiet metabolic work into DNA replication the moment tissue is lost. It has to be prepared. This is the priming phase. Priming is the stage in which injury, tissue loss, and inflammatory signaling alter the biochemical environment enough that hepatocytes become competent to hear growth signals. Cytokines such as tumor necrosis factor and interleukin-6 are important here, not because they force the cell to divide by themselves, but because they shift the hepatocyte out of its resting posture and make later proliferative signals effective. A quiescent cell is being told that the organ has lost enough mass that maintenance is no longer sufficient. The chemistry of ordinary function is being replaced by the chemistry of readiness.

Once primed, the hepatocyte can enter the proliferation phase. This is the stage most people think of when they hear the word regeneration, but it only works because the priming step happened first. Growth factors such as hepatocyte growth factor and epidermal growth factor now drive the cell deeper into the cycle. The hepatocyte moves through G1, commits to DNA synthesis, enters S phase, replicates its genome, and eventually proceeds toward mitosis so one working cell can become two. But proliferation in the liver is never just arithmetic. The cell is not reproducing in empty space. It is reproducing inside a high-flow exchange organ that still has to maintain blood handling, bile flow, membrane polarity, mitochondrial energetics, and extracellular architecture while the rebuilding is underway. That is why proliferation is so physiologically expensive. The liver is not simply making more cells. It is trying to restore enough parenchymal mass that glucose buffering, ammonia handling, albumin production, and bile secretion can move back toward adequate throughput. Proliferation is the visible growth phase, but underneath it the organ is trying to replace lost chemistry, not just lost volume.

Then the liver has to stop. That is the termination phase, and it matters just as much as the first two. Regeneration that cannot terminate is no longer clean repair. It becomes distorted growth inside a tissue that risks losing its architecture. Once enough functional mass has been restored, growth-promoting signals have to fall, inhibitory pathways have to rise, and hepatocytes have to return toward quiescence. The organ is not aiming for endless expansion. It is aiming for restored sufficiency. Termination means the liver has judged that the emergency of missing mass has been answered well enough that ordinary working identity can take over again. Cells stop cycling. Tissue remodeling settles. The regenerative program closes. This is one of the reasons regeneration and cancer are not the same thing, even though both involve cell division. Regeneration is burden-driven and self-limited. It starts because tissue was lost and stops because enough has been restored.

These phases are easiest to understand as a sequence of permission. Priming gives permission to leave rest. Proliferation gives permission to grow. Termination removes permission and restores restraint. If priming fails, the hepatocyte never becomes responsive enough to rebuild. If proliferation fails, tissue loss remains tissue loss. If termination fails, growth continues in a field that may no longer support normal architecture. Chronic liver disease often damages all three stages at once. Inflammation may be present, but it is no longer clean priming. Hepatocytes may proliferate, but they are doing so inside fibrosis and sinusoidal distortion. Termination may become disordered because the tissue keeps receiving injury signals before repair can settle. The liver still looks active, but activity is no longer the same thing as successful regeneration.

So priming, proliferation, and termination are not technical extras added onto the concept of liver regeneration. They are the actual logic of it. The liver first prepares the surviving hepatocyte to hear the need for growth, then drives expansion of functional cell mass, then shuts that expansion down once enough working tissue has been restored. Regeneration is not one event. It is a timed sequence of readiness, rebuilding, and restraint.

Why Preserved Architecture Matters

The liver can regenerate impressive amounts of tissue, but regeneration only becomes useful if the architecture those cells re-enter is still livable. That is why preserved architecture matters so much. A hepatocyte is not just a generic cell that can be dropped anywhere and expected to perform liver work. It has to sit in the right relation to sinusoidal blood, to neighboring hepatocytes, to the bile canaliculus, to the extracellular matrix, and to the oxygen and substrate gradients that normally organize the lobule. Blood has to be able to reach the cell. Plasma has to be able to cross the fenestrated sinusoidal lining and enter the space of Disse. The hepatocyte has to preserve polarity so one membrane domain faces blood and another faces bile. Canalicular flow has to remain possible. Central venous drainage has to remain open. If those structural relationships are preserved, new hepatocytes can step back into real function. If those relationships are lost, cell division alone does not rebuild a liver. It only fills damaged space with cells struggling inside the wrong design.

This is why acute injury and chronic injury feel so different in outcome. In an acute setting, a substantial number of hepatocytes can die and yet the remaining scaffold may still be open enough that regeneration means real replacement. Sinusoids still conduct blood. The matrix still gives positional guidance. Bile can still find a route out. The surviving hepatocytes can re-enter the cell cycle and repopulate a tissue whose deeper organization is still recognizable. In that setting the liver's regenerative reputation is justified in the strongest sense. The organ lost working mass, but it did not lose the blueprint. Chronic injury is more dangerous because it attacks the blueprint itself. Fibrosis thickens the space of Disse. Sinusoids lose their fenestrated permissive character and begin behaving more like ordinary capillaries. Portal and central flow patterns distort. Bile canaliculi are compressed or function less cleanly. The extracellular matrix becomes a field of restraint rather than a guide to orderly exchange. Now the hepatocyte may still divide, but the daughter cells are being asked to inhabit an increasingly unlivable architecture.

That is the deeper meaning of preserved architecture. It is not cosmetic shape. It is preserved function written into space. A healthy liver lobule is an arrangement of gradients, surfaces, and directional flows. Oxygen is higher near the portal inflow and lower near the central vein. Blood moves one way. Bile moves the other way. Hepatocytes near the portal region and those near the central region do not do identical chemistry because the blood reaching them is not identical. Kupffer cells and endothelial cells are positioned in specific places because portal blood must be read immunologically as well as metabolically. Stellate cells occupy the space of Disse because reserve, exchange support, and fibrogenic risk all live there. Destroy enough of this and you do not merely have a smaller liver. You have a liver that has forgotten how to be spatially intelligent.

This is why fibrosis is so devastating to regeneration. Collagen deposition does not only stiffen the tissue. It interrupts the relationships regeneration depends on. A newly divided hepatocyte still needs contact with blood, access to substrate, oxygen delivery, a functioning canalicular network, and a matrix environment that supports polarity rather than confusion. If fibrosis has turned the liver into a scar-partitioned organ with distorted sinusoids and nodular isolation, then growth becomes less like restoration and more like trapped survival. Cells may multiply. Nodules may enlarge. But nodular proliferation inside broken architecture is not the same as true regeneration. The liver may look active histologically while still moving farther from functional recovery because the tissue field can no longer host normal exchange.

So preserved architecture matters because regeneration is not a numbers game. The liver does not heal by replacing lost cells in the abstract. It heals by replacing them into a structure that still knows how to conduct blood, form bile, maintain gradients, and preserve cell polarity. When that structure is preserved, regeneration can mean return of real liver function. When that structure is lost, proliferation becomes increasingly compromised, and the organ may keep trying to rebuild itself inside a framework that no longer supports what a liver is supposed to do.

When Regeneration Succeeds and When It Stalls

Regeneration succeeds when the liver is asked to replace lost working mass inside a tissue field that still remembers how to be a liver. That is the real requirement. Hepatocytes can re-enter the cell cycle, growth factors can rise, and surviving cells can divide, but none of that matters if the blood cannot reach them properly, if bile cannot leave them properly, or if the extracellular scaffold that normally guides polarity and exchange has already been replaced by fibrosis. In successful regeneration, the injury is often substantial but the deeper architecture is still usable. Sinusoids still conduct blood in a way that lets plasma meet hepatocytes. The space of Disse is still permissive enough for exchange. Canaliculi still connect into a functioning biliary path. Portal and central flow relationships are still recognizable enough that new or surviving hepatocytes can regain real functional position rather than simply surviving in a disordered field. The cell division matters, but the preserved map matters more. The liver succeeds when growth is occurring inside a framework that can still host blood chemistry control, bile formation, protein synthesis, and nutrient handling once the new cells arrive.

It also succeeds when the injury is finite enough that the regenerative program has room to complete itself. A liver recovering from a partial resection or an acute reversible insult is not only replacing lost tissue. It is doing so under conditions where the main destructive force has already fallen. The burden is no longer arriving as fast as the organ is trying to rebuild. Priming can occur cleanly. Proliferation can proceed with enough mitochondrial energy, enough nutrient supply, and enough synthetic reserve to support DNA replication and membrane building. Termination can arrive at the right time because the tissue is no longer being hammered by the same force that triggered repair in the first place. This is one reason the liver's reputation is deserved in acute injury. It is not that the organ ignores damage. It is that, when the damaging event stops soon enough and the scaffold remains navigable, the surviving hepatocytes can repopulate an environment that still rewards correct rebuilding.

Regeneration stalls when the liver is trying to rebuild inside conditions that keep destroying the very logic of rebuilding. Chronic inflammation is one of the main reasons. If endotoxin delivery stays high, if steatotic stress keeps distorting the hepatocyte interior, if alcohol continues driving redox burden and toxic intermediates, if bile retention keeps bathing tissue in detergent chemistry, or if viral injury persists, the liver never gets a quiet enough interval to move cleanly from priming into restoration. It stays trapped in damaged readiness. Hepatocytes may still divide, but they are dividing inside oxidative stress, cytokine noise, and ongoing cell death. Stellate cells keep laying down collagen. Sinusoids become less open. Canalicular flow becomes less reliable. The exchange field narrows. Regeneration then starts losing the conditions it needs even while the organ is still trying to mount it.

Fibrosis is one of the clearest reasons this happens. A hepatocyte can divide in scarred tissue, but scar does not provide the same instructions as a preserved sinusoidal-canalicular architecture. Collagen thickens the space of Disse, distorts oxygen and substrate gradients, raises portal resistance, and turns a high-flow exchange organ into a mechanically stiffer and less readable one. Now proliferation starts looking less like restoration and more like compensation inside confinement. Nodules can form. Cell mass can increase locally. But a nodule is not proof of success. If the surrounding architecture is still fibrotic, congested, cholestatic, and immunologically activated, the new cells are being placed into a field that cannot fully support normal hepatic behavior. The liver is growing, but it is not necessarily recovering. That is the tragedy of chronic liver disease. The organ can remain full of regenerative intent while moving farther away from true regenerative outcome.

So regeneration succeeds when lost mass is being replaced inside preserved architecture, under a burden that has become finite enough for the program to complete itself. It stalls when division continues inside chronic injury, fibrotic distortion, and unresolved chemical load. In the successful liver, growth restores function. In the stalled liver, growth becomes trapped inside damage.

Chapter 35

Fibrosis

What Fibrosis Is at the Tissue Level

Fibrosis is what happens when the liver stops repairing injury by restoring open working tissue and starts repairing it by laying down structural restraint. At the tissue level, that means extracellular matrix begins accumulating where rapid exchange, soft flow, and close contact used to define the organ. Extracellular matrix is the protein-rich scaffold outside cells, made of collagen and other structural molecules that normally provide support without suffocating function. In a healthy liver that scaffold is present, but it is disciplined. It helps preserve form while still allowing portal blood to move through sinusoids, plasma to cross the fenestrated endothelial lining, hepatocyte microvilli to reach into the space of Disse, and bile to be secreted cleanly into canaliculi. Fibrosis begins when that restrained scaffold becomes excessive, thicker, stiffer, and placed in the worst possible locations for a high-flow exchange organ.

That placement is everything. Fibrosis in the liver is not like scar on the skin, where the surface can close and the body may still function. The liver survives through permeability, throughput, and contact. Blood has to reach hepatocytes. Oxygen and substrates have to move across a thin exchange surface. Hepatocytes have to maintain polarity so one side faces blood and the other faces bile. Once collagen and other matrix proteins begin accumulating in the space of Disse and around sinusoidal structures, the distance between blood and hepatocyte effectively increases. Fenestrations are lost or function less cleanly. Sinusoids become less like low-resistance exchange channels and more like ordinary capillary-like tubes. The tissue is not just getting scarred. It is being rebuilt into a form less capable of doing liver work. Fibrosis is architectural interference.

This is why the first consequences of fibrosis are often microscopic but physiological rather than dramatic and gross. Solute movement slows. Oxygen gradients become less favorable. Portal resistance rises. Hepatocytes receive blood less cleanly and work under greater strain. Bile-facing functions may begin suffering because the same cell losing blood-side efficiency is the cell that also has to conjugate bilirubin, secrete bile acids, and maintain canalicular transport. Kupffer cells continue reading injury. Stellate cells continue producing matrix if the signal persists. What begins as tissue reinforcement gradually becomes tissue isolation. Cells are still present, but they are no longer meeting the stream they were built to govern.

At the tissue level, fibrosis therefore should be understood as a wound-response pattern that became chronic enough to remodel the liver against its own design. The liver was trying to stabilize repeated injury, but stabilization in this organ quickly becomes dangerous because too much structure means too little exchange. The matrix that was supposed to support the tissue begins partitioning it. Blood must now move through a narrower, stiffer, more resistant field. Hepatocytes must now survive farther from clean access to portal chemistry and oxygen. The organ becomes harder, but hardness is not healing here. Hardness is loss of permeability. In the liver, stiffness is often the physical signature of throughput being buried under repair.

Stellate Cell Activation and Collagen Deposition

Stellate cell activation is the moment the liver's quiet reserve cell becomes a wound-response cell. In health, stellate cells sit in the space of Disse, the narrow exchange compartment between sinusoidal blood and the hepatocyte surface, holding vitamin A in lipid droplets and helping preserve the softness of the perisinusoidal environment. They are part of what makes the liver an open exchange organ rather than a stiff mechanical tissue. Chronic injury changes that identity. Repeated hepatocyte stress, oxidative burden, bile acid retention, alcohol metabolites, endotoxin arriving from the gut, and inflammatory cytokines released by Kupffer cells all begin telling the stellate cell the same thing. This tissue is under threat. The cell responds by losing its quiet storage phenotype, emptying much of its retinoid reserve, becoming more contractile, proliferating, and turning toward matrix production. Activation is not a decorative marker on a pathology slide. It is a change in job description.

The signals driving that change are local and cumulative. Injured hepatocytes release damage-associated molecules. Kupffer cells release cytokines such as transforming growth factor beta and platelet-derived growth factor, signals that push stellate cells toward a fibrogenic program. Transforming growth factor beta is especially important because it drives the cell toward collagen production and extracellular matrix expansion. Platelet-derived growth factor helps push proliferation and migration. Reactive oxygen species add another layer of pressure by damaging membranes and proteins in the local environment and making inflammatory signaling more persistent. The stellate cell is living exactly where all of these messages converge. It sits beside the hepatocyte, beside the sinusoid, beside the returning portal burden. Once those messages become repetitive enough, activation stops being a brief wound response and becomes the new baseline behavior of the cell.

Collagen deposition is the physical result of that activated state. The stellate cell begins laying down extracellular matrix proteins, especially collagen types I and III, in a part of the liver that was never meant to be heavily armored. The space of Disse is supposed to remain thin and permissive so plasma can move easily across the fenestrated sinusoidal endothelium and reach hepatocyte microvilli. Microvilli are membrane projections that increase the cell's exchange surface. Once collagen begins accumulating there, the distance between blood and hepatocyte effectively increases. Exchange becomes less efficient. Sinusoids begin losing their open, sieve-like behavior and drift toward capillarization, meaning they start behaving more like ordinary tighter capillaries than the specialized exchange channels the liver needs. The organ is still full of blood and still full of cells, but the intimacy between blood chemistry and hepatocyte surface is being buried under matrix.

That matrix does more than thicken the exchange surface. Activated stellate cells also develop contractile behavior, which means they help narrow the sinusoidal bed and increase resistance to portal flow. Now fibrosis is not only an exchange problem. It is a hemodynamic problem, a blood flow problem. Portal pressure rises upstream. Oxygen delivery becomes less reliable downstream. Hepatocytes work under increasing metabolic strain because the same scar that separates them from blood is also making blood harder to move through the tissue at all. This is why collagen deposition is so devastating in the liver. The organ does not fail because scar is visually ugly. It fails because scar is being laid exactly where flow and exchange were supposed to remain open.

So stellate cell activation and collagen deposition are the bridge between repeated cellular injury and true architectural disease. The liver may begin with fat stress, bile retention, alcohol burden, viral injury, or gut-derived inflammatory traffic, but once stellate cells activate, the tissue starts converting those insults into structure. The reserve cell becomes a scar-building cell. The exchange space becomes a matrix-rich barrier. The sinusoid becomes narrower and less permeable. At that point fibrosis is no longer a side effect of injury. It has become a new organizing force inside the liver, and it organizes the organ in the wrong direction.

Extracellular Matrix Expansion

Extracellular matrix expansion means the liver is beginning to fill its exchange spaces with structural material that was never supposed to dominate them. The extracellular matrix is the protein framework outside cells, built from collagens, glycoproteins, proteoglycans, and other scaffold molecules that normally provide support without obstructing function. In a healthy liver this matrix is present, but it is restrained. It helps hold the tissue together while still allowing portal blood to move through sinusoids, plasma to cross the fenestrated endothelial lining, hepatocyte microvilli to reach into the space of Disse, and bile-facing polarity to remain intact. Expansion begins when that restraint fails. More matrix is laid down, more collagen accumulates, and the thin exchange field starts being converted into a thicker mechanical barrier. This is not just more support. It is support invading function.

The problem is location. The liver is not a dense static organ that can tolerate extra scaffold without consequence. It is a high-flow exchange organ. Blood has to reach hepatocytes quickly. Oxygen and substrates have to cross a short distance. Waste has to move back. Bile has to be formed on the opposite side of the same cell. Once extracellular matrix expands in the space of Disse and around sinusoidal structures, the distance between blood and hepatocyte effectively widens. The endothelial surface loses some of its open, permissive character. Hepatocyte microvilli no longer meet the moving plasma with the same intimacy. Solute transfer slows. Oxygen delivery becomes less favorable. The tissue may still look full of cells, but those cells are now living farther from the stream they were built to govern. Matrix expansion therefore is not a cosmetic change in the background. It is a direct attack on throughput.

This expansion is driven mainly by activated stellate cells, but the full tissue environment participates. Injured hepatocytes release danger signals. Kupffer cells release cytokines. Oxidative stress alters local signaling and membrane integrity. Cholestatic retention, steatotic stress, alcohol burden, viral injury, and gut-derived inflammatory traffic all keep telling the liver that the tissue is under threat and needs reinforcement. The stellate cell answers by producing more collagen and other extracellular matrix components. At first this can look like an attempt at stabilization. A wounded organ is trying to brace itself. But the liver pays for that brace immediately because its work depends on openness. The more matrix accumulates, the more the organ begins protecting itself in the wrong language. It trades permeability for stiffness, and stiffness is a losing bargain in liver tissue.

As matrix expansion continues, blood flow changes as well. Sinusoids narrow and begin behaving less like specialized exchange channels and more like restricted vascular tubes. Portal resistance rises. Pressure builds upstream. Hepatocytes farther downstream receive less favorable oxygen and substrate delivery. Bile secretion becomes harder to sustain because the same cell that is now poorly supplied from the blood side still has to maintain ATP-dependent transport on the bile side. ATP, adenosine triphosphate, is the cell's immediately usable energy currency, and matrix-expanded tissue makes ATP harder to generate cleanly by worsening perfusion and raising cellular stress. So extracellular matrix expansion is never just a fibrosis marker. It is the beginning of a new operating state in which the liver is less able to meet moving blood, less able to preserve gradients, and less able to repair itself effectively.

This is why extracellular matrix expansion is one of the clearest turning points in chronic liver disease. Injury can be survived for a while. Inflammation can be survived for a while. Even hepatocyte loss can sometimes be replaced. But once the exchange field itself is being thickened and stiffened, the organ starts losing the physical conditions that made recovery possible. New cells entering that environment do not inherit a healthy liver. They inherit a scaffold already moving in the wrong direction. Matrix expansion is therefore the stage at which repair begins turning into architectural sabotage. The liver is not simply scarred. It is being rebuilt into a form less capable of being a liver.

Why Scar Changes Blood Flow and Exchange

Scar changes blood flow and exchange because the liver is not built as a dense solid organ that can tolerate extra structural material without consequence. It is built as a soft high-flow exchange field. Portal blood is supposed to enter through the portal tracts, spread into low-resistance sinusoids, slow down enough for plasma to meet hepatocytes across a thin permeable interface, and then drain toward the central veins. At the same time bile is supposed to move away from hepatocytes in the opposite direction through canaliculi and ducts. This design only works if the tissue stays open, thin, and permissive. Scar breaks that design by adding collagen and other extracellular matrix into spaces that were supposed to remain narrow enough for rapid transfer and loose enough for easy flow. The liver does not simply become stiffer. It becomes less readable to its own blood stream.

The first change is at the exchange surface. In healthy tissue, sinusoidal endothelial cells are fenestrated, meaning they contain pores, and the space of Disse between blood and hepatocyte is thin enough that plasma solutes can move quickly to the hepatocyte membrane. Hepatocyte microvilli project into that space and expand the contact area. Scar thickens this interface. Collagen accumulates in the space of Disse. Endothelial cells lose some of their fenestrated character and the sinusoid begins to undergo capillarization, which means it starts behaving more like an ordinary tighter capillary than a specialized exchange channel. Blood is still present. Hepatocytes are still present. But the intimacy between them is gone. Glucose buffering becomes less efficient. Bilirubin uptake becomes less efficient. Hormone handling, lipoprotein processing, ammonia clearance, and drug metabolism all begin paying for the extra distance and the less permeable surface. Scar changes exchange because it physically separates the moving blood from the cell that is supposed to govern it.

The second change is hemodynamic, which means it changes the behavior of the blood stream itself. A healthy sinusoidal network offers low resistance. Blood can spread, slow, and pass through. Scar narrows that field and makes it mechanically less compliant. Activated stellate cells add another layer by becoming contractile, so the tissue is not only being filled with matrix. It is also being squeezed. Portal blood now has to push through a narrower stiffer route, and the pressure required to do that rises. This is the beginning of portal hypertension. The important point is that scar does not merely sit beside the circulation. It becomes part of the circulation's resistance. Once that resistance rises, the problem is no longer microscopic only. The spleen enlarges. Collateral vessels form. Ascites becomes easier to generate. The liver has changed blood flow by changing the shape and compliance of the channels blood has to cross.

Oxygen delivery changes with it. The liver already runs on gradients. Blood arrives near the portal region richer in oxygen and leaves near the central vein after some of that oxygen has been spent. Scar makes that gradient harsher and less reliable because flow is more distorted and exchange is less efficient. Downstream hepatocytes, especially those nearer the central vein, are now living under thinner oxygen support while still being asked to run high-energy transport, protein synthesis, and detoxification chemistry. ATP, adenosine triphosphate, is the cell's immediately usable energy currency, and less oxygen means less clean mitochondrial energy production. Once ATP falls, bile secretion weakens, membrane polarity becomes harder to maintain, and the hepatocyte becomes less capable of compensating for the structural damage around it. Scar therefore changes blood flow and exchange in a way that becomes a metabolic injury inside the cell, not just a vascular abnormality outside it.

This is why fibrosis becomes a turning point rather than a background finding. A liver can survive inflammation, fat burden, toxins, and even substantial cell loss for a time if the exchange field remains open enough for the surviving cells to keep meeting blood. Once scar begins invading that field, every later step becomes harder. New hepatocytes divide into a distorted scaffold. Blood reaches them less cleanly. Bile leaves less cleanly. Portal pressure stays high. The organ becomes more dependent on compensation and less capable of real restoration. Scar changes blood flow and exchange because it replaces a living permeable architecture with structural restraint. In skin, that may close a wound. In liver, it buries the work.

Why Fibrosis Is Not Only a Cosmetic Tissue Change

Fibrosis is not a cosmetic tissue change because the liver is not an organ whose function can be separated from its architecture. In skin, a scar may change appearance and texture while still leaving the basic job of coverage intact. In liver, scar is laid into the operating surface itself. The organ survives by keeping blood moving slowly through sinusoids, by letting plasma cross a thin fenestrated endothelial barrier, by keeping the space of Disse open enough for hepatocyte microvilli to meet the passing chemistry, and by preserving the polarity that allows the same cell to face blood on one side and bile on the other. Once fibrosis begins, those relationships start changing. Collagen accumulates where exchange was supposed to remain thin. Sinusoids stiffen and lose openness. The distance between blood and hepatocyte increases. This is not a change in how the tissue looks. It is a change in how the tissue works.

That is why fibrosis damages function long before cirrhosis becomes visually dramatic. A scarred liver may still be large. It may still contain many hepatocytes. But if those hepatocytes are being separated from blood by expanding extracellular matrix, receiving less favorable oxygen delivery, and living inside a rising field of portal resistance, they are not doing the same job they once did. Glucose buffering becomes less efficient because substrate reaches the cell less cleanly. Bilirubin uptake and secretion worsen because both the blood-facing and bile-facing sides of hepatic work depend on intact membrane transport and adequate ATP. Lipoprotein handling suffers. Hormone transformation becomes less graceful. Ammonia clearance weakens because the cell has become harder to supply and harder to sustain. Fibrosis is therefore not scar sitting beside function. It is scar interrupting function molecule by molecule.

Blood flow changes at the same time. A healthy liver is a low-resistance exchange organ. Fibrosis narrows and stiffens that field, and activated stellate cells add contractile force on top of the collagen they lay down. Portal blood now has to push harder through a route that was built to remain soft. Portal pressure rises upstream. The spleen enlarges. Collateral vessels form. Ascites becomes easier to generate. None of that happens because the liver became cosmetically rough. It happens because scar has become part of the vascular problem. The liver is no longer only processing blood. It is obstructing the normal way blood is supposed to move through it.

The same is true for regeneration. A tissue that is merely scarred in appearance might still heal by replacing cells into a preserved scaffold. A fibrotic liver is harder to restore because the scaffold itself has been rewritten. New hepatocytes entering that environment do not inherit normal sinusoidal flow, normal exchange distance, or normal biliary architecture. They divide into a field already biased toward poor perfusion, altered gradients, and mechanical distortion. This is one reason fibrosis is such a turning point. The liver may still be trying to repair itself, but the conditions required for real repair are being buried under matrix. The organ keeps generating evidence of activity while losing the capacity for true recovery.

So fibrosis is not only a cosmetic tissue change because the liver is an organ of flow, contact, and permeability. Scar in this organ means blood meets cells less cleanly, bile leaves less cleanly, oxygen gradients worsen, portal resistance rises, and regeneration becomes less effective. The surface may look different. The deeper event is that the tissue has been rebuilt in a way that no longer suits liver physiology.

Reversible Versus Advanced Scarring

Not all scar in the liver means the same thing, and that distinction matters because fibrosis is a process before it is a permanent monument. Early scarring is still dynamic tissue. Activated stellate cells are laying down extracellular matrix, especially collagen, but the matrix has not yet fully reorganized the liver into a new hard architecture. The exchange field is injured, but not completely rewritten. Blood can still move through sinusoids with less distortion. Hepatocytes are still closer to the stream they need. Bile channels are still more likely to remain connected. The matrix itself is still under active turnover, which means the liver is not only making scar. It is also still capable of breaking some of it down. That breakdown depends on enzymes that digest extracellular matrix, especially matrix metalloproteinases, and on whether the inflammatory environment stops telling stellate cells to keep producing more collagen than the tissue can clear. In this stage, scar is real, but it is still participating in a live argument with repair. If the injuring burden falls soon enough, alcohol, cholestatic retention, viral injury, endotoxin load, steatotic stress, toxic exposure, the liver can begin tipping that argument back toward remodeling, thinning of matrix, and partial restoration of normal exchange.

This is why reversible scarring should not be imagined as scar simply vanishing overnight. Reversal means the liver has regained enough control over the wound response that matrix deposition slows, collagen breakdown becomes more favorable, stellate cells deactivate or disappear, and the tissue starts moving back toward softness, permeability, and usable flow. The important point is not that every collagen strand dissolves. The important point is that the organ becomes more liver-like again. Sinusoids reopen. Capillarization lessens. The space of Disse becomes less crowded. Hepatocytes meet blood more cleanly. Portal resistance falls enough that the circulation stops paying such a high mechanical price for the injury. Reversibility, then, is functional before it is cosmetic. The liver is recovering because the architecture is becoming more compatible with exchange.

Advanced scarring is different because by that stage the matrix has stopped behaving like a temporary wound response and started behaving like a new physical regime. Collagen is denser, more cross-linked, and more difficult to remove. Cross-linking means the collagen fibers are chemically tied together into a tougher network that resists degradation. Fibrous septa, thick bands of scar, begin partitioning tissue and separating hepatocyte groups into nodules. Sinusoids lose more of their normal fenestrated character. Portal and central venous relationships become distorted. Blood starts finding abnormal shortcut routes around functional parenchyma instead of through it. Bile flow becomes more vulnerable because the same architectural distortion that disrupts blood exchange also disrupts canalicular and ductal continuity. At this stage the scar is no longer just excess matrix in a previously healthy design. The scar is becoming the design.

That is why advanced scarring is so much harder to reverse. The liver may still retain some capacity for remodeling, but the problem is no longer only excess collagen. The problem is that collagen has reorganized flow, oxygen gradients, biliary drainage, and regeneration itself. New hepatocytes entering this field do not inherit normal lobular architecture. They inherit a scarred map. Portal hypertension becomes easier to sustain because the resistance is now built into the structure. Hypoxia becomes easier to provoke downstream. Nodules may enlarge, but nodular growth inside a fibrotic framework is not the same as restoration of normal liver tissue. The organ is still trying to survive, but survival now happens inside a mechanically and chemically altered scaffold that keeps rewarding distortion.

So reversible versus advanced scarring is really the difference between matrix that is still negotiable and matrix that has become law. In earlier fibrosis, the liver can still retreat from scar if the injuring pressure falls and the wound-response program is allowed to quiet down. In advanced scarring, the collagen has become thicker, more permanent, more cross-linked, and more deeply woven into blood flow and tissue geometry. The liver may improve, sometimes significantly, but the road back is longer because the problem is no longer just injury. The problem is architecture that has learned the wrong shape.

Chapter 36

Cirrhosis

What Cirrhosis Actually Means Structurally

Cirrhosis means the liver has stopped being organized as a normal lobular exchange organ and has been rebuilt into a scar-partitioned organ with regenerative nodules and distorted flow. That is the structural meaning. It is not merely "a lot of fibrosis," and it is not just a badly inflamed liver. In cirrhosis, chronic injury has driven enough fibrosis, nodule formation, and vascular reorganization that the normal lobular architecture is no longer intact. The portal tracts, sinusoids, central veins, and hepatocyte plates no longer relate to one another in the clean open pattern that makes normal liver work possible. Scar has become framework. Regeneration is still happening, but it is happening inside the wrong scaffold. That is why cirrhosis is a turning point rather than a heavier version of ordinary fibrosis. The organ has changed its shape at the microscopic level in a way that changes everything blood and bile are allowed to do.

The two defining structural features are diffuse fibrosis and regenerative nodules. Fibrosis means excess extracellular matrix, especially collagen, has been laid down through the parenchyma. Regenerative nodules mean surviving hepatocytes have proliferated into rounded islands or nodules of parenchyma surrounded by fibrous septa. Septa are the bands of scar tissue that bridge and partition the liver. That combination matters because the liver is still trying to replace lost cell mass, but it is replacing it inside a framework of scar rather than inside preserved lobular design. So cirrhosis is not the absence of regeneration. It is regeneration trapped inside architectural distortion. The liver is growing and scarring at the same time, and the result is not restored normal tissue. It is a nodular organ whose cells are increasingly separated from the open vascular and biliary relationships they need.

Once that architecture is distorted, blood flow changes immediately. Normal sinusoidal circulation is a low-resistance, high-exchange system. Cirrhosis replaces that with vascular reorganization, neoangiogenesis, and scar-based resistance. Blood is diverted, compressed, rerouted, and forced through a stiffer field. Portal hypertension follows because portal inflow is now trying to cross an organ whose microvascular structure no longer behaves like a permissive exchange bed. The significance here is structural, not only hemodynamic. Blood is still entering the liver, but it is no longer meeting hepatocytes under normal sinusoidal conditions. The organ is less able to buffer, extract, convert, and clear because the stream itself is being redirected around or through scarred tissue in abnormal ways.

This is why cirrhosis should not be pictured as a liver with surface scarring alone. The clinically important event is internal. The exchange surface between portal blood and hepatocytes has been altered, the drainage pattern has been altered, and the tissue's own repair program has become mechanically trapped. That is also why cirrhosis is generally treated as irreversible damage, even though earlier fibrosis may be reversible and treatment can still prevent worsening or improve outcomes. By the time cirrhosis is established, the liver is no longer just wounded. It has been structurally reauthored into a nodular fibrotic organ with disordered blood flow and impaired functional geography.

Nodules, Distorted Blood Flow, and Hardened Architecture

Once cirrhosis is established, the liver is no longer organized as a soft continuous exchange field. It has become a partitioned organ. Surviving hepatocytes keep trying to regenerate, but they do so inside fibrous bands that cut the tissue into nodules rather than restoring normal lobular structure. These nodules are not proof that the liver has healed well. They are proof that the liver is still trying to replace working mass inside a scaffold already deformed by scar. A regenerative nodule is a cluster of hepatocytes surrounded by fibrous septa, bands of collagen-rich matrix that now act like walls inside an organ that was built to remain open. In a healthy liver, hepatocytes are arranged in plates that let blood move through sinusoids with close contact, low resistance, and broad exchange surface. In cirrhosis, that plate-and-sinusoid logic is broken. Regrowth happens, but it happens in compartments.

That compartmentalization changes blood flow immediately. Portal blood is supposed to enter the liver and spread into a low-pressure sinusoidal network where plasma can cross a thin fenestrated endothelium and meet hepatocytes across the space of Disse. In cirrhosis, fibrous septa narrow, distort, and reroute that network. Sinusoids become less like specialized exchange channels and more like compressed or capillarized vascular paths. Capillarized means they lose the open sieve-like quality that normally makes the liver such an efficient exchange organ. Blood is still entering the liver, but it is no longer being distributed through tissue with the same intimacy or efficiency. Some areas are underperfused. Some are bypassed. Some receive blood under altered pressure gradients that favor survival of the tissue patch rather than clean control of blood chemistry. This is why cirrhosis is not just scar accumulation. It is vascular redesign in the wrong direction.

The nodules make this worse because they are not sitting inside normal inflow and outflow relationships. A hepatocyte cluster trapped inside fibrous distortion may still synthesize, still metabolize, and still divide, but it is no longer living in a clean portal-to-central gradient. Oxygen delivery becomes less predictable. Nutrient exposure becomes less predictable. Waste removal becomes less predictable. The hepatocyte has not simply lost neighboring tissue. It has lost positional logic. Zone 1 and zone 3 behavior become harder to preserve when blood is being rerouted around scar and through abnormal channels. A nodular liver is therefore not only lumpy in appearance. It is confused in physiology. Cells are present, but the map that told them what kind of blood they were standing in has been damaged.

Hardened architecture is the physical result of all of this. Collagen deposition and matrix cross-linking make the liver stiffer, less compliant, and more resistant to incoming portal flow. Compliance means the tissue's ability to deform and accommodate flow without becoming a pressure problem. A healthy liver yields enough to let blood spread. A cirrhotic liver resists it. Portal pressure rises because the inflow is meeting a field that has lost its softness. That increased resistance is not an abstract number. It is the beginning of portal hypertension, collateral vessel formation, splenic enlargement, and ascites. The hardened liver is no longer simply less efficient at exchange. It is actively obstructive to the circulation trying to cross it.

This is why nodules, distorted blood flow, and hardened architecture belong together as one structural event. The liver tries to regenerate, but scar partitions the tissue into nodules. Those nodules grow inside a distorted vascular map. The distorted vascular map forces blood through a harder, narrower, less permeable organ. The organ still contains hepatocytes, but it no longer behaves like a normal liver because the relationship between cell, blood, and bile has been rebuilt incorrectly. Cirrhosis is not just more scar. It is the loss of soft organized throughput.

Why a Cirrhotic Liver May Be Present in the Body but Functionally Unavailable

A cirrhotic liver can still be physically present, still contain hepatocytes, and still occupy its place under the diaphragm, yet be functionally unavailable because the body no longer has clean access to what those cells are supposed to do. That is the key distinction. Liver function is not created by tissue mass alone. It depends on architecture. Blood has to enter through a low-resistance sinusoidal network, slow down, cross a fenestrated endothelial surface, and meet hepatocytes across a thin exchange space. Bile has to leave in the opposite direction through intact canaliculi and ducts. In cirrhosis, diffuse fibrosis, regenerative nodules, altered lobular architecture, and intrahepatic vascular shunts break that arrangement. The cells may still exist, but the stream that was supposed to reach them has been rerouted, compressed, or separated from them by scar. The liver is present. Functional access to the liver is not.

This is why cirrhosis is not simply "less liver." It is liver tissue trapped inside the wrong geometry. Regenerative nodules mean hepatocytes are still trying to replace lost mass, but they are doing so inside fibrous septa rather than inside normal lobular design. Sinusoidal capillarization, loss of fenestrae and formation of a basement membrane-like barrier, makes the exchange surface less permissive, so plasma chemistry no longer reaches hepatocytes with the same intimacy or efficiency. The result is that glucose buffering, bilirubin uptake, ammonia handling, hormone transformation, lipoprotein processing, and protein synthesis all become harder to perform even when cells are still there. Cirrhosis makes the organ cellularly present but operationally distant.

Blood flow changes are what make this structural loss clinically brutal. Scar and nodules increase intrahepatic vascular resistance, portal pressure rises, and blood begins seeking easier routes around the liver through intrahepatic and portosystemic shunts. That means some of the blood that should have been processed by hepatocytes bypasses effective exchange altogether. The liver is still in the body, but the circulation is no longer passing through it in the way liver physiology requires. A portion of portal inflow is now functionally lost to shunting, and effective liver blood flow falls even while the organ itself remains anatomically present. The body is carrying a liver it can no longer use fully.

That is why a cirrhotic liver can look like organ persistence and behave like organ withdrawal. It still occupies space. It may still produce some bile, some albumin, some clotting factors, and some metabolic work. But it does so through a scar-partitioned, shunt-ridden, hardened architecture that no longer lets blood, bile, and hepatocytes meet under normal conditions. The deeper failure is not disappearance. It is unavailability. The organ remains in the body while more and more of its functional territory becomes unreachable to the bloodstream that needed it.

The Difference Between Compensated and Decompensated Cirrhosis

Compensated cirrhosis means the liver is already structurally cirrhotic, scar-partitioned, nodular, and hemodynamically abnormal, but the body is still extracting enough usable function from that distorted organ that the major overt failure events have not yet declared themselves. The architecture is badly altered, yet the person has not crossed into the stage marked by ascites, variceal hemorrhage, overt encephalopathy, or jaundice as a clinical consequence of cirrhosis. In this phase the liver is not healthy. It is still holding the line. Blood is still moving through a damaged organ with enough residual control that the wider organism has not yet been forced to display the full price of architectural collapse. Patients in the compensated stage are at highest risk of progressing once clinically significant portal hypertension develops, which is one reason this stage can look deceptively quiet while the structural burden underneath it is already severe.

Decompensated cirrhosis begins when that uneasy stability breaks and the distorted architecture starts producing overt downstream failures the body can no longer hide. The classic threshold events are ascites, variceal bleeding, hepatic encephalopathy, and jaundice. These are not random complications attached to cirrhosis from the outside. They are the visible consequences of a liver that has become structurally present but increasingly unavailable. Ascites means portal pressure, splanchnic vascular changes, and reduced effective circulatory stability have exceeded compensation. Variceal bleeding means portal blood has been forced through collateral channels fragile enough to rupture. Encephalopathy means the blood is no longer being cleared with enough precision to keep neuroactive burden away from the brain. Jaundice means bilirubin handling and biliary outflow are failing openly enough for pigment to rise in blood. Decompensation is therefore not "worse cirrhosis" in a vague sense. It is cirrhosis that has crossed from architectural distortion into clinically exposed organ failure.

The structural difference between the two states is not that fibrosis exists in one and scar exists in the other. Scar is present in both. The deeper difference is whether the remaining functional pathways through the organ are still sufficient to keep pressure, flow, exchange, and synthetic work from failing in public. In compensated cirrhosis, the body is still extracting enough throughput from nodular scarred tissue to maintain a tense balance. In decompensated cirrhosis, portal hypertension has become clinically consequential, shunting and vascular resistance have become more destructive, and functional reserve has fallen below the threshold needed to prevent overt manifestations. The liver is no longer only distorted. It is losing the ability to conceal that distortion.

This is why the transition matters so much prognostically. The compensated phase can remain clinically silent for years, but once the first decompensating event appears, the disease typically accelerates and the conversation changes from chronic architectural disease to survival with a failing organ, often including transplant evaluation. The liver did not become cirrhotic on the day ascites appeared. It became decompensated on the day the architecture could no longer keep that ascites from appearing. That is the cleanest way to hold the distinction. Compensation means distorted structure still governing enough of the body to keep failure mostly hidden. Decompensation means the same distorted structure has started losing control in ways the whole organism is now forced to feel.

Why Cirrhosis Is the Collapse of Organization, Not Just the Presence of Scar

Cirrhosis is what the liver becomes when scar stops being a wound response and starts becoming a new organizing principle. Scar matters, but scar alone does not explain the failure. The deeper event is that the liver loses the arrangement that made liver function possible in the first place. A healthy liver is not simply a mass of hepatocytes sitting in red tissue. It is a spatial system. Portal blood enters at the portal tracts, moves through low-resistance sinusoids, slows enough for plasma to cross a fenestrated endothelial surface, and meets hepatocytes across a thin exchange compartment called the space of Disse. Bile is formed on the opposite side of those same hepatocytes and moves away through canaliculi and ducts. Oxygen, substrate concentration, immune surveillance, and metabolic specialization all depend on that layout. Cirrhosis means that layout has been broken. The organ may still contain liver cells, but the map that let those cells behave like liver tissue has been lost.

This is why diffuse fibrosis by itself is not the full definition. In cirrhosis, fibrous septa, bands of collagen-rich scar, begin partitioning the parenchyma into regenerative nodules. Those nodules are the liver trying to replace itself, but the replacement is happening inside walls. Hepatocytes are no longer arranged in their normal plates with clean sinusoidal flow around them. They are grouped into islands surrounded by distorted matrix and abnormal vascular routes. Blood is rerouted. Some channels narrow. Some shunt. Some areas are underexposed to the stream they were built to govern, while other regions receive blood under abnormal pressure and abnormal timing. The liver is still present, but its cells are no longer positioned inside a coherent exchange organ. They are surviving inside a scarred maze.

Once that happens, every major liver function becomes harder for structural reasons before it becomes impossible for cellular reasons. Bilirubin uptake depends on blood reaching hepatocytes cleanly. Albumin synthesis depends on hepatocytes being adequately supplied with oxygen and substrate. Ammonia clearance depends on portal blood actually crossing functional parenchyma rather than bypassing it through shunts. Bile secretion depends on intact polarity and downstream drainage. Glucose buffering depends on hepatocytes standing in the path of incoming post-digestive blood. Cirrhosis weakens all of this at once because it turns a high-flow exchange organ into a nodular low-order organ with hardened internal boundaries. The problem is no longer merely damaged cells. The problem is that even surviving cells are harder to reach, harder to supply, and less able to hand their work back to the circulation in an orderly way.

This is also why portal hypertension becomes such a defining feature. Scar does not simply sit beside blood flow. It becomes part of the resistance blood has to cross. Sinusoids lose their softness. Stellate cell-driven matrix and contractile force narrow the vascular field. Portal inflow now meets an organ that has become mechanically less compliant and geometrically less open. Pressure rises upstream. Collaterals form. Splenic congestion develops. Ascites becomes easier to generate. These are not secondary decorations added to a scarred liver. They are what happens when organization collapses deeply enough that blood can no longer cross the organ the way the organ was built to receive it. The vascular problem is the architectural problem made visible.

That is why cirrhosis should be understood as collapse of organization rather than as presence of scar alone. Scar is one material inside the process. The real disaster is that flow, exchange, polarity, gradients, regeneration, and drainage have all been reorganized into the wrong shape. The liver still tries to repair itself. Hepatocytes still proliferate. Proteins are still synthesized. Bile is still formed in part. But the work now happens inside an organ that has forgotten its own geometry. A cirrhotic liver does not fail because collagen exists. It fails because collagen, nodules, shunts, and hardened tissue have replaced the ordered architecture that made liver physiology possible.

How Cirrhosis Changes Every Major Job the Liver Performs

Cirrhosis does not damage one liver function at a time. It damages the conditions that make all liver functions possible at once. That is why the disease feels so total once it declares itself. A healthy liver is a high-flow exchange organ with soft architecture, open sinusoids, preserved cell polarity, low-resistance portal inflow, orderly central venous drainage, intact canalicular bile flow, and enough hepatocyte mass standing in the right place to meet the blood that needs governing. Cirrhosis breaks that arrangement. Fibrous septa partition the parenchyma. Regenerative nodules trap hepatocytes inside distorted compartments. Sinusoids capillarize and lose permeability. Blood is shunted, compressed, or rerouted. The space between plasma and hepatocyte thickens. Portal pressure rises. Bile drainage becomes less clean. Once that happens, every major hepatic job begins failing for the same structural reason. The stream no longer reaches the cell normally, and the cell no longer sits inside a liver built to let it work.

Blood chemistry control suffers first because the cirrhotic liver is no longer seeing portal blood in the way a normal liver sees it. Glucose buffering becomes less precise because incoming post-meal substrate is not crossing a broad open sinusoidal network cleanly. Ammonia clearance weakens because blood is increasingly bypassing functional hepatocyte contact through intrahepatic and portosystemic shunts, and because the hepatocytes that remain are operating under worse oxygen delivery and less favorable substrate access. Bilirubin handling deteriorates because uptake, conjugation, and secretion all depend on preserved cell polarity, adequate ATP, and intact biliary flow. Hormone transformation becomes less reliable for the same reason. The liver may still contain hepatocytes capable of doing these tasks in principle, but principle is not enough. Function requires access. A cirrhotic liver still exists in the body, but more and more of the body's blood is no longer meeting that liver in a usable way.

Synthetic work also changes profoundly. Albumin falls not only because there are fewer healthy hepatocytes, but because the hepatocytes that remain are living inside a scarred field with worse perfusion, more inflammatory signaling, and less reserve. Clotting factor production becomes unstable for the same reason. The liver loses the ability to maintain plasma as a carefully built protein environment, and the bloodstream begins revealing that loss through edema, ascites, altered coagulation, and reduced transport stability. The same synthetic decline affects endocrine output as well. Insulin-like growth factor 1 falls. Thrombopoietin support weakens. Transport proteins that shape hormone movement can change. The liver stops being a quiet manufacturer of blood architecture and becomes a compromised factory operating under interrupted supply, distorted drainage, and chronic inflammatory pressure.

Bile physiology changes just as deeply. A cirrhotic liver does not only struggle with blood. It struggles with direction. Hepatocytes are supposed to face blood on one side and bile on the other, but fibrosis, nodular distortion, ATP depletion, and cholestatic stress all make that directional organization harder to preserve. Canalicular secretion becomes less reliable. Ductal flow becomes more vulnerable. Cholesterol, bilirubin, and bile acids are harder to move outward cleanly. That means digestion suffers downstream because the intestine receives less orderly biliary support, and the liver suffers upstream because compounds meant to leave now begin accumulating or refluxing. The same organ that once governed both blood chemistry and digestive outflow becomes less competent at both because the tissue architecture that separated and coordinated those two traffic systems has been damaged.

Immune behavior changes too. The healthy liver is built to tolerate low-grade portal microbial traffic without collapsing into constant inflammation. Cirrhosis weakens that balance. Kupffer cells remain activated more easily. Cytokine signaling stays louder. Endotoxin clearance becomes less efficient. The acute-phase state becomes easier to sustain. Stellate cell activation and extracellular matrix expansion, which helped create the cirrhotic organ in the first place, are kept alive by the very burden the cirrhotic organ is now worse at clearing. The liver therefore stops being merely an organ injured by inflammation and becomes an organ that helps perpetuate inflammatory distortion because it can no longer absorb the portal burden quietly. Tolerance weakens. Chronic activation becomes easier. The organ grows louder as it grows less capable.

Regeneration changes perhaps most tragically of all. The cirrhotic liver is often full of regenerative effort. Hepatocytes still proliferate. Nodules still enlarge. But growth inside cirrhosis is no longer the same thing as restoration. New cells are being laid into hardened architecture, distorted blood flow, thickened exchange spaces, and abnormal biliary pathways. The liver is still trying to replace what was lost, but it is doing so inside a scaffold that no longer rewards normal organization. That is why cirrhosis changes every major hepatic job at once. It does not merely remove function. It collapses the architecture that made integrated function possible. Blood handling, bile flow, protein synthesis, immune tolerance, endocrine signaling, detoxification, and regeneration all begin failing together because they were never separate jobs to begin with. They were different expressions of one organized organ, and cirrhosis is what happens when that organization gives way.

Part X

When Flow Fails

Chapter 37

Portal Hypertension

What Portal Pressure Is

Portal pressure is the force carried by blood as it tries to move through the portal venous system and then cross the liver. That definition has to stay connected to structure or it becomes too abstract. The portal vein is not an ordinary supply vessel feeding a passive organ. It is the main channel carrying blood from the stomach, intestine, spleen, pancreas, and related abdominal organs into the liver before that blood reaches the general circulation. This blood is rich in absorbed nutrients, microbial fragments, hormone signals, metabolic byproducts, and everything else the gut and spleen have handed forward. For the system to work, portal blood has to enter the liver, spread through a low-resistance sinusoidal network, exchange with hepatocytes, and then drain out through the hepatic veins into the systemic venous return. Portal pressure is therefore not just pressure inside a vein. It is the pressure required to move this entire upstream blood stream through the hepatic exchange field.

In a healthy liver that pressure remains relatively low because the organ is built as a soft, compliant, low-resistance bed. Sinusoids are open enough, the space of Disse is thin enough, the extracellular matrix is restrained enough, and the hepatocyte plates are organized in a way that lets blood move, slow, and exchange without meeting a hard mechanical barrier. The liver is not simply accepting blood. It is designed to let blood cross it. Portal pressure becomes clinically important when that crossing grows more difficult. If resistance rises anywhere between portal inflow and hepatic venous outflow, the upstream pressure required to maintain flow begins to increase. The body is now trying to push the same blood through a field that has become narrower, stiffer, more distorted, or more compressed than it was built to be.

That is why portal pressure is best understood as a relationship between flow and resistance rather than as a number floating in isolation. Blood is arriving continuously from the splanchnic circulation, the vascular territory of the gut, spleen, pancreas, and related abdominal organs. If the liver is soft and organized, that inflow meets manageable resistance. If fibrosis, nodules, stellate cell contraction, sinusoidal capillarization, thrombosis, or downstream venous obstruction distort the pathway, resistance rises and pressure backs up. The portal system then begins carrying more pressure than it was designed to hold. This matters because veins are not built to behave like arteries. The portal circulation is meant to deliver a high-volume nutrient and burden stream into a low-pressure exchange organ. Once the pressure rises, the entire upstream territory begins paying for the liver's loss of compliance.

So portal pressure is not merely a measurement taken to describe liver disease. It is the mechanical expression of whether the body can still push gut-derived blood through the liver in the way liver physiology requires. A healthy liver keeps that flow quiet. A diseased liver makes that flow expensive. When portal pressure rises, it means blood is having a harder time reaching and crossing the organ that was supposed to govern it.

Why Scarred Tissue Resists Portal Inflow

Portal blood is supposed to enter a soft organ, not a barricade. That is the first thing to see. The liver is designed as a low-pressure exchange bed where blood arriving from the intestine and spleen can spread through thin-walled sinusoids, slow down, cross a permeable endothelial surface, and meet hepatocytes across a very short distance. Scar changes the physical terms of that encounter. Collagen and other extracellular matrix accumulate in spaces that were meant to remain thin, open, and compliant. Compliance means the tissue can yield under flow instead of pushing back hard against it. A scarred liver yields less. The same volume of portal blood now meets a stiffer field, and pressure begins rising upstream because the organ no longer behaves like an easy place to cross.

The resistance is not only passive, like water meeting a narrower pipe. It is structural and dynamic at the same time. Structural resistance comes from fibrous septa, nodules, capillarized sinusoids, and thickened exchange spaces. Fibrous septa are bands of scar that partition the liver and interrupt the normal lobular path blood is supposed to take. Capillarization means the sinusoid loses its open fenestrated character and begins behaving more like an ordinary tighter capillary. The plasma can no longer reach hepatocytes with the same ease, and blood cannot spread through the sinusoidal bed with the same low resistance. Dynamic resistance comes from activated stellate cells. These cells do not only lay down collagen. They also become contractile, which means they can narrow the sinusoidal space further. So scarred tissue resists portal inflow because it is both physically narrowed and actively tightened.

Nodular distortion makes the problem worse because the blood is no longer entering an organized map. In a healthy liver, portal inflow moves through a predictable network toward the central veins. In cirrhosis, regenerative nodules and fibrous bands break that network into irregular compartments. Blood has to find its way through a rearranged organ in which some routes are compressed, some are blocked, and some are diverted into abnormal vascular channels. The result is not simply slower flow. It is disordered flow. Some hepatocytes are bypassed. Some regions are underperfused. Some vessels are exposed to pressure they were not built to hold. Portal inflow is still arriving, but the organ no longer offers a clean low-resistance pathway through which that inflow can be distributed.

This resistance also rises because scar changes what the vessel wall is pushing against. A healthy sinusoidal bed is soft enough that portal blood can enter without needing high driving force. A fibrotic liver is mechanically harder. The extracellular matrix is thicker, more cross-linked, and less deformable. Cross-linking means collagen fibers are chemically tied together into a more rigid network. That rigidity matters. Blood entering from the portal vein now meets a tissue that does not open under pressure the way healthy liver does. It behaves more like a compressed sponge than a living exchange field. Portal pressure rises because flow is trying to enter an organ that has lost its softness.

So scarred tissue resists portal inflow because scar does not sit harmlessly beside the circulation. It invades the route the circulation has to use. It narrows sinusoids, thickens the exchange surface, partitions the parenchyma, stiffens the organ, and adds contractile force through activated stellate cells. The portal vein is still delivering blood from the gut. The problem is that the liver has become a worse place for that blood to go.

Collateral Vessel Formation

Collateral vessel formation begins when portal blood can no longer cross the liver cheaply enough to keep using the normal route. The portal venous system is supposed to deliver blood from the intestine, spleen, pancreas, and stomach into a low-resistance hepatic bed where that blood can be processed before it enters the general circulation. Once fibrosis, nodules, stellate cell contraction, sinusoidal capillarization, and hardened architecture raise intrahepatic resistance, the same incoming volume now meets a liver that has become mechanically expensive to cross. Blood does not stop arriving just because the organ has become harder. The gut keeps draining. The spleen keeps draining. Portal inflow keeps pressing forward. So the circulation begins looking for alternate exits. Collateral vessels are those alternate exits, preexisting small vascular connections between the portal and systemic venous systems that enlarge under chronic pressure until they become major bypass routes.

This is the key point. Collaterals are not usually brand-new vessels appearing out of nowhere in a moment of crisis. They are often small, low-importance communications that become recruited, widened, and remodeled because the pressure gradient has changed. Blood follows resistance. When the liver offers too much of it, even minor side channels start becoming attractive. Veins that once carried little flow are exposed to sustained higher pressure and higher volume, so they dilate, their walls remodel, and over time they become visible evidence that the portal system is unloading itself around the liver instead of through it. The body is improvising a detour. That detour is protective in one narrow sense because it lowers some of the upstream pressure burden. It is also dangerous because it sends portal blood into vessels never designed to carry that much flow and because it allows gut-derived blood to bypass hepatic processing.

The classic sites of collateral formation are not random. They are places where portal and systemic venous territories already lie close enough to communicate. At the lower esophagus and upper stomach, portal tributaries from the left gastric system can connect with systemic veins draining toward the azygos system. In the rectal region, superior rectal venous drainage can communicate with middle and inferior rectal veins.

Around the umbilicus and abdominal wall, paraumbilical and superficial systemic veins can enlarge. Retroperitoneal channels can do the same deeper in the abdomen. These sites become important because the pressure does not distribute evenly. It exploits anatomical opportunities. The result is a network of bypass vessels carrying blood away from the congested portal system into systemic venous return without requiring that blood to cross the scarred liver first.

That change has two major consequences. The first is mechanical fragility. These collateral veins are thin-walled, low-pressure structures by origin, not reinforced conduits built for chronic high-flow stress. As they dilate under portal hypertension, they become varices, enlarged tortuous veins with walls that are vulnerable to rupture. Esophageal and gastric varices matter most because bleeding there can be sudden, massive, and life-threatening. The second consequence is functional bypass. Blood that should have entered the liver carrying ammonia, microbial products, hormone fragments, drugs absorbed from the intestine, and other gut-derived chemistry is now reaching the systemic circulation with less hepatic review. This is one reason portal hypertension is not only a vascular problem. It is also a detoxification and blood chemistry problem. The body has preserved venous return at the cost of skipping part of liver function.

So collateral vessel formation is the circulation's structural answer to a liver that has become too resistant to cross. Portal blood keeps coming. The scarred organ keeps refusing flow cheaply enough. Small portosystemic connections enlarge into major bypass channels. Pressure is partly relieved, but only by rerouting blood into fragile veins and around the very organ that was supposed to process it. The detour keeps the stream moving. It does not solve the disease.

Varices and Bleeding Risk

Varices are collateral veins that have been forced into a job they were never built to perform. They begin as part of the body's attempt to survive portal hypertension. Blood coming from the intestine, spleen, and stomach is supposed to cross the liver before returning to the systemic circulation. Once cirrhosis and fibrosis make that route too resistant, portal blood starts looking for alternate exits. Small venous connections between the portal system and the systemic veins enlarge under chronic pressure and become major bypass channels. At the lower esophagus and upper stomach, this produces esophageal and gastric varices, thin-walled distended veins carrying blood under a load far greater than their original design. The body has created a detour around the scarred liver. The price is that the detour is fragile.

The danger is mechanical first. A normal vein in that region is not reinforced to tolerate sustained high pressure and high flow. As the vessel enlarges, its wall stretches and thins. The amount of force pressing outward on that wall rises as the radius of the vessel increases, which means a larger varix is not just a bigger vein. It is a vein under more wall tension. Wall tension is the stress trying to pull the vessel open. If the wall becomes thinner while the vessel becomes wider, the structure becomes easier to rupture. This is why large varices matter so much clinically. They are living under a geometry that favors failure. The liver did not merely create collateral circulation. It created a set of thin pressurized tubes in one of the most vulnerable locations in the body.

The location makes the risk worse. The distal esophagus is constantly exposed to swallowing, reflux, mechanical irritation, and changes in luminal pressure. A dilated collateral vein running just beneath that mucosal surface has little margin for error. It does not need to be torn by some dramatic external injury. It only needs the wall to thin enough, the pressure to rise enough, or the local surface to become compromised enough that the vessel gives way. Once it ruptures, bleeding can be massive because the portal system is still trying to unload itself through that channel. The same adaptation that helped decompress the congested circulation becomes a direct hemorrhage risk the moment the wall fails.

This is why variceal bleeding is not simply a blood pressure problem and not simply a clotting problem. It is a structural rupture happening inside a person who often also has unstable coagulation, low platelet count, reduced clotting factor synthesis, and poor vascular reserve because of advanced liver disease. A healthy vessel with a healthy clotting system can often tolerate small defects and seal them. A large varix in cirrhosis is different. The wall is weak, the inflow is strong, and the blood may already be less capable of forming a stable hemostatic response once bleeding starts. Portal hypertension creates the vessel. Cirrhosis makes the whole situation harder to control.

So varices matter because they are the point where compensation turns into catastrophe. The body formed them to keep portal blood moving when the liver became too resistant to cross. But those same vessels are thin, superficial, enlarged, and exposed to forces they were never designed to bear. Bleeding risk is therefore built into their existence. The collateral saves flow until the day the wall can no longer hold it.

Splenomegaly and Blood Cell Sequestration

Splenomegaly in portal hypertension is not an isolated spleen problem. It is the upstream venous consequence of a liver that has become too resistant to cross. Portal blood coming from the intestine and spleen is supposed to pass through a low-resistance hepatic bed. Once cirrhosis and fibrosis distort that bed, pressure rises backward into the portal system, and the spleen, which drains into that same system through the splenic vein, begins living under chronic venous congestion. Over time it enlarges. That enlargement is not decorative. It changes how the spleen handles blood cells. The organ that normally filters aged or abnormal blood elements and serves as an immune reservoir begins trapping a larger share of the circulating blood cell mass than it should. This is the beginning of sequestration. The cells still exist in the body, but more of them are being held in the spleen instead of remaining available in the circulating pool.

Platelets show this most clearly. In cirrhosis with portal hypertension, thrombocytopenia often appears early, and one major reason is splenic sequestration from hypersplenism, the state in which an enlarged spleen retains and removes blood cells more aggressively than normal. Earlier reviews describe the redistribution as striking, with up to 90% of the circulating platelet mass located in the enlarged spleen in some cirrhotic patients, and more recent reviews still identify splenic sequestration as a central mechanism of thrombocytopenia in chronic liver disease. The important physiological point is that the marrow may still be making platelets, but fewer of those platelets remain available in peripheral blood because the spleen has become a larger, more demanding reservoir.

The same logic can extend beyond platelets. Red blood cells and white blood cells may also become more heavily pooled and removed, which is why hypersplenism can contribute to broader cytopenias rather than isolated thrombocytopenia alone. The enlarged spleen is not simply fuller of blood in a passive sense. It is filtering, retaining, and turning over formed elements under altered hemodynamic conditions. That is why splenomegaly in liver disease matters so much clinically. It is one of the places where a vascular consequence of cirrhosis becomes a hematologic consequence. Portal resistance rises in the liver, venous congestion enlarges the spleen, and the spleen then changes what fraction of the body's blood cells remain functionally available in circulation.

This also explains why low platelet count in cirrhosis is never only a spleen story. Splenic sequestration is major, but reduced hepatic thrombopoietin production also lowers marrow stimulation, so platelet loss is often coming from both sides at once: fewer platelets being produced and more being held or destroyed in the enlarged spleen. Splenomegaly therefore should be read as a structural sign that portal flow is failing and as a functional sign that the circulating blood cell pool is being redefined upstream of the marrow and downstream of the scarred liver. The cells are still present in the organism. More of them are simply trapped in the wrong reservoir.

Ascites as a Flow and Protein Problem at the Same Time

Ascites is not simply fluid in the abdomen. It is what happens when the liver loses control of where vascular water is supposed to stay. That failure begins as a flow problem and deepens as a protein problem. Portal blood is supposed to cross a soft low-resistance hepatic bed, exchange with hepatocytes, and drain onward without forcing the upstream venous system to live under chronic backpressure. Cirrhosis changes that. Fibrous septa, regenerative nodules, sinusoidal capillarization, and stellate cell contraction make the liver harder to cross, so portal pressure rises behind it. The splanchnic circulation, the venous territory draining the intestine, spleen, pancreas, and related abdominal organs, now sits under a hydraulic burden it was not designed to carry. More fluid is pushed out of congested capillary beds and into the interstitial space and peritoneal cavity. The abdomen begins filling because blood can no longer move through the liver cheaply enough to keep fluid where it belongs.

But portal pressure alone does not fully explain ascites. If it did, every rise in venous pressure would produce the same degree of abdominal fluid accumulation. The other half of the problem is plasma protein architecture, especially albumin. Albumin is one of the main proteins creating oncotic pressure, the inward pulling force that helps hold water inside the vascular compartment. In cirrhosis, albumin synthesis falls because the hepatocyte is no longer maintaining the same synthetic output. Now the circulation is being squeezed from both directions.

Portal and splanchnic hydrostatic pressure are pushing fluid outward harder, while plasma oncotic pressure is pulling fluid back inward less effectively. More pressure out. Less protein pull in. The peritoneal cavity becomes the place where this failed balance shows itself most clearly.

That is why ascites should be understood as a misdistribution problem before it is understood as a volume problem. A person with tense ascites can look fluid-overloaded and still have an underfilled effective circulation. Effective circulation means the fraction of blood volume that is actually being sensed as adequate by the arterial system and kidneys. In cirrhosis, blood is pooling in the splanchnic vasculature, leaking into the abdomen, and bypassing normal hepatic processing through collateral pathways. The kidneys read this as circulatory threat even when the abdomen is visibly swollen. They respond by retaining sodium and water through the renin-angiotensin-aldosterone system, sympathetic activation, and antidiuretic hormone signaling. The body is trying to defend perfusion. Instead it adds more fluid to a system that already cannot hold fluid in the right compartment. The retained water does not solve the architecture problem. It enlarges it.

The liver's lymphatic system is pulled into this too. A healthy liver makes lymph continuously, but portal hypertension and sinusoidal pressure increase hepatic lymph production dramatically. Once lymph formation exceeds drainage capacity, more fluid escapes into the peritoneal space. So ascites is not caused by one leak. It is the result of several fluid vectors converging. Congested portal flow raises hydrostatic pressure. Synthetic decline lowers oncotic restraint. Splanchnic vasodilation and effective arterial underfilling trigger renal sodium and water retention. Hepatic lymph production rises beyond what drainage can carry. The abdomen becomes the reservoir where all of those failures meet.

This is why ascites belongs in the language of flow and protein at the same time. It is a vascular problem because the scarred liver resists portal inflow and forces pressure backward into the abdominal venous system. It is a synthetic problem because low albumin weakens the blood's ability to retain its own water. It is a renal problem downstream because the kidneys answer the false signal of underfilling by retaining more salt and water. But the deeper truth is still hepatic. Ascites appears when the liver can no longer keep circulation soft enough, protein-rich enough, and structurally organized enough to hold fluid inside blood instead of surrendering it to the abdomen.

Chapter 38

Ascites, Edema, and Fluid Misplacement

How Portal Pressure and Low Albumin Work Together

Portal pressure and low albumin work together because they attack fluid balance from opposite sides of the same capillary wall. Portal hypertension raises the force pushing fluid outward. Low albumin weakens the force pulling fluid inward. That is the whole architecture in one sentence. The liver is supposed to sit between the splanchnic circulation and the rest of the body as a soft low-resistance exchange organ. Once cirrhosis and fibrosis distort that organ, portal blood coming from the intestine, spleen, pancreas, and stomach meets more resistance than it should. Pressure rises behind the liver, especially in the splanchnic venous system. Capillary hydrostatic pressure, the physical force of blood pressing outward against the vessel wall, increases in that congested territory and drives more water out of the vascular compartment into surrounding tissue and into the peritoneal cavity. At the same time, the same failing liver is often making less albumin. Albumin is the main plasma protein generating oncotic pressure, the protein-based inward pull that helps keep water inside blood. So the circulation is now being pushed toward leakage from both directions. More fluid is being forced out. Less fluid is being held in.

This is why ascites and peripheral edema are not simply signs of "too much fluid." They are signs of fluid in the wrong place. The person may indeed retain more total sodium and water later in the process, but the first major failure is distribution. Blood entering the congested splanchnic bed is exposed to a stronger outward push, and plasma that now contains less albumin has less power to reclaim that escaped water. In the abdomen, this becomes ascites, fluid accumulating in the peritoneal cavity. In peripheral tissues, especially the lower extremities where gravity adds its own hydrostatic burden, it becomes edema. The body has not only gained water. It has lost control over where that water belongs. This is why a swollen cirrhotic patient can still be effectively underfilled in the arterial circulation. The fluid is present in the organism, but too much of it has left the vascular space and become unavailable for clean circulatory work.

The kidneys then make the whole problem worse while trying to solve it. A circulation that is losing usable intravascular volume because of portal congestion, splanchnic vasodilation, and low oncotic pressure is sensed as threatened, even if the abdomen is visibly full of fluid. Renal perfusion and arterial stretch signals fall enough to activate the renin-angiotensin-aldosterone system, sympathetic outflow, and antidiuretic hormone release. Sodium and water retention increase. This would be adaptive if the main problem were simple dehydration. In cirrhosis it becomes maladaptive because the added volume is being poured back into a system that still has high portal resistance and weak plasma protein restraint. More retained fluid means more fuel for the same misplacement. The abdomen enlarges further. The tissues swell further. The effective circulation may still remain unstable because the structural problem in the liver and the protein problem in plasma have not been corrected.

Portal pressure is therefore not enough by itself, and low albumin is not enough by itself, but together they form one of the main hydraulic signatures of advanced liver disease. High portal pressure without severe synthetic decline can still push fluid outward, but stronger oncotic pressure may partly restrain the escape. Low albumin without major portal congestion can still favor edema, but the abdominal cavity will usually not be driven under the same forceful splanchnic hydraulic load. In cirrhosis, the two converge. Scar makes the liver harder to cross, so upstream hydrostatic pressure rises. Synthetic failure lowers albumin, so the vascular pull weakens. The same organ created both sides of the fluid problem. It became too resistant to flow and too weak to maintain plasma protein architecture.

That is why ascites and edema should be understood as fluid misplacement created by failed hepatic order. Portal hypertension drives water out of the vascular compartment. Low albumin makes it harder to bring that water back. The kidneys answer the resulting arterial underfilling by retaining even more sodium and water, which deepens the misplacement rather than correcting it. The swollen body is therefore not a body with too much usable circulation. It is a body whose circulation has lost its grip on its own fluid.

Sodium Retention and Neurohormonal Compensation

Sodium retention in cirrhosis is not the kidney acting irrationally. It is the kidney acting on the wrong map. The body does not judge circulatory adequacy by looking at how swollen the abdomen is or how much fluid has collected in the tissues. It judges adequacy by how full and pressurized the effective arterial circulation feels at the level of the baroreceptors, the pressure-sensing structures that monitor stretch in key vessels and help determine whether perfusion is secure. In cirrhosis, portal hypertension and splanchnic vasodilation distort that reading. Blood pools in the vascular territory of the gut and related abdominal organs, fluid escapes into the peritoneal cavity and tissues, and the circulation that remains inside the arterial tree is sensed as underfilled even when total body water is already high. That false underfilling is the trigger. The body thinks volume is threatened, so it begins defending sodium.

The first major response is activation of the renin-angiotensin-aldosterone system, often abbreviated RAAS. Renin is released from the kidney when perfusion pressure and sodium delivery fall in ways the organ interprets as circulatory shortage. Renin then drives formation of angiotensin II through the angiotensin system already stocked with liver-made angiotensinogen, and angiotensin II constricts vessels and stimulates aldosterone release from the adrenal cortex. Aldosterone acts mainly on the distal nephron, the later segments of the kidney tubule, increasing sodium reabsorption so more sodium is pulled back into the body instead of being lost in urine. Water follows sodium osmotically. This is the body trying to refill the arterial tree. The intent is survival. The effect in cirrhosis is expansion of a fluid burden that the circulation still cannot hold in the right place.

The sympathetic nervous system joins this response because arterial underfilling is also read as a threat requiring vascular defense. Sympathetic activation narrows certain vascular beds, supports renin release, and shifts renal handling toward conservation rather than excretion. At the same time antidiuretic hormone, also called vasopressin, rises and tells the kidney to retain more free water. This means the cirrhotic body is not simply holding salt. It is running a full emergency program designed for low-volume states such as hemorrhage or dehydration. But the problem is not true empty-body volume depletion. The problem is maldistributed fluid in a circulation distorted by portal resistance, low albumin, and splanchnic pooling. So the emergency program becomes self-defeating. More sodium is retained. More water is retained. The abdomen enlarges further. Edema worsens. The effective circulation may still remain unstable because the structural reason fluid is leaving the vascular space has not been solved.

This is why sodium retention in cirrhosis should be understood as compensation that has become part of the disease. The kidney is trying to preserve perfusion, not create ascites. But because the liver has become a high-resistance organ and because plasma oncotic pressure is often weakened by low albumin, the extra retained volume keeps feeding the same leak-prone system. Portal pressure continues pushing fluid outward. Low albumin continues weakening the inward pull that should help hold water in blood. Splanchnic vasodilation continues making the arterial circulation look emptier than total body volume would suggest. The kidney keeps hearing emergency. The abdomen keeps filling.

So sodium retention and neurohormonal compensation are the body's attempt to rescue circulation from a problem the liver has made impossible to solve by volume alone. The signals are real. The response is biologically logical. The outcome is misplacement made worse. The person does not retain sodium because the body has forgotten how to regulate fluid. The person retains sodium because the body is still trying to defend arterial life inside a vascular architecture that can no longer keep fluid where it belongs.

Why Fluid Leaves the Vasculature

Fluid leaves the vasculature because blood vessels are not sealed pipes. Capillaries are exchange surfaces. Water is always being pushed and pulled across their walls, and that movement is part of normal physiology before it ever becomes edema or ascites. Hydrostatic pressure inside the vessel pushes fluid outward. Plasma proteins, especially albumin, create oncotic pressure that pulls fluid inward. The capillary wall itself determines how permissive that movement can be, and the lymphatic system clears much of what leaves the vascular space but does not immediately return. In health, these forces remain in tension with one another. Some fluid leaves to bathe tissues and deliver dissolved solutes. Some is pulled back. Some is taken away by lymphatics and returned later to the circulation. The system works because the outward leak is limited, the inward pull is preserved, and the escape routes remain open enough that the interstitial space does not drown.

The problem begins when the outward force rises, the inward force falls, or the barrier itself becomes easier to cross. More hydrostatic pressure means more water is being driven against the capillary wall and out through it. This is what happens in venous congestion and portal hypertension. Blood is still arriving, but it is meeting downstream resistance, so pressure backs up into the upstream vascular bed. The capillary now behaves less like a quiet exchange surface and more like a pressurized seep. At the same time, if albumin falls, the plasma loses part of its protein-based inward pull. Fluid that leaves the vessel is less effectively drawn back. The water has not become more rebellious. The circulation has become worse at holding it.

Permeability matters just as much. A capillary wall with intact endothelial cells and disciplined junctions limits what can cross and how freely it can do so. Inflammation changes that. Cytokines, oxidative stress, endothelial injury, and local immune activation can make the wall leakier, which means not only more water but more protein-rich fluid can escape into the interstitial space. Once proteins leave with the fluid, oncotic pull within the vessel weakens further and the tissue space becomes more capable of holding on to what escaped. This is one reason inflamed tissue swells so easily. The capillary is no longer simply under more pressure. It has become more permissive to loss.

In cirrhosis and portal hypertension, all of these forces begin converging. Portal and splanchnic hydrostatic pressure rise because the scarred liver resists inflow. Albumin often falls because synthetic function declines. Splanchnic vasodilation makes the effective arterial circulation look underfilled, so the kidneys retain sodium and water through neurohormonal compensation. More volume is now being poured into a vascular system already biased toward leakage. Hepatic and intestinal lymph formation also rises, and once lymphatic return cannot keep up, the excess fluid finds the peritoneal cavity or soft tissues. The person then appears fluid-overloaded, but the deeper truth is misplacement. The vasculature has lost part of its ability to hold and reclaim its own water.

So fluid leaves the vasculature whenever the balance between outward push, inward pull, barrier restraint, and lymphatic recovery is broken in the wrong direction. Normal physiology allows controlled filtration. Disease exaggerates it. More pressure pushes fluid out. Less albumin pulls it back. Greater permeability opens the wall wider. Weak lymphatic recovery lets escape become accumulation. Edema and ascites are what that imbalance looks like once the circulation can no longer keep fluid where blood was meant to stay.

Ascitic Fluid as a Downstream Consequence of Failing Architecture

Ascitic fluid is not the original disease. It is what the abdomen becomes once the liver can no longer preserve the physical conditions that keep blood moving through the right channels and water staying in the right compartment. A healthy liver is a low-resistance exchange organ. Portal blood arrives from the intestine and spleen, spreads through a soft sinusoidal bed, crosses a thin permeable interface, and drains onward without forcing the upstream circulation into chronic backpressure. Plasma proteins, especially albumin, help keep vascular water from escaping too freely while lymphatic drainage quietly clears the fraction that leaves during normal exchange. Ascites appears when that architecture fails in enough places at once that the body can no longer keep fluid inside blood without losing it elsewhere. The fluid in the abdomen is therefore not just extra water. It is the visible residue of blood flow, protein synthesis, vascular pressure, and lymph handling all becoming less organized than the liver was built to maintain.

The first failure is hydraulic. Cirrhosis and fibrosis turn the liver from a soft exchange field into a scarred resistant organ. Portal inflow now meets fibrous septa, narrowed sinusoids, nodular distortion, stellate cell contraction, and hardened extracellular matrix. Pressure rises upstream because blood is still arriving from the gut and spleen but can no longer cross the liver cheaply enough. That increased hydrostatic pressure in the splanchnic circulation pushes more fluid out of capillary beds and into the interstitial space. Some of that excess filtration is picked up by lymphatics, but chronic portal hypertension drives lymph production so high that the drainage system eventually loses the race. Once lymphatic return is outpaced, the peritoneal cavity becomes one of the easiest spaces for that excess fluid to collect. Ascitic fluid is therefore partly failed portal flow made visible as misplaced water.

The second failure is synthetic. The same cirrhotic liver resisting flow is also often making less albumin. Albumin provides much of the plasma oncotic pressure that helps pull water back toward the vascular compartment. When albumin falls, the bloodstream loses part of its inward holding force. Fluid pushed out by high hydrostatic pressure is now less effectively reclaimed. The circulation becomes a worse container for its own water. This is why ascites cannot be explained by portal pressure alone. Pressure pushes fluid out, but low albumin makes it easier for that fluid to stay out. The abdomen fills because the liver has become both mechanically obstructive and synthetically weaker. It is resisting blood and failing to build part of the protein architecture that helps blood hold itself together.

Then the kidneys join the story, and what began as misplacement becomes self-amplifying retention. The arterial system reads the cirrhotic circulation as underfilled because blood is pooling in the splanchnic bed, bypassing the liver through collaterals, and leaking into tissues and the peritoneal cavity. Baroreceptors do not care that the belly is full of fluid. They care that effective arterial volume feels low. The kidney answers by retaining sodium. Water follows. Renin rises. Angiotensin II rises. Aldosterone rises. Antidiuretic hormone rises. This neurohormonal compensation would be useful in hemorrhage or dehydration. In cirrhosis it becomes part of the problem. More sodium and water are added to a vascular system that still cannot keep fluid in the right compartment. The extra volume does not repair the scarred liver, reopen the sinusoids, or restore albumin synthesis. It simply feeds the same leak-prone architecture more aggressively. Ascitic fluid is therefore not only what escaped. It is also what the body later retained while trying to rescue a circulation it could no longer stabilize normally.

This is why ascitic fluid should be read as a downstream consequence of failing architecture rather than as an isolated abdominal symptom. The peritoneal cavity is acting as the collection basin for a liver that no longer allows cheap portal flow, no longer preserves plasma protein support well enough, no longer keeps the splanchnic circulation hydraulically quiet, and no longer prevents renal compensation from becoming maladaptive. The fluid itself is often relatively simple in composition compared with blood, but the path that created it is not simple at all. Scar raised resistance. Nodules distorted flow. Albumin fell. Lymph overflowed. Kidneys retained more sodium and water. The abdomen became the place where those failures met. Ascites is therefore the geometry of cirrhosis turned into fluid.

Spontaneous Bacterial Peritonitis and Why Stagnant Fluid Becomes Dangerous

Spontaneous bacterial peritonitis, usually abbreviated SBP, is infection of ascitic fluid without an obvious surgically correctable intra-abdominal source such as a perforation or abscess. That definition matters because it tells you the problem is not dirty fluid from a ruptured bowel spilling into the abdomen in the usual sense. The problem is that cirrhotic ascites has become a compartment in which bacteria can arrive and survive because the normal barriers, flow patterns, and immune defenses have already been weakened. Bacteria or bacterial products translocate from the gut, meaning they cross the intestinal barrier, reach lymph or blood, and then seed the ascitic compartment. In cirrhosis this becomes easier because the gut barrier is more permeable, the microbiome is altered, local and systemic immune defenses are blunted, and portosystemic shunting lets more gut-derived microbial traffic bypass normal hepatic filtering. The ascitic fluid is therefore not dangerous just because it exists. It is dangerous because it is the downstream collection basin of a failing gut-liver immune axis.

The fluid itself becomes permissive because cirrhotic ascites is often immunologically poor. Low ascitic protein matters here. Protein in ascitic fluid is not just a lab number. It includes part of the defensive chemistry that helps kill bacteria or mark them for clearance. When ascitic protein is low, complement levels and opsonic activity fall. Opsonic activity means the fluid is worse at coating microbes in a way that makes them easier for immune cells to recognize and ingest. A low-protein ascitic compartment is therefore a badly defended space. Add stagnant flow, poor effective immune surveillance, and repeated bacterial translocation from the gut, and the fluid stops behaving like a neutral ultrafiltrate. It becomes a vulnerable culture medium sitting inside a host already weakened by cirrhosis.

This is why stagnant fluid becomes dangerous. In a healthy system, contaminated burden is cleared, diluted, or handled before it can establish itself. In advanced cirrhosis, ascites accumulates because portal flow is failing, albumin is low, sodium and water are being retained, and lymphatic overflow is pouring fluid into the peritoneal cavity faster than normal handling can contain it. That means the abdomen is now holding a large poorly defended fluid reservoir while the same patient has cirrhosis-associated immune dysfunction, altered hemodynamics, and more bacterial traffic arriving from the gut. The stagnant compartment does not simply sit there harmlessly. It lowers the distance between bacterial arrival and bacterial persistence. Once infection takes hold, SBP can rapidly trigger acute kidney injury, worsen circulatory dysfunction, precipitate hepatorenal syndrome and encephalopathy, and sharply worsen mortality.

So spontaneous bacterial peritonitis is best understood as infected ascites born from failed architecture, failed filtering, and failed local defense. The fluid collected because portal flow and protein balance broke down. It became dangerous because gut-derived bacteria had an easier route in and the fluid had too little opsonic and complement support to clear them back out. Ascites is the hydraulic consequence of cirrhosis. SBP is what happens when that hydraulic consequence turns into an infectious compartment.

What Fluid Accumulation Reveals About Systemic Liver Failure

Fluid accumulation in cirrhosis reveals that the liver is no longer failing in one isolated job. It is failing as an organizer of circulation. Ascites and edema mean the body has lost control over pressure, protein architecture, vascular tone, renal compensation, and lymphatic handling all at once. Portal hypertension has already made the liver too resistant to cross, so blood is backing up into the splanchnic circulation instead of moving quietly through a low-resistance hepatic bed. At the same time, cirrhosis drives splanchnic vasodilation, which lowers effective arterial blood volume even while total body water is rising. The kidneys read this as circulatory threat, not abundance, and respond by activating the renin-angiotensin-aldosterone system, the sympathetic nervous system, and antidiuretic hormone, all of which push sodium and water retention harder. Fluid accumulation therefore is not simply swelling. It is proof that the liver has lost command over the geometry of blood flow and the body has shifted into maladaptive rescue mode.

It also reveals synthetic failure. A cirrhotic liver is often making less albumin, and albumin is one of the main plasma proteins providing oncotic pressure, the inward pull that helps keep water inside the vascular compartment. Once portal and splanchnic hydrostatic pressure are pushing fluid outward and albumin is no longer supplying enough inward restraint, the circulation starts surrendering water to the wrong spaces. The abdomen fills. Peripheral tissues swell. The swollen patient may still be effectively underfilled in the arterial circulation because the fluid is present in the body but misplaced outside the vascular tree. That is why ascites is such an important threshold sign. It tells you the failing liver is now affecting not only detoxification or bile flow, but also the physical behavior of plasma itself.

Fluid accumulation also reveals that compensation has turned against the host. The sodium and water retention that would be protective in hemorrhage or dehydration becomes self-defeating in cirrhosis because the added volume is being poured back into a system whose portal resistance, splanchnic pooling, and low oncotic pressure have not been corrected. More retained sodium means more retained water. More retained water means more ascites and edema. As ascites becomes refractory, prognosis worsens sharply, which is why modern hepatology treats fluid accumulation as a hallmark of decompensation rather than as a cosmetic complication. The abdomen is showing that the liver's structural disease has become a whole-body circulatory disease.

So fluid accumulation reveals systemic liver failure because it proves the liver can no longer keep blood moving through itself cheaply enough, protein-rich enough, and coherently enough to hold water in the circulation. Portal pressure is high. Effective arterial volume is low. Neurohormonal rescue is chronically activated. Albumin support is weakened. The kidneys are retaining salt and water against a false map. The abdomen and tissues become the place where all of those failures finally become visible. Ascites is not just fluid. It is failed hepatic architecture made hemodynamic.

Chapter 39

Hepatorenal Syndrome

Why Liver Failure Can Destabilize Kidney Function Without Primary Kidney Disease

Hepatorenal syndrome begins as a circulation problem before it becomes a creatinine problem. The kidney may be structurally present, the nephron may not be the original site of injury, and yet renal function begins failing because the liver has already lost control over the vascular world the kidney depends on. In decompensated cirrhosis, portal hypertension drives profound splanchnic vasodilation, which means blood pools in the vascular territory of the gut instead of supporting the effective arterial circulation cleanly. The arterial tree is then sensed as underfilled even when total body water is high. The body answers with the same emergency systems it would use in hemorrhage: renin-angiotensin-aldosterone activation, sympathetic activation, and intense sodium and water retention. Those responses help preserve perfusion at first, but as cirrhosis worsens they progressively constrict the renal circulation, lower renal perfusion, and drive the kidney into a state of extreme sodium avidity and falling filtration. That is the core logic of hepatorenal syndrome. The liver fails first as an organizer of blood flow, and the kidney begins failing inside the distorted circulation that follows.

This is why the syndrome was historically described as "functional" renal failure. Older observations, including angiographic and postmortem findings, showed kidneys that could appear structurally intact even while renal function collapsed in advanced cirrhosis and ascites. The kidney was not initially the organ declaring war on itself. It was a tightly vasoconstricted organ receiving the message that perfusion had to be defended at all costs. Modern hepatology keeps that circulatory explanation at the center but adds an important refinement: hepatorenal syndrome is not always purely hemodynamic in a sterile sense. Systemic inflammation, oxidative stress, and bile-related tubular injury can participate as the syndrome deepens. So the cleanest way to say it is this: hepatorenal syndrome can arise without a primary intrinsic kidney disease setting the process in motion, but the kidney is still being injured by the circulatory and inflammatory environment created by advanced liver failure.

Ascites belongs in this mechanism because it marks the stage at which sodium retention and circulatory distortion have already become systemic. In cirrhosis, ascites develops when portal hypertension is established and the kidney has already shifted into chronic sodium conservation under the influence of splanchnic vasodilation and reduced effective arterial blood volume. That same sodium-retaining state is part of the path toward hepatorenal syndrome. The body keeps defending a circulation it cannot truly stabilize, and the kidney keeps sacrificing filtration to preserve pressure. This is why the syndrome often appears in decompensated cirrhosis with ascites rather than in quiet liver disease. By the time hepatorenal syndrome appears, the liver has already turned portal hypertension, fluid misplacement, and neurohormonal compensation into a whole-body hemodynamic disorder. The kidney is then pulled into failure not because it forgot how to be a kidney, but because the vascular conditions required for kidney function have been progressively taken away.

The syndrome is often tipped into the open by a precipitating event. Bacterial infection, especially spontaneous bacterial peritonitis, gastrointestinal bleeding, nonsteroidal anti-inflammatory drug use, or large-volume paracentesis without adequate albumin support can further destabilize the already fragile circulation and push renal perfusion below the threshold the kidney can tolerate. That is why hepatorenal syndrome should never be framed as an isolated renal mystery. It is the kidney reading liver failure through blood flow, vascular tone, inflammatory burden, and effective arterial underfilling. The kidney becomes the organ that finally declares the crisis, but the crisis was built upstream in the scarred, vasodilated, sodium-retaining world that cirrhosis created.

Circulatory Changes in Advanced Liver Disease

Advanced liver disease changes circulation by turning the liver from a low-resistance exchange organ into a high-resistance obstacle and then forcing the rest of the vascular system to improvise around that obstacle. Portal blood is still arriving from the intestine, spleen, pancreas, and stomach, but a cirrhotic liver no longer offers the same open sinusoidal bed for that blood to cross. Fibrous septa, regenerative nodules, sinusoidal capillarization, and stellate cell contraction all raise intrahepatic resistance. Portal pressure rises behind the liver because the inflow continues while the organ becomes harder to traverse. That is the first circulatory change. Blood that was supposed to move through the liver begins backing up in the splanchnic circulation, the vascular territory of the abdominal digestive organs, and the whole upstream system starts living under abnormal venous pressure.

The body does not tolerate that quietly. One of the paradoxes of advanced cirrhosis is that portal congestion and systemic vasodilation develop at the same time. The splanchnic circulation becomes abnormally dilated, driven in part by excess vasodilator signaling such as nitric oxide, so a larger fraction of the blood volume pools in the abdominal vascular bed rather than supporting the effective arterial circulation cleanly. Effective arterial circulation means the portion of blood volume the body actually senses as available to maintain organ perfusion and pressure. This is why a cirrhotic patient can be fluid-overloaded and still be treated physiologically as if they are underfilled. The abdomen may be swollen, the tissues may be edematous, yet the arterial tree is being sensed as inadequately filled because too much blood has pooled in the wrong place and too much fluid has escaped the vascular compartment. The circulation becomes hyperdynamic in one sense and underpowered in another. Cardiac output may rise. Systemic vascular resistance falls. Effective perfusion becomes less trustworthy.

That false underfilling triggers the same rescue systems the body would use in hemorrhage or dehydration. The renin-angiotensin-aldosterone system activates. The sympathetic nervous system activates. Antidiuretic hormone rises. Sodium retention increases. Water follows. The kidney begins conserving volume aggressively because it is being told the circulation is threatened. This compensation is logical at first, but in advanced liver disease it becomes part of the pathology. More retained sodium and water are poured back into a system still burdened by portal resistance, low albumin, and splanchnic pooling. Ascites grows. Peripheral edema worsens. Renal blood flow becomes more vasoconstricted and more precarious. The body is defending pressure by sacrificing distribution. The circulation is no longer simply transporting blood. It is fighting to keep enough of its own volume in the arterial tree to preserve organ survival.

Collateral vessels are another major circulatory adaptation. When portal blood cannot cross the liver efficiently, it looks for alternate exits into the systemic venous circulation. Small communications enlarge into major bypass routes. Varices form in the esophagus and stomach. Paraumbilical veins and other collateral channels dilate. This partly decompresses the portal system, but it does so by sending gut-derived blood around the liver instead of through it. That means the circulation is not only changing pressure. It is changing the route by which blood encounters hepatic processing. Ammonia, microbial products, hormone fragments, drugs absorbed from the intestine, and other portal burdens are now more likely to bypass the hepatocyte field that should have modified them. The circulatory adaptation protects flow at the cost of liver access.

The result is that advanced liver disease becomes a disease of vascular organization as much as a disease of hepatocytes. Portal pressure rises because the liver resists inflow. Splanchnic vasodilation lowers effective arterial filling. Neurohormonal compensation retains sodium and water. Collaterals bypass hepatic processing. Renal perfusion becomes harder to preserve. Ascites, varices, edema, and hepatorenal syndrome all emerge from this same circulatory logic. A failing liver does not only stop doing chemistry. It rewrites the map through which blood has to move, and once that map changes, every other organ starts living inside the consequences.

Renal Vasoconstriction and Falling Kidney Perfusion

In advanced cirrhosis the kidney does not usually fail first by becoming inflamed or structurally destroyed. It fails because the circulation surrounding it has become less trustworthy. The scarred liver raises portal resistance. The splanchnic circulation dilates and begins holding a larger share of the blood volume in the abdominal vascular bed. Fluid leaves the vasculature more easily because portal pressure is high and albumin is often low. The arterial system then reads this as underfilling. That reading matters more than the total amount of water in the body. The person may have ascites, edema, and expanded total body fluid, yet the kidney is being told that effective perfusion is threatened. Renal vasoconstriction is the body's attempt to defend arterial pressure under those conditions. The problem is that the defense eventually begins sacrificing the kidney itself.

The sequence is neurohormonal before it is renal. The renin-angiotensin-aldosterone system activates because the juxtaglomerular apparatus in the kidney senses reduced perfusion pressure and altered sodium delivery. Renin rises. Angiotensin II rises. Sympathetic nervous system output rises. Antidiuretic hormone rises. These are not minor adjustments. They are the same programs the body uses in hemorrhage or major dehydration. The circulation is behaving as though it has lost usable volume, and the kidney responds by narrowing its own vascular bed, especially the renal cortex and afferent circulation, in order to preserve systemic pressure and maintain filtration long enough to survive. At first this can be adaptive. Over time it becomes self-defeating. Blood flow to the kidney falls more and more, oxygen delivery worsens, filtration drops, and the nephron begins living under sustained ischemic stress even though the primary event began outside the kidney.

This is why renal vasoconstriction in cirrhosis is so dangerous. The kidney is a flow-dependent organ. Glomerular filtration depends on enough plasma arriving under enough pressure through a vascular network still open enough to sustain it. Once vasoconstriction becomes severe, the kidney retains sodium almost compulsively because it is trying to preserve intravascular volume. Water follows. Urine sodium falls. Circulating creatinine rises later as filtration declines further. The body is still trying to save circulation, but it is doing so by squeezing one of the main organs that depends on circulation the most. This is the logic behind hepatorenal syndrome. The kidney may be structurally intact at the beginning, but it is being starved of blood in a body that keeps calling that starvation a necessary sacrifice.

The deeper point is that falling kidney perfusion in advanced liver disease is not a kidney decision. It is the renal expression of a larger vascular collapse in which portal hypertension, splanchnic vasodilation, fluid misplacement, low effective arterial volume, and relentless neurohormonal compensation have all converged. The scarred liver changes the vascular map. The kidney reads the consequences and constricts. At first that preserves pressure. Eventually it preserves nothing.

Why the Problem Is Systemic, Not Local

Hepatorenal syndrome only looks like a kidney problem if the reader starts too late in the sequence. By the time creatinine rises, the real disorder has already spread across the circulation. Advanced cirrhosis raises portal resistance, distorts blood flow through the liver, and drives profound splanchnic vasodilation, which means the vascular bed of the gut and related abdominal organs begins holding and dissipating blood volume that should have been supporting the effective arterial circulation. The kidney then reads underfilling, not because the body lacks fluid in the simple sense, but because the arterial tree has lost secure access to usable volume. What follows is not local nephron disease declaring itself first. It is a whole-body hemodynamic failure in which the kidney becomes the organ forced to live at the end of the worst vascular compromise.

That is why the renal response is so dramatic and yet so misleading if it is read in isolation. The body answers this false underfilling with renin-angiotensin-aldosterone activation, sympathetic activation, and intense sodium and water retention. These are the same rescue systems used in hemorrhage and dehydration. They are trying to preserve pressure and survival. But in cirrhosis they act inside a circulation already warped by portal hypertension, splanchnic pooling, ascites, and low effective arterial volume, so the rescue becomes part of the injury. Renal vasoconstriction deepens, perfusion falls, filtration falls, and the kidney begins failing in a body that is actually fluid-expanded but functionally underfilled. The problem is therefore systemic because the kidney is obeying circulatory instructions written upstream by the failing liver, not inventing an isolated renal disease of its own.

Systemic inflammation makes the argument even stronger. Modern hepatology no longer treats hepatorenal syndrome as a purely sterile plumbing disorder. In advanced cirrhosis, bacterial translocation from the gut, endotoxemia, cytokine release, and infection-related decompensation intensify circulatory dysfunction and directly favor multiorgan failure. The kidney is one of the main organs exposed to that inflammatory field, but it is not alone. The same state that destabilizes renal perfusion also worsens vascular tone, cardiac performance, sodium handling, and immune dysregulation across the whole organism. This is why spontaneous bacterial peritonitis, gastrointestinal bleeding, or other decompensating events can precipitate hepatorenal syndrome so quickly. They do not merely injure the kidney secondarily. They destabilize the cirrhotic body as a system, and the kidney is one of the first organs to lose reserve inside that collapse.

So the problem is systemic, not local, because liver failure at this stage is no longer confined to hepatocytes or bile flow. It has become a disease of vascular organization, inflammatory signaling, and organ-to-organ dependence. The scarred liver changes the map of circulation. The arterial tree loses effective fill. Neurohormonal compensation turns chronic. Inflammation stays too high. Ascites and sodium retention declare the same disorder in fluid form. The kidney then fails inside that environment, not apart from it. Hepatorenal syndrome is the renal face of a body whose circulatory order has already been broken by advanced liver disease.

The False Appearance of a Normal Kidney Trapped in an Abnormal Circulation

Hepatorenal syndrome creates one of the most deceptive pictures in medicine because the kidney can appear structurally ordinary while function is collapsing. That is why older descriptions called it a "functional" renal failure. Histology may be normal or near-normal, proteinuria is typically absent or minimal, and the renal vasculature can look intact, yet filtration falls because the kidney is living inside a circulation that has become profoundly abnormal. In advanced cirrhosis, portal hypertension and marked splanchnic vasodilation make the arterial tree feel underfilled even when total body fluid is excessive. The body answers with renin-angiotensin-aldosterone activation, sympathetic activation, and intense renal vasoconstriction. The kidney is not failing because it forgot how to be a kidney. It is being starved of usable blood flow by a systemic hemodynamic disorder created upstream by liver failure.

That is the false appearance. On the slide or at autopsy, the kidney may not show the kind of primary parenchymal destruction that would explain the severity of the laboratory abnormality by local pathology alone. But structurally present is not the same thing as functionally supplied. A nephron can be intact and still underperform if perfusion pressure has been withdrawn, renal vasoconstriction has become extreme, and the surrounding circulation is redistributing blood away from effective renal support. The organ is present in the body, but the body is no longer delivering the vascular conditions the organ needs in order to work. This is why hepatorenal syndrome belongs to the language of circulatory failure more than to the language of intrinsic nephron disease.

This is also why reversal is tied so closely to restoring the circulation or replacing the liver rather than to treating the kidney as the primary source of failure. When hepatic and circulatory conditions improve enough, kidney function can improve because the main problem was never local structural collapse to begin with. The kidney looked normal because, in an important sense, it was normal tissue trapped in the wrong vascular world. The damage was real. The map around the organ was the original injury.

Why Liver Disease Has to Be Read as Multi-Organ Disease

Liver disease has to be read as multi-organ disease because the liver is not performing one isolated task that can fail in private. It is governing blood chemistry, portal blood handling, bile formation, plasma protein synthesis, hormone transformation, iron traffic, immune filtering, nutrient conversion, and vascular resistance all at once. Once that organ begins losing structure, every other organ is forced to live inside a different bloodstream. The kidney receives a circulation with worse effective arterial filling. The brain receives blood less cleanly cleared of ammonia and other gut-derived burden. The spleen begins draining into a portal system that can no longer empty through the liver cheaply enough. The intestine keeps sending portal blood into an organ that is now worse at filtering bacterial fragments and endotoxin. The endocrine system keeps releasing signals into a plasma environment whose transport proteins, conversion pathways, and steroid handling are no longer normal. The liver is not an accessory gland attached to digestion. It is one of the main organs that decides what kind of internal environment the rest of the body is allowed to inhabit.

That is why the first visible consequences of liver disease often appear somewhere else. A person develops ascites and edema because circulation can no longer hold fluid in the vascular space cleanly. Platelets fall because portal hypertension enlarges the spleen and the failing liver makes less thrombopoietin, the hormone that supports platelet production. Mental status changes because blood bypasses hepatic processing through collateral routes and neuroactive burden reaches the brain more freely. Renal function falls because portal hypertension and splanchnic vasodilation, dilation of the abdominal vascular bed, make the arterial tree look underfilled, so the kidney constricts and retains sodium in a world where perfusion is already unstable. These are not side effects sitting next to liver disease. They are the liver's structural failure being expressed through other organs.

The same is true for metabolism and immunity. A damaged liver does not simply stop clearing toxins in the vague popular sense. It loses the ability to buffer glucose, regulate iron traffic through hepcidin, maintain albumin, produce clotting factors, shape thyroid and steroid hormone handling, and keep low-grade portal microbial traffic from becoming systemic inflammatory pressure. The bloodstream then changes character. It becomes less stable as a transport fluid, less disciplined as a protein environment, more prone to fluid misplacement, more prone to inflammatory signaling, and more permissive to bypass chemistry coming from the gut. The rest of the organism is not looking at a diseased liver from the outside. It is bathing in the consequences of that liver's changed output every minute.

Portal hypertension makes the multi-organ logic even harder to ignore because once blood can no longer cross the liver normally, the whole upstream vascular territory begins remodeling around that fact. Collateral vessels form. Varices become fragile alternative drainage routes. The spleen enlarges and sequesters blood cells. Ascites fills the abdomen. The kidney interprets the circulation as threatened and begins conserving sodium and water in a way that deepens the problem. The brain becomes vulnerable because portosystemic shunting means more gut-derived and nitrogen-related chemistry reaches systemic blood without normal hepatic review. The liver may still be physically present in the body, but the body has already begun reorganizing itself around the fact that blood can no longer use that liver properly.

So liver disease has to be read as multi-organ disease because the liver is one of the main organs that makes a body-wide internal order possible. Once that order fails, no other organ gets to continue operating under normal conditions. The kidney is forced into vasoconstriction. The brain is exposed to altered blood chemistry. The spleen becomes congested. The gut-liver immune border weakens. Fluid leaves the vasculature. Hormone traffic changes. Coagulation destabilizes. The liver does not fail alone. It pulls the rest of the organism into a new physiological map, and that map is what advanced liver disease really is.

Chapter 40

Hepatic Encephalopathy

Ammonia and Other Neuroactive Compounds

Hepatic encephalopathy begins when the liver is no longer clearing portal blood cleanly enough, or when that blood is no longer passing through the liver cleanly enough, for the brain to be protected from gut-derived neuroactive burden. Ammonia is the central molecule in that story, but it is not the only one. Ammonia is generated continuously, especially from intestinal bacterial metabolism and nitrogen handling, and in health the liver converts much of that burden into urea so it can be excreted safely. In cirrhosis or portal-systemic shunting, that protection weakens. More ammonia escapes hepatic handling, more reaches the systemic circulation, and more is delivered to the brain. Hepatic encephalopathy is therefore not simply confusion in a person with liver disease. It is brain dysfunction caused by liver failure and or portal-systemic shunting, with ammonia as the major neurotoxic driver inside a broader toxic and inflammatory field.

The reason ammonia matters so much is that it is small, diffusible, and chemically disruptive once it accumulates beyond the body's ability to contain it. The gut keeps producing it. Muscle and other tissues can buffer some of it. But the liver is the main organ built to convert ammonia into a safer excretory form through the urea cycle. When hepatocyte function falls or portal blood bypasses hepatocytes through collateral vessels and shunts, the body loses that protective conversion step. Ammonia then circulates at higher levels and reaches the brain, where it contributes to astrocyte swelling through glutamine accumulation, alters neurotransmission, and helps push the brain into metabolic dysfunction. This is why a scarred liver and a shunted circulation can produce a neurologic syndrome even when the brain itself was not the original site of disease. The liver failed first as a blood-clearing organ. The brain becomes the organ that has to live with the consequence.

But ammonia alone does not explain the full syndrome. Modern hepatology treats hepatic encephalopathy as multifactorial. Other neuroactive compounds and conditions contribute, including inflammation, altered amino acid traffic, manganese accumulation in chronic liver disease, GABAergic tone, false neurotransmitters, gut-derived toxins, and the broader metabolic instability that follows liver failure. The important point is not to dilute ammonia's importance. It is to place ammonia in the correct setting. A cirrhotic or shunted body is not exposing the brain to one purified toxin. It is exposing the brain to a blood chemistry increasingly shaped by failed hepatic clearance, portal-systemic bypass, intestinal dysbiosis, and systemic inflammation. Ammonia is the main signal of that failure, but the syndrome becomes more dangerous because the rest of the internal environment is failing with it.

This is also why ammonia levels have to be interpreted with discipline. In most cases of hepatic encephalopathy ammonia is elevated, but a normal ammonia level does not fully exclude the diagnosis, and the degree of elevation does not map neatly onto clinical grade in every cirrhotic patient. The syndrome is clinical because the brain is responding to the whole disturbed internal environment, not to one laboratory value in isolation. Even so, ammonia remains central because it explains the core logic: the gut is generating neuroactive nitrogen burden, the failing or bypassed liver is no longer clearing enough of it, and the brain is paying for that escape. Hepatic encephalopathy is therefore the nervous system reading liver failure through blood.

Why Brain Symptoms Arise From Liver Failure

Brain symptoms arise from liver failure because the brain is being asked to function inside blood the liver no longer knows how to clean. That is the starting point. The liver normally receives ammonia-rich portal blood from the gut and converts much of that nitrogen burden into urea through the urea cycle before it can circulate freely. When liver function falls, or when portal blood is diverted around the liver through portosystemic shunts, ammonia and other gut-derived neurotoxic compounds escape that first-pass clearance and enter the systemic circulation in higher amounts. Hepatic encephalopathy is the brain reading that failure through blood. It is not a primary brain disease that happens to coexist with cirrhosis. It is neuropsychiatric dysfunction caused by a liver that can no longer keep portal chemistry from reaching the nervous system.

Ammonia matters most because it is continuously generated in the gut, small enough to circulate easily, and dangerous enough to distort brain physiology once it accumulates. In cirrhosis and shunting, more ammonia reaches the brain, where it contributes to astrocyte swelling, oxidative stress, and neurotransmitter imbalance. Astrocytes are the glial support cells that help maintain the brain's chemical environment, so when they swell and lose metabolic control, neural signaling becomes less stable even before the brain shows obvious structural damage. This is why early hepatic encephalopathy can look like sleep-wake reversal, slowed thinking, irritability, personality change, and impaired attention, and later deepen into lethargy, disorientation, and coma. The blood has changed first. The brain is displaying the result.

But ammonia is not acting alone. Modern hepatology no longer explains hepatic encephalopathy as a one-toxin story. Systemic inflammation, gut dysbiosis, altered neurotransmitters, manganese accumulation in chronic liver disease, and other neuroactive compounds all add weight to the syndrome. The liver fails as a clearing organ, the portal circulation bypasses it more readily, and the brain ends up exposed to a chemical mixture it was never meant to see in that concentration. That is why the symptoms can range from subtle cognitive slowing to profound neurologic collapse. The brain is not malfunctioning in isolation. It is functioning inside a bloodstream that has stopped being fully filtered by the organ built to protect it.

Confusion, Slowed Thinking, Sleep Reversal, Asterixis, Coma

These symptoms are not random neurologic curiosities attached to liver disease. They are the staged behavioral output of a brain trying to operate inside blood the liver has stopped clearing well enough. In hepatic encephalopathy, the liver is no longer converting enough ammonia into urea, and portal blood is increasingly bypassing hepatic processing through collateral vessels and shunts. The result is that ammonia and other gut-derived neuroactive compounds reach the systemic circulation more freely and arrive at the brain in concentrations the brain was not designed to tolerate. Astrocytes, the support cells that help stabilize the brain's chemical environment, begin swelling as ammonia is converted into glutamine inside them. That swelling does not need to produce dramatic structural destruction to cause trouble. It only needs to disrupt the fine control of neurotransmission, energy use, and ion balance that normal thought depends on. The brain becomes slower before it becomes visibly damaged. Function blurs before consciousness disappears.

Confusion is often one of the first ways this shows itself because organized thought depends on clean timing between cortical networks. A brain exposed to chronic neuroactive burden does not stop all at once. It loses sharpness first. Attention drifts. Conversation becomes less linear. The person can still speak, still recognize family, still answer questions, but the internal sequencing is less reliable. Slowed thinking belongs to the same stage. It is not just fatigue. It is cognitive throughput falling off. Mental tasks that once ran automatically now require effort, and even then they proceed more slowly. Questions are answered after a delay. Planning weakens. Short-term working memory becomes less trustworthy. The person is still present, but the speed and precision of thought have been blunted by blood chemistry the liver no longer controls tightly enough.

Sleep reversal is one of the clearest signs that the brain's regulatory rhythm is being disturbed before consciousness fully breaks down. The person becomes awake at the wrong hours, sleepy during the day, restless or more confused at night, and unable to maintain a normal sleep-wake pattern. This is not simply poor sleep hygiene in a sick person. The brain's timing systems are being pulled off their normal rhythm by the same metabolic and neurochemical burden driving the rest of the syndrome. The cortex is slower, but the regulatory architecture underneath that cortex is also becoming less stable. The body may still appear superficially alert at certain times of day, yet its internal schedule is no longer holding together.

Asterixis is what happens when the motor system can no longer sustain steady posture cleanly. It is often described as a flapping tremor, but that phrase can mislead because it sounds like a rhythmic tremor generated continuously by muscle contraction. Asterixis is better understood as intermittent failure of sustained tone. When the person extends the wrists or holds the arms out, there are brief lapses in posture, and the hands drop and recover in a flapping irregular motion. The brain is no longer maintaining continuous motor output with normal reliability. This is important because it shows that hepatic encephalopathy is not only a disorder of abstract thought. It is a disorder of cortical and subcortical control reaching into movement itself. The same chemistry slowing thought is also making posture harder to hold.

Coma is the far end of the sequence. By that point the brain is no longer merely slowed, mis-timed, or intermittently disorganized. It is globally unable to maintain wakeful consciousness. The progression from confusion to slowed thinking to sleep reversal to asterixis to coma is therefore not a list of disconnected symptoms. It is the arc of a brain losing its ability to think clearly, regulate time, sustain motor control, and finally remain awake at all. The liver failed first as a clearing organ. The blood changed next. The brain then displayed the entire collapse in stages the bedside can see.

Gut-Derived Nitrogen and the Liver's Failed Control

The gut is constantly generating nitrogen burden, and the liver is supposed to keep that burden from becoming brain chemistry. That is the starting logic. Dietary protein is digested into amino acids, intestinal bacteria metabolize nitrogen-containing substrates, bacterial urease liberates ammonia from urea, and enterocytes themselves contribute through glutaminase activity. The result is a continuous stream of gut-derived ammonia and related nitrogenous compounds entering the portal circulation. Under healthy conditions this is not a crisis because portal blood is supposed to deliver that burden directly to hepatocytes, where ammonia is converted into urea through the urea cycle before it can circulate systemically in dangerous amounts. The gut makes the burden. The liver is supposed to neutralize it.

That control is more important than it first appears because ammonia is not a harmless metabolic leftover. Blood ammonia has to stay low. Even modest elevations are neurotoxic. The liver's urea cycle is therefore not a side pathway for cleanup. It is one of the main chemical firebreaks between intestinal nitrogen production and brain dysfunction. Extrahepatic tissues such as skeletal muscle can buffer some nitrogen temporarily by incorporating it into glutamine and alanine, but that buffering is only partial. It is a support system, not the main disposal route. The liver is still the organ expected to take portal ammonia and turn it into something safely excretable.

When liver failure develops, or when portal blood is diverted around the liver through collateral vessels and portosystemic shunts, that control breaks in a very specific way. The gut keeps producing ammonia. Portal blood keeps carrying it forward. But hepatocytes no longer convert enough of it into urea, and more of that nitrogen burden escapes into the systemic circulation. The problem is therefore not simply that ammonia exists. It always existed. The problem is that the liver is no longer intercepting it. Blood that should have been cleaned is now moving onward with gut-derived nitrogen still in it. That is the basic metabolic failure underneath hepatic encephalopathy.

This is why gut-derived nitrogen belongs at the center of liver-brain physiology. The intestine is producing the load. The failing liver is no longer controlling the load. The brain then becomes one of the main organs forced to live with the result. Ammonia accumulates, other gut-derived neuroactive compounds contribute, and what began as a failure of nitrogen handling in portal blood becomes slowed thinking, sleep reversal, asterixis, and coma. The liver did not merely fail to detox in some vague sense. It failed to keep ordinary intestinal nitrogen from becoming systemic neurologic burden.

Why Encephalopathy Is One of the Clearest Demonstrations of Liver Necessity

Hepatic encephalopathy is one of the clearest demonstrations of liver necessity because it shows what happens when the liver stops protecting the brain from the chemistry of the gut. The intestine is always generating ammonia and other neuroactive burden. Portal blood is always carrying that burden forward. In health, the liver intercepts it, converts much of the ammonia into urea, and keeps the systemic circulation from becoming a direct extension of the bowel. In cirrhosis or portosystemic shunting, that protection fails. Blood either crosses a liver that can no longer process it well enough or bypasses the liver through collateral channels entirely. The brain then receives what the liver used to neutralize. Confusion, slowed thinking, sleep reversal, asterixis, and coma are the nervous system's proof that the liver was never optional.

This is why encephalopathy is such a ruthless teaching tool. Many liver functions can be hidden for a while by compensation. Albumin can fall gradually. Portal pressure can rise silently for years. Fibrosis can advance before the patient understands what architecture has been lost. But once the brain starts changing, the body is displaying the liver's necessity in public. The person is no longer simply carrying abnormal labs or scarred tissue. Thought itself is being altered because blood has lost one of its main clearing organs. The brain is exquisitely dependent on the liver's quiet work. It does not tolerate portal chemistry arriving unedited for long.

Encephalopathy also makes the broader point that liver disease is not a local organ problem. The liver fails first, but the brain becomes the tissue that reveals how much of normal consciousness depends on hepatic control of circulation. The syndrome is neuropsychiatric, but its origin is hemodynamic and metabolic. Gut-derived compounds are no longer being cleared properly. Portosystemic shunting lets more of them bypass the liver. The bloodstream changes. The brain pays. Few complications show so clearly that the liver is not just making bile, storing glycogen, or processing drugs. It is continuously keeping ordinary digestion from becoming neurologic injury.

How the Gut, Liver, and Brain Are Linked in Real Time

The gut, liver, and brain are linked in real time because they are sharing one moving chemical stream, not passing messages to one another only after damage is already established. The intestine is continuously generating ammonia through bacterial urease activity, glutamine metabolism, and broader nitrogen handling inside the lumen. That ammonia and other gut-derived compounds enter portal blood immediately. In health, the liver receives that portal stream and converts much of the ammonia into urea before the chemistry can spread systemically. In cirrhosis or portosystemic shunting, that filtration and conversion step weakens, so the brain begins seeing more of what the gut is making from moment to moment. This is why hepatic encephalopathy is not best understood as a late abstract complication. It is the nervous system living in real time with the quality of portal blood the liver failed to control.

The link is therefore not symbolic. It is physical. Gut microbes alter ammonia generation, intestinal permeability, inflammatory signaling, and the composition of small molecules moving into portal blood. The liver is supposed to stand in the middle and decide what remains local to the gut, what becomes urea, what is cleared, what is conjugated, and what is allowed into the wider circulation. Once that hepatic checkpoint weakens, blood reaching the brain is no longer the same blood. It carries more ammonia, more inflammatory burden, more gut-derived neuroactive traffic, and less evidence of normal first-pass control. The brain then responds not because it has become the primary diseased organ, but because it is exquisitely sensitive to changes in the chemistry of the blood bathing it.

Ammonia is still the clearest molecule in this sequence because it reaches the brain and is handled by astrocytes, the support cells that help stabilize the neural environment. Astrocytes convert ammonia into glutamine, and when that burden rises too high, glutamine accumulates osmotically inside the cell, drawing water with it and promoting astrocyte swelling. That swelling disturbs neurotransmission, energy handling, and network timing before it needs to produce dramatic structural destruction. This is why the earliest manifestations can be slowed thinking, poor attention, sleep-wake reversal, and subtle cognitive blunting rather than immediate coma. The brain begins losing precision because the blood arriving from the gut is no longer being edited properly by the liver.

But the real-time link is broader than ammonia alone. Gut dysbiosis changes what kinds of bacteria dominate, which changes what kinds of metabolites and inflammatory signals are produced. Increased intestinal permeability allows more endotoxin and microbial fragments into portal blood. Systemic inflammation then amplifies the neurologic consequences of hyperammonemia, making the brain more vulnerable to the same nitrogen burden than it would be in a quieter immune state. The liver is the middle organ in all of this. When it is healthy, the gut can remain colonized and chemically active without the brain paying such a high price. When it fails, the distance between bowel chemistry and mental status becomes alarmingly short.

So the gut, liver, and brain are linked in real time because the gut is continuously making chemical burden, the liver is continuously supposed to neutralize and filter that burden, and the brain is continuously exposed to the result. A healthy liver keeps the bowel from becoming a neurologic event. A failing liver lets intestinal chemistry become thought, mood, sleep, motor control, and eventually consciousness itself. That is the real-time axis. The gut writes the first draft. The liver is supposed to edit it. The brain reads whatever survives the edit.

Part XI

The Major Disease Patterns

Chapter 41

Viral Hepatitis

Hepatitis as Inflammation, Not a Single Disease

Hepatitis is a tissue event before it is a diagnostic label. The word means inflammation of the liver, not one disease with one cause, one time course, or one biological outcome. That distinction matters because the liver can become inflamed for many different reasons and still produce a superficially similar laboratory picture. Hepatocytes swell, die, or leak intracellular enzymes into blood. Kupffer cells become activated. Cytokine signaling rises. Sinusoidal flow becomes less quiet. Cholestatic features may appear if bile handling is disturbed. The tissue begins behaving like an injured organ. Viral hepatitis belongs inside that broader logic. A virus infects the liver or targets hepatocytes strongly enough that the liver becomes an inflamed battlefield. But the inflammation is the liver's tissue state. The virus is the trigger. Those are not the same thing.

This is why viral hepatitis should never be taught as though "hepatitis" names one unified disease entity the way measles or mumps names a more specific infection. Hepatitis A, B, C, D, and E do not behave identically because they are not the same biological problem wearing different letters. They differ in route of transmission, persistence, immune dynamics, chronicity, and how they injure the liver over time. Some are usually acute infections that the body clears. Some can become chronic and turn years of low-grade viral persistence into fibrosis, cirrhosis, portal hypertension, and liver cancer risk. Some are strongly shaped by fecal-oral spread and intestinal exposure. Others are blood-borne and build long-term residence inside the body. So when we say viral hepatitis, we are not naming one disease. We are naming a family of viral causes capable of pushing the liver into the inflammatory state called hepatitis.

The liver experiences that state through the same core tissue language regardless of which virus started it. Hepatocytes become the site of viral replication or immune attack. Cytotoxic immune cells begin identifying infected targets and killing hepatocytes in the effort to control infection. Cytokines rise and alter hepatic metabolism, protein synthesis, and bile handling. Bilirubin may accumulate if hepatocyte function or biliary secretion falls behind. The blood begins showing elevated aminotransferases because injured hepatocytes are leaking their enzymes. Fatigue, jaundice, right upper quadrant discomfort, dark urine, pale stool, and appetite loss may follow, but those symptoms are downstream. The primary event is that liver tissue has been pushed into inflammatory injury by a viral process, and the exact meaning of that injury depends on whether the infection is brief and self-limited or persistent enough to turn inflammation into remodeling and remodeling into scar.

That is the real reason the phrase hepatitis as inflammation, not a single disease, matters so much. It forces the reader to think structurally. The liver has only so many ways to display injury. Inflammation is one of them. A virus is one cause of that inflammation, but not the only cause, and not every virus does it in the same way. To understand viral hepatitis clearly, the reader has to separate the tissue state from the triggering agent. Hepatitis is the inflamed liver. Viral hepatitis is what happens when a virus drives the liver into that state. The word names the organ's response. The lettered viruses explain why that response began and what kind of future the tissue is now being pushed toward.

Hepatitis A, B, C, D, E From a Structural Point of View

From a structural point of view, these viruses are best separated by the kind of liver injury pattern they tend to impose, not just by the letter attached to them. Hepatitis A and hepatitis E are usually enteric infections, acquired mainly by the fecal-oral route, that drive an acute inflammatory event rather than a long remodeling project. The liver becomes inflamed, hepatocytes are injured, jaundice can appear, and the tissue can become sick enough to fail in severe cases, but the usual arc is a short-lived attack rather than years of architectural rewriting. Hepatitis A does not become chronic and does not typically leave the liver on a path toward cirrhosis. Hepatitis E is also usually self-limited, though chronic infection can occur in immunocompromised hosts. Structurally, A and most E infections are the viruses of acute hepatic stress, an inflamed liver trying to survive a sharp insult rather than a liver being slowly rebuilt into scar.

Hepatitis B is different because it can be either an acute inflammatory event or a chronic architectural disease. It is transmitted through blood and other body fluids, and once chronic infection is established the important question is no longer only whether hepatocytes are infected on a given day. The important question is what years of ongoing immune recognition, hepatocyte injury, and regenerative pressure are doing to the lobule. Chronic hepatitis B can push the liver from inflammation into fibrosis, then into cirrhosis and liver cancer risk. So from a structural standpoint, hepatitis B is a virus of dual possibility. It may pass through the liver as a finite inflammatory episode, or it may remain long enough that the organ starts trading open exchange architecture for scar and nodularity.

Hepatitis C is even more defined by chronicity. Acute infection is often quiet enough that the person may not know it happened, but the architectural burden accumulates later because a large share of infections become chronic if untreated. The liver then lives under a long low-grade inflammatory pressure in which hepatocytes are repeatedly injured, the immune system never quite closes the case, and stellate-cell-driven fibrotic remodeling becomes easier to sustain year after year. This is why hepatitis C is so strongly associated with progressive fibrosis, cirrhosis, and hepatocellular carcinoma. Structurally, hepatitis C is less the virus of dramatic acute collapse and more the virus of slow parenchymal attrition, a liver that keeps working while quietly being converted into scar.

Hepatitis D is structurally the amplifier. It cannot exist on its own. It requires hepatitis B virus infection, because it depends on HBV to complete its life cycle. That dependency is not a trivia detail. It means hepatitis D enters a liver that is already living inside hepatitis B biology and then intensifies the damage. CDC-associated sources describe hepatitis D as the most severe form of chronic viral hepatitis, with frequent progression toward decompensation and liver cancer. From a tissue perspective, D is the virus that pushes an already threatened liver harder toward architectural failure. It is not merely another hepatitis letter. It is a second viral burden riding on the back of B and accelerating the path from inflammation to collapse.

So the structural map is simple once the letters are translated into tissue behavior. Hepatitis A is usually an acute inflamed liver that recovers. Hepatitis E is usually the same pattern, though it can become chronic in the immunocompromised. Hepatitis B can be acute or can become a long war that reshapes the organ. Hepatitis C is the classic chronic remodeling virus, often quiet at entry and destructive over time. Hepatitis D is the severe intensifier that can only exist with B and tends to make the architecture fail faster. Same organ. Same word hepatitis. Very different futures for the tissue.

Acute Versus Chronic Injury

Acute liver injury is a hit the organ may survive without forgetting its shape. Chronic liver injury is a repetition that teaches the tissue the wrong shape. That is the cleanest structural difference. In an acute injury, hepatocytes may swell, leak enzymes, undergo necrosis or apoptosis, and push the organ into a visible inflammatory state, but the deeper framework of sinusoids, portal tracts, central veins, bile channels, and exchange surfaces may still remain preserved enough that regeneration means real replacement of working tissue. In chronic injury, the liver is not being asked to recover from one blow. It is being asked to keep repairing while the blows continue. The result is that inflammation stops being an event and becomes the background, stellate cells remain activated, extracellular matrix keeps expanding, and fibrosis begins rewriting the architecture itself. Acute injury threatens function. Chronic injury threatens organization.

This is why the time course matters less than the tissue logic. An acute injury can be severe, even catastrophic, but if the injuring force stops and the scaffold is still usable, the liver may rebuild enough parenchyma to restore real hepatic work. The hepatocyte re-enters the cell cycle, regeneration proceeds, and the organ can move back toward open exchange. Chronic injury is different because the liver never gets a clean interval in which repair can finish. Viral persistence, alcohol burden, steatotic stress, cholestatic injury, autoimmune attack, or repeated toxic exposure keep the tissue in a state of recurring wounding. Each round of injury recruits more inflammatory signaling and more matrix deposition, so the organ begins defending itself with scar rather than restoring itself with architecture. What was once repair becomes remodeling, and the remodeling is hostile to liver physiology.

The structural consequences follow from that difference. In acute injury, the liver may show hepatocyte death, cholestatic disturbance, or dramatic enzyme release into blood, yet still retain enough normal lobular organization that blood can reach surviving cells and bile can still find its way out. In chronic injury, repeated fibrosis, nodule formation, and vascular reorganization begin distorting the normal lobular map. The portal-to-central flow pattern becomes less clean. Sinusoids lose their permissive character. Exchange across the space of Disse becomes less efficient. Regeneration becomes trapped inside scar. This is why chronic injury is the real road to cirrhosis. The liver is not only losing cells. It is losing the physical arrangement that made those cells useful.

So acute versus chronic injury is not just a question of shorter versus longer. It is a question of whether the liver is still repairing inside preserved architecture or whether repeated injury has begun converting repair into scar. Acute injury can be violent and still reversible. Chronic injury can be quiet and still be the one that buries the organ.

Why Persistent Inflammation Changes the Tissue Future

Persistent inflammation changes the tissue future because the liver does not experience inflammation as a harmless background color. It experiences it as instruction. A short inflammatory response can help the organ survive injury. Kupffer cells clear debris, cytokines recruit defense, hepatocytes shift synthetic priorities, and damaged cells are removed so regeneration can begin. But when the signal does not shut off, the same machinery that once protected the tissue begins teaching it to rebuild in the wrong direction. Cytokines stay elevated. Reactive oxygen species remain high. Stellate cells do not return quietly to vitamin A storage. They stay activated and keep laying down extracellular matrix. Hepatocytes are pushed into repeated cycles of injury, compensation, and imperfect regeneration. The liver stops living in the rhythm of damage followed by recovery and starts living in the rhythm of damage during repair. That is the turning point. The organ is no longer choosing its future freely. Inflammation is choosing it.

This matters because the liver is an exchange organ, not just a mass of cells. Blood has to cross a soft sinusoidal bed. Plasma has to reach hepatocyte microvilli across a thin space of Disse. Bile has to move cleanly in the opposite direction. Persistent inflammation slowly dismantles all of those conditions. Matrix expands where openness should remain. Sinusoids lose their fenestrated character and become more capillary-like, which means less permeable and less suited for rapid exchange. Stellate cells contract as well as deposit collagen, so the organ becomes not only more scarred but more resistant to flow. The hepatocyte is now doing its work farther from the blood, under worse oxygen delivery, in a louder cytokine environment, with less room for clean polarity and clean secretion. The future of the tissue changes because the tissue itself is being rebuilt into a form less capable of being liver.

Regeneration changes with it. A healthy liver can repair itself impressively when injury is finite and the scaffold remains usable. Persistent inflammation ruins that advantage by making growth occur inside a bad field. Hepatocytes still divide, but the new cells inherit the same inflammatory pressure and the same matrix distortion that injured the old ones. That means proliferation no longer guarantees restoration. It may only create nodules inside scar. The liver remains active, but activity becomes deceptive. Cells are still turning over. Protein is still being synthesized in part. Blood is still entering. Yet the architecture that would let those processes become real recovery is slipping away. Inflammation therefore changes the tissue future not only by damaging cells in the present, but by corrupting what new tissue will be allowed to become.

This is why persistent inflammation is more dangerous than an isolated episode of severe injury. An acute insult may kill cells, but if the signal ends, the liver can often rebuild with some fidelity. Chronic inflammation keeps giving the wound-response program new orders before the old orders have resolved. Scar keeps being rewarded. Regeneration keeps being distorted. Portal resistance keeps rising. Bile flow keeps becoming more fragile. The organ does not merely stay inflamed. It learns inflammation structurally. The future liver is then built out of the memory of repeated injury, and that memory is written in collagen, nodules, shunts, and declining functional reserve.

Immune-Mediated Injury Versus Direct Cytopathic Damage

These are two different ways a liver cell can be injured, and the distinction matters because they produce different tissue futures. Direct cytopathic damage means the infectious agent or toxic exposure is injuring the hepatocyte by its own internal activity, replication, or chemistry inside the cell. The cell is damaged because the agent itself is structurally destructive. Immune-mediated injury means the infected hepatocyte becomes the target of the body's own defense system. Cytotoxic lymphocytes, cytokines, macrophages, and inflammatory signaling do the majority of the tissue damage while trying to control the threat. In viral hepatitis, especially hepatitis B and hepatitis C, the dominant pattern is not usually that the virus simply shreds the hepatocyte from within. Much of the injury is driven by the host immune response directed against infected liver cells.

That difference changes how the tissue should be read. In a strongly immune-mediated pattern, the severity of liver injury does not map neatly to how much virus is physically present inside the liver at a given moment. A person can carry viral material and still have relatively little injury if the immune system is not mounting a strong destructive response, while another person can have major hepatocyte death because the immune system is attacking infected cells aggressively. This is one reason hepatitis B has long been understood as a disease in which liver damage is largely tied to host response rather than direct viral cytotoxicity. It is also why chronic hepatitis C has classically been described as a condition in which liver lesions are mainly immune-mediated rather than directly cytopathic. The virus starts the conflict. The immune system often determines how much tissue is lost.

A directly cytopathic pattern feels different at the tissue level because the injuring force is more tightly coupled to the agent itself. The cell is being damaged by what the pathogen or toxic molecule is doing inside it, not only by what the immune system is doing around it. In an immune-mediated pattern, the liver can end up paying for defense as much as for infection. Cytotoxic T cells recognize infected hepatocytes. Kupffer cells and other immune populations amplify cytokine signaling. Hepatocytes die, enzymes spill into blood, and the tissue enters inflammation because the organ is under attack from the host side as well as the viral side. This is why persistent viral hepatitis is so dangerous. Repeated immune recognition keeps turning infection into chronic inflammation, and chronic inflammation keeps turning inflammation into fibrosis, nodules, and eventually cirrhosis.

So the cleanest way to hold the distinction is this. Direct cytopathic damage means the agent itself is the main destroyer of the hepatocyte. Immune-mediated injury means the hepatocyte is being damaged mainly because the immune system is trying to eliminate the infected cell. In viral hepatitis, the liver often suffers most not because the virus is a simple cellular poison, but because the body is fighting an infected tissue it cannot clear without wounding the organ that contains it.

How Chronic Hepatitis Feeds Fibrosis

Chronic hepatitis feeds fibrosis because the liver is being injured while it is trying to repair itself, and the repair program never gets permission to shut off. An acute inflammatory hit can kill hepatocytes and still leave the organ enough quiet time to rebuild working tissue. Chronic hepatitis is different. Viral persistence or ongoing immune-mediated attack keeps portal tracts inflamed, keeps hepatocytes dying at the interface between portal inflammation and lobular tissue, and keeps the wound-response machinery running long after it should have returned to baseline. The liver is therefore not healing from one event. It is living inside repeated injury. That is the condition in which fibrosis becomes the organ's default repair language.

The first structural changes are usually small but directional. Chronic viral hepatitis tends to begin with portal expansion and periportal inflammation, then move into periportal fibrosis and later septal or bridging fibrosis as bands of scar start linking portal areas to one another and eventually to central regions. That sequence matters because it shows how inflammation at the margins of the lobule gradually becomes a map-changing event. What began as a portal inflammatory process starts laying down collagen in places that blood and solutes have to cross. The exchange organ begins hardening from the edges inward.

The cell driving that shift is the stellate cell. In health it is a quiet vitamin A-storing cell in the space of Disse. Under chronic inflammatory pressure it is activated into a fibrogenic myofibroblast-like cell. Cytokines from damaged hepatocytes and immune cells, along with oxidative stress and ongoing matrix disruption, keep telling it the same thing: reinforce the tissue. The stellate cell answers by producing extracellular matrix, especially collagen, and by helping perpetuate a pro-inflammatory environment that makes later matrix deposition easier still. Chronic hepatitis therefore does not feed fibrosis only by killing hepatocytes. It feeds fibrosis by turning the liver's support cells into scar-building cells and then rewarding that behavior every day the inflammation continues.

Once that matrix begins accumulating, the future of the tissue changes. The space of Disse thickens, sinusoidal exchange becomes less efficient, blood meets hepatocytes less intimately, and regeneration starts happening inside a field that is already being distorted by scar. Hepatocytes may still divide, but repeated inflammation means new cells are being laid into worsening architecture rather than restored architecture. That is why chronic hepatitis is so dangerous even when the patient remains outwardly stable for years. The organ can still function while the inflammatory process quietly converts portal expansion into periportal fibrosis, periportal fibrosis into bridging fibrosis, and bridging fibrosis into cirrhosis. The inflammation does not merely accompany the scar. It teaches the scar how to grow.

Chapter 42

Alcohol-Associated Liver Injury

Why Alcohol Reaches the Liver in High Concentration From Portal Flow

Alcohol does not enter the body as a diffuse burden shared equally by every organ from the first second. It enters through the gut and is delivered to the liver through portal flow before the rest of the body has had much chance to dilute it. That positional fact is one of the reasons alcohol injury is so strongly liver-centered. Ethanol is absorbed rapidly across the stomach and especially the small intestine because it is a small water-miscible molecule that crosses biological membranes without needing complex digestion. Once absorbed, much of it enters the portal circulation directly. The portal vein then carries that absorbed alcohol straight to the liver, the organ sitting between the intestine and the general circulation. This means hepatocytes meet alcohol early, often at a higher effective concentration than many other tissues initially do, and they are asked to begin metabolizing it before the blood has been widely distributed through the rest of the body.

That early delivery matters because the liver is not meeting alcohol as one mild chemical among many. It is meeting it as a concentrated first-pass burden. The same portal blood that carries glucose, amino acids, microbial fragments, bile acid return, and other post-digestive chemistry now carries ethanol in a form that needs immediate handling. Hepatocytes begin oxidizing it mainly through alcohol dehydrogenase, converting ethanol into acetaldehyde. Acetaldehyde is more reactive and more damaging than ethanol itself, which means the first metabolic step does not simply neutralize the molecule. It creates a sharper intermediate the liver must then process further, mainly through aldehyde dehydrogenase, into acetate. So the liver is exposed not only to the incoming alcohol load, but also to the transitional chemical violence of turning that load into something less dangerous. Portal delivery places the liver first in line for both the parent compound and its reactive metabolite.

The other reason portal flow matters is that alcohol enters alongside the rest of intestinal chemistry, not in isolation. The liver is already handling nutrient traffic, hormone fragments, endotoxin, microbial metabolites, drugs, and foreign food compounds. Alcohol adds a redox burden to that existing workload. Redox burden means the cell's normal balance of electron-handling chemistry is being pushed in a direction that changes how other pathways behave. Ethanol oxidation increases the NADH to NAD+ ratio inside hepatocytes, and that shift alters fat oxidation, favors triglyceride accumulation, and makes steatosis easier to develop. So the liver is not merely the place where alcohol is detoxified. It is the place where alcohol arrives first, is metabolized first, and distorts existing metabolic traffic first.

This is why alcohol-associated liver injury begins as a structural consequence of route. The gut absorbs. Portal blood delivers. The liver receives the concentrated burden and has to pay the metabolic, oxidative, and inflammatory cost of handling it before the rest of the body has evenly shared that cost. The first pass becomes the first injury field.

Ethanol Metabolism and Acetaldehyde Production

Ethanol becomes dangerous in the liver not only because it arrives there first through portal flow, but because the first major step of handling it creates a more reactive molecule than the one that entered. Ethanol itself is a small alcohol that crosses the gut barrier rapidly and reaches hepatocytes early. Once inside the liver, it is oxidized mainly by alcohol dehydrogenase in the cytosol, producing acetaldehyde. That reaction also converts NAD+ to NADH, which immediately matters because the hepatocyte is not just processing a foreign compound. It is altering its own redox state while doing so. Redox state means the balance of electron carriers inside the cell, and changing that balance changes how the liver handles fat, glucose, and mitochondrial energy production. So the first act of alcohol metabolism is already two burdens at once. The liver has made a more reactive metabolite, and it has shifted its internal chemistry in a direction that makes normal metabolic control harder.

Acetaldehyde is the sharper problem. It is far more chemically reactive than ethanol and much more willing to bind proteins, lipids, and other cellular structures. That binding is not harmless decoration. It changes protein shape, interferes with enzyme function, damages membrane behavior, and creates molecular adducts, which are abnormal chemical attachments between acetaldehyde and cellular components. Once these adducts form, the hepatocyte is no longer only exposed to alcohol. It is carrying chemically altered versions of its own machinery. Mitochondria function less cleanly. Cytoskeletal proteins are distorted. Protein secretion becomes less orderly. The cell is now paying for the transitional chemistry required to turn ethanol into something more excretable. This is why alcohol metabolism has to be understood as a process of injury, not merely of clearance. The danger is not only the drink. The danger is what the liver has to create while trying to get rid of it.

The next step is oxidation of acetaldehyde into acetate, mainly by aldehyde dehydrogenase in mitochondria. This is a protective step because acetate is much less reactive and can be further metabolized into acetyl-CoA, entering broader energy and substrate pathways. But the fact that the pathway has a protective second step does not erase the first. Between ethanol and acetate lies acetaldehyde, and the hepatocyte must survive that interval every time alcohol is metabolized. If the burden is small and the cell is well supplied, this interval can be managed. If the burden is repeated, heavy, or layered on top of steatosis, inflammation, nutrient deficiency, or low glutathione reserve, the intermediate begins outrunning the cell's ability to contain it. The liver is then no longer metabolizing alcohol cleanly. It is accumulating the cost of doing so.

The redox shift intensifies everything. As NADH rises and NAD+ falls, beta-oxidation of fatty acids becomes less favorable, triglyceride synthesis becomes easier to sustain, and gluconeogenesis becomes less efficient. This is one reason fatty liver develops so readily with chronic alcohol exposure. The hepatocyte is being pushed into a biochemical state that favors fat retention while simultaneously enduring the reactive burden of acetaldehyde and the oxidative stress generated by downstream metabolism. Alcohol-associated liver injury therefore is not one insult layered on top of normal physiology. It is alcohol metabolism becoming a metabolic reprogramming event inside the very cell trying to detoxify it.

So ethanol metabolism and acetaldehyde production should be understood as a sequence in which the liver protects the body by creating a more dangerous intermediate and then trying to neutralize it before the tissue is injured. Ethanol is oxidized to acetaldehyde. Acetaldehyde is the reactive hinge point where protein binding, oxidative stress, and metabolic disruption begin. Acetate comes later, but the liver has to survive the chemistry in between. That is why alcohol damages the liver so efficiently. The organ is not merely exposed to the substance. It is forced to manufacture part of the danger itself.

Redox Shifts, Fat Accumulation, Inflammation, Fibrosis

Alcohol-associated liver injury becomes easier to understand once the reader sees it as a sequence rather than as a list. Ethanol enters the hepatocyte and is oxidized to acetaldehyde, then further toward acetate. In that process the cell spends NAD+, reducing it to NADH and shifting the internal redox state. That ratio change matters because the hepatocyte is not only clearing alcohol. It is changing the chemical environment in which fat oxidation, gluconeogenesis, and mitochondrial work are supposed to occur. As NADH rises, beta-oxidation of fatty acids becomes less favorable and triglyceride synthesis and retention become easier. The liver is pushed toward steatosis not by fat magically appearing, but by redox chemistry making incoming and newly formed lipid harder to burn and easier to store.

That fat accumulation is not a neutral storage event. At first, turning fatty acids into triglyceride droplets can look protective because neutral triglyceride is less immediately disruptive than free fatty acid burden. But once redox pressure remains high and lipid continues arriving or being synthesized, the droplet becomes evidence that throughput is failing. Mitochondria are oxidizing less cleanly, export cannot keep up, and hepatocytes begin filling with a burden they are no longer processing efficiently. Chronic alcohol exposure also increases oxidative stress through pathways such as CYP2E1-mediated metabolism, which generates reactive oxygen species and amplifies lipid peroxidation and protein damage. The steatotic hepatocyte is therefore not simply a cell holding fat. It is a cell living under altered fuel handling and rising oxidative strain at the same time.

Inflammation enters when that strained hepatocyte stops containing the cost quietly. Acetaldehyde forms adducts with proteins and other cellular components, reactive oxygen species damage membranes and mitochondria, and gut-derived endotoxin contributes additional inflammatory signaling through the liver. Kupffer cells are activated, cytokines rise, and the tissue shifts from metabolic distortion into immune injury. This is the point where alcohol-associated fatty liver begins feeding alcoholic steatohepatitis rather than remaining only a fat-storage problem. The liver is no longer just metabolically congested. It is inflamed because the chemistry of alcohol handling, oxidative injury, and gut-derived immune burden has become too loud to ignore.

Fibrosis follows when that inflammatory state stops being temporary. Stellate cells are activated by cytokines, oxidative stress, and hepatocyte injury, and they begin laying down extracellular matrix, especially collagen, in the space where blood and hepatocytes were supposed to meet across a thin exchange surface. The tissue responds to repeated alcohol-related injury by reinforcing itself structurally, but in the liver that reinforcement comes at the cost of function. The organ grows stiffer, blood flow grows more resistant, regeneration becomes less clean, and the future of the tissue changes from reversible metabolic injury toward scarred architecture. This is why the sequence matters so much. Redox shift feeds fat accumulation. Fat and reactive intermediates feed inflammation. Persistent inflammation feeds fibrosis. Alcohol injury is not one event. It is a chain.

Steatosis, Steatohepatitis, Cirrhosis

These are not three unrelated liver diagnoses stacked in a textbook list. They are three different operating states of the same organ as burden deepens and architecture begins to lose its way. Steatosis is fat accumulation inside hepatocytes. That is the first shift. The cell is receiving or making more lipid than it can oxidize or export, so triglyceride droplets begin filling the cytoplasm. At this stage the liver may still be quiet clinically and can remain reversible, because the dominant event is storage rather than inflammatory collapse. The hepatocyte is congested, but the organ has not yet fully turned that congestion into tissue-level destruction. This is why steatosis should be read as a traffic problem first. More fat is arriving or being built than the cell can clear cleanly.

Steatohepatitis is what happens when fat stops being a relatively contained storage burden and starts provoking cellular injury. Now the hepatocyte is not only holding lipid. It is swollen, stressed, and often ballooned, meaning the cell has lost part of its structural discipline under metabolic and oxidative pressure. Inflammation appears. Hepatocytes are injured or dying. Kupffer cells are activated. Cytokines rise. The liver has crossed from steatosis into steatohepatitis because fat is no longer just present. It is now feeding tissue damage. This is the moment when the future of the organ begins to change more seriously, because persistent hepatocyte injury is exactly the kind of signal that activates stellate cells and starts teaching the liver to repair with scar rather than with clean restoration.

Cirrhosis is the final architectural consequence when that cycle of injury and repair continues long enough to reorganize the organ itself. At this stage the liver is no longer simply fatty or inflamed. It is scar-partitioned, nodular, and structurally distorted. Fibrous septa divide the parenchyma, regenerative nodules form, and normal lobular blood flow is replaced by a harder, higher-resistance, less coherent vascular map. That is why cirrhosis is not just "more inflammation" or "more fibrosis." It is the collapse of normal organization. The liver may still contain hepatocytes, but they are now trapped inside an organ whose blood flow, exchange surfaces, and biliary architecture have been rewritten in the wrong direction.

So the sequence is not simply fat, then a worse fat disease, then scar. It is congestion, then injury, then architectural failure. Steatosis means the hepatocyte is storing more than it can process. Steatohepatitis means that storage burden has become inflammatory tissue injury. Cirrhosis means repeated injury and repair have hardened into a new structure the body can no longer use like a normal liver. The names change, but what is really changing is the organ's future.

Why Alcohol Injury Is Chemical, Energetic, and Inflammatory at the Same Time

Alcohol injures the liver on three fronts at once because the hepatocyte does not merely encounter ethanol as a passive poison sitting in blood. It has to metabolize it, and metabolism changes the chemistry of the cell while that cell is still trying to remain alive. The chemical injury begins immediately when ethanol is oxidized into acetaldehyde. Acetaldehyde is more reactive than ethanol itself and more willing to bind proteins, lipids, and other cellular structures in ways that distort their function. Those adducts are not innocent chemical decorations. They alter enzyme shape, interfere with membrane behavior, disrupt cytoskeletal organization, and make the hepatocyte carry chemically modified versions of its own machinery. The cell is not only exposed to alcohol. It is forced to manufacture part of the danger and then survive the intermediate it just created.

The energetic injury happens at the same time because alcohol metabolism rewrites the hepatocyte's redox state. As ethanol is oxidized, NAD+ is reduced to NADH, and the cell begins shifting toward a high NADH to NAD+ ratio. That matters because mitochondrial respiration, beta-oxidation of fatty acids, gluconeogenesis, and broader fuel handling all depend on a workable redox balance. Once that balance is pushed in the wrong direction, fat oxidation becomes less efficient, triglyceride accumulation becomes easier, and the liver starts storing lipid it would otherwise have processed more cleanly. ATP production also becomes harder to maintain under a growing oxidative and mitochondrial burden. ATP, adenosine triphosphate, is the cell's immediately usable energy currency, and the hepatocyte needs it constantly to maintain membrane polarity, run transporters, secrete bile, synthesize proteins, and defend itself against the very chemistry alcohol metabolism is generating. So alcohol does not just intoxicate the liver chemically. It drains the energetic order that lets the liver remain a liver while under chemical attack.

Inflammatory injury is then layered onto both of these. Acetaldehyde adducts, reactive oxygen species, damaged lipids, injured mitochondria, and hepatocyte death all become signals that the immune system can read. Kupffer cells are activated. Cytokines rise. Gut-derived endotoxin becomes more important because alcohol also destabilizes the intestinal barrier and increases the amount of microbial burden reaching portal blood. The liver is therefore being hit from both sides. Inside the hepatocyte, alcohol metabolism is generating reactive intermediates, redox imbalance, and oxidative stress. From the gut side, more inflammatory traffic is arriving through the portal vein. The result is that the liver does not experience alcohol as one clean toxic event. It experiences alcohol as chemical reactivity, metabolic distortion, and immune activation converging inside the same tissue at the same time.

This is why alcohol-associated liver injury progresses so easily from fatty liver into steatohepatitis and fibrosis. The fat accumulation comes from the energetic and redox distortion. The hepatocyte injury comes from acetaldehyde, oxidative stress, and membrane damage. The inflammatory amplification comes from Kupffer cell signaling, cytokines, and gut-derived endotoxin. Once that inflammatory state persists, stellate cells are activated and begin laying down collagen. So alcohol injury is never only chemical, never only metabolic, and never only inflammatory.

The same molecule forces the liver to create reactive intermediates, lose energetic balance, and recruit immune damage in one continuous sequence.

Why Some Tissue Recovers and Some Does Not

Tissue recovers when the injury ends before the architecture that supports function has been rewritten. That is the real dividing line. Cells can die and still be replaced if blood can still reach the surviving parenchyma, if oxygen can still cross the exchange surface, if bile can still leave in the right direction, if the extracellular scaffold still tells new cells where they belong, and if the inflammatory signal falls soon enough that repair is allowed to finish. Recovery is therefore not just about whether cells are capable of dividing. It is about whether the organ still remembers how to organize those cells into usable tissue. A liver can lose hepatocytes and still recover if the sinusoidal bed remains open, if the space of Disse remains thin enough for exchange, if stellate cells return to restraint instead of staying activated, and if regeneration occurs inside preserved structure rather than inside a field of permanent alarm. In that setting, new cells inherit a map that still makes sense. Blood meets them correctly. Substrates reach them. Waste leaves. The organ becomes itself again because its underlying logic survived the insult.

Tissue does not recover when injury keeps teaching the organ the wrong shape. That is what chronic inflammation, repeated toxic burden, cholestatic retention, viral persistence, and unresolved metabolic congestion do. They do not merely kill cells. They alter the rules of replacement. Stellate cells stay activated and lay down matrix. Sinusoids capillarize and lose permeability. The distance between blood and hepatocyte grows. Portal resistance rises. Oxygen delivery worsens. Regenerating hepatocytes divide into a scarred field and inherit distorted flow, abnormal gradients, and impaired biliary architecture. The organ may still look active under the microscope because cells are proliferating, cytokines are rising, and wound-response machinery is engaged, but activity is not recovery. A cell can be alive and still be placed in the wrong geometry. A liver can be regenerating and still be moving farther away from normal function because the architecture guiding that regeneration has already been corrupted. This is why some tissue returns and some only remodels. The deciding factor is not effort. It is whether the conditions required for true restoration are still present when the repair begins.

Chapter 43

Metabolic Dysfunction and Steatotic Liver Disease

Insulin Resistance, Excess Substrate Delivery, and Hepatic Fat Storage

Metabolic dysfunction-associated steatotic liver disease, MASLD, begins as a traffic problem before it becomes an inflammatory or fibrotic one. The liver is always handling lipid, but it is supposed to do so under conditions where incoming fatty acids, new fat synthesis, oxidation, and export remain in working balance. Insulin resistance breaks that balance from several directions at once. In adipose tissue, resistance to insulin's antilipolytic effect means stored triglyceride is broken down more freely, so more nonesterified fatty acids are released into blood and delivered to the liver. At the same time, hyperinsulinemia and excess carbohydrate exposure, especially in metabolically overfed states, continue pushing hepatic de novo lipogenesis, the synthesis of new fatty acids from nonfat carbon. The result is a liver receiving more fat from outside while also being told to make more fat inside. MASLD is the name given when hepatic steatosis develops in the setting of metabolic dysfunction, not as an isolated accident of fat presence but as the liver's visible record of disordered substrate handling.

This is why insulin resistance and steatotic liver disease are so tightly linked. A healthy adipose compartment buffers nutrient excess by storing fat under insulin's instruction and restraining lipolysis when fasting release is not required. Once that restraint weakens, the liver becomes the overflow site for a storage problem that adipose tissue is no longer containing well. The hepatocyte then has only a few options. It can oxidize incoming fatty acids in mitochondria, export triglyceride in very-low-density lipoprotein, VLDL, or re-esterify fatty acids into intracellular triglyceride droplets. When incoming fatty acid traffic and new lipogenesis outrun oxidation and export, triglyceride begins accumulating in the cytoplasm. The droplet is therefore not a random deposit. It is what remains after the liver has been handed more substrate than it can clear in real time.

Excess substrate delivery is broader than fatty acid release alone. In metabolic dysfunction the liver is often exposed to repeated carbohydrate surplus, increased fructose handling, and a hormonal environment that keeps lipogenic programs active even when the organ is already carrying fat. Hepatic triglyceride content is normally governed by the balance between fatty acid appearance, oxidation, and export, and de novo lipogenesis is now understood as a major contributor to increased hepatic fatty acid appearance during MASLD development. This means the liver is not merely a passive recipient of peripheral fat spillover. It is an active manufacturer of part of its own lipid burden when nutrient signaling keeps driving carbon into fatty acid synthesis. The steatotic hepatocyte is therefore being squeezed from both sides. More substrate is arriving. More substrate is being made. The cell stores what it cannot immediately burn or send away.

Hepatic fat storage in this setting should be understood as an early adaptation that can become the opening move in a more serious disease pattern. Triglyceride droplets are, in part, a way of packaging lipid into a less immediately reactive form than free fatty acids. But that adaptation does not solve the underlying traffic disorder. If insulin resistance persists, if substrate oversupply continues, and if oxidation and export remain insufficient, the liver stays in a state of chronic lipid congestion. That is the platform from which steatohepatitis and fibrosis become easier to generate later. The first visible event is fat in the cell. The deeper event is failed metabolic governance.

Why Metabolic Overload Converges on the Liver

Metabolic overload converges on the liver because the liver is the organ the body uses to absorb excess before excess is allowed to become blood chemistry. That is its normal job. After a meal, carbohydrate absorbed from the intestine reaches the liver first through the portal vein. Fructose reaches it in especially liver-centered fashion. Amino acids reach it. Lipid remnants return to it. When adipose tissue is insulin-resistant and begins releasing more fatty acids than it should, those fatty acids reach the liver too. So the liver sits at the crossing point where surplus from the gut, surplus from adipose tissue, and surplus generated by disordered hormone signaling all arrive in the same cell field. It is not only a storage organ. It is the first organ forced to decide what to do with incoming abundance. That is why overload keeps landing there. The liver cannot step out of the stream.

This becomes more obvious in insulin resistance. Insulin is supposed to help adipose tissue store fat and restrain lipolysis, which is the breakdown of stored triglyceride into fatty acids and glycerol. In metabolic dysfunction, that restraint weakens, so more fatty acids spill into blood and are delivered to hepatocytes. But the liver is hit from the other side at the same time. Hyperinsulinemia, the chronically elevated insulin state that often accompanies insulin resistance, can still keep hepatic fat-making pathways active even while adipose tissue is leaking more fat into circulation. That means the liver is being told to receive more substrate and to manufacture more lipid from carbohydrate at the same time. The contradiction is only apparent. It is the real chemistry of metabolic disease. Adipose tissue stops holding the line. The liver becomes the overflow chamber.

The liver is also the place where excess carbohydrate becomes a lipid problem. A muscle cell can take up glucose and burn or store part of it. An adipocyte can store incoming energy more directly. The hepatocyte has a harder assignment. It has to buffer blood sugar for the whole body, keep glucose available between meals, and handle large waves of portal substrate without letting the circulation swing too violently. When carbohydrate intake and insulin signaling remain chronically high, the liver begins converting more of that carbon into fatty acids through de novo lipogenesis, which means new fat synthesis from nonfat substrate. Fructose is particularly effective at feeding this burden because much of it is taken up and handled in the liver early. So metabolic overload converges on the liver not only because fat arrives there, but because excess sugar can be turned into fat there.

Once enough substrate converges on the hepatocyte, the exits start mattering more than the arrivals. The cell can oxidize fatty acids in mitochondria, package triglyceride into very-low-density lipoprotein, VLDL, for export, or store triglyceride in intracellular droplets. If delivery and synthesis outpace oxidation and export, storage wins by default. This is how steatosis begins. Not because the liver chose fat, but because the liver ran out of clean ways to move the fat through. Mitochondria become more stressed. Endoplasmic reticulum burden rises. Lipid droplets enlarge. Oxidative stress becomes easier to generate. The cell that was supposed to normalize metabolic excess now begins carrying that excess in its own cytoplasm.

There is another reason overload converges here. The liver is not just a fuel organ. It is an endocrine, immune, and traffic organ at the same time. It reads insulin, glucagon, nutrient arrival, inflammatory cytokines, microbial signals from the gut, and fatty acid delivery from adipose tissue all in one integrated stream. A metabolic disturbance in almost any upstream compartment eventually becomes a hepatic burden because the liver is where those compartments are reconciled. Too much carbohydrate, too much adipose lipolysis, too much gut-derived inflammatory traffic, too much alcohol, too much fructose, too much sedentary surplus all become one biochemical argument inside the hepatocyte. Other tissues can contribute to metabolic dysfunction. The liver is where that dysfunction is forced to become organized or fail.

That is the deeper answer. Metabolic overload converges on the liver because the liver is the organ that must make excess compatible with continued life. It receives the first post-meal wave from the intestine, the spillover from failing adipose restraint, and the hormonal commands that decide whether substrate is stored, burned, or exported. When the incoming pressure stays high enough for long enough, the organ built to buffer the surplus becomes the organ that visibly records it.

Progression From Steatosis to Inflammatory Injury

Steatosis becomes inflammatory injury when fat storage stops functioning as a tolerable containment strategy and starts exposing the hepatocyte to the cost of holding too much lipid for too long. A triglyceride droplet inside the cell is, at first, partly protective. The hepatocyte is taking incoming fatty acids and esterifying them into a more neutral storage form because free fatty acids and certain lipid intermediates are more chemically disruptive. They insert into membranes, distort signaling, stress mitochondria, and make the cell harder to keep organized. Triglyceride storage buys time. It does not solve the overload. If excess substrate delivery, de novo lipogenesis, reduced oxidation, and limited export continue, the droplet enlarges, the cytoplasm becomes crowded, and the cell begins living in a state of metabolic congestion rather than short-term adaptation. The droplet was the compromise. Inflammation is the bill.

The transition begins inside the hepatocyte before the immune system fully declares itself. Mitochondria, the organelles responsible for most aerobic energy production, are forced to oxidize heavy fatty acid loads while already working in a redox environment distorted by nutrient excess and insulin resistance. Redox refers to the balance of electron transfer reactions that determines how cleanly the cell can make energy without spilling reactive chemistry into itself. As mitochondrial stress rises, beta-oxidation becomes less efficient, reactive oxygen species increase, and ATP production becomes harder to sustain. ATP, adenosine triphosphate, is the cell's immediately usable energy currency, and the hepatocyte spends it continuously to maintain membrane polarity, transporters, protein synthesis, and bile secretion. At the same time, the endoplasmic reticulum is trying to manage lipid synthesis, membrane turnover, and protein folding inside a cell whose internal space is increasingly occupied by fat droplets. The hepatocyte is still alive, but it is no longer quiet.

Once that burden persists, the chemistry changes character. It is no longer mainly neutral triglyceride sitting in storage. Lipotoxic intermediates begin accumulating. Lipotoxic means the injury is being driven by particular lipid species and their downstream effects rather than by fat mass alone. Diacylglycerols, ceramides, oxidized lipids, and other reactive lipid derivatives begin interfering with insulin signaling, damaging membranes, and amplifying oxidative stress. Membrane lipids are oxidized. Proteins are modified. Mitochondrial function worsens further. Calcium handling becomes less stable. The hepatocyte starts releasing danger signals, fragments of its own damaged machinery that tell neighboring immune and stromal cells the tissue is no longer merely overloaded. It is injured.

That is the moment the liver shifts from steatosis to steatohepatitic injury. Kupffer cells, the resident macrophages sitting in the sinusoids, detect the rising burden of damaged-cell signals, oxidized lipids, endotoxin arriving from the gut, and inflammatory mediators generated inside the tissue. They respond by releasing cytokines, signaling proteins that amplify local inflammation and recruit broader defense pathways. The sinusoidal environment changes. Hepatocytes begin swelling and losing structural clarity, the ballooned cell of steatohepatitis. Ballooning means the cell has become enlarged and cytoskeletally unstable under metabolic and oxidative strain. Some hepatocytes die by apoptosis, which is controlled cell death, and others by necrotic injury when the damage is too severe and too disorganized for controlled exit. The liver is no longer storing excess quietly. It is now reacting to the consequences of that storage as a wound.

The gut helps push this transition forward. In metabolic dysfunction, intestinal permeability and microbial composition often change in ways that increase delivery of endotoxin and other inflammatory microbial products into portal blood. The liver therefore is not only handling lipid injury generated inside hepatocytes. It is doing so while receiving a dirtier portal stream from upstream. Kupffer cells read both messages at once. The hepatocyte is stressed from within, and the sinusoid is carrying more immune burden from the gut. This combination makes inflammatory progression much easier to sustain. The liver loses the ability to keep metabolic congestion local.

Once inflammation holds, fibrosis becomes easier to start. Stellate cells in the space of Disse stop behaving like quiet vitamin A reserve cells and begin producing collagen. The exchange surface thickens. Sinusoidal flow becomes less open. Regeneration begins to happen inside a field already being remodeled in the wrong direction. That is why progression from steatosis to inflammatory injury matters so much. The fat itself was the first sign that throughput had failed. Inflammation is the stage where failed throughput starts rewriting the tissue future.

Why This Disease Is Tied to Whole-Body Fuel Handling

A hepatocyte does not become fat because it independently decided to store more triglyceride. It becomes fat because the body's larger fuel-partitioning system has begun sending the wrong substrates, in the wrong amounts, under the wrong hormonal instructions. In healthy physiology, the work is divided. The intestine delivers absorbed fuel in pulses. Skeletal muscle takes up and burns a large share of glucose or stores it as glycogen. Adipose tissue takes surplus energy and locks it away as triglyceride. The liver buffers the first post-meal wave, stores some carbohydrate, packages some lipid, and keeps blood chemistry stable without becoming the long-term depot. In steatotic liver disease, that division of labor breaks. The liver becomes the organ forced to absorb what the rest of the system is no longer handling cleanly.

Adipose tissue is one of the first places this failure becomes visible. Insulin is supposed to restrain lipolysis, the breakdown of stored triglyceride into fatty acids and glycerol. In insulin-resistant adipose tissue, that restraint weakens. Fat cells keep releasing fatty acids into blood even when the body is not in a true fasting emergency. Those fatty acids travel to the liver bound to albumin, and the hepatocyte has to decide whether to oxidize them, export them, or store them. If delivery remains high, storage wins more often. This is one reason the disease belongs to whole-body fuel handling rather than to liver fat alone. The liver is carrying lipid traffic that adipose tissue failed to keep contained.

Muscle matters just as much. In insulin-resistant skeletal muscle, glucose uptake becomes less efficient, which means more carbohydrate remains in circulation and the pancreas often has to maintain higher insulin output to keep blood glucose under control. The liver is then exposed to two signals at once. More fatty acids are arriving from adipose tissue, and more carbohydrate pressure is arriving from the gut and bloodstream. Persistent insulin exposure, especially in this overfed setting, continues pushing hepatic de novo lipogenesis, the synthesis of new fat from nonfat carbon. The liver is therefore not only receiving excess lipid from outside. It is being instructed to manufacture more lipid from surplus substrate inside. This is the central metabolic contradiction of the disease. Too much is arriving, and too much is being made.

The liver itself then develops a particularly damaging kind of insulin resistance. In many people with steatotic liver disease, the organ becomes selectively resistant. It does a worse job suppressing glucose output, yet lipogenic pathways can remain relatively responsive enough to persistent insulin and substrate load that fat synthesis continues. In plain language, the liver can keep acting partly as if the body needs more glucose while also continuing to make fat. It is being pulled in opposite directions and doing both badly. This is one of the clearest demonstrations that the disease is systemic. A normal liver does not choose that state. It is driven into it by a whole-body fuel economy that has lost coordination between feeding, storage, oxidation, and release.

The gut is part of this too. Fructose and excess carbohydrate reach the liver directly through portal blood, and the liver is the first large organ forced to process that incoming carbon load. If meals repeatedly deliver more substrate than muscle can burn and adipose tissue can buffer safely, the hepatocyte becomes the place where excess carbon is converted into triglyceride. That triglyceride may then be exported in very-low-density lipoprotein, VLDL, or kept in intracellular droplets if oxidation and export cannot keep pace. This is why blood lipids often look abnormal in parallel with hepatic steatosis. The liver can be overexporting fat to the circulation and still remain fat-loaded itself because total substrate pressure is greater than both exits. The blood and the liver are showing the same disorder in two different compartments.

So this disease is tied to whole-body fuel handling because the fat droplet in the hepatocyte is only the local storage form of a body-wide traffic failure. Adipose tissue is leaking. Muscle is underclearing. The pancreas is signaling harder. The intestine is delivering repeated substrate excess. The liver is being asked to buffer all of it and eventually starts storing what it can no longer organize. The liver holds the triglyceride. The whole body wrote the route that put it there.

The Liver as the Organ That Exposes Prolonged Dietary and Metabolic Mismatch

The liver exposes prolonged dietary and metabolic mismatch because it sits where mismatch stops being abstract and becomes tissue. A bad pattern upstream can stay invisible for a while in appetite, habit, or blood sugar drift, but once that pattern keeps sending the wrong substrate mix, the wrong hormonal signals, and the wrong timing into portal blood and systemic circulation day after day, the liver is the organ forced to absorb the contradiction physically. It is receiving carbohydrate excess while already carrying fat. It is receiving fatty acids from adipose tissue that should have stayed stored. It is still being told by insulin-rich signaling to convert surplus carbon into new lipid even while oxidation and export are falling behind. That is why the liver becomes such a clear witness to long-term mismatch. It is not morally judging the diet. It is recording what the body could not successfully partition, burn, store, or release under the conditions it was given.

This matters because the liver is the first major organ asked to reconcile opposing metabolic instructions. In a healthier system, meals arrive, glucose is buffered, glycogen is stored, adipose tissue restrains lipolysis when feeding is abundant, and the liver can process incoming substrate without becoming the storage site for the failure of everything else. In prolonged mismatch, that order breaks. The intestine keeps delivering more digestible energy than the body is using. Adipose tissue becomes less responsive to insulin's restraint and releases more fatty acids into blood than it should. Muscle underclears glucose. Hyperinsulinemia persists. The liver is then forced to stand in the middle of a body that is simultaneously overfed and metabolically disorganized. It receives the excess and the confusion together. This is why the steatotic liver is so informative. It shows that the organism has not only eaten too much in some generic sense. It has failed to distribute fuel according to the logic of its own tissues.

The hepatocyte makes that mismatch visible because it can only solve the overload for so long by storage. It turns fatty acids into triglyceride to reduce immediate lipotoxicity. It packages some into VLDL for export. It oxidizes what it can in mitochondria. But when intake pattern, hormonal state, and substrate delivery keep pushing in the wrong direction, the exits stop matching the arrivals. Fat begins accumulating inside the cell. Mitochondria work under greater strain. Redox balance worsens. Endoplasmic reticulum stress rises. The cell is no longer simply doing liver work. It is carrying the metabolic history of the whole organism in its cytoplasm. The liver therefore exposes mismatch not because it is unusually fragile, but because it is unusually central. It is where prolonged excess, poor fuel partitioning, and failed adipose restraint are forced to become structural reality.

That is the deeper reason liver disease so often reveals what the rest of metabolism has been doing quietly for years. The liver is the organ that cannot fake harmony forever. If fuel handling is chronically wrong, the bloodstream may fluctuate, adipose tissue may enlarge, insulin may rise, muscle may become resistant, but the liver is one of the first places where the cumulative contradiction becomes visible as altered architecture. A fat-loaded liver is not just a local problem in a local organ. It is the body's long nutritional and hormonal argument written into tissue.

Why the Disease Is Common but Still Poorly Understood by the Public

This disease is common because the metabolic conditions that feed it are common. Excess digestible substrate, frequent feeding, poor adipose restraint, insulin resistance, altered muscle glucose handling, fructose-heavy intake, low movement, and chronic energy surplus all converge on the same organ. The liver receives the first wave from the intestine and the spillover from the rest of the body. It is therefore exposed to the whole argument long before the public sees a symptom dramatic enough to take seriously. Fat begins accumulating quietly. Redox balance shifts. Mitochondria work harder under worse conditions. Export and oxidation begin losing the race against delivery and synthesis. None of this has to hurt at first. The organ can keep functioning while carrying the burden. That is one reason the disease spreads so easily. The body can keep negotiating with it for years before the person realizes a negotiation is happening.

It is poorly understood because the public is usually taught to think in labels rather than in traffic. People are told the liver is fatty, but not why the fat stayed there. They are told the disease is about obesity, but not why a liver can become steatotic in a person who is not outwardly massive or why two people with similar body size can carry very different hepatic burdens. They are told it is a liver condition, which is true, but that phrasing hides the deeper point that the liver is often exposing a whole-body failure of fuel partitioning. Adipose tissue is leaking. Muscle is underclearing. Insulin is signaling in a distorted way. The intestine keeps delivering surplus. The liver is where that systemic mismatch becomes visible. The public sees a liver image with fat in it. The real disease is a broken fuel map.

It is also poorly understood because early disease is metabolically loud and symptomatically quiet. The organ can enlarge, store triglyceride, alter blood chemistry, and begin losing clean control of lipid and glucose handling without producing the kind of pain or immediate crisis that forces attention. The person may feel vaguely tired, or nothing at all. A disease that builds through silent congestion is always easier to underestimate than one that announces itself with fever, bleeding, or acute collapse. By the time inflammation, fibrosis, or cirrhotic architecture appear, the public often experiences the disease as if it arrived suddenly. It did not. The liver buffered the mismatch until it could not.

There is another reason the understanding stays shallow. Public language usually swings between two bad explanations. One is moral, too much food, too little discipline. The other is generic, inflammation, unhealthy lifestyle, poor metabolic health. Neither one shows the tissue event. The useful question is not whether the person was virtuous or careless. The useful question is what the hepatocyte was being asked to do every day and why that request became impossible to satisfy cleanly. More fat arrived than could be burned. More fat was made than should have been. Less was exported. The cell stored what it could not move, then began paying for that storage with stress, inflammation, and eventually scar. The disease is common because the modern substrate environment keeps creating that sequence. It is poorly understood because the public is usually given slogans where mechanism should have been.

Chapter 44

Autoimmune and Cholestatic Disorders

Autoimmune Hepatitis

Autoimmune hepatitis is hepatitis driven by the immune system treating hepatocytes as targets rather than tolerating them as self. That is the core event. The liver becomes inflamed not because a virus is replicating inside it or because bile is mechanically obstructed, but because the adaptive immune system is sustaining a misdirected attack against liver tissue. The histologic pattern classically centers on interface hepatitis, inflammation and hepatocyte injury at the boundary between portal tracts and surrounding parenchyma, often with a plasma-cell-rich infiltrate and evidence of hepatocyte destruction extending inward from the portal margin. This is why autoimmune hepatitis should be understood as a structural immune disease rather than as a mysterious elevated-liver-enzyme syndrome. The immune system is damaging the liver at the tissue edge where portal inflammatory activity meets the working parenchyma, and if that attack persists, fibrosis and cirrhosis can follow just as surely as they do in chronic viral or metabolic injury.

What makes the disease so important mechanistically is that the liver is normally built for tolerance. It sits downstream from the gut and sees constant antigenic and microbial traffic without defaulting into destructive immunity. Autoimmune hepatitis represents a failure of that restraint directed inward. The exact trigger is not fully defined, but the result is clearer than the cause. T-cell-mediated immune injury, autoantibody associations, cytokine signaling, and ongoing hepatocyte destruction produce a chronic inflammatory environment that keeps telling the liver it is under attack. Hepatocytes leak aminotransferases into blood. Immunoglobulin G, or IgG, is often elevated. Portal inflammation deepens. If left untreated, the repeated immune attack keeps activating stellate cells and feeding matrix deposition, so the disease can move from inflammation into fibrosis and cirrhosis. This is why autoimmune hepatitis is not just "autoimmunity in the liver." It is chronic immune-driven remodeling pressure on hepatic architecture.

Clinically the disease can be protean, ranging from asymptomatic transaminase elevation to acute hepatitis and even fulminant liver failure, which is one reason it is often missed or misunderstood. But the structural logic remains the same across presentations. The liver is inflamed because self-directed immunity is being sustained against the parenchyma. The longer that inflammatory conversation persists, the more likely the tissue future shifts toward scar. That is the danger. Autoimmune hepatitis is not just a liver enzyme abnormality with antibodies attached. It is a chronic wound-response state created by the immune system itself, and the liver pays for that misrecognition by slowly trading normal architecture for fibrosis if the attack is not brought under control.

Primary Biliary Cholangitis

Primary biliary cholangitis is a chronic autoimmune cholestatic disease in which the immune system targets the small intrahepatic bile ducts, the tiny ducts inside the liver that receive bile after hepatocytes have secreted it into canaliculi. That location is the whole key. The disease is not beginning as a generalized attack on hepatocytes in the same way autoimmune hepatitis does, and it is not beginning as a mechanical obstruction of the large extrahepatic ducts. It is beginning as progressive destruction of the small biliary channels embedded inside the liver itself. Once those channels are lost, bile can no longer leave the parenchyma cleanly. Cholestasis develops. Bile acids and other biliary constituents begin being retained in the wrong compartments. The liver is then injured not only by immune attack, but by the chemistry of its own trapped outgoing stream.

This is why primary biliary cholangitis has to be read structurally as a small-duct loss disease. The earliest injury is centered on the bile ductules in portal tracts. Portal inflammation develops, bile duct epithelial cells are damaged and gradually disappear, and the ductular network that was supposed to carry bile forward becomes less complete. The liver can still make bile. Hepatocytes can still conjugate bilirubin, synthesize bile acids, and pump biliary contents into canaliculi. But the route out is being erased one duct at a time. The consequence is cholestatic pressure inside the tissue. Bile accumulates where it should have moved. Over time this drives periportal fibrosis and eventually cirrhosis if the process is not controlled. So the disease is not merely autoimmune in some broad blood-test sense. It is autoimmune destruction of the organ's internal biliary outflow architecture.

That duct-centered injury also explains the symptom logic. Fatigue and itching are common because cholestasis is developing before massive hepatocyte collapse has to occur. Fat-soluble vitamin handling can eventually suffer. Cholestatic enzyme patterns, especially alkaline phosphatase elevation, become part of the laboratory signature because bile flow is impaired at the duct level. The liver is still present, and much hepatocyte work may remain intact for a long time, but the patient is living in a body whose bile stream is increasingly interrupted inside the organ itself. This is why primary biliary cholangitis should not be thought of as a vague inflammatory liver disease. It is a disease of progressive intrahepatic bile duct disappearance, and the fibrosis that follows is the liver trying to survive that loss in the wrong language, by scar.

Primary Sclerosing Cholangitis

Primary sclerosing cholangitis, or PSC, is a chronic cholestatic disease in which the bile ducts inside and outside the liver become inflamed, scarred, narrowed, and eventually obstructed. That location is what defines the disease structurally. Primary biliary cholangitis destroys the small intrahepatic bile ducts. PSC is broader and more architectural. It involves the larger ductal tree as well, including ducts outside the liver, so the organ is injured not only by inflammation but by progressive distortion of its whole drainage system. Bile is still being made, but the channels meant to carry it forward are becoming rigid, irregular, and obstructed. The liver is therefore damaged by its own trapped outflow as much as by the inflammatory process itself.

This is why PSC should be understood as a fibrosing duct disease before it is understood as a liver enzyme pattern. The bile ducts become scarred and narrowed, bile backs up, and cholestatic injury begins spreading through the parenchyma downstream of that failed drainage. Over time this can lead to cirrhosis and liver failure. The disease often progresses slowly enough that many people have no symptoms at diagnosis, but the tissue is already moving in the wrong direction. The liver is being reorganized around obstructed bile flow. Itching, jaundice, abdominal pain, and cholangitic episodes make sense later because the ductal architecture has already been quietly hardening and narrowing for years.

PSC also sits in a different biologic neighborhood from many other liver diseases because it is strongly linked to inflammatory bowel disease, especially ulcerative colitis. That connection matters for structure as much as for epidemiology. The gut and biliary tree appear to be participating in the same broader immune and inflammatory story, which is one reason PSC belongs naturally inside the gut-liver axis rather than only inside hepatology. The exact cause remains uncertain, but immune dysregulation, genetic susceptibility, gut microbial factors, and bile-duct injury all appear to play roles. The disease is therefore not just a plumbing problem and not just a simple autoimmune label. It is chronic biliary inflammation feeding fibrosis in a ductal network that the liver cannot function without.

So the cleanest way to see PSC is this. It is a chronic inflammatory scarring disease of the bile ducts in which the liver keeps making bile while the duct system becomes progressively less able to carry it. The organ is then injured by retained biliary chemistry, recurrent obstruction, and the long structural consequences of chronic cholestasis. It is the disease of a drainage tree turning rigid inside an organ that survives by flow.

Why the Immune System Can Target Either Hepatocytes or Bile Ducts

The immune system can target either hepatocytes or bile ducts because the liver is not one uniform tissue. It is a composite organ made of different cellular compartments doing different jobs and presenting different molecular surfaces to the immune system. Hepatocytes are the main parenchymal cells, the high-volume metabolic workers handling portal blood, making bile acids, synthesizing plasma proteins, and transforming hormones and xenobiotics. Cholangiocytes, the epithelial cells lining the bile ducts, are different cells with different membrane proteins, different antigenic exposure, and a different structural role. If immune tolerance fails, the attack can be directed mainly against the hepatocyte field or mainly against the biliary tree, and the tissue consequence changes accordingly. Autoimmune hepatitis is the classic hepatocyte-targeted pattern, while primary biliary cholangitis and primary sclerosing cholangitis are bile-duct-centered diseases, with primary biliary cholangitis focused on small intrahepatic ducts and primary sclerosing cholangitis affecting larger intrahepatic and extrahepatic ducts.

That difference in target changes the whole shape of the disease. When the immune attack is directed mainly at hepatocytes, the dominant tissue picture is hepatitis: interface activity, hepatocyte injury, enzyme leakage into blood, and over time fibrosis driven by repeated parenchymal loss and regeneration under inflammatory pressure. When the immune system targets the bile ducts instead, the first major problem is cholestatic rather than purely hepatocellular. The liver may still make bile, but the ducts that are supposed to carry it become inflamed, narrowed, or destroyed. Bile then begins to stagnate, and retained bile acids and bilirubin create a second layer of injury downstream of the immune attack itself. In other words, targeting hepatocytes injures the workers. Targeting bile ducts injures the drainage system and then forces the liver to bathe in the chemistry it failed to move out.

This is why autoimmune liver disease is better understood as compartment-selective immune misrecognition rather than as one generic liver-directed error. The immune system is not merely "attacking the liver." It is attacking a particular structural component of the liver, and that component determines what kind of failure follows. Hepatocyte-targeted injury leads first toward hepatitis and parenchymal damage. Small-duct injury leads toward chronic cholestasis and duct loss. Large-duct scarring leads toward obstructed flow, biliary fibrosis, and secondary parenchymal injury from retained bile. The organ is the same. The attacked compartment changes the future.

How Biliary Disease Creates a Different Pattern From Hepatocellular Disease

Biliary disease and hepatocellular disease do not damage the liver in the same compartment, so they do not produce the same physiological pattern even when both eventually end in fibrosis, cirrhosis, and liver failure. Hepatocellular disease begins in the main working cell. The hepatocyte is injured directly, whether by virus, alcohol, steatotic stress, immune attack, ischemia, or toxin burden. When that cell is damaged, the dominant early consequences come from failed parenchymal work. Aminotransferases rise because injured hepatocytes leak intracellular enzymes. Glucose handling, ammonia conversion, albumin synthesis, clotting factor production, bilirubin uptake and conjugation, and xenobiotic metabolism begin to suffer because the cell that performs those jobs is under attack. The tissue picture is one of hepatocyte stress, inflammation, cell death, and compensatory regeneration. Blood chemistry changes because the worker is failing.

Biliary disease begins somewhere else. The hepatocyte may still be capable of making bile, but the channels meant to carry that bile forward are inflamed, narrowed, destroyed, or obstructed. The first problem is therefore not loss of synthetic labor in the same way. The first problem is outflow failure. Bile acids, conjugated bilirubin, cholesterol, and other biliary constituents are still being pushed into a drainage system that is becoming less able to move them onward. The liver is now injured by retention. Cholestasis develops. Biliary pressure rises. Ductal epithelium is damaged. Bile begins accumulating in compartments where it does not belong. The same bile acids that are useful in the intestinal lumen become detergent stress inside the liver. Conjugated bilirubin refluxes into blood. Fat digestion worsens because bile acids are no longer reaching the intestine properly. So biliary disease creates a combined excretory and digestive pattern from the beginning. The worker may still be present. The drainage system is failing.

That difference is why the laboratory and symptom profile often look different. Hepatocellular disease usually declares itself first through hepatocyte leakage and parenchymal dysfunction, which is why alanine aminotransferase and aspartate aminotransferase often rise more prominently early in that pattern. Biliary disease tends to produce a cholestatic pattern, with alkaline phosphatase and gamma-glutamyl transferase rising more strongly as the duct system becomes stressed or obstructed. The symptom logic follows the same structural difference. Hepatocellular disease may first look like fatigue, malaise, impaired synthetic function, inflammation, and later jaundice if bilirubin handling deteriorates. Biliary disease more readily produces dark urine, pale stool, itching, and fat malabsorption early in the course because the problem is already centered on failed bile flow. One disease is injuring the maker. The other is injuring the route out.

The fibrosis pattern can differ for the same reason. Hepatocellular disease feeds fibrosis through repeated parenchymal death, inflammatory signaling, and stellate-cell activation in response to hepatocyte injury. Biliary disease feeds fibrosis through chronic cholestatic injury, duct loss or duct narrowing, retained bile, and the inflammatory environment that develops around a liver trying to secrete into a blocked or damaged outflow system. In one case the scar follows repeated damage to the working cell. In the other it follows repeated damage to the drainage architecture and the chemical toxicity of trapped bile. Both can end in cirrhosis, but the road there is not the same.

So biliary disease creates a different pattern from hepatocellular disease because it begins with failed flow rather than failed labor. Hepatocellular disease injures the cell that performs the liver's main metabolic and synthetic work. Biliary disease injures the channels that carry the liver's outgoing stream. One primarily disrupts production and processing first. The other primarily disrupts excretion and delivery first. The liver may eventually fail in both cases, but it does not fail by the same route.

Why Symptom Patterns Follow the Injured Compartment

Symptoms follow the injured compartment because the liver is not one undifferentiated mass performing one generic job. It is a composite organ made of compartments with different traffic, different surfaces, and different responsibilities. Hepatocytes govern blood chemistry, nutrient conversion, bile formation, plasma protein synthesis, and detoxification. Bile ducts carry the outgoing stream. Sinusoids handle exchange between portal blood and hepatocytes. The portal venous system manages inflow. The hepatic veins manage outflow. Stellate cells and matrix determine whether exchange remains soft and open or becomes scarred and resistant. When disease strikes one compartment more than another, the first failures appear in the functions that compartment was built to carry. The symptom pattern is therefore not random decoration on top of liver disease. It is the body revealing which piece of the architecture failed first.

If hepatocytes are the main injured compartment, the symptoms and laboratory signals tend to reflect failed cellular work. Aminotransferases rise because damaged hepatocytes leak enzymes. Fatigue, malaise, impaired glucose handling, jaundice from failed bilirubin processing, coagulopathy from reduced clotting factor synthesis, and encephalopathy from poor ammonia control all belong to that kind of injury pattern. The worker is failing, so the outputs of the worker begin to fail. If the bile ducts are the main injured compartment, the pattern shifts. Now the liver may still be making bile, but it cannot move that bile properly. Conjugated bilirubin backs up into blood and darkens the urine. Less pigment reaches the gut, so stool pales. Retained bile acids and related biliary compounds contribute to itching. Fat digestion weakens because bile acids are no longer reaching the intestine in adequate quantity. The route out is failing, so symptoms cluster around retention and malabsorption rather than around primary parenchymal collapse.

The same logic applies to vascular compartments. If portal inflow is the main mechanical problem because cirrhosis has made the liver too resistant to cross, the symptom pattern shifts toward portal hypertension. Splenomegaly appears because the spleen is draining into a congested system. Varices form because blood is being rerouted through collateral veins. Ascites develops because hydrostatic pressure in the splanchnic circulation rises and fluid is pushed outward while albumin support is falling. If hepatic outflow is impaired, congestion dominates differently. If sinusoidal exchange is the main compartment being buried under fibrosis, the person may begin showing diffuse decline in multiple liver functions because blood can no longer meet hepatocytes cleanly even though both blood and hepatocytes are still present in the organ. The symptom pattern follows the traffic failure.

This is why good liver reasoning begins with asking where the problem sits physically. Is the injured compartment the hepatocyte. The bile duct. The sinusoid. The portal circulation. The outflow tract. The matrix around the exchange surface. The answer changes what the patient looks like, what the labs show, and how the disease will progress. The liver does not generate one symptom language for all injury. It speaks through the compartment that was hit. When the worker fails, the outputs fail. When the duct fails, bile is trapped. When flow fails, pressure rises and blood reroutes. When the exchange surface fails, everything grows less efficient at once. Symptom patterns follow the injured compartment because function was always compartmentalized before disease ever arrived.

Chapter 45

Genetic and Storage Disorders

Hemochromatosis

Hemochromatosis is not a disease of "too much iron" in the vague sense. It is a disease of failed iron restraint. The body has no efficient regulated way to dump excess iron once it has been absorbed, so long-term iron balance depends mainly on controlling entry from the intestine and release from storage and recycling cells. In hereditary hemochromatosis, that control is weakened. The usual result is inappropriately increased intestinal iron absorption and progressive iron loading of tissues, especially the liver first, then the pancreas, heart, joints, skin, and endocrine organs. The liver matters most because it is both a major iron storage site and the organ that makes hepcidin, the peptide hormone that normally restrains iron export into plasma by acting on ferroportin. When hepcidin is too low or the ferroportin-hepcidin relationship is otherwise disrupted, iron keeps moving when it should have been stopped.

That is why hemochromatosis is best understood as a traffic disorder rather than a simple storage disorder. Iron enters the enterocyte, leaves through ferroportin, binds transferrin in plasma, and is delivered to tissues. In the healthy body, hepcidin rises when iron stores are sufficient and shuts part of that traffic down. In hemochromatosis, the brake is too weak, so iron continues entering and circulating beyond what tissues can safely hold. The liver becomes one of the first major organs to absorb the consequence because hepatocytes readily store iron, but storage here is not harmless reserve. Iron is chemically useful and chemically dangerous at the same time. Once it accumulates beyond controlled ferritin storage, it drives oxidative injury, membrane damage, mitochondrial stress, fibrotic signaling, and eventually cirrhosis risk. The disease is therefore not that iron exists in the body. The disease is that the body keeps admitting iron after safe storage has already been exceeded.

The classic hereditary form is most often linked to HFE mutations, especially C282Y, and Merck notes that hereditary hemochromatosis is among the more common genetic disorders in people of northern European ancestry. But the structural logic matters more than the letter code. Different inherited forms affect different parts of the hepcidin-ferroportin system. Some reduce hepcidin production. Some alter proteins involved in iron sensing. Some affect ferroportin itself. The common endpoint is the same: too much iron is allowed into plasma or kept moving through plasma for too long, and tissue storage becomes tissue injury. This is why the disease can remain quiet for years. Iron accumulates slowly. The liver buffers it first. Symptoms often appear only after enough storage has been converted into organ damage. By the time fatigue, arthropathy, hypogonadism, diabetes, cardiomyopathy, skin pigmentation, or cirrhosis become obvious, the liver has usually been recording the overload silently for a long time.

So hemochromatosis belongs in a chapter on genetic and storage disorders because it is both. It is genetic failure of iron restraint, and it is storage turned toxic. The liver is the central witness to that failure. It makes the hormone that should have limited iron traffic, it stores much of the metal that should not have kept arriving, and it becomes one of the first organs scarred by the chemistry of excess. Hemochromatosis is not merely iron overload. It is the body losing the right to say no to iron soon enough.

Wilson Disease

Wilson disease is a genetic failure of copper handling in which the body keeps taking copper in but loses the ability to move enough of it safely out through bile. The liver is the first major organ forced to absorb that mistake because copper absorbed from the intestine reaches hepatocytes through portal blood, and hepatocytes are the cells normally responsible for packaging copper into ceruloplasmin and secreting excess copper into the biliary stream for elimination. In Wilson disease, that export step fails. Copper begins accumulating inside the liver first, then later in the brain, eyes, and other organs as storage capacity is exceeded and the metal starts escaping controlled handling. That is why this disease belongs in a chapter on genetic and storage disorders. It is not simply "high copper." It is a defect in the liver's authority over where copper is allowed to go and how it is allowed to leave.

Copper is useful and dangerous in the same atom. It is required for enzyme systems involved in energy production, antioxidant defense, connective tissue formation, and iron metabolism. But free copper is chemically reactive. If it is not bound tightly and trafficked correctly, it begins damaging proteins, membranes, and mitochondria. The healthy liver solves this by buffering copper inside hepatocytes, loading some onto ceruloplasmin for controlled circulation, and pushing excess copper into bile. Wilson disease interrupts that discipline. Copper keeps entering the hepatocyte, but too little leaves through the biliary route. The liver first looks like a storage organ under strain. Later it becomes an injured organ because stored copper is no longer quiet reserve. It becomes oxidative burden. Over time that burden can drive hepatitis, fibrosis, cirrhosis, acute liver failure, and extrahepatic injury, especially in the nervous system.

This is why Wilson disease can look like a liver disease, a neurologic disease, or both, depending on how long the storage problem has been allowed to progress. Early on, the liver often carries the burden silently because it is the first reservoir. Later, once copper handling fails badly enough, the disease becomes more systemic. Kayser-Fleischer rings, neurologic symptoms, and psychiatric changes make sense only after the liver has already lost enough control that copper is no longer being contained locally. The deeper logic never changes. The intestine keeps delivering copper. The liver is supposed to decide whether that copper stays safely bound, enters the bloodstream in controlled form, or leaves through bile. In Wilson disease, the exit route is broken, and the organ meant to govern copper traffic becomes the first place that traffic turns toxic.

Alpha-1 Antitrypsin Deficiency

Alpha-1 antitrypsin deficiency is a genetic liver disease built around retention, not simply absence. That distinction is what makes it unusual. Alpha-1 antitrypsin, often shortened to AAT, is a serine protease inhibitor made mainly by hepatocytes and normally secreted into blood, where it helps protect the lung from neutrophil elastase. In the deficient state, especially with severe disease-causing variants, the problem is not only that there is too little functional AAT reaching the circulation. The mutated protein misfolds inside the hepatocyte, polymerizes, and becomes trapped in the endoplasmic reticulum instead of being secreted normally. The lung suffers from too little circulating antiprotease. The liver suffers because the cell that made the protein is forced to store the malformed product it could not export. This is why the disease injures two organs in opposite ways at once, deficiency in the blood and toxic retention in the liver.

From a structural point of view, the liver lesion begins as a protein-trafficking disorder inside the hepatocyte. The cell is still synthesizing AAT, but the protein is not folding into a form that can pass cleanly through the secretory system. It accumulates, and that accumulation produces endoplasmic reticulum stress, proteotoxic burden, and eventually hepatocyte injury. Proteotoxic means the cell is being damaged by misfolded or aggregated protein rather than by a classic external toxin. The liver is therefore not injured because it lacks alpha-1 antitrypsin locally. It is injured because it cannot get rid of the abnormal alpha-1 antitrypsin it keeps producing. This is why alpha-1 antitrypsin deficiency belongs naturally among the genetic storage and trafficking disorders. The retained molecule is the disease mechanism.

The tissue future then follows the familiar liver logic. Injured hepatocytes trigger inflammation, stellate-cell activation, extracellular matrix deposition, and over time fibrosis and cirrhosis in some patients. NIDDK identifies alpha-1 antitrypsin deficiency as the most common genetic cause of liver disease in children and an important cause of chronic liver disease and liver cancer risk in adults. That progression makes sense once the mechanism is clear. A hepatocyte forced to live with retained misfolded protein is not just metabolically burdened for one day. It is under chronic intracellular stress, and chronic intracellular stress is exactly the kind of condition that teaches the liver to repair with scar.

So alpha-1 antitrypsin deficiency should be understood as a disease in which the liver is harmed by what it cannot secrete. The protein is made. The protein folds badly. The protein stays. The cell pays. The lung reveals the deficiency. The liver reveals the retention.

Glycogen Storage and Urea Cycle Disorders as Teaching Examples

These disorders matter because they turn invisible liver physiology into something the reader can finally see. In ordinary health, the liver stores glycogen quietly, breaks it back down quietly, converts ammonia into urea quietly, and the body moves through the day without ever noticing how much chemistry was being held together in the background. Glycogen storage disorders and urea cycle disorders strip that quiet away. They are inherited defects in specific metabolic pathways, but they should not be read only as rare pediatric syndromes or exam questions. They are teaching models. They show, with unusual clarity, what the liver is actually doing when everything appears normal. If glycogen cannot be built properly, mobilized properly, or trimmed properly, blood sugar control changes, hepatocytes swell with stored material, and the liver begins revealing that glycogen was never just backup fuel. It was a timing device for the whole organism. If ammonia cannot be converted cleanly into urea, the brain begins paying for a defect that started in hepatic nitrogen handling. What looked like one enzyme problem becomes proof that the liver normally protects consciousness itself.

Glycogen storage disorders are especially useful because they expose the liver's role in fuel timing rather than in crude calorie storage. A healthy liver takes incoming glucose after a meal, converts part of it into glycogen, then opens that reserve when the intestine goes quiet and the blood still needs glucose. That sequence feels simple until one of the enzymes inside it breaks. Depending on the specific defect, the liver may store glycogen but fail to release glucose efficiently, or may accumulate abnormal glycogen structure that enlarges and distorts hepatocytes while still leaving the bloodstream unstable between meals. The result can be fasting hypoglycemia, hepatomegaly, altered lactate handling, hyperuricemia, hyperlipidemia, or progressive structural liver burden. These findings are not random. They reveal that glycogen was doing two jobs at once. It was storing carbohydrate inside the hepatocyte and stabilizing blood chemistry outside the hepatocyte. The child with a glycogen storage disorder is therefore teaching a liver principle the healthy adult rarely sees. Hepatic glycogen is not inert reserve. It is controlled delay between intake and need.

Urea cycle disorders do the same thing for nitrogen. The body is always producing ammonia through amino acid turnover, intestinal microbial activity, and the broader chemistry of protein metabolism. In health, hepatocytes capture that burden and move it through the urea cycle so nitrogen can leave the body in a safer, water-soluble form. If one of the key enzymes in that pathway is absent or impaired, ammonia rises because the liver can no longer perform one of its most life-protective conversions. The result may look neurologic, irritability, vomiting, lethargy, seizures, coma, but the origin is hepatic chemistry. The brain is reacting to the liver's failure to neutralize portal and systemic nitrogen burden. This is why urea cycle disorders are such good teaching examples. They show that the liver is not merely one organ helping with protein metabolism. It is the organ standing between ordinary nitrogen turnover and toxic brain exposure. Remove one step in that sequence and the nervous system reveals how necessary the hepatic step always was.

These disorders also teach something deeper about storage and throughput. In glycogen disease, excess substrate may remain trapped in the hepatocyte because release is defective. In urea cycle disease, toxic burden may remain in circulation because conversion is defective. One disorder traps useful fuel in the wrong compartment. The other fails to trap dangerous nitrogen before it escapes. The liver is therefore doing more than storing and more than clearing. It is assigning molecules to the correct fate. Glucose must sometimes stay. Ammonia must not. A broken enzyme exposes the traffic rule that enzyme had been enforcing all along. This is one reason rare inherited disorders are so valuable conceptually. They make normal physiology visible by removing one quiet permission the liver usually grants.

So glycogen storage disorders and urea cycle disorders should be read as teaching examples of hepatic authority. The first shows what happens when the liver can no longer control when carbohydrate is released. The second shows what happens when the liver can no longer control whether nitrogen is allowed to remain dangerous. Both convert invisible daily physiology into visible disease. That is their value. They prove that the liver is not an organ that merely participates in metabolism. It is an organ that decides whether metabolism remains livable.

Why Inherited Defects Reveal Normal Physiology by Breaking It

Inherited defects reveal normal physiology by removing one quiet step the body usually performs so well that no one notices it is happening. In health, the liver stores glycogen, releases glucose, converts ammonia into urea, binds and moves metals, packages lipids, conjugates bilirubin, transforms hormones, and clears foreign compounds without drawing attention to any of those acts. The reader sees only the finished stability. Blood sugar stays usable between meals. Ammonia stays low enough for the brain to think clearly. Iron moves or stays trapped according to need. Bile reaches the intestine. Plasma proteins remain in circulation. The invisible success of normal physiology is what makes the organ seem simple from the outside. An inherited defect breaks that illusion because it deletes one enzyme, one transporter, one folding step, one export pathway, one storage mechanism, and suddenly the quiet order becomes visible as crisis.

That is why these diseases are such powerful teaching tools. They show that normal function is not a broad vague "support" of the body. It is a sequence of exact permissions. A glycogen storage disorder reveals that the liver is not merely holding carbohydrate. It is deciding when carbohydrate may become blood sugar. A urea cycle defect reveals that the liver is not merely helping with protein metabolism. It is deciding whether nitrogen remains a brain poison or becomes an excretable molecule. Hemochromatosis reveals that iron control is not passive mineral balance. It is a gated traffic system that must know when to say no. Wilson disease reveals that copper handling is not just storage. It is binding, routing, and biliary exit under strict control. Alpha-1 antitrypsin deficiency reveals that secretion is not guaranteed just because synthesis occurred. A protein can be made, misfold, remain trapped, and turn the cell that made it into the site of disease. These disorders are valuable because they do not add exotic physiology on top of the normal body. They subtract a normal protection and let the reader watch what that protection had been preventing.

The liver is especially vulnerable to this kind of teaching because so much of its work is transitional. It takes a molecule from one state to another, from ammonia to urea, from unconjugated bilirubin to conjugated bilirubin, from vitamin D precursor to 25-hydroxylated intermediate, from hydrophobic toxin to conjugated excretable product, from portal substrate surge to glycogen reserve, from fatty acid burden to ketone export. Transitional physiology is easy to underestimate when it succeeds because the body quickly forgets the unstable form once the stable form has been made. An inherited defect traps the process in the unstable middle. The disease then reveals not only what the liver normally produces, but what dangerous intermediate states it usually spares the organism from having to live through.

This is why breaking physiology often teaches more clearly than describing it. A healthy liver is so effective that it hides its own necessity. An inherited defect removes that cover. The body then starts declaring, in blood chemistry, tissue injury, neurologic failure, or organ overload, exactly what the missing hepatic function had been holding together all along. Normal physiology becomes visible the moment it can no longer complete itself.

How Storage and Transport Errors Expose the Deeper Logic of Liver Function

Storage and transport errors are so revealing because they do not simply produce "metabolic disease." They expose the exact point where the liver normally keeps chemistry from becoming structurally dangerous. In health, the liver does not just store molecules and move them around like a warehouse with pipes. It changes their form, binds them to the right carriers, places them into the right compartment, and decides when they are allowed to stay, circulate, or leave. Iron must be stored in a buffered form and released under hormonal control. Copper must be bound, distributed, and excreted through bile without being allowed to remain chemically loose. Bilirubin must be conjugated before it can leave. Lipids must be packaged into lipoproteins before they can travel through plasma. Vitamin A must be held in stellate cells rather than left freely active in circulation. B12 must be stored long enough to defend tissue function across interruptions in intake. When one of those storage or transport steps fails, the resulting disease is often more educational than the healthy state, because it reveals what the liver was quietly preventing all along.

A storage error shows what happens when a useful compound is allowed to remain in the wrong place or in the wrong amount. Hemochromatosis teaches that iron is not dangerous only when absent. It is dangerous when the body keeps admitting it without enough restraint. Wilson disease teaches that copper cannot simply be absorbed and trusted to behave. If biliary export fails, the liver becomes the first site where a useful metal turns into oxidative burden. Alpha-1 antitrypsin deficiency teaches that even a protein made for export can become toxic to the liver if folding and trafficking fail. In each case the liver is not merely "affected." It is revealing its deeper role as a governor of chemical placement. The disease appears because the organ no longer decides correctly where the molecule should reside. The storage site becomes the injury site. The reserve becomes the burden.

Transport errors reveal the same logic from another angle. A molecule may be chemically valuable and still unusable if it cannot move in the correct fluid. Fat cannot travel through plasma unprotected, so the liver packages it into lipoproteins. Unconjugated bilirubin cannot be excreted cleanly, so the liver changes its chemistry and routes it into bile. Thyroid hormone does not circulate mainly as free signal, so the liver makes transport proteins that buffer and distribute it. Free fatty acids use albumin. Copper uses ceruloplasmin. Bile acids are secreted into a detergent stream and then reclaimed through enterohepatic circulation. These are not decorative details. They are the real mechanics of what it means for a molecule to be biologically usable. A transport error shows that nutrition, hormones, metals, and waste are not governed only by how much exists. They are governed by whether the body can bind them, carry them, release them, and clear them in the right form.

This is why storage and transport disorders feel so clarifying. They strip away the public illusion that the liver is simply a place where substances pass through. The truth is harder and more precise. The liver is one of the organs that decides whether a molecule remains tolerable by assigning it to the right compartment and the right route. If iron is stored wrong, tissues oxidize. If copper is exported poorly, hepatocytes become poisoned. If triglyceride cannot be exported, the droplet remains in the cell and congestion becomes steatosis. If bilirubin cannot be conjugated or secreted, it stays in blood and stains the body. If albumin falls, fluid and drug transport become less stable. If bile acids are not reclaimed, the digestive detergent pool shrinks and nutrient extraction changes. The same organ keeps blood, bile, storage, and tissue chemistry from dissolving into each other. Break one storage rule or one transport rule and the whole hidden architecture comes into view.

That is the deeper logic these diseases expose. The liver is not just processing molecules. It is deciding where they are allowed to exist, in what form, for how long, and at what cost. Storage and transport errors reveal normal liver physiology by proving that presence alone is never enough. A compound can be inside the body and still be in the wrong state, the wrong compartment, the wrong carrier, or the wrong direction. The liver's job is to prevent that mismatch. When it fails, the disease teaches the rule.

Chapter 46

Drug-Induced Liver Injury

Predictable Versus Idiosyncratic Injury

Drug-induced liver injury divides first into two structural logics. Predictable injury, often called intrinsic DILI, is dose-related, reproducible, and easier to understand mechanistically because enough of the compound will injure the liver in most people if exposure is high enough. Acetaminophen is the classic example. The liver metabolizes the drug, generates a reactive intermediate, glutathione is exhausted if the dose is large enough, and hepatocyte injury follows in a pattern the toxicology world can anticipate. Idiosyncratic injury is different. It is not clearly dose-dependent in the same straightforward way, it occurs in only a small fraction of exposed people, and it often depends on host-specific susceptibility such as immune factors, genetics, metabolic variation, or the way a particular liver handles the compound under that person's specific conditions. The same drug may be tolerated by thousands of patients and still produce severe hepatic injury in one. That is why idiosyncratic injury is harder to predict and much harder to reduce to simple exposure arithmetic.

The tissue logic follows from that difference. In predictable injury, the hepatocyte is overwhelmed by a burden the drug itself or its metabolites impose directly. The mechanism is often chemically legible: too much reactive intermediate, too much oxidative stress, too much mitochondrial failure, too much depletion of protective systems such as glutathione. The liver is injured because the dose crossed a threshold the tissue could not survive. In idiosyncratic injury, the mechanism is usually less clean and often more biologically entangled. A therapeutic dose can still trigger hepatocellular, cholestatic, or mixed injury patterns because the problem lies partly in who the host is, not only in how much drug arrived. Immune activation, unusual metabolic handling, HLA-associated susceptibility, and interactions with inflammation or prior liver state can all matter. So one form of DILI teaches toxic dose logic. The other teaches susceptibility logic. Both damage the liver, but they do not begin from the same kind of risk.

That distinction matters because it changes how the liver should be read. Predictable injury reveals what happens when chemical burden is large enough to defeat ordinary hepatic defenses in almost anyone. Idiosyncratic injury reveals what happens when a liver-specific or host-specific vulnerability turns a tolerated exposure into an inflammatory, cholestatic, or hepatocellular event in the wrong person. One says the dose itself was enough to injure. The other says the person and the drug met in a biologically unlucky way.

Dose-Dependent Toxicity and Susceptibility-Based Toxicity

Dose-dependent toxicity means the compound itself becomes injurious once enough of it reaches the liver, and this happens in a way that is broadly reproducible across people. The classic pattern is intrinsic drug-induced liver injury. Acetaminophen is the teaching example because a toxic enough dose reliably overwhelms hepatic defenses, generates the reactive metabolite NAPQI, depletes glutathione, and produces hepatocellular necrosis in a way clinicians and toxicologists can predict. In this form of injury, the key variable is burden. Enough compound arrives, enough reactive intermediate is generated, and the liver is damaged because the chemistry crossed a threshold most bodies cannot safely absorb. The mechanism is legible. The dose-response relationship is visible. This is why intrinsic injury is often described as predictable.

Susceptibility-based toxicity is different because the same ordinary-looking dose can injure one liver badly while leaving many others apparently unharmed. This is the logic of idiosyncratic drug-induced liver injury. Here the drug is not simply toxic because the amount was large enough in every exposed person. The drug becomes dangerous because it meets a host with the wrong vulnerability. That vulnerability may involve immune recognition, HLA-associated risk, unusual metabolic handling, impaired adaptation to reactive intermediates, mitochondrial sensitivity, transporter differences, or the way inflammation and prior liver state change the response. The hepatocyte is still being injured, but the decisive factor is no longer dose alone. It is dose meeting predisposition. That is why idiosyncratic injury is often much harder to forecast and can occur at doses that are therapeutic for most people.

The tissue logic follows from that difference. In dose-dependent toxicity, the liver is overwhelmed by volume of chemical burden. In susceptibility-based toxicity, the liver is overwhelmed by mismatch between the compound and the person. One teaches threshold chemistry. The other teaches host biology. Both can produce hepatocellular, cholestatic, or mixed patterns of injury, but they do not begin from the same kind of danger. In one, the compound would injure most people if the dose were high enough. In the other, the compound injures only the wrong liver under the wrong internal conditions.

Why the Same Metabolic Machinery That Protects the Body Can Injure the Liver

The liver protects the body by taking compounds that cannot safely remain in their incoming form and forcing them through chemistry the rest of the body would rather not perform for itself. That is the first reason the same machinery can become dangerous. A foreign molecule that is too lipid-soluble, too stable, too membrane-compatible, or too chemically awkward to leave through urine or bile has to be altered before it can be cleared. Hepatocytes do that work through oxidation, reduction, hydrolysis, conjugation, packaging, and export. But those steps are not gentle. Phase I metabolism, especially through cytochrome P450 systems, often exposes a molecule by making it more chemically active than it was when it arrived. The liver is not always subtracting danger in a single motion. It is often creating a sharper intermediate so that later detoxifying steps can finally grab it. The body is protected only if the sequence is completed. If it stalls in the middle, the liver becomes the tissue first exposed to the hazard it created.

This is why metabolic defense is inseparable from metabolic risk. Cytochrome P450 enzymes sit largely in the smooth endoplasmic reticulum and use oxygen and reducing equivalents to alter compounds in ways that make them more polar, more reactive, or more recognizable to Phase II systems. That may be exactly what the body needs. A drug may need to be activated. A steroid may need to be inactivated. A hydrophobic xenobiotic may need a hydroxyl group added before glucuronidation can make it excretable. But every one of those steps carries a cost. Reactive oxygen species can be generated during the process. Unstable electrophilic intermediates can appear. Membrane lipids can be oxidized. Proteins can be modified. Glutathione has to be spent to neutralize what the cell just created. ATP has to be available to keep transporters running and maintain the polarity that separates blood-facing and bile-facing traffic. The hepatocyte is therefore protecting the organism by doing dangerous chemistry in a controlled space. If control weakens, the chemistry turns inward.

The cell's own design makes this paradox unavoidable. The liver is where orally absorbed compounds arrive first in meaningful concentration. Portal blood delivers alcohol, drugs, food compounds, microbial metabolites, and countless other molecules directly to hepatocytes before the rest of the circulation sees their full burden. That is protective at the organism level because the liver can intercept, transform, and reduce systemic exposure. But it also means hepatocytes stand in the highest local concentration gradient of what the body is trying to neutralize. The same cell that is buffering glucose, converting ammonia into urea, synthesizing albumin, making bile acids, and packaging lipids is also being asked to oxidize foreign compounds, conjugate reactive intermediates, and survive the redox shifts required to do it. A quiet hepatocyte can absorb some of that cost. A steatotic hepatocyte, a hypoxic hepatocyte, a cholestatic hepatocyte, or a glutathione-depleted hepatocyte cannot absorb it as well. The machinery did not become evil. The reserve became too thin for the work.

This is why classic liver toxins are often really failures of sequence rather than simple poison events. Acetaminophen is dangerous not mainly because the parent molecule is intrinsically catastrophic at therapeutic handling, but because enough of it can be driven through a pathway that creates NAPQI, a reactive intermediate normally neutralized by glutathione. Ethanol injures not only because alcohol is present, but because alcohol dehydrogenase and aldehyde dehydrogenase drive a redox shift while creating acetaldehyde, which is more reactive than ethanol itself. Many other xenobiotics follow the same logic. The liver begins a protective transformation, but the transitional chemistry is harsher than the starting material, and if glutathione, mitochondrial function, membrane integrity, or Phase II capacity cannot keep up, the intermediate binds proteins, disrupts mitochondria, distorts membranes, and turns the detoxifying cell into the casualty.

So the same metabolic machinery that protects the body can injure the liver because protection at this level is not passive filtering. It is controlled chemical violence. The hepatocyte takes molecules the body cannot safely keep, makes them more exposed, more manageable, and ultimately more excretable, but it has to survive the unstable middle state every time. When the burden is modest and reserve is intact, the organism benefits and the liver absorbs the cost quietly. When the burden stays high or the cell's defenses are thin, the machinery still runs, but now the price is paid in oxidative stress, mitochondrial injury, inflammation, and cell death. The body is protected. The liver pays first.

Patterns of Injury: Hepatocellular, Cholestatic, Mixed

Drug-induced liver injury and many other hepatic insults become easier to read when the damage is sorted by compartment rather than by cause alone. The three major patterns are hepatocellular, cholestatic, and mixed. These are not abstract laboratory categories invented to make charts cleaner. They are structural clues. They tell you whether the main injured compartment is the hepatocyte, the bile-forming and bile-draining system, or both at once. That distinction matters because the liver can fail through different routes, and the symptoms, laboratory behavior, and future of the tissue follow the route that was hit first.

A hepatocellular pattern means the main injury is centered on the hepatocyte. The working cell is being damaged directly or through immune-mediated attack, toxic intermediates, ischemia, viral burden, or other forms of parenchymal stress. The blood reflects that by showing stronger release of aminotransferases, especially alanine aminotransferase and aspartate aminotransferase, because those enzymes live inside hepatocytes and leak out when the cell membrane and internal structure are damaged. The physiology underneath this pattern is one of failed labor. Glucose buffering, ammonia control, bilirubin conjugation, albumin synthesis, clotting factor production, and xenobiotic handling all sit inside the hepatocyte, so when the hepatocyte is the primary compartment under attack, the liver begins losing the jobs only that cell can do. Jaundice can appear later if bilirubin handling falls behind, but the main pattern begins as cellular injury first.

A cholestatic pattern means the dominant failure is in bile formation, canalicular secretion, small-duct flow, or larger biliary drainage. The hepatocyte may still be present and capable of much of its work, but bile is not leaving properly. The blood then begins reflecting retention of biliary constituents and stress on the ductal system rather than mainly leakage from hepatocyte death. Alkaline phosphatase and gamma-glutamyl transferase tend to rise more prominently because the injury is centered on the biliary compartment. Symptom logic follows that same route. Dark urine appears because conjugated bilirubin backs into blood and reaches the kidney. Stool pales because less bilirubin pigment reaches the intestine. Itching develops because retained biliary chemistry begins circulating systemically. Fat digestion worsens because bile acids are no longer reaching the lumen in adequate quantity. In this pattern the liver is not only failing as a processor. It is failing as an outflow organ.

A mixed pattern means the insult has not respected one compartment cleanly. The hepatocyte is injured enough to leak cellular enzymes, but biliary flow is also impaired enough that cholestatic chemistry begins accumulating. This is common in drug-induced liver injury because the same compound can damage the hepatocyte directly or through reactive intermediates while also disturbing canalicular transport, bile acid handling, or small-duct physiology. Mixed injury often feels harder to read because the liver is now speaking in two dialects at once. Aminotransferases rise because cells are being damaged. Alkaline phosphatase rises because bile traffic is also being impaired. The symptom pattern can therefore include fatigue, malaise, and enzyme leakage typical of hepatocellular injury together with dark urine, itching, and more obvious cholestatic features. The organ is revealing that both the worker and the route out have been hit.

This is why these patterns matter so much clinically and mechanistically. A hepatocellular pattern tells you to think first about injury to the parenchymal worker. A cholestatic pattern tells you to think first about bile traffic and retention. A mixed pattern tells you the insult is crossing compartments. The liver is one organ, but it is not one tissue job. Its disease patterns follow the compartment that failed.

Why Medication History Matters

Medication history matters because the liver usually cannot announce "this drug did it" on its own. The tissue can show hepatocellular injury, cholestatic injury, or a mixed pattern, but those patterns are not specific enough to name the culprit without the timeline of exposure, dose, latency, and withdrawal. Drug-induced liver injury is, in large part, a diagnosis built from sequence. A medication was started, changed, increased, combined with something else, or taken in overdose. Liver tests changed after a plausible interval. Competing causes were excluded. The injury improved, or failed to improve, after the suspected agent was stopped. Without that chronology, the liver's pathology remains real but its cause remains blurred. This is why major liver references and clinical guidance emphasize that detailed history is crucial in DILI evaluation and that the latency between exposure and injury may range from days to many months depending on the agent and mechanism.

The timing matters because predictable and idiosyncratic injuries do not behave the same way. An intrinsic toxic injury such as acetaminophen overdose can follow a large dose in a relatively direct, more reproducible sequence. Idiosyncratic injury is harder. A perfectly ordinary therapeutic dose may injure one liver and not another, and the onset can be delayed enough that the patient no longer associates the drug with the symptoms. Some agents produce hepatocellular patterns. Others produce cholestatic patterns. Others mimic autoimmune hepatitis. That is why the history has to include more than prescription drugs taken this week. It has to include over-the-counter medications, herbals, supplements, recent antibiotics, antidepressants, anticonvulsants, anabolic agents, alcohol use, and any drug stopped in the recent past. The liver is often reacting to a sequence the patient no longer sees as one sequence.

Dose matters too, but so does context. A medication history is not only a list of names. It is a map of chemical burden. The relevant questions are what was taken, how much, for how long, with what dose changes, with what co-exposures, and on what background liver. A drug that is usually tolerated may become dangerous in a steatotic liver, an inflamed liver, a cholestatic liver, or a liver already handling alcohol and multiple other compounds. A medication that is safe at one dose may become toxic at another. A therapeutic drug may look harmless until the history reveals a supplement, another prescription, or a pattern of repeat dosing that changed metabolism or depleted reserve. This is why causality in liver injury relies so heavily on history and exclusion rather than on one pathognomonic blood test. The drug rarely leaves a signature. The sequence does.

So medication history matters because the liver speaks in injury patterns, but history tells you which chemical story produced the pattern. The hepatocyte can reveal that it was hurt. The bile system can reveal that flow was impaired. Only the exposure timeline explains whether the wound came from the medicine cabinet, the supplement shelf, the bottle, or the interaction between them.

How Clinicians Think Through Causal Patterns

Clinicians do not usually diagnose drug-induced liver injury by finding one lab value that announces the culprit. They diagnose it by reconstructing a sequence. The liver has only a limited number of ways to display injury. Hepatocytes can leak enzymes. Bile flow can stall. Bilirubin can rise. Coagulation can worsen. Symptoms can appear days later or months later. None of that is specific enough by itself to name the offending compound. So the first real question is not "is the liver abnormal." The first real question is "what pattern of injury appeared, when did it appear, and what exposures were active when that pattern began." A causal pattern is a timeline joined to a compartment. The clinician is asking whether the drug history, the latency, the laboratory profile, and the course after stopping the agent all fit the same mechanism.

The pattern of injury comes first because it tells the clinician which part of the liver was hit hardest. A hepatocellular pattern means the main worker cell, the hepatocyte, is being injured, so alanine aminotransferase and aspartate aminotransferase rise more prominently. A cholestatic pattern means bile formation or bile flow is being disrupted, so alkaline phosphatase and bilirubin often matter more. A mixed pattern means both compartments are being hit at once. This matters because different drugs and toxins tend to injure the liver in different ways. Some are classic hepatocyte poisons. Some behave more like cholestatic disruptors. Some mimic autoimmune hepatitis. Some create a predictable dose-related injury. Others create an idiosyncratic pattern, meaning the dose may be ordinary but the host response is not. The clinician therefore begins by asking what kind of liver failure language the tissue is speaking before asking which drug speaks that language most often.

Time is the next filter. Causality in liver disease is often hidden inside latency, the interval between exposure and injury. A large acetaminophen overdose can produce a hepatocellular pattern on a relatively short timeline because the chemistry is direct and dose-related. An antibiotic, anticonvulsant, antidepressant, herbal product, or supplement may provoke injury only after days or weeks of exposure, sometimes longer, because the mechanism is idiosyncratic, immune-mediated, or dependent on cumulative stress. This is why clinicians ask not only what the patient is taking today, but what was started recently, what was stopped recently, what dose changed, what was added on top of something else, and what over-the-counter or "natural" compounds were used outside the prescription list. A perfect liver injury pattern attached to the wrong time course is weak evidence. A strong time relationship turns suspicion into a much more serious claim.

Then comes exclusion, which is where clinical reasoning becomes disciplined rather than impressionistic. The liver can be injured by viruses, alcohol, ischemia, biliary obstruction, autoimmune disease, metabolic dysfunction, congestion, pregnancy-related disorders, and systemic infection, and many of those can imitate drug injury convincingly. So the clinician has to ask what else could produce this exact pattern in this exact patient. Could a viral hepatitis explain the enzyme rise. Could gallstones or a blocked duct explain the cholestatic picture. Could shock, heart failure, or sepsis have changed hepatic perfusion. Could alcohol or steatotic liver disease already have placed the hepatocyte under stress before the medication arrived. Drug-induced liver injury is often strongest when the pattern fits the exposure and the main competing explanations fall away. The diagnosis is not made by loving one theory. It is made by surviving the alternatives.

Finally, clinicians watch what happens after the suspected compound is removed. This is called dechallenge. If liver tests begin improving after the medication stops, that improvement supports the idea that the exposure mattered. It does not prove it absolutely, because some other conditions improve with time as well, but it strengthens the pattern. Rechallenge, the return of injury after the same drug is reintroduced, is even stronger evidence, but it is rarely sought intentionally because repeating the exposure may be dangerous. So the clinician often has to reason with imperfect evidence. The injury pattern, the latency, the dose relationship, the known behavior of the drug, the exclusion of alternatives, and the response after withdrawal are assembled into one causal narrative. That is how clinicians think through drug-induced liver injury. They do not ask whether the liver is sick. They ask whether this chemical history explains the exact way in which it became sick.

Chapter 47

Liver Cancer

Hepatocellular Carcinoma

Hepatocellular carcinoma, or HCC, is the major primary cancer of hepatocytes, the main working cells of the liver. It does not arise because the liver simply became inflamed once or because one scar patch appeared on a scan. It arises when repeated injury, repeated regeneration, and repeated inflammatory signaling keep forcing hepatocytes to divide inside a damaged architectural field for years. Cirrhosis is the most important soil for that process, because cirrhosis means the organ has already become nodular, fibrotic, hypoxic in places, and structurally disordered in the way blood reaches cells and the way cells repair themselves. That is a carcinogenic environment. The liver keeps trying to replace lost tissue, but the replacement is happening inside oxidative stress, cytokine pressure, altered signaling, DNA damage, and scar-partitioned regeneration. HCC is what can happen when repair stops being clean enough to remain only repair.

This is why HCC is best understood as the cancer of chronic architectural injury rather than as an isolated tumor event. Viral hepatitis, alcohol-associated liver disease, and metabolic dysfunction-associated steatotic liver disease all increase HCC risk because all three can keep the liver in the same dangerous loop: hepatocyte injury, inflammation, stellate-cell activation, fibrosis, regenerative nodules, and ongoing cell turnover. Hepatitis B deserves special attention because it can increase HCC risk even without cirrhosis, in part because HBV can integrate into the host genome and promote genomic instability directly. Hepatitis C, by contrast, is more strongly linked through the long path of chronic inflammation and cirrhotic remodeling. MASLD is now a major and growing HCC background as well, and a meaningful minority of MASLD-related HCC cases can occur without cirrhosis, which shows again that the deeper issue is not scar alone but a liver kept under long-term metabolic and inflammatory stress.

From a tissue point of view, HCC is the moment a regenerating hepatocyte population stops obeying the organ's normal growth limits. The liver is one of the few organs that can regenerate impressively, but that gift becomes dangerous when the stimulus to regenerate never really ends. Cells keep entering the cell cycle. DNA is copied again and again under conditions of oxidative burden and inflammatory signaling. Fibrotic septa distort blood flow and oxygen delivery. Growth factor pathways remain active. Over time, the chance that one clone of hepatocytes acquires enough survival and proliferative advantage to escape normal control rises. That is why HCC belongs naturally at the end of chronic liver disease, not as a separate surprise. It is the malignant version of a liver that has been trying to rebuild itself for too long inside the wrong environment.

So hepatocellular carcinoma is not just liver cancer in the generic sense. It is the cancer of the main liver cell, usually born from years of repeated damage and repeated imperfect repair. The liver's regenerative power is part of what keeps people alive through chronic injury. HCC is one of the prices the organ can pay when that regenerative pressure is sustained for too long.

Cholangiocarcinoma

Cholangiocarcinoma is a cancer of the bile duct epithelium, not of the hepatocyte. That distinction matters because it means the disease begins in the liver's drainage system rather than in its main metabolic worker cell. Bile ducts are lined by cholangiocytes, and cholangiocarcinoma arises from malignant transformation of that epithelial compartment anywhere along the biliary tree. It is commonly divided by location into intrahepatic cholangiocarcinoma, which arises within the liver's bile ducts, and extrahepatic cholangiocarcinoma, which includes perihilar and distal tumors outside the liver parenchyma proper. From a structural point of view, this is a cancer of outflow architecture. The organ may still be able to make bile, but the channels meant to carry that bile are now becoming the site of malignant obstruction, invasion, and progressive cholestatic injury.

This makes cholangiocarcinoma feel different from hepatocellular carcinoma even when both are called liver cancers. Hepatocellular carcinoma grows out of the parenchymal worker cell and is usually rooted in chronic regenerative pressure inside scarred tissue. Cholangiocarcinoma grows out of the ductal epithelium and damages the liver by narrowing, infiltrating, and obstructing the biliary tree. The resulting physiology is therefore more cholestatic from the start. Bile flow becomes interrupted, conjugated bilirubin backs up, jaundice and itching become easier to provoke, and the liver begins suffering not only from the mass itself but from retained biliary chemistry and repeated duct-centered injury. The tumor is not just occupying space. It is deforming the route out.

That is why cholangiocarcinoma should be understood as a cancer of a transport network. The bile ducts are not decorative tubing. They are the path by which the liver's outgoing stream reaches the intestine. Once the epithelium lining that path becomes malignant, the consequences are mechanical, chemical, and architectural all at once. Flow is impeded. Pressure rises upstream. Cholestasis deepens. Fibrosis and parenchymal injury can accumulate around the failing outflow system. The tumor therefore injures the liver not only by being a cancer, but by corrupting one of the organ's central directional systems.

Why Chronic Injury Raises Cancer Risk

Chronic injury raises cancer risk because the liver is one of the few organs that keeps trying to rebuild itself while the damage is still ongoing. A single acute insult can kill hepatocytes and still leave enough preserved architecture that regeneration restores working tissue. Chronic injury is different. Viral hepatitis, alcohol-associated injury, cholestatic disease, and steatotic injury keep hepatocytes dying, keep inflammatory signaling active, and keep stellate cells laying down scar. The organ is forced into repeated cycles of cell death and cell replacement inside a field that is becoming progressively more fibrotic, more hypoxic in places, and more structurally disordered. That is carcinogenic soil. The danger is not only that cells are injured. The danger is that cells are being asked to divide again and again inside an inflammatory, scarred, growth-factor-rich environment where the normal rules of repair are already under strain.

This is why cirrhosis matters so much. Cirrhosis is not just scar. It is fibrosis plus regenerative nodules, vascular reorganization, and loss of normal lobular order. In that setting, hepatocytes are no longer regenerating inside a clean exchange organ. They are regenerating inside nodules surrounded by fibrous septa, with altered blood flow and chronic inflammatory pressure. The liver is still trying to repair itself, but the repair is happening in the wrong geometry. Over time that repeated regeneration increases the chance that one clone of hepatocytes acquires enough genetic and signaling advantage to stop obeying the normal limits of tissue repair. This is one reason hepatocellular carcinoma generally arises from chronically injured and often cirrhotic liver rather than from a liver that was hurt once and allowed to heal cleanly.

The sequence is therefore structural as much as molecular. Chronic inflammation feeds fibrosis. Fibrosis distorts blood flow and exchange. Distorted architecture forces more imperfect regeneration. Imperfect regeneration under persistent oxidative and inflammatory stress raises malignant risk. The liver's regenerative power, which is one of the reasons people survive chronic injury for years, becomes part of the hazard when the stimulus to regenerate never truly ends. Repair keeps being called for, but the tissue field in which repair happens becomes less and less normal. Cancer emerges from that long argument between injury and rebuilding.

There is one important qualification. Chronic liver injury usually raises liver cancer risk most strongly by driving cirrhosis, but hepatitis B can raise hepatocellular carcinoma risk even without established cirrhosis because HBV, as a DNA virus, can integrate into the host genome and promote genomic instability more directly. That exception actually sharpens the rule. Most chronic injury raises cancer risk by years of inflammation, fibrosis, and regeneration. HBV can also add a more direct genomic route on top of that.

Regeneration Under Chronic Inflammatory Stress

Regeneration under chronic inflammatory stress is not the same thing as healthy repair repeated many times. It is repair happening inside a tissue that never returns to baseline. That difference is what makes it dangerous. In the healthy liver, hepatocytes can re-enter the cell cycle after injury and rebuild lost working mass inside a scaffold that still supports normal blood flow, normal bile flow, and normal exchange across the sinusoid. Chronic inflammation changes the terms of that recovery. Cytokines remain elevated. Kupffer cells stay activated. Reactive oxygen species remain high. Stellate cells continue laying down extracellular matrix. Blood reaches the regenerating parenchyma through a field that is becoming more distorted, more hypoxic in places, and less permissive to clean exchange. The hepatocyte is therefore being asked to divide in an organ that is still signaling danger. It is not regenerating inside quiet tissue. It is regenerating inside a wound that never closes.

That matters because cell division under stress is not neutral. Every round of regeneration requires DNA replication, membrane synthesis, mitochondrial support, protein production, and re-establishment of polarity. Under chronic inflammatory pressure, all of that is happening in the presence of oxidative damage, altered growth signaling, inflammatory mediators, and repeated hepatocyte death. The liver is trying to replace what was lost, but the replacement is occurring in a biochemical environment more likely to generate replication errors, more likely to select for survival advantages that would not matter in healthy tissue, and more likely to reward cells that can keep dividing under hostile conditions. Fibrosis makes this worse by partitioning the parenchyma into nodules and distorting the normal lobular map. Regeneration is still occurring, but it is increasingly happening inside scar-bounded compartments rather than inside preserved architecture. The organ remains active. The activity stops meaning restoration and starts meaning adaptation to chronic damage.

This is why regeneration under chronic inflammatory stress becomes part of carcinogenic risk rather than part of clean recovery. The liver survives chronic injury for years because it keeps trying to rebuild itself. That same persistence becomes dangerous when the inflammatory field never quiets. Cells are lost and replaced. Lost and replaced again. Each cycle occurs in a tissue already shaped by cytokines, matrix expansion, vascular distortion, and oxidative burden. Over time the regenerative program stops being purely protective. It becomes one of the forces that pushes the liver toward dysplasia, nodular transformation, and eventually malignancy. Chronic inflammation does not only scar the liver. It teaches regeneration to occur in the wrong atmosphere, and that changes the future of every new cell that enters the tissue.

Cirrhosis as a Carcinogenic Soil

Cirrhosis is carcinogenic soil because it forces hepatocytes to keep trying to survive and regenerate inside an organ that has already lost the structural conditions of normal repair. Scar partitions the parenchyma into nodules, distorts blood flow, alters oxygen delivery, thickens the exchange surface, and keeps inflammatory signaling active long after any single injury should have resolved. In that environment, hepatocytes are not dividing inside a quiet, well-perfused, correctly organized liver. They are dividing inside fibrous septa, altered gradients, oxidative stress, cytokine pressure, and chronic wound chemistry. That matters because cancer risk does not rise only when cells are damaged. It rises when damaged cells are repeatedly forced to replicate under conditions that reward survival at the expense of order. Cirrhosis creates exactly that condition. The liver is still trying to rebuild itself, but it is rebuilding inside a field that has become mechanically distorted, immunologically loud, and metabolically hostile. Regeneration continues, but the blueprint is gone.

This is why cirrhosis does more than scar the liver. It changes what kind of cell can succeed there. A hepatocyte that can tolerate oxidative stress better, ignore growth restraints more easily, survive under poorer oxygen delivery, or proliferate inside fibrotic distortion now has an advantage it would not have needed in healthy tissue. Over years, repeated cell death, repeated regeneration, and repeated inflammatory signaling create selection pressure. DNA damage accumulates. Repair becomes less faithful. Growth factor pathways stay more active than they should. The nodular cirrhotic liver becomes a place where abnormal clones can emerge and persist because the surrounding tissue is no longer enforcing normal architecture strongly enough to suppress them. That is the meaning of carcinogenic soil. The scar does not simply sit beside the future cancer. It prepares the ecological conditions in which malignant transformation becomes easier to sustain. A healthy liver can regenerate and return to order. A cirrhotic liver regenerates inside disorder, and that disorder is fertile ground for cancer.

Why Cancer Emerges From Distorted Architecture and Repeated Repair

Cancer emerges from distorted architecture and repeated repair because the liver is one of the few organs that keeps pushing damaged cells back into the cell cycle while the environment around those cells is becoming less and less normal. In a healthy liver, regeneration happens inside a soft low-resistance exchange field where blood reaches hepatocytes cleanly, bile leaves cleanly, oxygen gradients remain ordered, and the extracellular matrix still provides guidance rather than obstruction. Chronic injury changes all of that. Fibrous septa divide the parenchyma, nodules trap hepatocytes in abnormal compartments, sinusoids lose permeability, inflammatory cytokines remain elevated, and oxidative stress becomes background chemistry rather than temporary crisis. The cell that now divides is not being born into a normal liver. It is being born into a scarred, hypoxic, inflamed, and structurally distorted organ. That is why repeated repair becomes dangerous. It is not repair in a quiet tissue. It is replication under pressure.

That pressure matters because every round of repair requires DNA duplication, protein synthesis, membrane expansion, mitochondrial support, and a decision that this cell should keep living and proliferating inside the field it inherited. If the field is chronically inflamed, the hepatocyte is copying itself while surrounded by reactive oxygen species, growth factors, cytokines, bile-related stress, and distorted vascular supply. Oxidative injury damages DNA. Cytokines alter growth signaling. Fibrosis changes which cells receive blood and substrate cleanly enough to survive. Over time this creates selection pressure. Cells that can divide under bad conditions, ignore inhibitory signals more effectively, resist cell death, or adapt to metabolic stress begin gaining an advantage. The liver is not trying to produce cancer. It is trying to survive. But survival in a distorted architecture rewards traits that move cells away from normal restraint and toward clonal independence.

The architecture itself deepens that risk. A healthy liver enforces order through pattern. Portal blood enters, crosses a low-resistance sinusoidal bed, and exits through central veins. Hepatocytes occupy defined positions inside those gradients. Bile flows the other way. Cirrhosis erases that map. Blood is rerouted. Nodules enlarge inside fibrous boundaries. Exchange becomes patchy. Regeneration becomes nodular instead of restoring normal plates and lobules. This matters because tissue architecture is one of the ways the body keeps cell behavior disciplined. Once that architecture is distorted, the cell is less constrained by normal neighborhood logic. It is easier for a clone with abnormal growth behavior to persist when the surrounding tissue is no longer organized strongly enough to suppress it. The scar is not only damage left behind. It is a new landscape in which abnormal cells can survive more easily than they could have in the original organ.

So cancer emerges from distorted architecture and repeated repair because the liver is being forced to regenerate in a place that no longer rewards normal regeneration. Cells keep dividing because injury keeps demanding replacement. The environment keeps pushing those cells toward adaptation under inflammatory, fibrotic, and oxidative stress.

Eventually one clone gains enough advantage to stop behaving like a regenerative hepatocyte and start behaving like a malignant one. The cancer does not appear beside the repair process. It grows out of the repair process once the architecture around that repair has been changed enough to turn regeneration into selection.

Part XII

Reading the Liver Through Tests, Scans, and Patterns

Chapter 48

What Liver Enzymes Actually Mean

ALT and AST

ALT and AST are not liver function in the strict sense. They are injury signals. Alanine aminotransferase, ALT, and aspartate aminotransferase, AST, are intracellular enzymes that become more visible in blood when hepatocytes are damaged enough to leak their contents. ALT is more concentrated in the liver and is therefore somewhat more specific for hepatocellular injury, while AST is also found in skeletal muscle, heart, kidneys, red blood cells, and other tissues, which means an elevated AST does not automatically belong to the liver until the pattern and the clinical setting make that claim believable. This is why aminotransferases should be read as evidence that cells have been hurt, not as proof that liver function has already failed. A liver can leak ALT and AST and still maintain albumin, bilirubin handling, and coagulation for a time. The cell membrane has been injured. The whole organ may still be compensating.

That distinction matters because people hear "liver enzymes" and imagine the liver itself is becoming overactive or underactive in some vague way. What these enzymes really tell you is that hepatocytes are under stress, dying, or becoming permeable enough that the bloodstream can see what was supposed to stay inside the cell. When ALT and AST rise out of proportion to alkaline phosphatase and bilirubin, the pattern is called hepatocellular because the main injured compartment is the hepatocyte rather than the bile duct system. But even here the numbers do not tell the whole story by themselves. ALT usually exceeds AST in many forms of chronic hepatocellular injury, yet patterns can shift with alcohol-related disease, advanced fibrosis, muscle injury, or ischemic and toxic insults. Marked aminotransferase elevations can point toward acute hepatic injury, including viral, toxic, or ischemic causes, but the enzyme height alone is still not the diagnosis. It is the shape of the wound seen through blood.

This is why clinicians use ALT and AST as entry points rather than as final answers. Mild isolated elevations may normalize on repeat testing, and even chronic liver disease can exist with only modest aminotransferase changes or sometimes normal values when the disease is relatively quiet. A cirrhotic liver can be structurally advanced and still show only modest AST and ALT elevation because scar and lost organization are not measured directly by these enzymes. So ALT and AST should be understood as the bloodstream's early report that liver cells have been injured, not the bloodstream's full explanation of what the liver can still do. They tell you that the worker has been struck. They do not, by themselves, tell you how much of the factory still functions.

ALP and GGT

Alkaline phosphatase, ALP, and gamma-glutamyl transferase, GGT, are not mainly signals of hepatocyte death in the way ALT and AST are. They are more useful as signs that the biliary side of the liver is under stress. That distinction matters because the liver is not one compartment. A hepatocyte pattern tells you the working cell is being injured. An ALP and GGT pattern tells you to think about bile flow, bile duct epithelium, cholestasis, or pressure on the biliary system. ALP is found in bile duct epithelium, but it is also found in bone, placenta, and some other tissues, which means an elevated ALP by itself does not automatically belong to the liver. GGT is more helpful in that setting because it is more strongly associated with hepatobiliary tissue. When ALP rises and GGT rises with it, the pattern becomes more convincingly biliary. The bloodstream is then telling you that bile formation, bile movement, or bile duct integrity is being disturbed rather than simply that hepatocytes are leaking.

This makes physiologic sense. Bile ducts and canalicular transport systems are stressed when bile cannot move forward cleanly. Cholestasis, whether from intrahepatic bile formation failure, small-duct injury, or larger duct obstruction, places pressure on the biliary compartment and alters the epithelial surfaces meant to carry that outgoing stream. ALP becomes more prominent in blood under those conditions because the bile-facing architecture is being irritated, compressed, inflamed, or obstructed. GGT behaves similarly enough that it helps confirm the story. In plain terms, ALT and AST tell you the worker may be injured. ALP and GGT tell you the route out may be injured.

This is why the pattern matters more than any single number. A high ALP with a normal GGT should make you stop and ask whether the source might be bone rather than liver. A rise in both ALP and GGT points more strongly toward cholestatic or biliary disease. The patient with dark urine, pale stool, itching, and rising ALP and GGT is usually not giving you the same tissue story as the patient with very high ALT and AST after a hepatocellular insult. One pattern points toward trapped bile and injured ducts. The other points toward damaged hepatocytes. The liver is speaking in compartments, and ALP with GGT is the language of bile flow trouble.

Bilirubin

Bilirubin is one of the most useful liver tests because it does not simply tell you that tissue was injured. It tells you that a transport sequence may have broken. Bilirubin begins as a heme breakdown product, mostly from aging red blood cells, and reaches the liver in an unconjugated form bound to albumin. Hepatocytes are supposed to take it up, conjugate it into a more water-compatible form, secrete it into bile, and move it onward toward the intestine. A bilirubin result therefore sits at the intersection of blood-cell turnover, hepatic uptake, hepatocyte conjugation, biliary secretion, and downstream bile flow. That is why it is so valuable. It is not merely a liver pigment. It is a marker of whether the whole route from heme breakdown to excretion is still working.

This is also why bilirubin has to be interpreted by form and by context. Unconjugated bilirubin points more strongly toward an upstream burden or an early hepatic processing problem. Too much bilirubin may be arriving from increased heme breakdown, or the liver may be failing to take it up or conjugate it efficiently. Conjugated bilirubin tells a different story. The liver has already processed the pigment into a more water-soluble form, but that form is now backing up into blood because bile secretion or bile flow is impaired. Once conjugated bilirubin enters the bloodstream abnormally, the kidney can filter it, which is why dark urine often accompanies cholestatic or obstructive patterns. At the same time, if less bilirubin reaches the intestine, stool loses pigment and becomes pale. Bilirubin therefore is not just a number. It is a map. It shows whether the problem is before the liver, inside the hepatocyte, or after the liver in the biliary path.

That makes bilirubin different from ALT and AST. Aminotransferases tell you the hepatocyte is being injured. Bilirubin tells you the body is losing control over one of the liver's major routes of chemical disposal. A person can have a very high ALT and still maintain relatively normal bilirubin if the hepatocyte injury is acute and the excretory sequence is not yet overwhelmed. A person can have more modest enzyme elevations with a striking bilirubin rise if cholestasis or biliary obstruction is the dominant problem. This is why jaundice matters so much clinically. Once bilirubin rises enough to stain sclera and skin, the liver is no longer simply inflamed or fatty or scarred in the abstract. It is failing to move pigment out of the body cleanly.

So bilirubin is best understood as a test of flow and processing rather than of injury alone. It reflects whether the liver can still receive a heme-derived burden from blood, chemically modify it, and push it into bile for removal. When bilirubin rises, the body is telling you that this sequence has slowed, failed, or been obstructed.

Albumin

Albumin is not an injury marker. It is a production marker. That distinction is crucial. ALT and AST rise when hepatocytes are being damaged badly enough to leak their intracellular contents into blood. Albumin tells a different story. It tells you whether the liver is still making one of the main proteins that allows plasma to behave like plasma. Hepatocytes synthesize albumin continuously and release it into circulation, where it helps maintain oncotic pressure, the inward protein-driven force that keeps water inside the vascular compartment, and also binds many drugs, fatty acids, bilirubin, and other circulating molecules during transport. When albumin is low, the question is not usually "how much acute hepatocyte injury happened today." The question is "how long has synthetic function been inadequate, or what other process is lowering this protein in blood."

That is why albumin is more useful as a marker of chronic hepatic reserve than of sudden liver damage. The protein has a relatively long half-life, which means blood levels do not collapse the moment hepatocytes are injured. A person can have strikingly elevated aminotransferases from acute hepatitis and still have a normal albumin at first because the liver has not yet spent through the existing plasma pool. In chronic liver disease, the pattern is different. Month after month of reduced synthesis gradually lowers the level until the blood begins showing that the liver is no longer maintaining one of its most important structural proteins. Albumin therefore tells you less about the sharpness of today's insult and more about the depth and duration of lost synthetic competence.

This is also why low albumin has to be interpreted with discipline. The liver is a major source, but not every low albumin is caused by liver failure. Inflammation can suppress albumin because hepatocytes shift synthetic priorities toward acute-phase proteins. Protein loss through the kidney or gut can lower it. Severe malnutrition can reduce the amino acid supply needed for synthesis. So albumin is a liver test in the sense that the liver makes it, but it is not a liver-only fact. The number becomes most meaningful when read beside bilirubin, PT and INR, ascites, edema, nutritional state, inflammatory burden, and the broader architecture of the disease. A low albumin in cirrhosis means more than a low albumin in isolation because the failing organ, the swollen abdomen, and the altered coagulation profile are all telling the same story.

That story is physical as much as biochemical. When albumin falls, blood loses part of the protein force that helps keep water inside the circulation. Fluid leaves capillaries more easily. Edema and ascites become easier to sustain. Drug binding becomes less predictable because more free drug may be present for the same total concentration. The circulation becomes a worse transport medium. So albumin is not just another protein the liver happens to make. It is one of the main reasons blood can hold volume and carry cargo in an orderly way. When it falls, the lab is not merely showing low protein. It is showing that the liver is losing one of its largest construction jobs.

PT and INR

Prothrombin time, PT, and the international normalized ratio, INR, are not injury markers in the way ALT and AST are. They are synthesis markers. More specifically, they are rough blood-based tests of whether the liver is still making enough clotting proteins to let plasma clot at something close to a normal pace. PT measures how long it takes plasma to clot after tissue factor is added in the lab, and INR is the standardized way of expressing that result across laboratories. The reason these tests belong in liver physiology is simple. The liver synthesizes many of the clotting factors required for this pathway, especially factors II, V, VII, and X. When hepatocyte synthetic function declines, clot formation in the test tube slows, PT lengthens, and INR rises.

This makes PT and INR useful, but only in a narrow way. They are rough windows into hepatic synthetic function because they reflect one important class of liver-made proteins, not the whole organ and not the whole coagulation system. In acute liver injury, PT and INR can worsen quickly because some clotting factors, especially factor VII, have very short half-lives, so a failing liver is exposed fast. In chronic liver disease, a rising PT or INR often signals progression toward more advanced synthetic failure. But mild hepatocellular dysfunction can leave PT and INR normal, and compensated cirrhosis can still exist before these tests become markedly abnormal. The liver may already be structurally damaged while the clotting-time signal still looks deceptively quiet.

They are also poor as stand-alone bleeding tests. This is one of the most important corrections in modern hepatology. A high INR in cirrhosis does not mean the patient is "auto-anticoagulated," because the liver also makes natural anticoagulant proteins, and those can fall too. The result is an unstable but rebalanced coagulation state rather than a simple one-direction bleeding disorder. That is why AASLD emphasizes that elevated INR alone does not accurately reflect bleeding risk in cirrhosis and should not automatically be "corrected" before procedures. PT and INR show that part of hepatic protein synthesis has weakened. They do not capture the whole balance between clotting and anti-clotting forces in the cirrhotic bloodstream.

So PT and INR matter because they tell you whether the liver is still maintaining enough of the clotting-factor side of plasma architecture to support normal assay clotting. But they should be read as rough windows, not as complete explanations. They can show synthetic decline. They cannot, by themselves, tell you the whole truth about bleeding, clotting, or the total remaining functional reserve of the liver.

Why Injury Markers Are Not the Same as Function Markers

Liver injury markers and liver function markers answer different questions, and confusion between them is one of the biggest reasons people misunderstand liver disease. Injury markers ask whether liver cells or bile-facing tissue are being damaged right now. Function markers ask whether the liver can still perform the synthetic and excretory work the body depends on. ALT and AST rise when hepatocytes are injured enough to leak intracellular enzymes into blood. ALP and GGT rise when the biliary compartment is stressed or bile flow is impaired. These tests tell you the tissue is being hit. They do not, by themselves, tell you how much usable liver function remains.

Function markers are different because they reflect what the liver is still capable of building or clearing. Albumin reflects hepatic protein synthesis over time. PT and INR reflect whether the liver is still making enough clotting factors for plasma to clot normally in the assay. Bilirubin reflects whether the liver can take a heme-derived waste product, process it, and move it out through bile. These are not perfect tests, but they are closer to the question of whether the organ is still doing its job rather than simply whether the tissue is inflamed or damaged.

This is why a person can have very high aminotransferases and still have relatively preserved liver function, at least for a time. Acute hepatitis, ischemic injury, or toxin exposure can injure many hepatocytes quickly enough to drive ALT and AST very high while albumin is still normal and PT or INR are only mildly changed because the liver has not yet exhausted its synthetic reserve or because the plasma pool of those proteins has not yet turned over. The opposite can also happen. A person with advanced cirrhosis may have only modest enzyme elevations because there are fewer actively injured cells leaking enzymes, yet albumin is low, bilirubin is rising, and PT or INR are abnormal because the scarred organ can no longer synthesize and clear what the body needs. Injury and function are therefore related, but they are not the same thing and they do not fail on the same schedule.

The cleanest way to hold the distinction is this. Injury markers tell you the liver is being wounded. Function markers tell you whether the liver is still winning. A wounded liver can still function for a while. A failing liver may not leak dramatically if little active parenchyma remains to leak. That is why clinicians read patterns, not isolated numbers. The bloodstream can show tissue damage without organ failure, and it can show organ failure without spectacular enzyme release. The liver has to be read through both languages at once.

Chapter 49

Reading Patterns Instead of Single Numbers

Hepatocellular Pattern

A hepatocellular pattern means the main injured compartment is the hepatocyte, the liver's main working cell. That is the core idea. When hepatocytes are damaged, their intracellular enzymes leak into blood, which is why aminotransferases, especially ALT and AST, rise more prominently than alkaline phosphatase. In practical liver reading, a predominant AST and ALT elevation points first toward hepatocellular injury rather than toward a primary biliary or cholestatic process. The pattern does not tell you the cause by itself, but it tells you where the wound is centered.

That distinction matters because the liver has more than one kind of failure language. A hepatocellular pattern means the worker is being injured. Viral hepatitis, ischemia, alcohol-related injury, steatohepatitis, autoimmune hepatitis, and many forms of drug-induced liver injury can all produce this kind of signal because they damage hepatocytes directly or through immune or toxic mechanisms. The liver is then telling you that its main metabolic and synthetic cell is under attack, not primarily that bile flow is obstructed.

This is why a hepatocellular pattern should never be reduced to "high liver enzymes" in the vague sense. ALT and AST are injury markers, not clean function markers. They can rise dramatically while albumin, bilirubin handling, and PT or INR remain relatively preserved for a time, because the liver can still compensate even while hepatocytes are actively being wounded. The pattern tells you the cells are being hit. It does not, by itself, tell you how much total hepatic reserve remains.

The point of reading a hepatocellular pattern is therefore architectural. The blood is showing you that the main parenchymal cell is leaking, inflamed, dying, or both. Before you know whether the culprit is viral, toxic, ischemic, autoimmune, metabolic, or drug-related, you already know which compartment failed first. The worker is wounded. The rest of the workup begins there.

Cholestatic Pattern

A cholestatic pattern means the main injured compartment is the bile-forming or bile-carrying side of the liver rather than the hepatocyte alone. In bloodwork, this usually appears as a disproportionate rise in alkaline phosphatase, ALP, often with bilirubin elevation and frequently with gamma-glutamyl transferase, GGT, helping confirm that the source is hepatobiliary rather than bone. That distinction matters because the pattern is telling you the route out is under stress. Bile is not being formed, secreted, or moved normally, and the chemistry of retained bile is beginning to shape the lab profile.

This is why a cholestatic pattern should be read as a flow problem first. The hepatocyte may still be present and capable of much of its synthetic and metabolic work, but bile acids, conjugated bilirubin, cholesterol, and other biliary constituents are not leaving cleanly. The cause may be intrahepatic, such as small-duct disease or impaired canalicular secretion, or extrahepatic, such as a larger duct obstruction. Either way, the liver is beginning to speak in the language of trapped outflow rather than leaked cell contents. That is also why ultrasound and other biliary imaging become so important once the pattern is recognized. The clinician is no longer only asking whether liver cells are injured. The clinician is asking where bile is getting stuck.

The symptom logic follows the same compartment. A cholestatic pattern is more likely to travel with dark urine, pale stool, itching, and fat malabsorption because these symptoms belong to retained bile and reduced biliary delivery into the intestine. Conjugated bilirubin backs into blood and can reach the urine. Less pigment reaches the gut, so stool lightens. Bile acids and related compounds circulate abnormally and itch appears. Less bile reaches the duodenum, so fat handling weakens. The bloodstream is not only showing an abnormal number. It is showing that bile is no longer moving through the correct route.

So a cholestatic pattern is the liver's way of telling you that the drainage side of the organ has become the main site of stress. Hepatocytes may still be wounded secondarily, but the primary story is retained outflow. The worker may still be standing. The stream out has begun to fail.

Mixed Pattern

A mixed pattern means the liver is speaking in two injury languages at once. The hepatocyte is being damaged enough to leak aminotransferases, but bile flow is also being disturbed enough that alkaline phosphatase, bilirubin, and cholestatic physiology begin rising with it. This is not a vague "a little of everything" category. It is a real structural pattern. The worker is hurt, and the route out is hurt at the same time. That can happen because one insult crosses compartments, as in many forms of drug-induced liver injury, or because a primary injury in one compartment begins dragging the other down with it. The important point is that the blood is no longer showing a clean hepatocellular wound or a clean cholestatic blockade. It is showing both.

This matters because mixed injury is often harder to interpret if the clinician reads single numbers instead of tissue logic. ALT and AST may be high enough to show hepatocyte injury, but ALP and often bilirubin are also elevated enough to show impaired bile movement or biliary stress. That means the underlying insult has probably affected both the blood-processing side and the bile-handling side of hepatic work. The hepatocyte may be inflamed, swollen, oxidatively stressed, or dying, and the same hepatocyte may also be failing to maintain canalicular secretion cleanly, or the small ducts may be reacting secondarily to the same injury. A mixed pattern is therefore not just a numerical blend. It is the blood reflecting a liver whose two main directional systems, blood in and bile out, are both being disrupted.

The symptom logic follows that dual injury. A person with a mixed pattern may present with fatigue, malaise, and aminotransferase elevation typical of hepatocellular stress, but also dark urine, itching, and bilirubin elevation more typical of cholestatic disease. The reason is simple. The hepatocyte is wounded enough to leak enzymes, and bile chemistry is being retained enough to spill into blood and alter excretion. This is why mixed patterns often feel messier clinically. The liver is not failing through one clean compartment first. It is failing across the line that normally separates cellular injury from outflow injury.

So a mixed pattern should be read as the liver telling you that the insult has not stayed politely in one domain. The worker is injured. The stream out is impaired. The blood shows both.

Acute Injury Versus Chronic Decline

Acute injury and chronic decline do not represent the same liver failing at different speeds. They are different operating states with different tissue logic. Acute injury means the liver has been hit hard enough that hepatocytes are leaking enzymes, bilirubin handling may be disturbed, and the bloodstream is suddenly seeing the chemistry of cell damage. The cell membranes have been injured, reactive intermediates may be high, blood flow may have been interrupted, a virus may have triggered hepatitis, a toxin may have overwhelmed hepatic defenses, or bile flow may have failed abruptly. The important point is that the event is sharp. The organ is being wounded in the present tense. Chronic decline is different. The liver may not be leaking dramatically on any given day. Instead it is losing reserve, losing architecture, losing synthetic power, and losing clean access between blood and hepatocyte over months or years. Acute injury is a blow. Chronic decline is a reorganization in the wrong direction.

That is why the lab pattern can mislead if the reader thinks only in terms of magnitude. A person with massive acute aminotransferase elevation may still have a liver capable of real recovery if the insult stops and the underlying scaffold remains usable. The enzymes are high because cells have been hurt. The organ may still retain substantial synthetic reserve and regenerative capacity. A person with cirrhosis or long-standing chronic liver disease may show only modest enzyme elevations and yet be much closer to liver failure, because the problem is no longer mainly fresh leakage from wounded hepatocytes. The problem is that scar, nodules, portal resistance, and synthetic decline have been building quietly for years. Blood is no longer meeting the remaining hepatocytes cleanly. Albumin may be low. Bilirubin may be rising. PT and INR may be abnormal. Ascites or encephalopathy may be present. The liver is not necessarily louder. It is weaker.

Acute injury therefore is best read as a question of damage intensity and immediacy. Chronic decline is best read as a question of remaining functional reserve. In acute injury the organ may still know how to be a liver once the hit ends. In chronic decline the organ may be forgetting how to be a liver even on days when no single insult looks dramatic.

This is why injury markers and function markers have to be separated. ALT and AST may spike in acute injury while albumin and coagulation remain relatively preserved at first. In chronic decline, the enzyme picture can flatten while the function picture worsens because there are fewer healthy hepatocytes left to leak and less intact architecture left to support the work those cells are supposed to do.

The cleanest way to hold the difference is this. Acute injury tells you the liver is being struck. Chronic decline tells you the liver is running out of ways to compensate. One is the chemistry of present damage. The other is the slow disappearance of usable organ.

Why Numbers Have to Be Read With Symptoms, Timeline, and Mechanism

A liver number never explains itself. It is only a fragment of a physiological event, and if it is read alone it usually misleads. ALT may be high because hepatocytes are acutely injured, but that does not tell you whether the cause is viral, toxic, ischemic, autoimmune, or metabolic. Bilirubin may be high because of hemolysis, hepatocyte failure, cholestasis, or obstruction, and the same number can sit inside very different tissue stories. Albumin may be low because the liver is losing synthetic reserve, but it can also fall with inflammation, kidney loss, gut loss, or malnutrition. PT and INR may be prolonged because the liver is failing to produce clotting factors, but those tests do not tell you the whole coagulation state and they do not explain whether the problem is acute collapse, chronic decline, or fat-soluble vitamin malabsorption from cholestasis. A number therefore tells you where to look, not what the whole disease is.

Symptoms matter because they show what compartment the body is already losing control over. Dark urine and pale stool tell you bilirubin is failing to reach the intestine and is backing into blood. Itching tells you retained biliary chemistry is circulating. Ascites tells you portal pressure, oncotic decline, and renal compensation have already converged. Confusion tells you gut-derived neuroactive burden is reaching the brain. Right upper quadrant pain may suggest capsule stretch or gallbladder involvement rather than quiet chronic fibrosis. The same bilirubin value means something very different in a jaundiced itchy patient with pale stool than it does in a patient with hemolysis and no cholestatic symptoms. The number is one piece. The symptom pattern tells you which traffic system has already broken loudly enough for the person to feel it.

Timeline matters because liver disease speaks differently when it is acute than when it is chronic. A sudden spike in aminotransferases after a medication overdose tells a different story than a modest elevation present for years in a patient with steatotic liver disease. A normal albumin in the setting of very high ALT may still fit acute injury because synthetic pools have not yet fallen. A low albumin with only mild AST and ALT elevation may point toward chronic architectural disease rather than fresh massive cell death. A bilirubin rise over hours suggests a different mechanism from a bilirubin rise over months. The liver is always changing in time, and laboratory patterns only make sense if they are placed on a clock. A high value without a time course is only half a clue.

Mechanism matters because the liver is not one tissue job. Hepatocellular injury, cholestatic injury, portal hypertension, synthetic failure, and encephalopathy are not the same failure wearing different names. A hepatocellular pattern means the worker is being hit. A cholestatic pattern means the route out is being hit. A rising creatinine in decompensated cirrhosis means the kidney is now living inside a circulation the liver has already broken. A low platelet count may reflect splenic sequestration, reduced thrombopoietin, or both. The same number can belong to a different mechanism depending on what part of the organ or circulation is failing first. This is why pattern recognition matters more than isolated thresholds. The liver has compartments, and numbers become meaningful only when tied back to the compartment that produced them.

So numbers have to be read with symptoms, timeline, and mechanism because the liver is an organ of sequences, not isolated measurements. The bloodstream gives clues. The body gives context. The tissue architecture gives explanation. A number alone is a signal. Put it inside the right story, and it becomes physiology.

The Danger of Treating One Abnormal Lab as a Diagnosis

One abnormal liver lab is not a diagnosis because the liver only has a limited number of ways to leak its distress into blood. ALT can rise when hepatocytes are injured, but that same rise can come from viral hepatitis, ischemia, steatohepatitis, autoimmune injury, alcohol, drug toxicity, or other hepatocellular stress. Bilirubin can rise because too much pigment is arriving from heme breakdown, because hepatocytes are failing to process it, or because bile flow is obstructed after conjugation has already occurred. Alkaline phosphatase can point toward cholestasis, but not every alkaline phosphatase belongs to the liver, and even when it does, the next question is still where the bile pathway failed and why. Albumin can be low because hepatic synthesis is declining, but also because inflammation has shifted protein priorities, the kidney is losing protein, the gut is losing protein, or intake has been poor long enough to thin the substrate available for synthesis. The blood is showing a disturbance. It is not naming the cause for you.

This is why single-number thinking is so dangerous. It seduces the reader into treating chemistry as a label instead of as a clue. A person with a high AST is said to "have liver damage" as though that statement were already complete. A person with a high bilirubin is said to "have jaundice" as though the pigment itself were the disease rather than the visible result of a broken route. A person with an elevated alkaline phosphatase is said to "have cholestasis" without asking whether the failure sits in canalicular secretion, small-duct injury, large-duct obstruction, or some nonhepatic source entirely. The liver is an organ of compartments and sequences. One laboratory number can tell you that one part of that sequence is stressed. It cannot tell you by itself what the whole tissue event is.

The deeper problem is that isolated labs blur the difference between injury and function. ALT and AST can be strikingly high in an acute hepatocellular injury while albumin, PT, and bilirubin remain relatively preserved for a time because the organ still has reserve. A cirrhotic liver can show only modest aminotransferase elevation while albumin is low, INR is high, bilirubin is rising, and ascites or encephalopathy are already present. Treat the enzyme number as the diagnosis and you will miss the organ that is quietly failing behind it. Treat the bilirubin as the diagnosis and you may miss whether the burden is prehepatic, hepatocellular, or obstructive. Treat the INR as the diagnosis and you may miss that the liver's synthetic decline is only one part of a larger unstable coagulation state. A lab value is a window into a process. Confusing the window with the process is how shallow medicine happens.

So the danger of treating one abnormal lab as a diagnosis is simple. It stops the reasoning too early. The number should start the work, not end it. The real question is always what tissue, what route, what timing, what burden, what pattern. The liver does not speak in single words. It speaks in sequences.

Chapter 50

Imaging the Liver

Ultrasound

Ultrasound is usually the first imaging tool used in liver and biliary work because it is fast, noninvasive, relatively inexpensive, and especially useful for the structures that matter early in hepatology: the gallbladder, bile ducts, liver surface, spleen, ascites, and the larger vessels around the liver. It gives structural information rather than full biochemical explanation, but structure matters because so much liver disease is a problem of flow, outflow, distortion, and pressure before it is a problem of a dramatic mass lesion. Ultrasound is first-line for visualizing the biliary system, for evaluating right upper quadrant pain, for separating intrahepatic from extrahepatic causes of jaundice, and for screening for liver masses.

What the machine is really doing is simple. It sends sound waves into the tissue and reads how those waves reflect back depending on density, interface, and fluid content. That makes fluid black, dense interfaces bright, and moving blood readable when Doppler is added. In practical liver language, this means ultrasound is good at showing whether the liver looks enlarged, nodular, fatty, congested, or mass-bearing, whether bile ducts are dilated, whether the gallbladder contains stones or sludge, whether ascites is present, and whether the spleen is enlarged enough to suggest portal hypertension. It can also detect diffuse disorders such as cirrhosis and fatty liver, though it does not explain them at molecular depth.

This is why ultrasound belongs at the front of liver imaging rather than at the end. A cholestatic patient with dark urine and pale stool needs a first look at the biliary tree. A patient with suspected portal hypertension needs a first look at spleen size, ascites, and vascular directionality. A patient with metabolic liver disease needs a first look at whether the parenchyma already looks echogenic enough to suggest steatosis or irregular enough to suggest chronic architectural change. Ultrasound can localize focal lesions and can guide biopsy, but its deeper value is that it gives the clinician the first structural map of the problem. The blood may already have shown the pattern. Ultrasound shows where that pattern is living.

Doppler ultrasound extends that map from shape into flow. It can assess the direction of blood flow and vessel patency around the liver, especially in the portal vein, hepatic veins, and shunts. That matters because liver disease is often a disease of circulation before it is a disease of complete organ disappearance. Doppler can help detect portal hypertension through collateral flow patterns and portal vein enlargement, evaluate portal and hepatic vein patency, assess liver shunts, and identify unusual vascular structures such as cavernous transformation. In a cirrhotic patient, that means ultrasound is not only asking what the liver looks like. It is asking whether blood is still moving through it in the direction liver physiology requires.

Modern ultrasound also includes elastography, which is one of the most useful examples of imaging becoming more physiological. Elastography measures liver stiffness by tracking how shear waves move through the tissue. A stiffer liver conducts that wave differently, and stiffness becomes a noninvasive index of fibrosis. This does not replace all other methods, and it does not reduce fibrosis to one number, but it gives the clinician something closer to architectural burden than a simple grayscale image alone. That is why elastography is often paired with blood tests in chronic hepatitis and steatotic liver disease. The machine is no longer only showing anatomy. It is estimating how much scar has changed the feel of the organ.

Ultrasound also has limits, and those limits matter because the liver should never be read through one modality alone. It is operator-dependent, it can be more difficult in patients with obesity or intestinal gas, and its ability to characterize lesion vascularity and tissue composition is limited enough that it cannot definitively diagnose hepatocellular carcinoma on its own. It is excellent as a first structural examination, but not as the final word on every mass, every diffuse disorder, or every subtle vascular question. That is why CT, MRI, and more specialized imaging often follow when the first picture is not enough. Ultrasound opens the case. It does not always close it.

Computed Tomography

Computed tomography, CT, becomes especially useful when the clinician needs more than a first structural look. Ultrasound is excellent for triage, bile ducts, gallbladder, ascites, and a first look at liver texture and flow. CT is the next step when the question becomes sharper: Is there a mass. Is there vascular invasion. Is there active bleeding. Is the liver surface nodular because of cirrhosis or distorted by something focal. Are the bile ducts dilated because of a mechanical obstruction. CT gives a more detailed cross-sectional map of the liver, surrounding organs, vessels, and peritoneal cavity than ultrasound can usually provide. It is particularly valuable for detecting and characterizing focal lesions, staging malignancy, and evaluating complications of portal hypertension or trauma.

What makes CT so powerful in liver imaging is not just that it shows anatomy. It shows anatomy through time after contrast enters the bloodstream. Liver lesions are often defined by how they take up and release contrast across phases rather than by how they look on a single still image. Multiphasic CT is therefore central to liver work. The study may include an unenhanced scan, an arterial phase, a portal venous phase, and sometimes a delayed phase, depending on the clinical question. This matters because the liver has a dual blood supply and focal lesions often reveal themselves by how they behave during those different windows. Hypervascular lesions may stand out in the arterial phase. Other lesions become clearer in the portal venous phase. Washout, the loss of contrast relative to surrounding liver on later phases, can be a critical feature in characterizing hepatocellular carcinoma. CT is therefore not only looking at shape. It is watching blood move through abnormal tissue.

That phase logic makes CT especially important in liver cancer work. For hepatocellular carcinoma, modern liver imaging depends heavily on multiphasic contrast behavior, especially arterial phase hyperenhancement and later phase washout. This is one reason ACR and AASLD-linked guidance emphasize multiphase CT or MRI rather than single-phase studies when a suspicious liver lesion is being worked up. A lesion is not defined only by being present. It is defined by how it lives in blood. CT lets the clinician watch that vascular personality directly.

CT also matters because it sees beyond the liver itself. A liver disease patient may need assessment of portal vein patency, varices, splenomegaly, ascites, lymph nodes, metastatic spread, bowel changes, or postoperative anatomy. Cross-sectional imaging captures these relationships in one study. In trauma, arterial and venous phases can help identify active bleeding and solid organ injury. In chronic liver disease, CT can show nodular surface contour, portosystemic collaterals, splenic enlargement, and other signs that the organ has become a hemodynamic problem as much as a parenchymal one. The liver is rarely failing alone, and CT is one of the best ways to see that wider anatomical field.

But CT has limits, and those limits matter because the liver should never be read through one image modality alone. CT uses ionizing radiation. It often requires intravenous contrast, which may not be ideal in some patients with kidney impairment or contrast allergy. Some small lesions remain indeterminate and still need MRI for better characterization. Diffuse liver disease can alter CT appearance, but fibrosis, inflammation, fat, and malignancy do not always separate cleanly on CT alone. This is why CT is best understood as a high-resolution structural and vascular tool, not as a total explanation. It shows the organ in cross-section and in contrast phases. The physiology still has to be read through the pattern.

Magnetic Resonance Imaging

Magnetic resonance imaging, MRI, becomes most useful when the clinician needs more than a first structural survey and more than the density-based cross-sections CT provides. MRI is especially strong for liver work because it can image hepatic tissue, blood vessels, and bile ducts with excellent soft-tissue contrast, and it is better than CT or ultrasound for many diffuse liver disorders and for clarifying focal lesions such as tumors and hemangiomas. In other words, MRI is not just a prettier picture. It is a way of separating tissue types and fluid compartments that the liver often blurs on other studies.

That matters because the liver fails through architecture as much as through chemistry. Fatty change, iron overload, fibrosis, vascular distortion, nodules, bile duct disease, and focal masses can all alter the organ in ways that require more than a grayscale look. MRI is particularly valuable when the question is not simply "is there a lesion" but "what kind of tissue is this lesion behaving like." A hemangioma, a hepatocellular carcinoma, a metastasis, an abscess, or focal nodular hyperplasia may all occupy space in the liver, but they do not handle signal and contrast the same way. MRI is useful because it lets the clinician read tissue character, not only location.

MRI also extends beyond ordinary anatomic imaging into specialized forms that make liver physiology more visible. Magnetic resonance cholangiopancreatography, MRCP, images the biliary and pancreatic ducts noninvasively and is more sensitive than CT or ultrasound for common bile duct abnormalities, especially stones. It has largely replaced older nuclear biliary scans in many obstructive settings and is the screening test of choice for primary sclerosing cholangitis because it can map the biliary tree without entering it invasively. Magnetic resonance elastography measures liver stiffness and provides a highly accurate noninvasive estimate of fibrosis. Magnetic resonance proton density fat fraction can quantify hepatic fat content across the whole liver, reducing the sampling error that can affect biopsy. MRI-based iron quantification can estimate hepatic iron concentration when iron overload is a concern. These are not minor add-ons. They are ways of turning imaging into a direct readout of bile flow, scar burden, fat burden, and iron burden.

This is why MRI is so often the study that follows uncertainty. Ultrasound can show a mass, bile duct dilation, splenomegaly, or steatosis. CT can stage anatomy and vascular behavior with contrast phases. MRI is often where the liver's tissue identity becomes clearer. It sees blood flow, ducts, fat, iron, fibrosis, and focal lesions with a level of contrast and characterization that makes it especially valuable when the first scan showed that something was wrong but could not yet say exactly what kind of wrong it was.

Elastography and Fibrosis Assessment

Fibrosis used to hide behind the word chronic until the tissue was scarred enough to declare itself through portal hypertension, thrombocytopenia, ascites, or a biopsy needle. Elastography changed that by asking a more physical question much earlier. Not what the liver enzymes look like. Not what the liver says about itself in blood alone. What does the tissue feel like. A healthy liver is soft, compliant, and built to let blood spread through a low-resistance sinusoidal field. A fibrotic liver becomes stiffer because extracellular matrix expands, collagen cross-links, and the exchange architecture hardens. Elastography measures that change in stiffness. It does not look for scar by color or by gross nodules. It measures how easily a mechanical wave can move through the organ and uses that to estimate how much the liver has stopped behaving like a soft exchange tissue.

That matters because fibrosis is an architectural disease before it is a surface disease. Collagen is being laid into the space of Disse, around sinusoids, through portal tracts, and eventually across parenchyma in bridging bands. The liver may still look only mildly abnormal on routine ultrasound, and enzymes may be quiet enough to mislead, but the tissue is already harder than it should be. Elastography makes that hidden change more visible without needing to remove tissue. The basic logic is simple. Softer tissue deforms differently under force than stiff tissue does. The stiffer the liver, the faster or more forcefully the wave travels, depending on the specific technique. That stiffness then becomes a surrogate for fibrosis burden. It is not scar photographed directly. It is scar inferred through the mechanical behavior scar creates.

This is why elastography is so useful in chronic liver disease. Fibrosis does not develop all at once, and biopsy has limits even when it is done well. A biopsy samples only a tiny cylinder of tissue from an organ whose disease may be patchy, and that means sampling error is built into the method. Elastography reads a larger field. It is not perfect, but it gives the clinician something closer to a whole-organ mechanical impression rather than a single microscopic corridor. In viral hepatitis, steatotic liver disease, alcohol-related injury, and other chronic processes, that matters because the central question is often whether inflammation has already become scar and whether scar is still limited or beginning to distort architecture more seriously. Elastography helps answer that question by turning fibrosis into a physical measurement rather than waiting for the body to reveal it through later decompensation.

Different forms of elastography do this with different hardware, but the tissue logic is the same. Transient elastography uses a mechanical impulse and ultrasound-based wave tracking. Shear-wave ultrasound elastography folds the measurement into a conventional ultrasound platform. Magnetic resonance elastography uses the same principle inside MRI and can map stiffness across larger regions of the liver with greater detail. The tools differ. The physiology does not. All of them are trying to detect when the liver has stopped being soft enough to function like a healthy flow organ. That is why elastography belongs to fibrosis assessment rather than merely to imaging. It is not only showing where the liver is. It is telling you what kind of tissue the liver has become.

The limits are important because stiffness is not scar alone. Inflammation, congestion, cholestasis, and even recent meals in some settings can change measurements enough to confuse the picture. A very inflamed liver can feel stiffer than its true long-term fibrosis burden would suggest. A congested liver can do the same. So elastography has to be read with the pattern, not as a number that explains itself. But when used correctly, it gives something blood tests cannot give directly. It gives a readout of architecture. Not by cutting tissue out, but by asking the organ how hard it has become.

That is the deeper value of elastography. Fibrosis is the liver forgetting how to stay soft enough for exchange. Elastography measures that forgetting.

What Imaging Can Show and What It Cannot

Imaging can show structure, flow, stiffness, duct dilation, splenomegaly, ascites, nodules, masses, and vascular distortion. It can show what shape the liver has become, how fat-filled it looks, whether bile ducts are enlarged, whether portal hypertension has begun remodeling the circulation, whether a focal lesion is present, and, with elastography, whether the tissue has become mechanically stiffer in a way that suggests fibrosis. Ultrasound is often the first structural look. CT and MRI add cross-sectional detail, vascular behavior across contrast phases, and better lesion characterization. Elastography adds a physical estimate of fibrosis burden by measuring stiffness. In practical terms, imaging is excellent for answering questions of anatomy, flow, obstruction, and architecture.

What imaging cannot do reliably by itself is name every cause at the cell level. A scan can show fat in the liver, but routine imaging cannot tell you by itself whether that fatty liver is simple steatosis or steatohepatitis with active inflammation. It can show cirrhotic nodularity, but not always the precise mechanism that created the scar. It can detect a lesion, but not every lesion can be definitively classified without the right contrast pattern, follow-up, or sometimes tissue. It can suggest fibrosis through stiffness, but stiffness can also be influenced by inflammation, congestion, and cholestasis, not scar alone. Imaging therefore shows consequences and patterns better than it shows exact molecular causality.

That is why liver imaging has to be read beside symptoms, labs, and mechanism rather than treated as a final answer generator. A scan may show a dilated biliary tree, but the clinician still has to decide whether the underlying problem is stone, stricture, tumor, or inflammatory duct disease. It may show a stiff liver, but the reader still needs the timeline, enzyme pattern, bilirubin behavior, and synthetic markers to know whether the disease is acute, chronic, cholestatic, metabolic, vascular, or malignant. Imaging is strongest when it is used to answer the right kind of question. It shows what the liver looks like, how blood and bile may be moving, and how architecture may already have changed. It does not replace physiology. It gives physiology a map.

Why Structure, Stiffness, Flow, and Obstruction Each Require Different Tools

The liver does not fail in one dimension, so it cannot be imaged honestly with one kind of question. Structure, stiffness, flow, and obstruction are not four names for the same thing. They are four different physical problems. Structure asks what the organ and its surrounding anatomy look like. Is the liver enlarged, nodular, fatty, mass-bearing, cirrhotic, shrunken, or distorted by tumor, cyst, abscess, or vascular remodeling. Stiffness asks what the tissue feels like mechanically. Has fibrosis changed the liver from a soft exchange organ into a harder more resistant one. Flow asks how blood is actually moving through the portal vein, hepatic veins, collaterals, and shunts. Is the stream still crossing the organ in the direction and pattern liver physiology requires. Obstruction asks whether bile or blood is being physically blocked in a duct or vessel. Is the route out narrowed, compressed, occluded, or backed up. These are different questions, and different imaging tools exist because the body does not reveal all of them through one visual language.

A standard grayscale ultrasound is good at the first kind of work. It can show size, contour, obvious masses, steatosis, ascites, gallstones, splenomegaly, and ductal dilation. That makes it useful for structure and for a first look at obstruction. But ultrasound alone is not the best tool for everything that matters. A liver can look roughly present on grayscale imaging and still be fibrotic in a way the eye cannot grade well. Blood can be moving in the wrong direction and not announce itself on a static image. A bile duct can be enlarged, but the cause of that enlargement may remain uncertain. So the first tool opens the structural map. It does not solve every physiological question hidden inside it.

Stiffness needs a different kind of measurement because fibrosis is not only a visible scar. It is a mechanical change in how the tissue transmits force. Elastography exists for that reason. It asks what the liver feels like to a wave. A healthy liver is soft and compliant. A fibrotic liver is harder, less deformable, and more resistant. That change may precede the kind of obvious nodular architectural collapse that can be seen clearly on routine imaging. So stiffness needs a tool that reads mechanics, not only appearance. If the question is whether fibrosis has already buried the exchange architecture, a standard structural image may suggest it, but elastography is better designed to measure it directly.

Flow needs another tool because blood movement is not a still photograph. Doppler ultrasound, contrast-enhanced CT, MRI, and related vascular techniques are useful here because they can show whether portal blood is moving toward the liver or away from it, whether the portal vein is patent, whether hepatic veins are draining normally, whether collaterals have formed, and whether shunts are bypassing functional parenchyma. A cirrhotic liver can be structurally abnormal, but the real physiological danger often depends on what flow has become inside and around that structure. Flow therefore requires tools that can read direction, velocity, enhancement, and vascular timing rather than tissue shape alone.

Obstruction requires its own approach because blocked ducts and blocked vessels are route problems. The question is not only what the liver looks like. The question is where the stream stops. Ultrasound may show bile duct dilation and suggest there is a block downstream. MRCP, magnetic resonance cholangiopancreatography, is better for mapping the biliary tree in detail without entering it invasively. CT may identify a mass compressing the ducts. MRI may clarify whether a lesion is invading the hilum. A vascular obstruction may require venous-phase imaging or dedicated vascular study. The route out is a different anatomical problem from parenchymal texture or tissue stiffness, so the imaging has to follow the route itself.

That is why structure, stiffness, flow, and obstruction each require different tools. The liver is a shape, a texture, a hydraulic field, and a drainage system at the same time. No single image answers all four questions equally well. A good clinician does not ask the machine to explain the whole organ in one glance. The clinician asks a better question first, then chooses the tool that can actually see that kind of failure.

Chapter 51

Biopsy, Histology, and What Tissue Reveals

Why Tissue Sampling Is Still Sometimes Necessary

Blood tests can tell you that liver cells are being injured, that bile flow is disturbed, that synthetic reserve is falling, or that bilirubin is no longer moving correctly. Imaging can show contour, masses, duct dilation, stiffness, ascites, splenomegaly, and vascular distortion. But neither blood nor imaging can always tell you exactly what the tissue has become. That is why biopsy is still sometimes necessary. A liver disease can look similar in blood while being very different under the microscope. Steatosis can be present with little inflammation, or it can already have crossed into steatohepatitis with ballooned hepatocytes, inflammatory infiltrates, and early fibrosis. Cholestasis can be due to duct injury, canalicular dysfunction, obstruction, or mixed parenchymal disease. Autoimmune injury can mimic viral or drug-related injury in enzymes and symptoms. A scan can suggest cirrhosis, but only tissue can sometimes show how active the inflammatory process still is, how much fibrosis is present, and whether the parenchyma is being lost through one mechanism or another. Blood shows consequence. Tissue shows the event.

Histology matters because the liver is an architectural organ. The pathologist is not only looking for "damage." The pathologist is looking at where the damage sits, how the cells are responding, and what kind of repair is replacing the original structure. Are hepatocytes swollen, ballooned, necrotic, apoptotic, fatty, cholestatic, iron-loaded, copper-loaded, or full of retained abnormal protein. Are portal tracts inflamed. Are bile ducts lost, injured, or surrounded by fibrosis. Is the space of

Disse thickened. Are fibrous septa beginning to bridge portal areas and central veins. Are regenerative nodules already forming. Is there tumor. Tissue sampling becomes necessary when the clinician needs that level of spatial truth. It is the difference between knowing the liver is abnormal and knowing which compartment failed first, which direction the disease is moving, and what kind of scar is being built in response.

Biopsy is therefore not the first answer to every liver problem and should not be romanticized. It samples only a tiny core from a large organ, which means sampling error is real. Some diseases are patchy. One tract can look worse than another. Some questions are answered better by imaging or bloodwork than by tissue. But when the diagnosis remains uncertain, when staging matters, when competing mechanisms look plausible, or when treatment decisions depend on what the parenchyma actually shows, biopsy remains one of the few ways to stop guessing about architecture. The needle does not merely collect cells. It collects pattern, and in liver disease pattern is often the real diagnosis.

Steatosis, Inflammation, Necrosis, Fibrosis, Cirrhosis

These are not five separate liver words standing next to each other on a pathology slide. They are five different tissue states in a sequence of failure. The sequence matters because the liver does not move from health to cirrhosis in one jump. It is pushed there through changes in what the hepatocyte is storing, what the tissue is reacting to, what the immune system is amplifying, and what kind of repair the organ begins choosing. Steatosis is the earliest visible concession. The hepatocyte is receiving or synthesizing more lipid than it can oxidize or export cleanly, so it begins converting part of that burden into triglyceride droplets and holding them in the cytoplasm. At first that can look deceptively quiet. The cell is still alive, still performing much of its work, and still trying to buffer the burden rather than collapse under it. But the droplet is already telling the truth. Traffic has failed. More substrate is arriving than the cell can move through normal exits.

Inflammation begins when storage stops being a quiet containment strategy and starts becoming a tissue problem. The hepatocyte is no longer simply fat-loaded. It is now stressed enough to distort mitochondria, strain the endoplasmic reticulum, generate oxidative stress, and release danger signals that neighboring cells can read. Kupffer cells begin responding more aggressively. Cytokines rise. Sinusoidal immune tone changes. The tissue is no longer saying, "I am overloaded."

It is now saying, "I am being injured." This is the point where the disease stops being only a metabolic storage disorder and becomes an inflammatory organ disorder. The fat is still there, but the chemistry around that fat has changed. Lipotoxic intermediates, oxidized lipids, mitochondrial dysfunction, and inflammatory signaling are now part of the same field. A fatty liver at this stage is not just holding excess fuel. It is reacting to the cost of holding it.

Necrosis means the burden has become severe enough that hepatocytes are no longer merely stressed or ballooned. They are dying. The membrane loses integrity, the intracellular contents spill, and the surrounding tissue is exposed to the biochemical violence of cell death. This matters because a living but dysfunctional hepatocyte still offers the possibility of metabolic recovery. A necrotic hepatocyte is gone. The liver now has to replace parenchyma rather than merely calm it. Necrosis is therefore a threshold event. It tells you the tissue is no longer handling the burden at the level of reversible strain alone. The organ is losing working mass. Inflammation becomes more dangerous here because dead and dying cells amplify the wound signal, and the wound signal keeps telling surrounding tissue that the response can no longer remain temporary.

Fibrosis is the liver's attempt to survive that repeated injury by reinforcing itself structurally. Stellate cells in the space of Disse stop behaving like quiet vitamin A reserve cells and become matrix-producing cells. Collagen begins accumulating in the exchange spaces, around sinusoids, and through portal regions. At first that can be subtle enough to miss clinically, but it changes the tissue future immediately. The liver is an exchange organ. It survives by keeping blood close to hepatocytes, by keeping the space of Disse thin, by letting sinusoids remain low-resistance channels, and by preserving the polarity that separates blood-facing from bile-facing work. Fibrosis attacks all of that. The tissue may still contain many hepatocytes, but blood no longer meets them as cleanly, oxygen does not move as well, and the organ begins trading permeability for stiffness. Scar, in the liver, is never just a patch. It is an interference pattern laid into function.

Cirrhosis is what happens when fibrosis and regeneration stop being local and become the dominant architecture of the organ. The liver is now partitioned into nodules by fibrous septa. Blood flow is distorted. Portal resistance is high. Bile flow is less reliable. The lobular map that once told hepatocytes what blood they were standing in has been broken. This is why cirrhosis is not just "advanced fibrosis" in a casual sense. It is the collapse of organization. The liver still has cells. It may still make protein, process bilirubin, and conjugate drugs in part. But those cells are now trapped inside an organ that no longer gives them normal access to blood, normal gradients of oxygen and substrate, or normal routes of outflow. The person is no longer only carrying scar. The person is carrying a liver that has been rebuilt into the wrong shape.

So the sequence has to be read as a deepening argument between burden and structure. Steatosis means the cell is storing what it cannot process. Inflammation means the tissue is reacting to the consequences of that storage or other chronic injury. Necrosis means working cells are being lost. Fibrosis means repair has begun speaking in collagen. Cirrhosis means collagen and regeneration together have rewritten the organ into a nodular high-resistance field. These are not neat boxes. They overlap, feed each other, and often coexist in the same liver. But the direction is clear. What begins as excess substrate or repeated injury ends, if unchecked, as a liver that is still present anatomically and less and less available physiologically.

Histology as Architecture Made Visible

Histology matters because the liver is an organ whose logic lives in arrangement. Blood enters in one place, bile exits in another, oxygen falls along a gradient, hepatocytes change their behavior according to where they sit inside that gradient, and the exchange surface between plasma and cell has to remain thin, open, and properly polarized if the organ is going to function. A gross specimen can show size, color, nodularity, congestion, or a mass. Blood tests can show that cells are leaking, bile is retained, or synthesis is failing. Histology is where the actual arrangement becomes visible. It lets the reader stop guessing about whether the problem is mainly hepatocellular, biliary, vascular, inflammatory, fatty, fibrotic, or neoplastic and instead see where in the tissue the event is sitting. A histologic section is not interesting because it is microscopic. It is interesting because it turns the hidden architecture of function into something the eye can finally inspect.

This is why histology is never just a search for "abnormal cells." The pathologist is reading pattern. Are hepatocytes arranged in preserved plates or trapped in regenerative nodules. Are portal tracts expanded with inflammation. Is injury concentrated at the interface between portal tracts and parenchyma. Are bile ducts present, damaged, or disappearing.

Are canaliculi plugged with retained bile. Is the space of Disse still permissive or thickened with matrix. Are sinusoids open, congested, capillarized, or compressed by scar. Is fat present only as quiet steatosis, or are hepatocytes ballooned and inflamed in a way that says the droplet has already become injury. The slide answers these questions because it preserves spatial relationships. Histology is not merely listing what molecules are present. It is showing who is next to whom, what route has narrowed, what compartment has swollen, and what kind of repair the liver has started choosing.

That is what makes histology architecture made visible. In a healthy liver, the tissue reads like ordered traffic. Portal triads sit where inflow and biliary outflow belong. Blood moves through soft low-pressure sinusoids toward central veins. Hepatocytes face blood on one side and bile on the other. Zonation creates different metabolic neighborhoods along the portal-to-central axis. Stellate cells remain quiet in the space of Disse. In disease, the slide shows exactly how that order is being lost. Viral hepatitis widens portal tracts and drives interface activity where immune attack spills into parenchyma. Cholestatic disease injures ducts and leaves bile where it should not be. Steatohepatitis balloons hepatocytes, draws inflammatory cells into the lobule, and begins converting metabolic congestion into structural damage. Fibrosis lays collagen into the exchange field and eventually links one vascular territory to another in bridging bands. Cirrhosis converts the whole organ into nodules cut by septa. The beauty of histology is that all of these processes stop being abstract mechanisms and become visible geography.

It also matters because architecture explains function better than numbers do. A bilirubin can rise and tell you pigment is being mishandled, but it cannot show whether bile ducts are vanishing, whether hepatocytes are failing to conjugate and secrete, or whether fibrosis has begun partitioning the organ into nodules that no longer drain cleanly. ALT and AST can tell you cells are being injured, but they cannot show whether the injury is periportal, centrilobular, interface-driven, fatty, ischemic, or drug-related in pattern. Imaging can show stiffness, nodularity, masses, duct dilation, or portal hypertension, but histology can still reveal what kind of tissue event created those consequences. This is why biopsy remains necessary in selected cases. The liver often needs to be read not only through chemistry and scans, but through the arrangement of cells, ducts, vessels, matrix, and inflammatory infiltrate. Function leaves clues in blood. Architecture reveals the mechanism.

The deeper reason histology matters so much in the liver is that the liver can keep functioning while its architecture is quietly changing for years. A person can have fatty change, interface inflammation, portal fibrosis, duct injury, or early bridging scar before the organ has fully declared failure in a dramatic clinical way. The microscope can catch the tissue at that stage where the future is already visible but not yet final. It can show whether the liver is likely moving toward recovery, chronic inflammatory persistence, cholestatic remodeling, or cirrhotic collapse. Histology, then, is not only a way to confirm diagnosis. It is a way to see the future shape of the organ in the present tense. That is why it deserves to be understood not as tiny details on glass, but as liver architecture made visible enough to read.

What Pathologists Look For

A pathologist does not look at liver tissue the way a casual reader looks at a damaged organ and asks whether it seems generally inflamed or generally scarred. The pathologist is reading pattern, compartment, and sequence. The first question is where the injury lives. Is it centered in hepatocytes, in portal tracts, in bile ducts, around central veins, in the sinusoids, or in the matrix between blood and cell. That matters because the liver has more than one kind of architecture, and disease does not strike all of them equally. A hepatocyte disease looks different from a bile duct disease. A portal inflammatory process looks different from centrilobular toxic injury. Steatosis looks different from ballooning degeneration. Bridging fibrosis looks different from cirrhosis. So the pathologist begins by orienting the tissue spatially. Where is the blood supposed to be moving. Where is the bile supposed to be moving. Which compartments still look ordered, and which have started speaking the language of injury.

After that comes the question of what the cells themselves are doing. Are hepatocytes swollen, pale, ballooned, fatty, pigmented, cholestatic, iron-loaded, copper-laden, or packed with abnormal retained protein. Are they dying by apoptosis, where single cells shrink and fragment in a more controlled way, or by necrotic collapse, where larger areas lose membrane integrity and spill their contents into the tissue. Are there regenerative changes suggesting the organ is trying to replace lost mass. Are the cells atypical in a way that raises concern for dysplasia or malignancy. The pathologist is not only asking whether cells look sick. The pathologist is asking what kind of biochemical burden the cells appear to be carrying and what kind of response they are mounting against it. A fatty hepatocyte is telling one story. A ballooned hepatocyte with Mallory-Denk type injury is telling another. A hepatocyte packed with copper-associated protein or alpha-1 antitrypsin globules is revealing a very different mechanism again.

Inflammation is read with the same precision. Not all inflammatory infiltrates mean the same thing. A portal tract rich in lymphocytes and plasma cells pushes the mind toward a different disease pattern than a lobular inflammatory infiltrate with scattered dying hepatocytes, and both differ again from neutrophil-rich injury clustered around damaged fatty cells. Interface activity matters because it shows inflammation spilling from portal tracts into adjacent parenchyma, which is a structurally important clue in chronic hepatitis and autoimmune-pattern injury. Bile ducts are examined carefully. Are they intact, inflamed, disappearing, proliferating, or surrounded by destructive immune activity. Is bile accumulating in canaliculi or trapped in hepatocytes. A biopsy can therefore distinguish whether jaundice and cholestatic enzymes reflect duct injury, hepatocellular bile secretion failure, or mixed disease. The pathologist is reading the location and the character of the inflammatory response because inflammation in the liver is never just "present." It has a distribution, and that distribution usually says something about cause.

Fibrosis is read the same way, as architecture rather than as background scar. The question is not simply whether collagen exists. The question is where it is being laid down and how far that process has advanced. Is fibrosis confined to portal expansion, or has it started extending outward as delicate septa. Are portal areas beginning to bridge to one another or to central veins. Has the space of Disse begun thickening in a way that would make sinusoidal exchange less efficient. Are nodules forming in a way that means the organ is no longer only scarred but cirrhotic. This is one of the most important things tissue can reveal, because fibrosis is the liver teaching itself to survive by building the wrong kind of support. The pathologist is looking to see whether the repair response is still local and potentially reversible or whether the liver has already begun converting repeated injury into a new permanent map.

The deeper truth is that pathologists are looking for mechanism made visible. They are not simply classifying damage. They are asking what sequence of injury, adaptation, failed adaptation, and repair could have produced this architecture. A biopsy is valuable because it lets the liver stop speaking only through blood and imaging and start speaking through its own arrangement of cells, ducts, vessels, pigments, droplets, scars, and infiltrates. The pathologist is listening for that arrangement.

Limits of Sampling and Interpretation

A liver biopsy shows real tissue, but it does not show the whole liver. That is the first limit, and it is large enough to mislead if the reader forgets it. The pathologist is usually looking at a narrow cylinder of parenchyma taken from an organ weighing several pounds, an organ whose disease may be patchy, zonal, duct-centered, portal-centered, or unevenly scarred from one region to another. A core can pass through a tract rich in inflammation and make the disease look more active than the rest of the organ. A different core can miss the worst septa, the worst steatohepatitic change, the small duct lesion, or the edge of a nodule and make the disease look quieter than it truly is. This is why tissue is powerful and limited at the same time. It is direct evidence, but it is local evidence. The microscope sees what the needle captured, not automatically what the entire liver has become.

That matters especially in diseases defined by distribution. Fibrosis does not always advance evenly. Steatohepatitis can be more severe in one region than another. Bile duct loss may be focal enough that one sample contains the clue and another does not. Early cirrhosis can hide in a liver that still has broad regions of less distorted tissue. Tumor can be missed if the core never touched it, or overcalled if the sample comes from a particularly atypical regenerative edge that is not representative of a larger benign pattern. Even a very good pathologist cannot read tissue that was not sampled. The biopsy may reveal a mechanism clearly, but it can also under-stage or overemphasize part of the process simply because the needle entered one neighborhood and not another.

Specimen quality creates a second limit. A liver core has to be long enough, preserved well enough, and contain enough portal tracts and parenchyma to let architecture be read with confidence. Crush artifact, fragmentation, hemorrhage, poor fixation, tangential sectioning, and scant tissue can all damage the very patterns the pathologist needs to see. A distorted portal tract may look more inflamed than it really is. A broken core may make bridging fibrosis harder to judge. A tiny sample may show fat but not enough architecture to decide whether the disease has already become cirrhotic. Histology feels definitive because it is tissue, but tissue can still be technically compromised before interpretation even begins. A poor specimen does not simply produce a weaker answer. It can produce the wrong question.

Interpretation itself has limits because the liver does not always divide itself into textbook-pure diseases. Many patients carry overlap. A steatotic liver may also have alcohol-related injury, cholestatic stress, mild drug injury, and chronic congestion. Viral hepatitis can coexist with metabolic dysfunction. Autoimmune-pattern inflammation can be mimicked by drug injury. Cirrhosis can flatten older patterns and leave the pathologist reading the remains of multiple earlier events layered into one scarred organ. Histology therefore still requires clinical context. The pathologist may be able to say what the tissue pattern most resembles, how active the inflammation looks, how advanced the fibrosis appears, and what compartments are being hit, but the final interpretation is strongest when it is tied back to symptoms, laboratory patterns, imaging, exposure history, and time course. Tissue speaks clearly. It does not always speak alone.

This is why biopsy should be respected without being worshiped. It can reveal architecture in a way blood and imaging cannot. It can show ballooned hepatocytes, duct injury, interface activity, iron or copper loading, retained abnormal protein, and the actual placement of scar. But it is still a sample, still a moment in time, and still an interpretation built from a partial window into a large changing organ. The best use of biopsy is not to pretend it abolishes uncertainty. The best use is to let it narrow uncertainty by showing which tissue event is real enough to be seen directly.

Why Seeing the Tissue Changes How Disease Is Understood

Seeing the tissue changes how disease is understood because the liver stops being an abstraction and becomes an arrangement. Blood tests can tell you that hepatocytes are leaking, bilirubin is rising, bile is being retained, albumin is falling, or clotting factor synthesis is weakening, but none of those numbers can show you where the injury is sitting or what kind of architecture has begun replacing normal architecture. A slide can. It can show whether the problem is mainly fat inside hepatocytes, inflammation spilling across the portal-parenchymal boundary, duct loss in a cholestatic pattern, centrilobular necrosis from ischemic or toxic injury, copper or iron packed into storage cells, alpha-1 antitrypsin trapped inside hepatocytes, collagen thickening the space of Disse, or cirrhotic nodules trapped inside fibrous septa. Once the reader sees that, disease is no longer a list of elevated values. It is a spatial event.

The liver is not just abnormal. It is being damaged in a particular compartment, in a particular pattern, by a mechanism that now has visible shape.

That shift matters because most misunderstanding in liver disease comes from flattening unlike processes into the same word. Hepatitis sounds like one thing until the tissue shows the difference between acute lobular injury, interface activity, chronic portal inflammation, autoimmune-pattern plasma cell infiltrates, or steatohepatitic ballooning. Cholestasis sounds like one thing until the tissue shows whether bile is plugging canaliculi, whether small ducts are disappearing, whether hepatocytes are retaining pigment, or whether fibrosis is beginning from chronic bile injury rather than from primary hepatocyte death. Cirrhosis sounds like "advanced scar" until the microscope shows nodules, bridging septa, capillarized sinusoids, lost lobular order, and the collapse of the normal portal-to-central map. Tissue changes understanding because it breaks broad labels back into the actual architecture those labels were trying to compress.

It also changes how the future of the organ is read. A blood test can tell you that something is wrong now. Histology can often show what kind of future that wrongness is building. Simple steatosis and steatohepatitis are not the same future. Portal expansion with mild fibrosis and bridging fibrosis are not the same future. Quiet retained protein in a hepatocyte and active necroinflammatory injury driving stellate-cell activation are not the same future. Once the slide shows fat alone, inflammation with ballooning, active bile duct injury, portal-based immune attack, or collagen already beginning to partition the parenchyma, the clinician is no longer only naming the present disease. The clinician is seeing the direction the tissue is moving. The liver begins revealing not only what happened, but what it is becoming.

There is another reason tissue changes understanding so powerfully. It restores proportion. A dramatic enzyme elevation can come from a liver with real regenerative capacity still intact. A modest enzyme elevation can belong to a liver already deeply scarred and structurally reorganized. Without tissue, it is easy to confuse biochemical noise with total organ threat. With tissue, the difference becomes harder to miss. The reader can see whether the organ is mainly inflamed, mainly congested, mainly scarred, mainly cholestatic, or already nodular and cirrhotic. The disease stops being read as "high numbers" and starts being read as preserved architecture, strained architecture, or lost architecture. That is a much deeper understanding because the liver lives or dies by arrangement, not by enzyme values alone.

So seeing the tissue changes how disease is understood because it forces the mind out of shorthand and back into structure. The liver is revealed as an organ of compartments, gradients, routes, and repair programs. Once those are visible, diagnosis becomes less like naming a syndrome and more like recognizing a physical process. The slide does not merely confirm disease. It teaches what kind of liver is left, what kind of injury is active, and what kind of repair the organ has already begun choosing.

Part XIII

The Liver Across the Whole Body

Chapter 52

The Liver and the Brain

Ammonia

Ammonia is one of the clearest examples of why the liver is not optional to the nervous system. The brain does not make much use of metabolic rough drafts. It needs blood that has already been stripped of certain kinds of chemical danger, and ammonia is one of those dangers. Nitrogen is constantly moving through the body because protein is constantly being digested, amino acids are constantly being broken down and rebuilt, and intestinal microbes are constantly acting on nitrogen-containing material in the gut. That chemistry generates ammonia. In a healthy body, portal blood carries much of that burden straight to the liver, and hepatocytes convert it into urea through the urea cycle before the rest of thc circulation has to live with it. That step is not decorative housekeeping. It is a gate between ordinary digestion and abnormal brain chemistry. A person can eat protein, turn over tissue, and host a colonized intestine without becoming confused only because the liver is standing in the middle and rewriting nitrogen into a safer form.

The brain is unusually vulnerable here because ammonia is small, diffusible, and chemically disruptive in ways that do not need dramatic structural damage to produce dramatic functional change. Once ammonia escapes hepatic control, either because hepatocytes are failing or because portal blood is bypassing them through shunts, it reaches the systemic circulation and begins entering the brain. There the burden falls heavily on astrocytes, the support cells responsible for helping maintain the neural chemical environment. Astrocytes try to defend the tissue by combining ammonia with glutamate to form glutamine. That sounds protective, and at first it is. But glutamine is osmotically active. If too much accumulates inside the astrocyte, water follows it. The cell swells. A brain does not have much spare room for swelling, and even before intracranial pressure becomes a dramatic issue, the local chemistry begins to blur. Neurotransmitter balance changes. Energy handling becomes less reliable. Cortical timing becomes less precise. The person slows down. Attention loosens. Sleep shifts. Motor control begins to flutter. The brain is not failing mysteriously. It is trying to think inside blood the liver should have cleaned.

Ammonia also matters because it reveals that the liver's relationship to the brain is continuous rather than occasional. This is not the chemistry of rare poisoning. It is the chemistry of daily life, barely kept under control. The intestine produces ammonia every day. Skeletal muscle can buffer some of it temporarily by incorporating nitrogen into amino acids such as glutamine, but muscle is only support here, not the final solution. The liver remains the main organ built to take circulating ammonia and move it into the urea pathway at scale. That means the brain is depending on hepatic nitrogen control every hour, not only during overt liver failure. A healthy person does not feel this because the protection is so quiet. A cirrhotic person or a heavily shunted person begins revealing it because the protection is no longer strong enough. What looks like a neurologic complication is really a blood chemistry problem with a cerebral victim.

This is also why ammonia should not be flattened into a lone villain while still being treated seriously. The failing liver rarely sends the brain only one abnormal molecule. The same person who has hyperammonemia often also has portosystemic shunting, altered amino acid patterns, systemic inflammation, gut dysbiosis, retained manganese over time, disturbed energy metabolism, and a broader toxic field arriving from the bowel with less hepatic editing than before. But ammonia remains central because it shows the route most cleanly. Gut-derived nitrogen crosses into portal blood. The liver fails to contain it. The brain begins metabolically improvising against it. The symptoms of hepatic encephalopathy then appear not because the brain suddenly became diseased in isolation, but because the chemistry feeding the brain was no longer being filtered through the organ built to make it safe.

So ammonia is not just one lab abnormality in liver disease. It is one of the most direct demonstrations of what the liver does for the rest of the body. The gut makes a burden the brain cannot tolerate. The liver normally intercepts that burden and changes its form. When that control fails, thought itself becomes unstable. That is why ammonia belongs at the beginning of any serious account of the liver and the brain. It makes the dependency visible.

Glucose Stability

The brain lives closer to glucose than most people realize. It is not because the brain is uniquely fragile in some sentimental way. It is because neural tissue depends on a bloodstream that stays chemically predictable from minute to minute. Neurons can store very little fuel locally. They are built for signaling, not for warehousing. That means the brain is heavily dependent on what the liver is doing with incoming and outgoing carbohydrate all day long. In the fed state, the liver receives portal glucose from the intestine and absorbs enough of that surge to keep systemic blood sugar from becoming the raw extension of a meal. In fasting, the same liver opens glycogen stores, then shifts toward gluconeogenesis, the synthesis of glucose from lactate, glycerol, and certain amino acids, so the bloodstream does not suddenly starve the tissues that still require glucose. This is why the liver and brain are linked through blood sugar before they are linked through any dramatic liver disease state. The brain thinks clearly only because the liver is buffering time.

That buffering matters because glucose is not just another nutrient in the circulation. For the brain it is part of the immediate operating environment. A rapid fall in blood glucose does not remain a quiet metabolic detail. It becomes a problem of cognition, attention, coordination, mood, and eventually consciousness. The liver prevents that by standing between intermittent feeding and continuous neural demand. After a meal, it takes in glucose and stores some of it as glycogen instead of allowing the full absorbed load to strike the circulation all at once. Hours later, when the gut has gone quiet, the liver begins releasing glucose back into blood from that stored reserve. If the fast deepens further, it makes new glucose to preserve systemic supply. The intestine works in episodes. The brain cannot. The liver is the organ that converts episodic intake into continuous availability.

This is why the liver's role in glucose stability is already a brain-protection role even before encephalopathy enters the picture. A failing liver, or even a severely metabolically disordered liver, does not only lose some abstract control over fuel. It begins losing one of the body's main defenses against unstable neural supply. If glycogen handling is poor, if gluconeogenesis is impaired, if hepatocyte reserve is badly reduced, or if the organ is no longer reading fed and fasting signals coherently, the bloodstream becomes less reliable as a glucose source. The brain then pays quickly because it is one of the least tolerant tissues to abrupt swings in usable fuel. Confusion, weakness, slowed thinking, irritability, sweating, tremor, seizure, and loss of consciousness can all sit downstream of severe glucose instability. The liver therefore protects the brain not only by clearing ammonia, but by preserving one of the most basic conditions for normal neural work, a blood glucose level that does not collapse between meals.

Ketones can reduce some of this dependency during prolonged fasting, and that matters, but even then the liver remains central. The same organ that maintains glucose also makes ketones from fatty acid-derived acetyl-CoA when fasting deepens. The brain can adapt to using more ketones over time, which lowers but does not erase its need for glucose. That means the liver still has to manage both sides of the transition. It has to preserve enough glucose for tissues that cannot do without it while also shifting more of the organism onto fat-derived fuel. The brain is therefore tied to liver glucose stability in two ways at once. It depends on the liver to prevent glucose collapse, and it depends on the liver to reduce glucose demand by supplying alternate fuel when fasting goes long enough.

So glucose stability is one of the quietest and most constant ways the liver protects the brain. Most people notice the liver only when bilirubin rises, ammonia escapes control, or cirrhosis becomes visible. But every ordinary day of clear thinking already depends on the liver having turned meals into glycogen, glycogen into blood glucose, and fasting substrate into new glucose before the brain had to ask what happened to breakfast.

Ketones

Ketones matter to the brain because they are the liver's way of turning stored fat into a circulating fuel the nervous system can actually use when glucose supply becomes more limited. The brain cannot simply switch from glucose to long-chain fatty acids in the way muscle can, because fatty acids are not the right kind of fuel to move cleanly across that system in large enough quantity or in the right metabolic form. The liver solves that problem during fasting. Fatty acids released from adipose tissue enter hepatocytes, are broken down through beta-oxidation in mitochondria, and generate large amounts of acetyl-CoA. Once fasting deepens, some of that acetyl-CoA is diverted into ketogenesis rather than being burned to completion in the tricarboxylic acid cycle. The result is a new fuel class, mainly acetoacetate and beta-hydroxybutyrate, that can circulate in blood without lipoprotein packaging and reach tissues that need an alternative to constant glucose dependence. Ketones are therefore not incidental byproducts of starvation. They are a deliberate hepatic export program.

This matters because the liver is not making ketones for itself in the main. Hepatocytes generate them, but they do not use them efficiently in the same way many extrahepatic tissues do. That is part of the design. The liver can oxidize fatty acids directly for much of its own energetic work. The ketones are for distribution. Heart, skeletal muscle, kidney, and, during longer fasting adaptation, the brain can take up these molecules and turn them back into acetyl-CoA for energy production. The liver is therefore not just surviving the fast. It is reorganizing the fuel economy of the whole organism. It receives fat from adipose tissue, changes its chemical form, and sends it outward as a shared fasting currency. The brain benefits because this lowers how much glucose the liver has to keep supplying continuously, and that spares body protein from being dismantled too quickly for gluconeogenesis.

The neurologic importance of this shift is easy to underestimate because it feels less dramatic than ammonia toxicity or hepatic encephalopathy. But it is one of the great quiet services the liver provides to the brain. In a short fast, the liver holds glucose up with glycogen and then with gluconeogenesis. In a deeper fast, that alone would become increasingly expensive if the brain kept demanding glucose at fed-state levels. Ketones reduce that burden. They allow the brain to move part of its fuel use away from glucose and toward a liver-made substrate derived from stored fat. This does not make glucose irrelevant. The brain still needs some. But ketones reshape the ratio, and that change protects both blood chemistry and body structure. The liver is not only preserving fuel. It is preserving how that fuel is allocated across tissues.

That is why ketones belong in any serious discussion of the liver and the brain. The liver protects the brain not only by clearing gut-derived nitrogen and other neuroactive burden, but also by making sure the brain has a usable fuel strategy when feeding stops. A healthy liver can shift from post-meal glucose buffering to fasting glucose release to ketone export in an ordered sequence. A damaged liver loses some of that flexibility. The problem then is not only that toxins may accumulate. It is that fuel transitions become less graceful. The brain is one of the tissues most exposed to that failure because it depends so heavily on the liver to convert intermittent nutrition and stored fat into a steady, usable internal supply. Ketones are one of the clearest examples of the liver acting not merely as a processor of nutrients, but as an organ that decides what kind of fuel the brain will be allowed to live on.

Inflammation and Sickness Behavior

Sickness behavior is not weakness, poor attitude, or vague malaise. It is a coordinated metabolic and neurologic state in which the body deliberately lowers appetite, slows movement, blunts motivation, increases sleep drive, narrows attention, and changes social and cognitive behavior because inflammatory signaling has told the brain that normal outward life is temporarily too expensive. The liver belongs inside that sequence because inflammation is not only a local tissue event. Cytokines released from injured tissue, activated immune cells, the gut barrier, or circulating microbial burden reach hepatocytes and change what the liver releases into blood. Acute-phase proteins rise. Iron handling shifts through hepcidin. Albumin falls in relative priority. Glucose handling and amino acid traffic are pulled into a more defensive pattern. The bloodstream itself becomes a different medium, more oriented toward containment, defense, and resource reallocation than toward ordinary transport and performance. The brain does not create sickness behavior in isolation. It reads a body whose chemistry has already been rewritten, and the liver is one of the main organs doing the rewriting.

This is why inflammation reaches behavior so quickly. Cytokines such as interleukin-1, interleukin-6, and tumor necrosis factor do not stay trapped at the site of injury. They change vascular signaling, alter hepatic protein synthesis, shift endocrine tone, and communicate with the nervous system through both humoral and neural routes. The brain then changes priorities. Appetite falls because digestion and feeding are no longer being treated as immediate investment targets. Activity falls because locomotion, exploration, and outward effort cost energy that the immune system is now demanding elsewhere. Sleep drive rises because the organism is moving into a state built more for defense and repair than for hunting, working, building, or competing. This is not random fatigue. It is behavioral energy triage. The liver contributes by making the blood that carries and amplifies the inflammatory message. Once hepatocytes shift into acute-phase output, the whole body begins living inside an inflammatory plasma environment, and the brain responds accordingly.

The gut-liver axis makes this even more important. A compromised intestinal barrier, altered microbiota, endotoxin delivery, cholestatic burden, or steatotic inflammation can all increase the inflammatory traffic reaching the liver. The liver then stops being merely an organ processing nutrients and becomes a major amplifier of systemic illness behavior. More cytokine signaling. More acute-phase protein production. More altered iron and amino acid handling. More metabolic pressure toward conservation. In that state the brain is not only receiving signals that something is wrong. It is receiving a full-body instruction that fuel, attention, mood, and behavior should be reorganized around threat. This is one reason chronic liver disease and chronic inflammatory burden so often carry fatigue, apathy, slowed thinking, altered sleep, and depressed drive even before overt encephalopathy appears. The nervous system is living in a body that has already shifted its chemistry away from ordinary vitality.

So inflammation and sickness behavior belong in a liver chapter because the liver is one of the organs that turns local injury into whole-body physiology. The immune system sends the first warning, but the liver helps convert that warning into a blood state the brain can feel. Appetite changes. Motivation changes. Sleep changes. Movement changes. The person feels "sick" because the body has entered a coordinated inflammatory mode, and the liver is one of the main organs writing that mode into circulation.

Why the Brain Depends on Hepatic Order More Than Most People Realize

Most people think the brain depends on the heart for blood flow and the lungs for oxygen, which is true, but they underestimate how much the brain also depends on the liver for chemical order. The brain does not merely need blood. It needs blood that has already been edited. Glucose has to arrive in a usable range. Ammonia has to be kept low. Bile-derived waste products have to stay out of the circulation. Hormones have to be transformed and buffered. Drugs, alcohol, and gut-derived compounds have to be handled before they accumulate into a neuroactive burden. The brain is a high-demand tissue with very little tolerance for unstable chemistry. It can keep working through fatigue, poor sleep, and stress better than most organs can keep working through bad blood. That is why hepatic order matters so much. The liver is one of the organs that makes blood safe enough for the brain to think inside.

This dependence begins with fuel. Neurons do not keep large local reserves. They rely on a bloodstream that does not swing violently between post-meal excess and fasting scarcity. The liver absorbs the first glucose wave after a meal, stores part of it as glycogen, and later releases glucose back into blood when the intestine has gone quiet. As fasting deepens, the liver also shifts toward ketone production, which gives the brain another usable fuel and lowers how much glucose must be supplied continuously. So the brain is not just dependent on calories reaching the bloodstream. It is dependent on the liver turning irregular intake into continuous availability. A person can skip breakfast and still think because the liver is covering the gap. When hepatic order over fuel begins to fail, the brain feels it quickly as slowed thinking, irritability, weakness, poor concentration, and eventually altered consciousness if the failure is severe enough.

The dependence is even clearer with nitrogen. The gut is constantly producing ammonia through bacterial metabolism and protein breakdown, and portal blood is constantly carrying that burden toward the liver. The brain cannot tolerate free circulation of that chemistry for long. Hepatocytes convert ammonia into urea before the wider bloodstream has to live with it. When that control weakens, or when portal blood bypasses the liver through shunts, the brain begins receiving blood that is chemically wrong for neural tissue. Astrocytes swell. Neurotransmission becomes less stable. Thought slows. Sleep reverses. Motor control falters. Coma becomes possible. This is why hepatic encephalopathy is so revealing. It proves that the brain is not only dependent on what the body supplies. It is dependent on what the liver prevents.

The same dependency extends into inflammation and endocrine traffic. The liver changes plasma composition during inflammation by making acute-phase proteins, restricting iron through hepcidin, altering albumin synthesis, and shifting the body toward defense chemistry. The brain then feels that shift as sickness behavior, lower drive, more sleep need, less appetite, and slower cognition. The liver also helps shape thyroid hormone conversion, steroid hormone clearance, vitamin D metabolism, and the transport environment for many circulating signals. This means the brain is not only dependent on hepatic detoxification in the crude popular sense. It is dependent on hepatic refinement of the entire internal environment. Mood, clarity, motivation, sleep rhythm, and mental sharpness are all being influenced by whether the liver is keeping signal traffic clean enough and chemical burden low enough for the nervous system to remain stable.

That is why the brain depends on hepatic order more than most people realize. The liver is not beside the brain in function, but beneath it in chemistry. The brain can only be as clear as the blood it is given, and the liver is one of the main organs deciding what kind of blood that will be.

The Neurologic Face of Liver Breakdown

The neurologic face of liver breakdown is not subtle once the chemistry begins reaching the brain. A person who once looked "only liver sick" starts looking mentally slowed, inattentive, sleep-shifted, irritable, forgetful, poorly coordinated, or frankly altered. In early hepatic encephalopathy the changes may be quiet enough to pass for fatigue or mood disturbance, but the pattern is already specific if you know what you are seeing. Attention weakens. Working memory and psychomotor speed slow. Sleep-wake rhythm begins reversing. As the burden deepens, the person can become disoriented, somnolent, asterictic, stuporous, and eventually comatose. That progression is not a side note to liver failure. It is one of the most visible ways liver failure announces that blood has stopped being safe for the brain to inhabit.

What makes this the neurologic face of liver breakdown rather than a primary brain disease is route. The gut keeps generating ammonia and other neuroactive burden. Portal blood keeps carrying that burden forward. The liver is supposed to neutralize much of it before the bloodstream reaches the wider circulation. Once hepatocytes can no longer clear enough of that chemistry, or once portosystemic shunts let blood bypass them, the brain begins reading the consequence in real time. The person is not confused because the brain suddenly developed an independent lesion. The person is confused because the liver has stopped preserving the chemical conditions that normal cognition requires. Reduced alertness, impaired judgment, personality and behavior change, poor concentration, asterixis, and coma are therefore best understood as blood chemistry becoming neurologic function.

This is why the neurologic picture can look so broad. The liver is not failing in one neurotransmitter pathway. It is failing as an organizer of the bloodstream. The result reaches cognition, mood, sleep, motor tone, and consciousness itself. Asterixis shows that posture can no longer be held steadily. Sleep reversal shows that even circadian patterning is being pulled off course. Confusion and slowed thinking show cortical processing struggling under metabolic burden. Coma shows the far end of a sequence in which the brain has lost the ability to maintain wakeful organized function inside blood the liver no longer knows how to clean. The neurologic face of liver breakdown is therefore not one symptom. It is the entire nervous system revealing how dependent it was on hepatic order all along.

Chapter 53

The Liver and the Gut

Bile Flow and Digestion

Bile flow is the clearest way to see that the liver and gut are not separate subjects joined loosely by a duct. They are one digestive circuit. The liver makes bile continuously because hepatocytes are always synthesizing bile acids from cholesterol, conjugating bilirubin, secreting phospholipids, moving cholesterol into the biliary stream, and building an outgoing fluid that will later enter the intestine. But the intestine is where that fluid proves what it is for. A liver can make bile, but if bile does not reach the duodenum in the right amount, at the right time, and in the right chemical balance, digestion changes immediately. The meal may still enter the lumen. Pancreatic enzymes may still be present. The intestinal surface may still be intact. But the physical conditions of fat handling will already have deteriorated. This is why bile belongs in digestive physiology as centrally as acid or enzymes do. It is not a supporting fluid added after the main work is done. It is one of the main conditions that make the work possible.

The first problem bile solves is physical, not nutritional in the abstract. Dietary fat enters a watery lumen, and fat does not behave well in water. It aggregates into larger hydrophobic masses with relatively little exposed surface area. Hydrophobic means poorly compatible with water. Pancreatic lipase cannot solve that by itself because it works at the surface of lipid droplets, not inside a large oil mass. Bile acids are the molecules that change the geometry of the meal. They are amphipathic, meaning one side interacts with water and the other interacts with fat. When they enter the duodenum, they coat and disperse large lipid globules into much smaller droplets, increasing total surface area and making triglyceride hydrolysis far more efficient. This is emulsification. It is not the same as digestion, but without it digestion remains physically handicapped. The gut does not merely need enzymes. It needs the liver's detergent chemistry to make those enzymes useful at scale.

Then comes the second bile-dependent step, which is just as important and often less appreciated. Once lipase has acted, the products of lipid digestion still do not move well through the watery intestinal contents. Fatty acids, monoglycerides, cholesterol, and fat-soluble vitamins are still hydrophobic enough that they require a transport arrangement to reach the absorptive surface. Bile acids help form micelles, small mixed assemblies that keep these lipid products suspended and mobile in the lumen long enough to carry them through the unstirred water layer to the brush border of the enterocyte. The enterocyte is the absorptive cell lining the small intestine. So bile does not only help break fat apart. It helps escort the products of that breakdown to the membrane where absorption can finally occur. This is why impaired bile flow changes the nutritional result of a meal so profoundly. The fat may still be present in the intestine. The body loses access to part of it because the physical and transport conditions of absorption have failed.

The relationship between bile flow and digestion also explains why cholestatic disease creates symptoms that look intestinal even though the primary lesion may be in the liver or bile ducts. If bile acids do not reach the duodenum properly, fat digestion weakens, fat absorption falls, and vitamins A, D, E, and K become harder to absorb because they ride with the lipid phase. Stool can become pale not only because bilirubin pigment delivery is reduced, but because the meal's fat phase is now being handled poorly. Steatorrhea, greasy fat-rich stool, is the digestive face of failed bile flow. The liver is still upstream, but the intestine is where the failure becomes visible. This is one of the deeper reasons the liver and gut have to be read together. The liver does not digest food directly, yet it determines whether the intestine receives one of its main tools for doing so.

The timing of bile flow matters just as much as its chemistry. Hepatocytes make bile continuously, but the intestine does not need full-strength biliary delivery continuously. The gallbladder solves that mismatch by storing and concentrating bile between meals, then contracting when cholecystokinin signals that fat and protein have entered the duodenum. That means bile flow is not only a structural fact of ducts and secretion. It is meal-synchronized physiology. The liver writes the fluid. The gallbladder sharpens its timing and concentration. The intestine triggers its release. This is not three separate organs doing three independent jobs. It is a single digestive sequence distributed across three tissues. If any part of that timing breaks, the meal changes. A weak trickle is not the same as a coordinated detergent pulse delivered when fat actually arrives.

The deeper point is that bile flow shows how the gut depends on the liver for more than detoxification or nutrient storage. The liver is sending a working fluid into the intestine that decides whether the fat phase of food remains chemically stranded or becomes absorbable. It is also sending out the route by which bilirubin, cholesterol, and other selected burdens leave the body. Digestion and excretion therefore share one stream. The gut receives bile because the meal needs it. The body needs bile because certain compounds must leave through it. That is why bile flow and digestion belong together. The intestine is not merely downstream from the liver. It is the place where the liver's outgoing stream becomes usable biology.

Portal Exposure to Gut-Derived Chemistry

Portal blood is not just the bloodstream after digestion. It is the bloodstream after the intestine has decided what may cross. That distinction is the beginning of the gut-liver relationship. The gut lumen contains food fragments, bacterial metabolites, bacterial structural debris, endotoxin, bile acids returning from the distal intestine, drug molecules, alcohol, ammonia precursors, and all the chemistry produced when a colonized digestive tube processes a meal. Some of that chemistry is useful. Some of it is neutral until concentration changes. Some of it is inflammatory. Some of it is burdensome from the moment it enters blood. Once these compounds cross the intestinal barrier, they do not go first to the heart or brain. They go to the liver through the portal vein. The liver is therefore the first major organ exposed to what the gut has allowed through, and that exposure is broader than nutrition. It is exposure to the whole chemical personality of the intestine.

This matters because the intestine is not a perfect sorting machine. It is a selective membrane system operating under heavy traffic, not a sterilizing chamber. It can absorb glucose, amino acids, minerals, and water-soluble vitamins efficiently while still allowing smaller amounts of microbial fragments, lipopolysaccharide, fermentation products, food antigens, and xenobiotics to cross into portal blood. Xenobiotics are compounds foreign to the body's own structural chemistry. The liver has to live with this ambiguity. It receives a stream that contains fuel and burden in the same fluid. A meal, from the liver's point of view, is not just nourishment entering the body. It is a mixed report containing substrate, signaling molecules, immune triggers, and waste precursors. That is why portal exposure to gut-derived chemistry is such a central concept. The liver is downstream from digestion, but it is also downstream from the microbial and inflammatory consequences of digestion.

The exposure is continuous, not occasional. Even between meals, the gut is still metabolically active, bacteria are still producing compounds, epithelial turnover is still releasing debris, bile acids are still cycling, and low-level antigenic traffic is still reaching portal blood. The liver therefore cannot treat portal chemistry as if it were only a post-meal nutrient surge. It has to operate as a standing checkpoint. Kupffer cells sample that blood for bacterial fragments and endotoxin. Hepatocytes take up, transform, store, and excrete what they can. Sinusoidal endothelial cells participate in how that traffic is presented to the tissue. This is not a side function. It is built into the organ's position. The gut is always sending chemistry forward. The liver is always reading it.

This is why portal exposure becomes so important in disease. If the intestinal barrier grows more permeable, if dysbiosis changes the microbial output of the gut, if bile acid handling changes luminal chemistry, or if alcohol and dietary burden amplify endotoxin movement, the composition of portal blood changes before symptoms become obvious elsewhere. The liver then starts receiving a dirtier stream. More inflammatory signals arrive. More microbial residues arrive. More detoxification burden arrives. More ammonia and nitrogen handling pressure arrive. The liver can buffer some of this, sometimes for a long time, but it pays for the buffering. Kupffer cell activation rises. Acute-phase signaling rises. Hepatocytes work under greater oxidative and inflammatory burden. Stellate cells become easier to activate. Portal exposure is therefore not simply a background fact of anatomy. It is one of the main ways gut dysfunction becomes liver disease.

So portal exposure to gut-derived chemistry should be understood as the liver's daily reality. The organ is not waiting for rare toxins to appear. It is continuously receiving the molecular results of having a digestive tract, a microbiome, and an absorptive surface open to the outside world. Nutrients arrive in that stream, but so do inflammatory burdens, microbial traces, hormone metabolites, bile acid return, alcohol, drugs, and metabolic waste. The liver stands between that complexity and the rest of the body. It is the first organ forced to decide what from the gut becomes systemic truth.

Microbial Metabolites and Immune Signaling

The gut does not send the liver only intact nutrients and bacterial fragments. It also sends a dissolved chemical commentary on what the intestinal ecosystem has been doing. Microbes ferment carbohydrates, degrade amino acids, modify bile acids, split host- and food-derived compounds, and generate a large secondary chemistry that enters portal blood alongside absorbed substrate. Some of these molecules are relatively useful. Short-chain fatty acids produced from fermentation can act as fuel and signaling compounds. Others are more ambiguous or burdensome. Ammonia, amines, phenols, indoles, secondary bile acid derivatives, and other microbial products can change epithelial behavior, immune tone, and hepatic workload depending on their amount and context. The liver therefore receives not just the products of digestion, but the products of digestion after microbial interpretation. That matters because the gut is not only breaking food down. It is chemically rewriting part of it before the liver ever sees the stream.

This is one of the reasons the liver has to behave like an immune-reading organ and not just a metabolic one. Portal blood entering the sinusoids carries these microbial metabolites directly to Kupffer cells, sinusoidal endothelial cells, and hepatocytes. Kupffer cells do not merely ask whether bacteria are present in some dramatic infectious sense. They read pattern and burden. A small amount of ordinary portal microbial traffic can be handled through quiet clearance and tolerance. A larger or more inflammatory signal changes the tissue response. Endotoxin, bacterial DNA fragments, and some microbial metabolites can activate pattern-recognition pathways, leading to cytokine release and a shift in hepatic behavior. Cytokines are signaling proteins that tell the tissue whether it should remain metabolically quiet or move into an inflammatory state. The liver is therefore using microbial chemistry as information. It is asking, through the portal stream, what kind of intestinal world is upstream of it right now.

The metabolites themselves help determine that answer. A short-chain fatty acid arriving from fermentation is not read the same way as excess ammonia or a rising burden of endotoxin-associated chemistry. Modified bile acids arriving from microbial metabolism are not read the same way as a low-grade background signal from a stable intestinal barrier. The liver is not only clearing these compounds. It is integrating them into its decisions about acute-phase protein production, bile acid synthesis, glucose handling, immune tone, and fibrotic risk. This is where gut-liver communication becomes very literal. The intestine and its microbes create molecules. Those molecules enter portal blood. The liver changes its behavior in response. If the microbial output is relatively disciplined, the hepatic response can remain mostly tolerant and metabolic. If the microbial output becomes more inflammatory, more toxic, or more excessive, the liver starts reallocating itself toward defense. The same organ that was buffering glucose and making bile is now also shifting cytokines, complement proteins, and immune cell behavior because of what the microbiome upstream has generated.

This is why microbial metabolites belong inside any serious account of liver disease progression. They are one of the bridges between dysbiosis, intestinal permeability, steatosis, cholestasis, and chronic inflammation. A gut producing a harsher portal chemistry increases Kupffer cell activation, makes hepatocyte stress more inflammatory, and gives stellate cells a stronger reason to leave quiet reserve mode and begin building matrix. Over time, the liver stops experiencing the gut merely as a source of nutrients and starts experiencing it as a source of chronic immune instruction. The tissue future changes from there. Microbial metabolites, then, are not side chemistry riding along with digestion. They are part of the real-time signal traffic by which the gut tells the liver whether today's portal blood should be treated as ordinary life or as the beginning of injury.

Malabsorption and Cholestasis

Cholestasis turns the same liver problem into a digestive problem because bile is not only an excretory fluid. It is one of the intestine's main working tools for handling fat. When bile flow falls, bile acids no longer reach the lumen in the same quantity, and the physical chemistry of the meal changes immediately. Large lipid droplets remain less effectively emulsified, micelle formation weakens, and the products of fat digestion move less efficiently through the watery intestinal contents toward the absorptive surface. That means the person can still be eating fat, cholesterol, and fat-soluble vitamins, but the body is extracting less of them. Cholestasis therefore changes the nutritional result of a meal, not just the bilirubin pattern in blood.

This is why malabsorption belongs naturally inside cholestatic disease. The intestine is no longer receiving enough bile salts to support normal absorption of dietary fat and vitamins A, D, E, and K. Merck's clinical summaries state this directly: inadequate bile in the gastrointestinal tract leads to malabsorption of fat and fat-soluble vitamins, and chronic cholestatic states commonly require fat-soluble vitamin supplementation for exactly this reason. The chemistry underneath that clinical fact is simple. Bile acids are the detergent molecules that make hydrophobic food chemistry usable in a water-based lumen. When the detergent stream weakens, the meal's fat phase becomes less accessible to the body.

The consequences then spread beyond the gut because the missing absorbed nutrients were not trivial. Less fat absorption means less usable energy from the meal and, over time, poorer nutritional status. Less vitamin K absorption can prolong PT and raise INR through a fat-malabsorption route rather than through hepatocyte synthetic failure alone. Less vitamin D absorption can contribute to bone loss. Less vitamin A and vitamin E absorption weakens epithelial and antioxidant support. The liver created the problem by failing to deliver bile, but the intestine reveals it first as nutritional loss.

So malabsorption and cholestasis belong together because cholestasis is not only retained bile. It is absent bile where the meal needed it most. The liver's outgoing stream is supposed to carry digestive detergent into the intestine and excretory burden out of the body. When that stream stalls, the body is injured twice. Waste is retained upstream, and nutrition is lost downstream.

Small Intestine, Colon, and Liver as One Connected Axis

The small intestine, colon, and liver should not be imagined as three organs standing beside each other performing separate jobs in sequence. They are one metabolic axis with different territories. The small intestine is the main absorptive surface. It takes apart the meal, exposes nutrients to transporters, and transfers glucose, amino acids, minerals, water, and many other compounds into portal blood. The colon does something different but no less important. It receives what escaped digestion upstream, exposes that residue to dense microbial fermentation, recovers water and electrolytes, modifies bile acids, and generates a second wave of chemistry from material the small intestine did not fully settle. The liver sits downstream from both, receiving the blood-borne consequences of absorption from the small intestine and the microbial consequences of fermentation and barrier contact from the colon. The result is one continuous system. The small intestine decides what enters. The colon decides what is transformed later. The liver decides what form of that combined chemistry the rest of the body will be allowed to live with.

This is why portal blood should never be described as "nutrient blood" in the simple sense. Blood leaving the small intestine is carrying freshly absorbed substrate, but blood draining the broader intestinal system is also carrying microbial metabolites, ammonia precursors, bacterial fragments, modified bile acid species, short-chain fatty acids, and other colon-derived signals that can only be understood in the context of the microbiota. The liver receives both classes of input in the same stream. It buffers post-meal glucose, sorts amino acids, handles fructose, recycles bile acids, and converts ammonia into urea, but it is also sampling endotoxin, reading inflammatory traffic, conjugating compounds that came from microbial transformation, and deciding whether the portal burden still belongs to ordinary physiology or has crossed into pathologic load. This is the deeper meaning of the gut-liver axis. The liver is not merely downstream from nutrients. It is downstream from digestion and fermentation at the same time.

The small intestine and colon also divide labor in a way that makes the liver's role unavoidable. In the small intestine, bile acids emulsify fat, pancreatic enzymes reduce macronutrients into absorbable fragments, and enterocytes move those fragments into blood or lymph. In the colon, the problem is no longer primary nutrient extraction from the meal's original structure. The problem becomes salvage, fermentation, and barrier control. Bacteria produce short-chain fatty acids from fermentable carbohydrate, generate nitrogenous compounds from protein residues, transform primary bile acids into secondary forms, and produce signaling molecules that affect intestinal epithelium, immune tone, and hepatic metabolism. The colon is therefore not simply a water-recovery organ. It is a chemical reactor. The liver has to stand downstream because this reactor is not making compounds in a sealed container. It is making them one epithelial layer away from portal access.

This is why barrier integrity along the axis matters so much. A healthy small intestine absorbs aggressively without becoming an open door to inflammatory burden. A healthy colon contains dense microbial life without allowing that microbial world to pour into blood unchecked.

When these barriers hold, the liver receives a demanding but manageable stream. When they weaken, the liver receives a different blood. More endotoxin, more bacterial fragments, more inflammatory metabolites, more ammonia burden, more bile acid distortion, more work for Kupffer cells, more cytokine pressure on hepatocytes, more incentive for stellate cells to leave quiet reserve mode and move toward matrix production. The liver then starts paying for gut dysfunction in real tissue currency. Steatosis becomes easier to sustain. Cholestatic stress becomes harder to tolerate. Fibrosis becomes easier to trigger. This is why the axis should be understood as one system under load rather than as three organs each failing independently.

Bile closes the loop and makes the connection even tighter. The liver synthesizes bile acids and sends them into the small intestine to make fat digestion possible. The terminal ileum reclaims most of those bile acids and returns them through portal blood. What escapes into the colon is then exposed to microbial conversion, which changes bile acid chemistry further and can alter signaling and toxicity depending on the pattern. The colon therefore is not downstream from bile in a passive sense. It is modifying one of the liver's own major secretions and sending part of the consequences back into the liver's world through microbial metabolism and barrier interaction. The small intestine uses liver output. The colon rewrites some of that output. The liver then receives the rewritten chemistry. That is not a loose association. It is a circulating loop of digestion, salvage, fermentation, and reprocessing.

So the small intestine, colon, and liver are one connected axis because food, microbes, bile, barrier function, and portal blood all move through them as one integrated event. The small intestine extracts. The colon transforms and recovers. The liver filters, converts, stores, excretes, and decides. None of the three can be understood clearly in isolation. The meal becomes blood in the small intestine, becomes microbial chemistry in the colon, and becomes systemic reality only after the liver has finished reading both.

Why Gut Symptoms and Liver Symptoms Often Reflect the Same Upstream Disorder

Gut symptoms and liver symptoms often reflect the same upstream disorder because the liver is not downstream from a clean digestive machine. It is downstream from a living barrier, a microbial ecosystem, a bile-dependent absorptive surface, and a portal drainage system that carries whatever the intestine decided to let through. If that upstream system is functioning well, the liver receives a chemically mixed but manageable stream. Nutrients arrive in usable form, microbial burden stays low enough for quiet clearance, bile acids cycle efficiently, and the bloodstream that leaves the liver is more stable than the portal blood that entered it. But if digestion, barrier integrity, bile flow, microbial balance, or nutrient handling begin to fail, the gut and liver start displaying the same disturbance in two different languages. The intestine shows it through bloating, diarrhea, fat malabsorption, altered stool, pain, or irregular transit. The liver shows it through endotoxin burden, cholestatic stress, steatosis, inflammation, altered bilirubin handling, or rising portal pressure. One upstream disorder. Two organ faces.

Bile is one of the clearest examples. If bile does not reach the small intestine properly, the gut loses one of its main tools for handling fat. Large lipid droplets remain poorly dispersed, micelle formation weakens, fat-soluble vitamin absorption falls, and the meal becomes less biologically usable even if it looked nutrient-dense on paper. The person may develop greasy stool, bloating, pale stool, or nutritional deficiency. At the same time the liver is paying for the same failed stream from the other side. Bile acids and conjugated bilirubin are being retained instead of moving outward. Cholestatic pressure rises. Biliary epithelium becomes stressed. Itching, dark urine, and cholestatic enzymes appear. The gut symptom and the liver symptom are not separate diseases living side by side. They are the digestive and hepatic consequences of the same failed outflow path.

The intestinal barrier creates the same kind of overlap from the opposite direction. A weakened barrier does not merely produce vague gut inflammation. It changes portal blood. More lipopolysaccharide, more bacterial fragments, more inflammatory microbial metabolites, and more chemically disruptive luminal products reach the liver directly. Kupffer cells begin clearing a dirtier stream. Hepatocytes begin working under stronger inflammatory pressure. Acute-phase signaling rises. Steatosis becomes easier to sustain. Fibrosis becomes easier to trigger if the burden persists. The person may complain first of bloating, irregular stool, abdominal discomfort, or food intolerance, but the liver is already reading the same event as microbial translocation, inflammatory burden, and altered blood chemistry. The gut is showing local failure. The liver is showing what that local failure becomes once it enters circulation.

This is also why metabolic disease so often looks intestinal and hepatic at the same time. Excess carbohydrate, fructose-heavy intake, poor adipose restraint, dysbiosis, slower or abnormal transit, and altered bile acid signaling do not stay confined to one compartment. The intestine delivers more substrate and different microbial chemistry. The colon modifies bile acids and generates a different fermentation profile. Portal blood carries that altered chemistry to the liver. The liver begins storing more fat, shifting acute-phase output, handling more endotoxin, and losing clean control over blood chemistry. The patient may experience reflux, bloating, constipation, loose stool, or post-meal discomfort while the liver is accumulating triglyceride and moving toward inflammatory injury. Again, the body is not producing two unrelated disorders. It is expressing one broken nutrient-handling and barrier system through two organs joined by direct blood flow.

That is the deeper reason gut symptoms and liver symptoms so often belong to the same story. The small intestine, colon, bile system, and liver are not loosely associated structures. They are one axis of digestion, absorption, microbial interpretation, and chemical review. When the axis is working, the gut extracts and the liver refines. When it breaks, the gut becomes less able to contain or process what is in the lumen, and the liver becomes the first organ forced to absorb the cost. The bowel reveals the disorder locally. The liver reveals what the disorder looks like once it becomes blood.

Chapter 54

The Liver and the Kidney

Urea Generation and Renal Excretion

The liver and the kidney are linked by one of the body's most important chemical handoffs. The liver takes nitrogen that is too dangerous to leave in circulation and converts it into a form the kidney can safely excrete. The kidney then removes that form from the body in urine. That sequence sounds simple until you see what it is protecting the organism from. Protein is constantly being digested, amino acids are constantly being broken down and rebuilt, and gut microbes are constantly generating nitrogenous compounds in the lumen. All of that chemistry produces ammonia and related nitrogen burden. Ammonia is small, diffusible, and toxic enough that it cannot be allowed to remain in blood for long without threatening brain function and broader cellular stability. The liver solves that problem through the urea cycle. Hepatocytes take ammonia and related nitrogen, spend energy to incorporate that burden into urea, and release urea back into the bloodstream as a safer, water-soluble waste molecule. At that point the job is only half done. The liver has neutralized the chemical threat. The kidney must now carry the burden out.

This is why urea generation is not just another metabolic pathway buried in biochemistry. It is the liver's way of making blood compatible with protein metabolism. A body that digests protein but cannot convert ammonia into urea is not merely inefficient. It is chemically unstable. The brain becomes one of the first tissues forced to reveal that instability, which is why ammonia control belongs as much to neuroprotection as it does to nitrogen disposal. But the liver cannot finish the job alone. Urea is safer than ammonia because it can travel through plasma without the same toxic effect and because the kidney can filter it into urine. The kidney therefore is not just receiving generic waste from the liver. It is receiving a molecule the liver specifically manufactured for renal removal. This is a division of labor so precise that neither organ makes full sense without the other. The liver turns danger into transportable waste. The kidney turns transportable waste into excretion. The bloodstream is the handoff lane between them.

The chemistry also shows why the liver-kidney axis is about more than "detoxification." The liver is not only removing something bad. It is changing the form of nitrogen so the kidney can work within the physical rules of filtration. The kidney handles water-soluble compounds well. Ammonia in excess is too toxic and too disruptive to leave unmanaged. Urea is the compromise molecule. It lets nitrogen stay mobile in blood long enough to reach the nephron without turning that blood into a neurotoxic stream. Once it reaches the kidney, urea is filtered at the glomerulus, moves through the tubular system, and is excreted in urine, though some of it is reabsorbed depending on hydration state and renal handling. That nuance matters because the kidney is not simply a drainpipe either. It is a selective regulator of water and solute movement. Still, the broader logic holds. The liver makes the waste fit for the kidney. The kidney makes the waste leave the body.

This is why failure in either organ changes the meaning of the other organ's work. If the liver fails, ammonia rises because urea generation is inadequate, and the kidney is left receiving less of the safe excretory form and more of a burden the body should never have circulated freely in the first place. If the kidney fails, the liver may still generate urea correctly, but the molecule now accumulates because excretion is impaired. The body then lives with rising blood urea nitrogen not because the liver did the wrong chemistry, but because the second half of the pathway failed. The two organs are therefore linked by sequence. Liver function creates a renal task. Kidney function completes a hepatic intention. When the sequence holds, nitrogen leaves the body quietly. When it breaks, either through failed hepatic conversion or failed renal excretion, the bloodstream starts revealing how dependent normal physiology was on this partnership all along.

Bilirubin Metabolites

Bilirubin does not finish its life when the liver conjugates it. Conjugation is only the point where the molecule becomes fit to enter the next environment. Unconjugated bilirubin is hydrophobic, albumin-bound, and moving through blood on its way to the liver. Hepatocytes take it up and attach glucuronic acid, creating conjugated bilirubin, a form much more compatible with biliary secretion. At that point the liver has not destroyed the pigment. It has changed its transport fate. Conjugated bilirubin is secreted into bile, delivered into the intestine, and handed off to a different chemical world, one shaped not by hepatocyte enzymes but by luminal flow and bacterial metabolism. The gut is where bilirubin stops being a liver problem and starts becoming a microbial substrate.

Once conjugated bilirubin reaches the intestine, bacteria begin stripping and reducing it into urobilinogen. This is the key intermediate metabolite in the downstream story. Urobilinogen is not the final pigment seen in stool or urine. It is the branching point. Some of it stays in the intestinal lumen and continues moving distally, where it is further oxidized into stercobilin and related brown fecal pigments. That is why normal stool has its characteristic color. The liver sends bilirubin into bile. The intestine converts that bilirubin through bacterial action. The colon finishes the pigment story in a way the eye can finally see in stool. The brown color is not cosmetic. It is proof that biliary pigment reached the bowel, survived the trip, and was metabolized all the way down the line.

Not all urobilinogen stays in the gut. A portion is reabsorbed through the intestinal wall and returns through the portal circulation to the liver. This is one more example of the liver and intestine functioning as a loop rather than as separate organs standing beside each other. The liver is not only sending bilirubin outward. It is receiving part of its intestinal metabolite traffic back. Some of that reabsorbed urobilinogen is taken up again by hepatocytes and resecreted into bile, continuing the enterohepatic cycle. A smaller fraction escapes hepatic uptake, enters the systemic circulation, and reaches the kidney. There it can be filtered into urine and oxidized into urobilin, which contributes to the normal yellow color of urine. So the two visible pigments, brown stool and yellow urine, both arise from the same original heme burden after the liver has conjugated bilirubin and the intestine has metabolized it into downstream forms.

This is why bilirubin metabolites are such useful physiologic clues. If conjugated bilirubin cannot reach the intestine because bile flow is blocked, less urobilinogen is formed in the bowel, less stercobilin reaches the stool, and stool becomes pale. At the same time more conjugated bilirubin may back into blood and be filtered directly by the kidney, darkening the urine. The route has changed, so the pigments change with it. If bilirubin production is high upstream from hemolysis or other increased heme breakdown, more substrate may reach the intestine and urinary urobilinogen patterns can shift accordingly. The body is therefore constantly writing the status of bilirubin handling into its waste streams. Blood carries the precursor. The liver changes its form. The intestine makes the metabolite. Stool and urine display the result.

The deeper point is that bilirubin metabolites reveal the full architecture of the liver-gut axis in motion. A red blood cell is dismantled. Heme becomes bilirubin. The liver conjugates it and pushes it into bile. The intestine converts it into urobilinogen. The colon oxidizes part of it into stool pigment. The circulation and kidney carry another fraction toward urinary pigment. One molecule passes through macrophages, hepatocytes, bile, intestinal bacteria, colon, portal blood, and kidney before its story is finished. That is why bilirubin metabolism is so useful as a teaching sequence. It makes the whole route visible.

Fluid Balance and Cirrhosis

Fluid balance in cirrhosis fails because the liver is one of the organs that normally helps keep blood inside blood. Once cirrhosis develops, that control is lost from several directions at the same time. The scarred liver resists portal inflow, so pressure rises behind it in the splanchnic circulation, the vascular territory draining the intestine, spleen, pancreas, and related abdominal organs. That higher hydrostatic pressure pushes more fluid out of capillary beds and into the interstitial space and peritoneal cavity. At the same time, the cirrhotic liver often makes less albumin, and albumin is one of the main plasma proteins generating oncotic pressure, the inward pull that helps keep water in the vascular compartment. More pressure is pushing fluid out. Less protein force is pulling it back in. The result is not simply "extra fluid." It is fluid leaving the wrong compartment and collecting where it should not.

This is why cirrhosis produces misplacement before it produces simple overload. A person can be swollen, distended with ascites, and edematous in the legs while still having an effectively underfilled arterial circulation. The body does not judge volume by looking at the abdomen. It judges volume by what the arterial tree feels like to pressure sensors and to the kidney. In cirrhosis, a large fraction of blood volume becomes trapped in the dilated splanchnic vascular bed, while more fluid leaks outward because portal pressure is high and plasma protein restraint is weaker. The circulation becomes inefficient. Blood is present, but too much of its water is outside the vascular space and too much of its volume is pooled where it does not support systemic perfusion cleanly.

The kidney answers this as if the body were dehydrated or bleeding. Renin rises. Angiotensin II rises. Aldosterone rises. Antidiuretic hormone rises. Sodium retention increases, and water follows. This compensation is logical in a true low-volume state. In cirrhosis it becomes part of the disease. More sodium and water are added to a system whose basic hydraulic problem has not been corrected. The liver is still resistant to portal flow. The splanchnic circulation is still dilated. Albumin is still low or low relative to the need. So the newly retained fluid is again pushed toward the abdomen and the tissues. Ascites enlarges. Edema worsens. The body keeps trying to rescue circulation by adding volume, but the architecture now misplaces much of what it retains.

The liver also contributes through lymph formation. A healthy liver generates lymph continuously, but cirrhosis and portal hypertension can greatly increase hepatic and splanchnic lymph production. Once lymphatic drainage can no longer keep pace, the excess fluid finds the peritoneal cavity. This is one reason ascites can become such a dominant feature of advanced disease. It is not one leak. It is the combined result of increased hydrostatic pressure, reduced oncotic restraint, neurohormonal sodium retention, and overwhelmed lymphatic return. The abdomen becomes the reservoir where all of those failures meet.

So fluid balance in cirrhosis is really circulatory balance lost. The liver no longer permits cheap portal flow, no longer maintains plasma protein architecture strongly enough, and no longer protects the body from turning volume compensation into volume misplacement. Blood pressure, protein synthesis, renal behavior, and abdominal vascular tone all begin telling the same story. The person is not merely retaining fluid. The person is losing the ability to keep fluid where circulation needs it most.

Hepatorenal Syndrome

Hepatorenal syndrome is what kidney failure looks like when the kidney is being trapped inside the wrong circulation rather than destroyed first as a primary renal disease. The liver fails as an organizer of blood flow before the kidney fails as a filter. In advanced cirrhosis, scar, nodules, and rising intrahepatic resistance create portal hypertension, and the splanchnic arterial circulation becomes profoundly vasodilated. Blood volume is then redistributed into the abdominal vascular bed instead of supporting the effective arterial circulation cleanly. The body senses underfilling even when total body water is high and ascites is already present. That false signal is the beginning of the syndrome. The kidney is not responding to a healthy map. It is responding to a cirrhotic one.

Once that underfilling signal is established, neurohormonal compensation hardens around it. Renin rises. Angiotensin II rises. Aldosterone rises. Sympathetic tone rises. Sodium and water retention deepen. Most importantly for the kidney, renal vasoconstriction intensifies in an effort to preserve arterial pressure in the rest of the body. At first this is protective. Eventually it becomes self-defeating. Renal blood flow falls, glomerular filtration falls, and the kidney begins failing in a body that is actually fluid-expanded but functionally underfilled. This is why hepatorenal syndrome is not just "a bad kidney complication of cirrhosis." It is the renal face of a whole-body hemodynamic disorder created by advanced liver disease.

The deceptive part is that the kidney can look structurally ordinary while this is happening. The nephron may not show the kind of primary inflammatory or destructive lesion that would explain the severity of the renal failure by local pathology alone. The organ is present. The blood chemistry is abnormal. The circulation around it is worse. The kidney becomes functionally unavailable because perfusion has been withdrawn and vasoconstriction has become extreme. That is why older descriptions called hepatorenal syndrome a functional renal failure, even though modern understanding recognizes that systemic inflammation and other secondary tubular stresses can deepen the injury as the syndrome progresses. The original problem still remains circulatory.

This is also why hepatorenal syndrome so often appears in decompensated cirrhosis with ascites and why it is frequently precipitated by events such as bacterial infection, especially spontaneous bacterial peritonitis, gastrointestinal bleeding, or other episodes that further destabilize the already fragile circulation. The liver has already lost enough architectural control that the body is surviving on compensation. A new hit then pushes the kidney below the threshold it can tolerate. So hepatorenal syndrome should be read as proof that liver disease has become systemic disease. The liver changed the vascular map, the arterial tree lost effective fill, neurohormonal rescue became chronic, and the kidney was forced to fail inside that environment.

Why Nitrogen Disposal Is Shared Labor Between Organs

Nitrogen disposal is shared labor because the body cannot let protein metabolism end where it begins. Amino acids are too valuable to waste, so the body is constantly stripping nitrogen from some of them while saving the carbon skeleton for fuel, glucose production, or other chemistry. That deamination event, the removal of the amino group, creates a problem immediately. Free ammonia is chemically small, diffusible, and dangerous enough that it cannot be allowed to accumulate in blood for long. But the body does not solve that problem in one place. The gut generates part of the burden, the liver converts much of it into urea, muscle helps buffer overflow temporarily, and the kidney performs the final excretion. Nitrogen disposal is therefore not one pathway sitting inside one organ. It is a relay system moving a toxic burden through different tissues, each handling the part of the problem it is built for.

The gut is where the burden begins becoming real. Dietary protein is digested into amino acids, intestinal cells metabolize some of those amino acids locally, and gut microbes act on protein residues, peptides, and urea in the lumen. Bacterial urease splits urea and liberates ammonia. Other microbial reactions produce amines and related nitrogenous compounds that can feed the same broader burden. This is why portal blood is never carrying nutrients alone. It is also carrying the chemical consequences of intestinal nitrogen metabolism. The liver sits directly downstream because the body cannot permit gut-derived nitrogen to move into the general circulation unchecked. A colonized intestine makes that impossible. Someone has to receive the first load and change its form before the brain and the rest of the body are forced to live with it.

The liver solves the central chemical problem through the urea cycle. Hepatocytes take ammonia and related nitrogen, spend ATP, adenosine triphosphate, the cell's immediately usable energy currency, and incorporate that burden into urea, a far safer and far more water-soluble molecule. This is one of the most important conversions in the body because it turns diffuse neurotoxic nitrogen into a transportable waste form the blood can carry without the same danger. But even here the liver is not doing everything alone. Peripheral tissues, especially skeletal muscle, can help buffer ammonia by incorporating nitrogen into glutamine and alanine. This is temporary protection, not final disposal. Muscle buys time. The liver is still the major organ that turns dangerous nitrogen into a molecule meant for exit. That is why hepatic failure so quickly becomes neurologic. The body loses the organ responsible for finishing the main protective conversion.

The kidney then completes the sequence. Urea leaving the liver re-enters blood, reaches the nephron, and is filtered into urine. The kidney is therefore not receiving generic waste. It is receiving a liver-made solution to a liver-sized problem. At the same time the kidney contributes its own nitrogen handling through ammonium excretion and acid-base regulation. In states of acidosis, the kidney can increase ammoniagenesis from glutamine, generate more ammonium, and trap acid in urine through that route. So the kidney is not only the drainpipe at the end of hepatic chemistry. It is an active nitrogen-handling organ in its own right. The point is that the two organs are solving different parts of the same burden. The liver neutralizes and packages. The kidney removes. If the liver fails, ammonia rises because the package is never made properly. If the kidney fails, urea and other nitrogenous waste accumulate because the package is made but not cleared.

This shared labor is why nitrogen disorders reveal the architecture of the whole body so clearly. A urea cycle defect looks neurologic because the liver failed to neutralize the burden. Hepatic encephalopathy looks cerebral because portal ammonia escaped liver control and reached the brain. Renal failure raises blood urea nitrogen because the kidney failed to remove what the liver had already prepared. Muscle wasting worsens hepatic encephalopathy because one of the body's temporary buffering tissues has been reduced. The intestine, liver, muscle, kidney, and brain are all standing in the same nitrogen story, but they are not doing the same job. The body survives protein turnover because it divides the work. The gut and tissues create the burden. Muscle buffers part of it. The liver makes it transportable. The kidney makes it leave. That is why nitrogen disposal is shared labor. No one organ can carry the whole problem without the others.

How Liver Injury Changes Renal Workload and Circulation

The kidney does not meet liver injury as a distant observer. It meets it as a changed blood supply, a changed solute load, and a changed hormonal environment all at once. That is the first principle. Once liver architecture stiffens and portal resistance rises, the circulation stops distributing blood normally. Portal hypertension develops, the splanchnic vascular bed begins pooling blood, and effective arterial filling falls even when total body fluid is already expanding through ascites and edema. The kidney reads that as underperfusion, not abundance. It answers by activating renin, angiotensin, aldosterone, sympathetic tone, and vasopressin, all of which increase sodium and water retention and narrow the renal circulation in an attempt to preserve pressure. What begins as liver injury therefore becomes renal workload immediately because the kidney is forced to defend a circulation the liver no longer organizes properly.

That circulatory shift changes the kidney's job before it changes the kidney's structure. A healthy kidney filters in the setting of adequate perfusion and balanced volume signaling. In advanced liver disease, the same kidney has to work inside arterial underfilling, portal hypertension, splanchnic vasodilation, and chronic sodium-retaining commands. Filtration becomes harder to maintain because renal blood flow is increasingly sacrificed to support systemic pressure. Sodium handling becomes more conservative because the body is convinced volume is threatened. Water follows sodium, which enlarges ascites and edema but does not solve the original problem of fluid misplacement. The kidney is not misbehaving. It is obeying a bad map. Liver disease changes renal workload by converting the kidney from a regulator of balance into an organ of emergency retention.

The chemical burden changes too. The liver normally converts ammonia and related nitrogen burden into urea that the kidney can safely excrete. When hepatic function falls or portal blood bypasses the liver through shunts, the kidney receives a different nitrogen landscape. More ammonia escapes control upstream, while the overall hemodynamic state makes filtration less reliable downstream. Drug handling changes in parallel because liver disease alters intestinal absorption, plasma protein binding, hepatic blood flow, portal-systemic shunting, biliary excretion, and even renal clearance. The kidney is therefore not just filtering the same blood more slowly. It is filtering different blood under worse perfusion. Both workload and working conditions have changed.

This is why advanced liver disease so often turns into kidney disease without the kidney being the original organ that failed. The scarred liver changes flow, volume signaling, and blood composition. The kidney then has to live inside that altered physiology every minute after. Renal vasoconstriction deepens, sodium retention becomes chronic, effective perfusion falls, and what finally looks like a renal problem is really the renal expression of a body whose circulatory order has already been broken by the liver.

Chapter 55

The Liver and the Endocrine System

Thyroid Handling

The thyroid gland does not finish the whole job of thyroid hormone physiology by secreting T4 and T3 into blood. The liver takes that signal and begins editing it. Much of what the thyroid releases is thyroxine, T4, a prohormone with less direct hormonal force than triiodothyronine, T3. The liver helps determine how much of that circulating pool remains stored as precursor, how much is converted toward active hormone, and how much is pushed toward inactivation and clearance. It also manufactures part of the transport environment that keeps thyroid hormone moving through plasma in a buffered state rather than as a wildly fluctuating free signal. This means the liver is not a passive organ sitting downstream from the thyroid. It is one of the places where thyroid hormone becomes usable, measurable, and removable.

That handling begins in blood. Thyroid hormone does not circulate mainly as naked free molecule. Most of it is bound to transport proteins, and the liver is one of the organs that helps maintain that transport architecture through synthesis of proteins such as thyroxine-binding globulin, transthyretin, and albumin. Binding matters because it stabilizes the circulating reservoir, limits sudden swings in free hormone, and determines how much hormone remains available to enter tissues at any given moment. The bloodstream is not carrying only hormone concentration. It is carrying hormone under transport control, and the liver is one of the organs shaping that control.

The liver also participates directly in thyroid hormone conversion. Deiodination is the most important metabolic pathway in thyroid hormone handling, and type 1 deiodinase in the liver helps convert T4 into the more active T3 while also participating in reverse T3 metabolism. That means the liver is contributing to the balance between active signal, stored precursor, and inactive thyroid-related forms moving through blood. At the same time, thyroid hormones undergo conjugation in the liver through glucuronidation and sulfation, which alters their solubility, recirculation, and excretion. The liver is therefore doing with thyroid hormone what it does with many other compounds: changing form, changing fate, and deciding whether the molecule keeps circulating as endocrine message or moves toward disposal.

This is why liver disease can change thyroid physiology without the thyroid gland itself being the first organ at fault. If transport protein synthesis changes, total hormone values can shift even when free hormone behaves differently. If hepatic deiodination becomes less effective, circulating T3 can fall and reverse T3 can rise under illness conditions. If conjugation and clearance change, thyroid-related compounds may remain in circulation differently than they would in a healthy liver. The result is that thyroid status in liver disease can become harder to read unless the clinician remembers that the bloodstream is carrying a liver-edited hormone system, not just gland output. The gland writes the first message. The liver helps decide how long that message stays active, how much of it is sharpened, and how much is sent away.

Sex Hormone Metabolism

Sex hormones do not simply circulate until target tissues are finished with them and then vanish on their own. They have to be transformed, buffered, carried, and cleared, and the liver sits at the center of that work. Estrogens, androgens, progesterone-derived compounds, and cortisol-family steroids all pass through a tissue that is built to decide how long a signal remains active, what chemical form it will take next, and whether it will stay in circulation, move toward bile, or be prepared for urinary loss. That is why sex hormone metabolism belongs to liver physiology rather than floating out on its own as an endocrine footnote. The gonads and adrenal glands write part of the hormonal message, but the liver helps decide how long that message survives in blood and what shape it has while it survives.

This happens through transformation first and through transport second. Hepatocytes receive steroid hormones from the circulation and subject them to oxidation, reduction, hydroxylation, glucuronidation, and sulfation. Oxidation and reduction change the chemical state of the steroid ring or side groups. Hydroxylation adds a hydroxyl group and often creates a better handle for later processing. Glucuronidation and sulfation are conjugation reactions, meaning the liver attaches new chemical groups that make the hormone more water-compatible and easier to excrete. This does not always mean the molecule is destroyed in one step. It means the hormone is being moved away from its original active form and toward a state more suited to controlled clearance. Estrogens are hydroxylated and conjugated. Androgens are reduced, oxidized, and conjugated. The liver is therefore not just cleaning up after endocrine signaling. It is actively rewriting steroid chemistry so the body does not live under endless unregulated hormone exposure.

Transport matters just as much, because sex hormones do not circulate mainly as free unrestricted molecules. The liver makes sex hormone-binding globulin, SHBG, and also supplies albumin, both of which help determine how much hormone remains bound in plasma and how much remains free and biologically available to enter tissues. A change in liver function can therefore alter endocrine physiology even when glandular production is not the original problem. The ovaries, testes, or adrenal glands may still be releasing hormones, but if hepatic binding protein production changes, or if hepatic transformation and clearance slow down, the pattern of free versus bound hormone changes with it. The blood is no longer carrying the same endocrine message even if secretion upstream has not changed proportionally. This is one reason liver disease can create hormonal symptoms that look reproductive, dermatologic, or metabolic before the person ever thinks of the liver as an endocrine organ.

The liver also links sex hormone metabolism to bile and excretion. Once estrogens and androgens are conjugated, they can be secreted into bile or returned to blood for renal clearance. That means hormone metabolism is tied directly to hepatic outflow and to the chemistry of the intestine. Some conjugated steroids entering bile can later be modified by intestinal bacteria, and part of that downstream chemistry can influence what is reabsorbed or lost. The liver is therefore not just reducing hormone activity. It is deciding route. A steroid signal that should no longer remain fully active is transformed into something the body can carry toward elimination. This routing function is one of the reasons cholestatic disease, cirrhosis, and broader hepatic dysfunction can all distort sex hormone balance. The signal is no longer being processed and removed with the same discipline.

That is why sex hormone symptoms in liver disease often feel systemic rather than neatly glandular. When hepatic metabolism falls behind, estrogens may remain active longer, androgen balance may shift, transport protein patterns may change, and the body begins displaying the result in tissues far from the liver. In men this can contribute to gynecomastia, spider angiomas, testicular atrophy, reduced libido, and altered hair patterning. In women menstrual irregularity and broader reproductive endocrine instability can appear. The deeper point is not the symptom list. The deeper point is that the liver is one of the organs deciding whether a sex hormone remains a circulating signal, becomes a conjugated waste, or changes into another form first. It is an endocrine editor. When that editor weakens, the body keeps the hormones longer, keeps them differently, or clears them poorly, and the endocrine picture begins changing even before the glands themselves are asked to explain it.

Cortisol and Steroid Metabolism

Cortisol does not simply circulate until tissues are finished with it and then disappear. It has to be transformed, buffered, and cleared, and the liver is one of the main organs performing that work. In broad terms, hepatic cortisol metabolism is a sequence of reduction, oxidation, and conjugation reactions that turn an active steroid signal into more water-compatible metabolites suited for excretion. Clinical endocrine references describe the enzymatic system for cortisol metabolism as primarily hepatic, and they also note that corticosteroid metabolites are then excreted mainly in urine. This is why cortisol physiology cannot be reduced to adrenal secretion alone. The adrenal gland writes the first signal. The liver helps decide how long that signal remains chemically active in blood.

One important part of that handling is the interconversion between cortisol and cortisone. The 11β-hydroxysteroid dehydrogenase system performs this shuttling, and classic endocrine literature describes these enzymes as major regulators of active versus inactive glucocorticoid forms, with hepatic expression playing an important role in the broader balance of glucocorticoid availability. In liver and adipose tissue, 11β-HSD1 acts mainly as a reductase, favoring conversion of inactive cortisone into active cortisol under physiologic conditions. The same logic explains why prednisone, an inactive glucocorticoid precursor, is primarily converted in the liver into prednisolone, its active form. The liver therefore is not only a place where steroid hormones are inactivated. It is also one of the places where inactive glucocorticoid forms are reactivated. That makes hepatic steroid handling a two-way editorial process rather than a simple disposal route.

The conjugation side matters just as much. Once cortisol and related steroids have been reduced, oxidized, or otherwise reshaped, the liver attaches sulfate or glucuronide groups to many of those metabolites. Glucuronidation is a major phase II pathway for endogenous substances as well as drugs, and this is one of the main ways steroid molecules become hydrophilic enough to move cleanly toward urinary excretion. In plain physiological language, the liver is taking a lipid-friendly endocrine signal and turning it into something the kidney can carry away. That is why steroid metabolism belongs to the liver's larger logic of conversion. The organ does not only respond to hormones. It determines whether those hormones stay active, become inactive, are reactivated locally, or are pushed toward exit.

This is also why liver disease changes cortisol biology even when the adrenal gland itself is intact. Older and newer literature both note that chronic liver disease can impair cortisol metabolism, alter steroid handling, and change sensitivity to administered glucocorticoids. Once hepatic transformation and conjugation slow down, the bloodstream is no longer carrying the same steroid pattern it carried in health. Some signals last longer. Some conversions become less efficient. Some drug precursors are activated differently. The deeper point is simple. Cortisol is not only an adrenal hormone. It is also a liver-processed hormone, and the body feels the difference when that processing falls behind.

IGF-1 Output

IGF-1 output is one of the clearest examples of the liver acting like an endocrine organ in its own right. Growth hormone is released by the pituitary, but a major share of the circulating insulin-like growth factor 1, IGF-1, that carries growth and anabolic signaling through the body is produced by hepatocytes in response to that upstream signal. The liver is therefore not just listening to growth hormone. It is translating it into a second blood-borne message with different reach, different stability, and different tissue consequences. Circulating IGF-1 is predominantly liver-derived, even though many tissues also make IGF-1 locally for paracrine and autocrine use.

That output matters because growth hormone is not simply a direct command to bone and muscle. Much of its systemic effect depends on the liver receiving the signal and releasing IGF-1 into plasma. IGF-1 then helps support longitudinal growth, protein anabolism, tissue maintenance, and broader metabolic adaptation. It also participates in feedback control of the growth hormone axis, helping regulate the very signal that stimulated its production. In that sense the liver is not a passive relay. It is part of the architecture that determines how strong the growth signal becomes and how long it remains active in circulation.

The liver's IGF-1 output is also conditional. It is influenced not only by growth hormone, but by nutrition and insulin status, which means the hepatocyte does not produce endocrine growth signal in a vacuum. A body that is inflamed, malnourished, or metabolically disordered may still have growth hormone present while hepatic IGF-1 output falls because the liver is no longer in a state that can support full anabolic translation of that signal. This is one reason IGF-1 is such a useful physiologic clue. It reflects not only pituitary instruction, but also whether the liver is in a condition to answer that instruction.

So IGF-1 output should be understood as the liver taking a hormonal command and turning it into a whole-body growth environment. The pituitary writes the first signal. The liver decides how much of that signal becomes circulating anabolic reality. When hepatic output is strong, growth hormone becomes systemic tissue instruction. When hepatic output weakens, the body may still have the glandular message but lose much of the bloodstream's ability to carry it forward.

Vitamin D Conversion

Vitamin D does not enter the body in its final usable endocrine form. Whether it is made in skin under ultraviolet B light or absorbed from food and supplements, it begins as a precursor. The liver is the first major organ that changes that precursor into a circulating intermediate the rest of the body can actually work with. Hepatocytes perform 25-hydroxylation, adding a hydroxyl group and converting vitamin D into 25-hydroxyvitamin D, written as 25(OH)D. That step matters because 25(OH)D is the main circulating storage and transport form of vitamin D, the form that remains in blood in meaningful quantity and the form clinicians usually measure when they want to know what the body's vitamin D status really is.

This means the liver is not a side station in vitamin D physiology. It is the organ that turns a newly arrived precursor into an endocrine resource. Raw vitamin D from skin or diet is not yet the major blood-borne pool the body will draw from over time. Once the liver hydroxylates it, the molecule becomes a more stable circulating reserve, one that can move through plasma, remain available over days rather than moments, and supply the next activation step when the body needs it. That is why the standard blood test does not focus on the original parent compound. It focuses on the liver-made intermediate. The question is not simply whether vitamin D recently entered the body. The question is whether it successfully became 25(OH)D.

The kidney then performs the next major transformation, converting 25(OH)D into 1,25-dihydroxyvitamin D, the more tightly regulated active hormonal form. That two-step sequence is crucial because it separates arrival, storage, and final activation into different compartments. The liver creates the main circulating substrate. The kidney refines that substrate into the strongest endocrine signal. This division of labor is one reason vitamin D physiology is really a liver-kidney axis rather than a simple skin-or-diet story.

It also explains why liver disease can distort vitamin D biology before the kidney ever gets its turn. If hepatic 25-hydroxylation falls behind, or if the transport and nutritional conditions surrounding vitamin D handling deteriorate, the body can have lower 25(OH)D even when sunlight exposure or intake look reasonable on the surface. The precursor may have entered. The liver may have failed to turn enough of it into the main circulating form. Vitamin D deficiency in chronic liver disease is therefore not only a dietary or sunlight problem. It can also be a conversion problem.

So vitamin D conversion belongs in liver physiology because the liver is where vitamin D first becomes blood chemistry with endocrine meaning. Skin and diet provide the raw molecule. The liver gives it durable circulatory form. The kidney gives it final hormonal force. Miss the hepatic step, and the whole pathway is earlier, weaker, and less usable than it appears.

Why Hormone Symptoms Can Begin With Hepatic Dysfunction

Hormone symptoms can begin with hepatic dysfunction because the liver does not only receive hormones after they have done their work. It determines how those hormones travel, how long they remain active, what fraction stays free in plasma, what fraction is converted into a stronger or weaker form, and when the signal is finally pushed toward excretion. That means the endocrine system is not only gland secretion. It is gland secretion plus hepatic editing. A thyroid gland can release T4, but the liver helps convert part of that pool into T3 and helps maintain the transport proteins that buffer thyroid signal in blood. The gonads and adrenals can release steroids, but the liver oxidizes, reduces, conjugates, and clears those molecules so the body does not remain trapped under yesterday's signal. Growth hormone can be present, but a large share of circulating IGF-1 depends on the liver answering that signal. Vitamin D can enter from skin or diet, but the main circulating intermediate is made in the liver. So when hepatic function falls behind, the body does not only lose detoxification and bile flow. It loses endocrine finishing work. Hormone symptoms can therefore appear before anyone thinks to blame the liver, because the liver was shaping the hormonal environment quietly the whole time.

This becomes visible first in transport. Hormones do not move through plasma as naked messages in most cases. They travel bound to specific proteins, and the liver makes many of those proteins. Thyroxine-binding globulin, transthyretin, albumin, and sex hormone-binding globulin all change how much hormone remains buffered in blood and how much is immediately free to enter tissue. A change in hepatic protein synthesis can therefore alter hormone behavior even when glandular production has not changed proportionally. Total hormone levels may shift because binding capacity changed. Free hormone fractions may become less stable. Drug-hormone interactions may intensify because plasma binding is different. The patient then develops fatigue, low drive, reduced libido, menstrual disturbance, altered body composition, or other endocrine-looking symptoms, while the deeper problem is that the bloodstream is no longer carrying signals in the same way. The gland still speaks. The transport environment has changed the meaning of what it said.

Conversion is the next layer. The liver is one of the major organs taking hormone precursors and changing them into more useful circulating forms. T4 to T3 conversion is one example. Vitamin D 25-hydroxylation is another. Steroid metabolism is another still. In each case the liver is not merely cleaning up after hormones. It is deciding whether a signal becomes sharper, weaker, more water-compatible, more transportable, or more excretable. That is why hepatic dysfunction can create hormonal symptoms that look diffuse and hard to classify. The person may not present with one classic endocrine syndrome caused by one damaged gland. Instead the person begins drifting into flatter thyroid signaling, weaker anabolic tone through lower IGF-1, altered estrogen-androgen balance because steroid clearance changed, poorer vitamin D status because hepatic conversion weakened, and broader fatigue because the bloodstream is now carrying a less coherent endocrine pattern. The symptom picture can look mixed because the liver was participating in multiple hormone systems at once.

Sex hormone metabolism makes this especially obvious. The liver normally transforms estrogens and androgens through oxidation, reduction, glucuronidation, and sulfation, then routes the resulting metabolites toward bile or urine. If that handling slows, sex steroids remain in circulation longer or in altered ratios. In men this can show up as gynecomastia, spider angiomas, reduced libido, testicular atrophy, and changes in hair patterning. In women it can contribute to menstrual irregularity and broader reproductive endocrine instability. The point is not only that "hormones are out of balance." The point is that the liver has stopped imposing normal timing and clearance on circulating steroid signals. A hormone that should have been reduced, conjugated, and removed is still present. A transport protein that should have held the pool in better balance is lower or altered. The symptom appears in breast tissue, skin, gonads, or mood, but the failure began in hepatic chemistry.

So hormone symptoms can begin with hepatic dysfunction because the liver is one of the organs that turns endocrine secretion into endocrine reality. Glands release signals. The liver converts them, binds them, restrains them, extends them, or clears them. When hepatic function weakens, the person does not only lose liver-specific functions. The person begins living in a different hormone environment, and the body often reveals that long before anyone says the word cirrhosis.

Chapter 56

The Liver and the Blood

Plasma Protein Production

The liver does not only receive blood and alter what is dissolved in it. It helps build the fluid itself. That is the first thing the reader has to see. Plasma is not just water with nutrients drifting through it. It is a protein-built transport medium whose behavior depends heavily on what hepatocytes are continuously manufacturing and releasing into circulation. Albumin, clotting factors, complement proteins, transport proteins, acute-phase proteins, and a wide range of other circulating molecules are not passive decorations dissolved in blood after the important work is done. They are part of what makes blood capable of carrying fluid, hormones, fatty acids, bilirubin, metals, inflammatory signals, and coagulation potential in an organized way. The liver is therefore not merely processing the bloodstream after it arrives. It is one of the organs deciding what kind of bloodstream the body will have tomorrow.

This synthetic work begins inside hepatocytes as a matter of organelle architecture and directional secretion. Rough endoplasmic reticulum builds the protein backbone. The Golgi modifies, packages, and sorts the product. Then the hepatocyte releases that protein across its blood-facing membrane into the sinusoidal side of the tissue. That direction matters. A hepatocyte is not only pulling compounds out of blood for metabolism. It is also writing new proteins back into blood as part of normal circulation. Albumin enters plasma this way and helps maintain oncotic pressure while carrying fatty acids, bilirubin, and many drugs. Fibrinogen and prothrombin enter plasma this way and help determine whether blood can clot when a vessel is injured. Complement proteins enter plasma this way and help make blood immunologically active. Transferrin and other carrier proteins enter plasma this way and determine how metals and hormones move. The liver is not just making proteins because tissues need protein in the abstract. It is making the specific proteins that convert plasma into a functioning transport environment.

This is why plasma protein production has to be understood as a whole-body construction job rather than a liver-specific detail. Albumin affects whether water stays in the vascular compartment or leaks into tissue. Clotting proteins affect whether bleeding stops or continues. Acute-phase proteins affect how inflammation changes blood behavior. Binding proteins affect how much hormone remains free, how much drug remains active, and how long different molecules stay buffered in circulation before they are cleared. A change in hepatic protein synthesis therefore does not stay local to the liver. It changes fluid balance, coagulation, immunity, endocrine signaling, nutrient transport, and vascular stability across the body. The bloodstream becomes a different fluid because the liver changed what it was making.

This is also why chronic liver disease becomes visible in plasma before the organ is physically absent. A cirrhotic liver may still be present on imaging, still contain nodules of hepatocytes, and still process some bilirubin or some drug burden, yet if albumin production falls, ascites and edema become easier to generate. If clotting factor synthesis falls, PT and INR begin changing. If thrombopoietin falls, platelet production support weakens. If acute-phase signaling dominates, albumin may fall further in relative priority while inflammatory proteins rise. The body is then not only dealing with damaged liver tissue. It is dealing with a blood compartment that has lost part of the protein architecture the liver used to provide. Plasma protein production is therefore one of the clearest ways the liver reaches every organ without physically touching them. It builds the medium that touches them for it.

So the liver and the blood are linked at a deeper level than most people are taught. The liver does not merely filter a completed fluid. It manufactures much of the protein framework that gives that fluid its carrying power, its clotting power, its immune tone, and its ability to hold water where it belongs. The person with liver disease is not only losing a detox organ or a bile organ. The person is losing one of the main tissues that keeps plasma from becoming chemically and physically unstable. That is what plasma protein production really means. The liver is not outside the bloodstream looking in. It is one of the organs writing blood from within.

Clotting and Anticoagulation

The liver does not merely help blood clot. It helps decide whether blood remains fluid in motion or becomes solid at the right place and right time. That distinction matters because coagulation is not a one-direction system built to stop bleeding at any cost. It is a balance system. The bloodstream has to move through miles of vessels without turning into a thrombus, then harden within seconds when a vessel wall tears, then stop hardening before the repair itself becomes an obstruction. The liver sits near the center of that balance because hepatocytes synthesize most of the soluble proteins that make clot formation possible and many of the proteins that keep clot formation from running too far. Fibrinogen, prothrombin, and multiple other clotting factors are made there. Antithrombin, protein C, and protein S are made there too. This means the liver is not building a clotting machine. It is building a clotting field with brakes already mixed into the fluid.

The procoagulant side is easier to picture first. A vessel is injured, platelets adhere, coagulation factors are activated in sequence, thrombin is generated, and fibrinogen is converted into fibrin to reinforce the platelet plug. None of this happens if the proteins are missing. The liver writes most of those proteins into plasma before any injury occurs, so the blood is always carrying latent clotting potential in reserve. But that reserve would be catastrophic without restraint. Thrombin cannot be allowed to wander freely through circulation, and activated factors cannot be permitted to spread the clotting signal indefinitely beyond the wound. That is why the same liver that makes the accelerator also makes the braking proteins. Antithrombin helps neutralize activated clotting enzymes. Activated protein C, supported by protein S, helps shut down important amplification steps. The blood stays usable because these opposing systems are built together from the start.

This is why liver disease destabilizes coagulation rather than simply weakening it in a straight line. A failing liver often makes fewer clotting factors, which is real and clinically important. But it also makes fewer anticoagulant proteins. The person is not merely losing the ability to clot. The person is losing the ability to regulate clotting. This is one of the reasons cirrhosis produces such misleading bedside intuitions.

The patient may bruise, bleed, or show a prolonged PT and INR, yet still remain vulnerable to thrombosis because the anti-clotting side has been thinned as well. The circulation is no longer strongly defended in either direction. It becomes unstable, easier to tip toward bleeding in one setting and toward thrombosis in another. The liver has not created "thin blood." It has created blood with fewer margins.

Vitamin K sharpens this point further because several major clotting factors depend on vitamin K-mediated modification before they can function properly. The liver may still synthesize the protein backbone, but without enough vitamin K, factor II, factor VII, factor IX, and factor X are not finished correctly for normal coagulation chemistry. Since vitamin K is fat-soluble, cholestasis and poor bile delivery can impair its absorption and create a coagulation defect even before hepatocyte synthesis has fully collapsed. That means clotting and anticoagulation in liver disease are not only about whether the liver can make proteins. They are also about whether bile is reaching the intestine, whether fat-soluble vitamin handling is intact, and whether the bloodstream is still receiving fully functional versions of the proteins being made.

This is why clotting tests must be interpreted with discipline. A prolonged PT or elevated INR can reveal loss of clotting factor synthesis, but these numbers do not capture the whole balance between procoagulant and anticoagulant systems. They show one face of the problem. They do not fully describe whether the patient is safer from clot or closer to it. The deeper physiology is more complicated. The liver-made coagulation field has become thinner on both sides. Platelets may be reduced by splenic sequestration and lower thrombopoietin. Fibrinolysis may be altered. Endothelium may be more reactive. Portal flow may already be abnormal enough to favor thrombosis in one territory while varices threaten bleeding in another. The liver is therefore best understood as the organ that makes fluid blood possible by manufacturing both the proteins that stop hemorrhage and the proteins that prevent that solution from overshooting.

So clotting and anticoagulation belong together because the liver built them together. A healthy liver writes a plasma environment that can harden locally and remain fluid elsewhere. A failing liver does not simply take clotting away. It removes the order that kept clotting under control.

Iron Regulation

Iron regulation is not built around easy excretion. That is the first structural fact that makes the whole system different from sodium, potassium, or water. The body can lose some iron through bleeding, epithelial shedding, menstruation, and slow background turnover, but it does not possess a large fast regulated iron-dumping valve it can open when stores rise. That means control has to happen at the level of movement, not disposal. The real question is never simply how much iron exists in the body. The real question is whether iron is being allowed to enter the circulation, whether it is being held in storage, whether it is being handed to the marrow for erythropoiesis, and whether it is being trapped away from microbes or reactive chemistry when that is safer. The liver sits at the center of this because it is both a storage organ and an endocrine control organ. It does not merely hold iron. It decides whether iron traffic is open or closed.

The traffic begins in the gut but it is not finished there. Dietary iron crosses the intestinal epithelium, mainly in the duodenum, and enters the enterocyte. At that point the body has not yet truly committed to systemic iron gain. The enterocyte can store some iron transiently, but for that iron to become blood iron it has to leave the cell through ferroportin, the major cellular iron exporter. That same exporter matters in macrophages that recycle iron from aging red blood cells and in hepatocytes and other storage cells that release reserve iron back into plasma. Ferroportin is therefore the gate at which intracellular iron becomes circulating iron. The liver controls that gate hormonally through hepcidin. When hepatocytes release hepcidin into blood, hepcidin binds ferroportin, drives it off the cell surface, and causes it to be degraded. The door disappears. Iron remains in the enterocyte, the macrophage, or the storage cell instead of entering plasma. This is why iron regulation should be understood as permission, not abundance. Iron can be present in the body and still be unusable if the liver has decided it should stay trapped.

That decision is made against two opposite risks. Too little circulating iron starves erythropoiesis, weakens hemoglobin production, and constrains mitochondrial and enzymatic work across the body. Too much circulating or poorly contained iron creates a different danger because iron is chemically reactive. In the wrong redox state it participates in free radical chemistry that can damage membranes, proteins, and nucleic acids. The body therefore has to strike a narrow balance between sufficiency and restraint. When iron stores are high, hepcidin rises and ferroportin-mediated export falls. Less dietary iron enters blood. Less recycled iron leaves macrophages. Less reserve iron leaves the liver. When iron demand rises, or when oxygen-carrying capacity has to be rebuilt, that brake is lowered. More ferroportin remains on the cell surface. More iron enters plasma and is carried largely by transferrin toward the marrow and other tissues. The system is elegant because it solves a dangerous problem without needing a large excretory escape hatch. The liver reads the state of the body and controls the gates.

Inflammation shows how strategic this control really is. During inflammatory states, the liver raises hepcidin even when total body iron is not excessive. This is not a mistake. It is defense. Many microbes need iron, and the body responds by withholding it. Enterocytes absorb less usable iron into blood. Macrophages hold on to recycled iron instead of releasing it. Plasma iron falls. From the outside this can look like deficiency, but the deeper problem is restriction, not absence. Iron may still be present in storage while the marrow experiences scarcity because the liver has hormonally decided that free movement is too dangerous under inflammatory conditions. This is why anemia of inflammation is so revealing. The body is not necessarily out of iron. It has been instructed not to share it. Once again the liver is not just storing a mineral. It is adjudicating who gets access to it and under what circumstances.

So iron regulation is best understood as a liver-governed movement problem. The intestine takes iron in, macrophages recover it, storage cells hold it, transferrin carries it, and the marrow spends it, but the liver decides whether that traffic proceeds. It does this not by flushing iron away, but by controlling ferroportin through hepcidin and by serving as one of the major storage sites where the consequences of excess or restraint become visible. Iron biology therefore is not just nutrition. It is controlled circulation of a dangerous useful metal, and the liver is the organ that makes that circulation livable.

Platelet Production Signals

Platelets do not appear in blood simply because bone marrow happens to be active. Their production is driven by a signal, and one of the liver's clearest endocrine jobs is making that signal. The hormone is thrombopoietin, usually abbreviated TPO. It is produced predominantly in the liver and stimulates the bone marrow to expand megakaryocyte development and platelet production. Megakaryocytes are the large marrow cells that shed platelets into circulation. This means platelet biology is not only a marrow story. The marrow performs the manufacture, but the liver provides much of the instruction that keeps the production line moving.

That matters because the bloodstream is always consuming platelets quietly. Small vessel injury, endothelial wear, and ordinary hemostatic surveillance constantly use and clear them. The body therefore needs a standing growth signal that keeps production matched to loss. Thrombopoietin provides that signal. It circulates in blood and acts on megakaryocytic cells in the marrow, telling them that platelet support must be maintained. In a healthy system this remains mostly invisible because the signal and the output stay in balance. But the balance depends on hepatic output more than most people realize. If the liver stops making enough thrombopoietin, the marrow can still exist and still be structurally normal, yet platelet production begins losing one of its main hormonal drivers.

This is why liver disease so often shows itself through low platelet counts before the public fully understands what is happening. In advanced cirrhosis, thrombocytopenia is not only a spleen problem from portal hypertension and splenic sequestration, though that is real. Modern clinical references note that altered or reduced thrombopoietin production by the liver contributes directly to thrombocytopenia in cirrhosis. The platelet count falls from both directions at once. The enlarged spleen traps more platelets, and the failing liver provides less of the signal that tells the marrow to replace them. The low platelet count therefore is not only a secondary accident of portal congestion. It is also evidence that the liver's endocrine support for hemostasis has weakened.

So platelet production signals belong in liver physiology because the liver is not merely making clotting proteins and albumin. It is also shaping whether the marrow is told to keep supplying the blood with platelets at all. The marrow makes the platelets. The liver helps decide how strongly that production is commanded. When hepatic function declines, one of the quietest but most important blood-support signals begins to thin, and the circulation starts revealing that loss through thrombocytopenia.

Splenic Consequences of Portal Hypertension

The spleen is one of the first organs forced to reveal that portal blood can no longer cross the liver cheaply enough. It drains into the portal system through the splenic vein, which means a scarred liver does not trap only intestinal blood behind it. It traps splenic venous outflow too. Once intrahepatic resistance rises, pressure backs up through the portal circulation and the spleen begins living under chronic venous congestion. This is why splenomegaly, enlargement of the spleen, is such a dependable structural sign of portal hypertension. The liver has become a harder organ to cross, and the spleen is one of the upstream organs that pays the mechanical price.

That enlargement is not just a size change. It changes what the spleen does with blood. A healthy spleen already filters aged or abnormal blood cells and participates in immune surveillance. A congested enlarged spleen begins retaining and clearing more of the circulating formed elements than it should. This is called hypersplenism. The blood cells are still in the body, but more of them are being held in the splenic reservoir instead of remaining available in the bloodstream. Platelets usually reveal this earliest and most clearly, which is why thrombocytopenia is such a common blood finding in portal hypertension. Red blood cells and white blood cells can also be affected, so the pattern can widen into broader cytopenias as the splenic burden deepens.

This is why splenomegaly in cirrhosis is never just an incidental enlargement beside the liver. It is one of the main ways vascular failure becomes a hematologic problem. The portal system is congested. The spleen enlarges under that congestion. The enlarged spleen then changes how many platelets and other blood cells remain in circulation. The clinician may first notice a falling platelet count, but the deeper event is structural. Blood is no longer draining through the liver normally, and the spleen has become part of the body's attempt to live with that abnormal route.

The splenic consequence also matters because it compounds other liver-related blood problems. A cirrhotic liver may already be making less thrombopoietin and fewer clotting proteins. Add hypersplenism on top of that and the bloodstream loses both platelet support and platelet availability at the same time. This is one reason portal hypertension has to be read as a whole-body circulatory disorder rather than as pressure in one vessel alone. The scarred liver changes the route. The spleen enlarges to absorb part of that change. The blood then begins showing the cost.

The Blood as One of the Main Places Liver Function Becomes Visible

The liver becomes visible in blood because the liver is one of the organs that writes blood continuously. It does not simply receive a finished circulating fluid and clean a few mistakes out of it. It changes what enters, what stays, what leaves, and what the plasma itself is made of. Glucose arriving from the gut is buffered there before the rest of the body sees it in full. Ammonia is converted there into urea so the bloodstream does not become a neurotoxic nitrogen stream. Bilirubin is taken up, conjugated, and pushed toward bile so pigment does not remain circulating. Albumin is synthesized there and helps determine whether plasma can hold water inside the vascular space. Clotting factors, complement proteins, transport proteins, hepcidin, thrombopoietin, and multiple hormone-related intermediates are all written into blood by the liver. This is why the bloodstream is one of the main places liver function becomes visible. Blood is not just passing by the liver. Blood is being edited by it.

That makes blood chemistry one of the clearest mirrors of what kind of liver is still present. A healthy liver produces blood that is more stable than the portal blood it receives. Glucose is less violent after a meal because the liver absorbed part of the surge. Ammonia stays lower because the liver converted it. Bilirubin stays lower because the liver moved it into bile. Albumin remains adequate because hepatocytes keep making it. PT and INR remain within a usable range because clotting proteins are being supplied. Iron moves or stays trapped according to hepcidin. Platelet support continues through thrombopoietin. Thyroid, steroid, and vitamin D physiology remain more coherent because the liver is converting, binding, and clearing their signals in the background. The blood of a healthy person is not merely blood that circulates. It is blood that has already passed through hepatic judgment.

This is why liver disease often becomes visible in blood before it becomes obvious anywhere else. A damaged hepatocyte leaks ALT and AST. A cholestatic liver allows bilirubin and bile-related chemistry to back into circulation. A cirrhotic liver produces less albumin and fewer clotting factors, so fluid balance and coagulation begin changing even if the person still looks outwardly intact for a time. A liver under inflammatory pressure changes plasma composition by making acute-phase proteins and altering iron handling. A liver failing to clear ammonia allows neurologic chemistry to drift. A liver receiving more fatty acid burden than it can organize begins exporting different lipid traffic and storing the rest. The body can therefore read the liver through blood long before the liver becomes a visibly shrunken or nodular organ in the public imagination. The circulation starts telling the story first.

But blood does not only reveal that the liver is injured. It reveals which job is failing. A rising ALT says the worker is being hit. A rising bilirubin says processing or outflow is breaking. A falling albumin says the protein factory is tiring. A prolonged PT or rising INR says synthetic reserve for coagulation is thinning. A falling platelet count may reflect portal hypertension and splenic sequestration, reduced thrombopoietin, or both. Hyperammonemia says the nitrogen gate is failing. Iron restriction with inflammation says hepcidin is high and traffic is being withheld. The blood becomes a map of compartments and sequences. This is why numbers have to be read as physiology rather than as labels. The liver speaks through different changes in blood depending on which compartment is injured and which function is failing first.

So the blood is one of the main places liver function becomes visible because the liver is one of the main organs that determines what blood is allowed to contain. It takes a chemically rough portal stream and turns it into something more usable for the rest of the body. When that work succeeds, the liver stays hidden because the blood looks normal enough to live in. When that work fails, the bloodstream begins revealing exactly how much of its stability had been hepatic all along.

Part XIV

Clinical Logic for the Reader Who Wants to Think Like an Expert

Chapter 57

How to Approach a Liver Problem

Start With Structure, Then Flow, Then Synthesis, Then Excretion

The first mistake in liver thinking is to start with the lab value, the diagnosis label, or the symptom name as though those were the organ itself. They are not. The liver is an arrangement before it is a number. It is a vascular organ receiving portal blood from the gut, arterial oxygen from the hepatic artery, and sending bile out through a separate directional system at the same time. It contains hepatocytes that face blood on one side and bile on the other, sinusoids that are supposed to stay open and permeable, stellate cells that are supposed to remain quiet, and a portal-to-central geometry that determines which cells see which chemistry first. So the expert approach begins with structure. What kind of organ am I dealing with right now. Is the architecture still soft and exchange-ready. Is there steatosis, fibrosis, nodularity, duct injury, vascular distortion, mass effect, congestion, or cholestatic remodeling. Before asking what the liver test means, ask what kind of liver could have produced it. The organ always speaks through its architecture first, even when the clinician is only hearing bloodwork.

Once structure is in view, the next question is flow. Blood has to reach the hepatocyte. Bile has to leave it. Portal blood has to cross a low-resistance sinusoidal bed. Hepatic venous blood has to drain. Bile has to move from canaliculus to duct to intestine. A surprising amount of liver disease is really a failure of movement rather than a failure of cellular intelligence in the abstract. Fibrosis raises resistance and changes portal pressure. Collaterals appear when blood can no longer cross the liver cheaply enough. Cholestasis develops when bile can no longer leave in the direction the cell intended. Ascites appears when the circulation can no longer keep fluid inside blood. Hepatorenal syndrome appears when the whole vascular map has changed enough that the kidney begins living in an underfilled circulation. So after structure, think flow. Where is blood being slowed, shunted, backed up, or deprived of access. Where is bile being trapped, diluted, blocked, or withheld from the gut. A liver can still contain cells and still be functionally unavailable if the routes through it have become wrong.

Only after structure and flow should synthesis come fully into focus. The liver is not important merely because it processes incoming burden. It is important because it keeps making the bloodstream and the endocrine environment the rest of the body depends on. Albumin, clotting factors, complement proteins, transport proteins, thrombopoietin, hepcidin, IGF-1, and many other plasma molecules are part of hepatic output. This means the question is not only whether the liver is inflamed. The question is whether it is still manufacturing. A high ALT tells you the worker has been hit. A low albumin tells you the factory is slowing. A prolonged PT or elevated INR tells you clotting-factor synthesis is no longer keeping up. Low platelets may reflect splenic sequestration, reduced thrombopoietin, or both. The expert does not confuse injury with productive capacity. A dramatic injury can occur in a liver that still has reserve, and a quiet cirrhotic liver can be failing synthetically while leaking little. So after you ask what kind of structure is present and how blood and bile are moving through it, ask what this liver is still building well and what it has stopped building well enough.

Then comes excretion, and here the expert keeps the route in mind instead of treating waste as one generic concept. The liver does not only destroy things. It changes their form and assigns them to the right exit. Bilirubin must be conjugated and secreted into bile. Ammonia must be turned into urea so the kidney can take over. Bile acids and cholesterol must move outward through the biliary tree. Drugs and xenobiotics may need phase I exposure chemistry and phase II conjugation before they are fit for renal or biliary elimination. So when bilirubin rises, do not merely say jaundice and stop. Ask whether too much burden is arriving, whether hepatocytes are failing to conjugate and secrete, or whether the route out is blocked. When encephalopathy appears, ask whether nitrogen disposal failed at the liver, whether portosystemic shunting bypassed hepatocytes, or both. When fat-soluble vitamin deficiency appears, ask whether the problem is intake or whether bile never reached the meal. Excretion is never just a final waste chapter. It is one of the clearest places the liver's logic becomes visible, because the body is forced to reveal what the organ could no longer move out.

This sequence, structure, then flow, then synthesis, then excretion, protects the reader from shallow pattern matching. It keeps you from calling every enzyme rise hepatitis, every bilirubin rise jaundice as if that were enough, every low albumin a "liver number," and every cirrhotic patient simply "end-stage" without asking which part of the architecture failed first and which downstream organs are now paying for it. It also trains the eye to think mechanistically across diseases that look unrelated on the surface. Viral hepatitis, steatotic liver disease, alcohol injury, cholestatic disease, autoimmune disease, portal hypertension, ascites, hepatorenal syndrome, and encephalopathy are not separate islands once you read them this way. They are different ways structure, flow, synthesis, and excretion can become disordered in the same organ.

That is how an expert begins. Not by memorizing which lab goes with which disease, but by asking a harder and more useful question. What kind of liver is left, how is blood moving through it, what is it still making, and what is it failing to move out. Once those questions are answered in the right order, the diagnosis usually becomes much easier to see because the organ has already told you the logic of the failure.

Ask Whether the Main Problem Is Hepatocyte Injury, Bile Flow Obstruction, Vascular Distortion, or Metabolic Overload

A liver problem becomes easier the moment the reader stops asking for the disease name first and starts asking which part of the organ has failed first. That is how expert reasoning begins. The liver can be injured as a cell field, blocked as an outflow system, distorted as a vascular organ, or overwhelmed as a metabolic buffer. Those are not tiny distinctions inside one broad category called liver disease. They are different kinds of failure with different consequences, different blood patterns, different imaging patterns, different symptoms, and different futures for the tissue. If the main problem is hepatocyte injury, the working cell is being struck. If the main problem is bile flow obstruction, the outgoing stream is being trapped. If the main problem is vascular distortion, blood can no longer cross the organ cleanly. If the main problem is metabolic overload, the liver is being forced to store and process more substrate than it can move through its normal exits. One organ. Four very different first questions.

Hepatocyte injury is the pattern to think of when the worker is failing before the route out or the circulation has fully collapsed. Viral hepatitis, toxin injury, ischemia, autoimmune attack, and many forms of drug-induced liver injury live here first. The hepatocyte swells, leaks enzymes, loses membrane integrity, fails to conjugate and secrete bilirubin cleanly, and begins falling behind on glucose buffering, ammonia control, and protein synthesis if the burden is severe enough or sustained enough. The blood usually shows this through a hepatocellular pattern, especially rising aminotransferases. The tissue shows it through ballooning, necrosis, apoptosis, interface activity, or other signs that the cell itself is under attack. So if the worker is wounded, think first about causes that directly injure the parenchyma and only then ask what secondary consequences followed from that.

Bile flow obstruction is different because the liver may still be making bile while failing to move it. The problem here is not primarily that hepatocytes cannot do chemistry. The problem is that the chemistry they are sending outward is not reaching the intestine. That can happen at the canalicular level, in the small intrahepatic ducts, or in the larger extrahepatic tree. The pattern it creates is cholestatic. Conjugated bilirubin backs into blood. Urine darkens. Stool pales. Itching appears because retained biliary compounds are now circulating systemically. Fat digestion worsens because bile acids are not reaching the meal. Alkaline phosphatase and GGT rise more prominently because the biliary side of the organ is under pressure. So if the route out is trapped, think first about cholestasis, duct injury, or obstruction and let the hepatocyte injury that follows remain secondary until proven otherwise.

Vascular distortion belongs to another class again. A liver can still contain many hepatocytes and still be functionally unavailable because blood is no longer meeting them in the right way. Fibrosis, cirrhosis, sinusoidal capillarization, regenerative nodules, hepatic vein outflow problems, and portal vein abnormalities all belong here because they change how the stream moves through the organ. Once blood cannot cross the liver cheaply enough, portal pressure rises, collaterals form, the spleen enlarges, ascites develops, and the kidney begins conserving sodium and water inside a circulation it now reads as underfilled. This is why a patient with portal hypertension, varices, splenomegaly, edema, or hepatorenal syndrome is often declaring a flow problem more than a fresh cell-death problem. The organ has become harder to cross. The circulation has reorganized around that hardness. If you miss that and keep asking only about enzymes, you will miss the mechanism.

Metabolic overload has its own logic. Here the liver is not first inflamed, blocked, or fibrotic. It is first overfed and overburdened. Too much fatty acid arrives from adipose tissue. Too much carbohydrate, especially fructose and glucose, is delivered through portal blood. Hyperinsulinemia continues driving lipogenesis. Oxidation cannot keep up. Export cannot keep up. The hepatocyte begins storing triglyceride because that is the only remaining safe holding pattern. Steatosis appears. If the pressure persists, oxidative stress rises, inflammation follows, and fibrosis becomes easier to trigger. This means a liver can begin as a metabolic traffic jam and only later become an inflammatory or cirrhotic liver. If you start with the wrong question, you may see the fat and think the disease is only about storage when the deeper problem is whole-body fuel handling converging on the liver.

So the right approach is not to memorize dozens of liver diseases as separate islands. It is to ask which kind of failure came first. Was the hepatocyte struck. Was bile trapped. Was blood rerouted or obstructed. Was substrate overload the opening move. Once that first structural question is answered, the rest of the disease usually becomes easier to read because the labs, symptoms, imaging, and long-term consequences begin arranging themselves around the same compartment. That is how an expert narrows chaos. Not by knowing more names than everyone else, but by asking the organ what kind of problem it is having before asking what that problem is called.

Why Numbers Have to Be Read With Symptoms, Timeline, and Mechanism

A liver number never explains itself. ALT may be high because hepatocytes are being injured, but that does not tell you whether the cause is viral hepatitis, ischemia, steatohepatitis, autoimmune injury, or drug toxicity. Bilirubin may be elevated because too much pigment is arriving from heme breakdown, because hepatocytes are failing to conjugate and secrete it, or because bile flow is obstructed after conjugation has already occurred. ALP and GGT may point toward cholestasis, but that still leaves open whether the failure sits in canalicular secretion, small-duct injury, or extrahepatic obstruction. Even the basic distinction between injury markers and function markers has to be kept straight. ALT, AST, ALP, and GGT are markers of liver injury, while albumin, clotting factors measured through PT and INR, and bilirubin are read more as markers of liver function and handling. A number therefore is a clue to compartment, not a diagnosis by itself.

Symptoms matter because they tell you which route has already broken loudly enough for the body to feel it. Dark urine and light-colored stool point toward bilirubin being diverted away from the intestine and back into blood, which is a different problem from an isolated AST rise. Itching points you toward retained bile-related chemistry. Confusion, slowed thinking, and asterixis point you toward ammonia and hepatic encephalopathy rather than toward simple inflammation. Fatigue, anorexia, and nausea are common but nonspecific, which means they gain meaning only when paired with the rest of the pattern. The bloodstream can show injury. Symptoms tell you how that injury is being translated into lived physiology.

Timeline matters because acute injury and chronic decline do not look the same even when a number overlaps. Very high aminotransferases can point toward acute viral hepatitis, acetaminophen toxicity, or hepatic ischemia, but a cirrhotic liver with advanced architectural loss may show only modest enzyme elevations while albumin, bilirubin, and PT or INR are already worsening. A cholestatic pattern developing over weeks after a medication change tells a different story from a bilirubin rise that follows years of progressive bile duct disease. The liver always has to be read in time, because the same value can mean a fresh wound in one patient and exhausted reserve in another.

Mechanism matters because the liver fails by compartment. A hepatocellular pattern means the worker cell is being struck. A cholestatic pattern means bile flow or bile duct integrity is the dominant problem. A mixed pattern means both compartments are involved. Portal hypertension, hepatic encephalopathy, and cirrhosis then add another layer, because now the issue is not only cell injury or bile retention but altered blood flow, shunting, and loss of architectural order. This is why a good reader never asks only "what is the number." The better question is "what kind of liver would produce this pattern, in this person, on this timeline, with these symptoms." The number starts the reasoning. Symptoms, chronology, and mechanism make the reasoning honest.

Recognize When the Liver Is Primary and When It Is Absorbing Damage From Elsewhere

One of the most important habits in liver thinking is learning to ask whether the liver is the origin of the problem or the organ forced to carry someone else's failure. The liver sits in the middle of too many systems to be read lazily. It receives portal blood from the gut, venous return from the spleen and pancreas, arterial oxygen from the hepatic artery, hormone signals from the endocrine system, inflammatory signals from injured tissues, fatty acid traffic from adipose tissue, and drug, alcohol, and xenobiotic burden from the outside world. That means the liver can be diseased in its own right, viral hepatitis, steatohepatitis, autoimmune hepatitis, cholestatic duct disease, fibrosis, cirrhosis, primary liver cancer. But it can also look abnormal because the gut barrier is leaking, because adipose tissue is releasing too much fat, because red blood cells are breaking down too quickly, because the heart is backing blood up into the venous system, because the kidney is failing to excrete what the liver already prepared, because systemic inflammation is rewriting hepatic protein output, or because shock has cut perfusion badly enough that hepatocytes are now ischemic casualties. The blood may still say "liver." The tissue story may say "downstream victim."

This is why pattern and compartment matter more than reflex. If hepatocytes are the main compartment under attack, the liver is more likely to be primary. Viral injury, immune-mediated injury, toxic injury, alcohol injury, steatohepatitis, and inherited metabolic disease all begin with the worker cell or its immediate environment. The liver is then generating the pathology from within its own structure. But if bilirubin is rising because hemolysis is flooding the liver with pigment burden faster than it can conjugate it, the liver is not necessarily the original disease. If alkaline phosphatase is high because infiltrative bone disease is present, that number is not hepatic truth at all. If AST rises from heavy muscle injury, the liver may be accused by a laboratory pattern it did not create. If ascites and edema are driven partly by severe right-sided heart failure congesting the hepatic venous side, the liver may become enlarged, congested, and biochemically abnormal while still acting mainly as the organ being compressed by a cardiac problem. The liver has one of the highest burdens of secondary interpretation in medicine because it is downstream from so much of the body.

The circulation is one of the best places to learn this distinction. A primary liver disease usually changes blood because hepatocytes, bile ducts, or hepatic architecture are failing first. A secondary hemodynamic problem changes the liver because blood is arriving or leaving under the wrong pressure. In shock, the liver may suffer ischemic hepatitis because perfusion collapsed systemically. In right heart failure, central venous congestion can enlarge the liver, impair sinusoidal drainage, and elevate liver tests because the organ is drowning in backed-up blood rather than being primarily inflamed. In severe portal inflow abnormalities, the liver may become a high-pressure victim of altered vascular route rather than the original mechanical culprit. The expert question is always the same. Is this liver test abnormal because the hepatocyte initiated the failure, or because the blood reaching or leaving the hepatocyte has already been made abnormal by another organ.

The same reasoning applies to metabolism. A fatty liver can be primary in the sense that the pathology now lives in the hepatocyte, but the fuel mismatch that created it is often systemic. Adipose tissue is leaking fatty acids. Muscle is underclearing glucose. Hyperinsulinemia is driving lipogenesis. The intestine is delivering repeated substrate surplus. The liver is the organ where those failures are forced into structure. It becomes the visible victim of a whole-body fuel disorder. The same is true for inflammatory disease. A damaged gut barrier, dysbiosis, infection elsewhere, autoimmune activation, or chronic tissue injury can send cytokines and microbial burden into the bloodstream, and the liver may respond by raising acute-phase proteins, altering hepcidin, shifting albumin output, and becoming inflamed under a pressure it did not author alone. A liver under inflammatory stress may therefore be primary or secondary depending on who wrote the signal first. The tissue may be real liver disease either way. The causal logic is different.

This is why experts keep asking what direction the burden is traveling. Did the problem begin in the hepatocyte and spread outward into blood, bile, and whole-body physiology. Or did the problem begin in the gut, muscle, spleen, bone marrow, heart, kidney, immune system, or circulation and then arrive at the liver as imposed work. The organ gives clues. A strongly hepatocellular pattern with compatible symptoms and timing often points inward. A bilirubin rise dominated by unconjugated pigment in the setting of hemolysis points upstream. Congestive hepatopathy points backward to the heart. Hyperammonemia with severe cirrhosis points to failed hepatic nitrogen handling, but elevated urea with renal failure may reflect a kidney that failed after the liver did its part. The liver always tells the truth about the burden it is carrying. It does not always mean the truth began there.

So the discipline is simple, though not easy. Do not ask only whether the liver is abnormal. Ask whether the liver is the first damaged organ or the organ forced to absorb damage generated elsewhere. That one distinction protects the reader from treating every abnormal liver number as a primary liver disease and from missing the opposite error, blaming another system when the liver is the true origin. The liver is too central to be read superficially. Sometimes it is the author of the disorder. Sometimes it is the first witness. Sometimes it is both.

Build the Habit of Mechanism-First Interpretation

A good liver reader does not begin with the label. He begins with the event. What is the tissue doing wrong right now. Is a hepatocyte being injured. Is bile failing to leave. Is blood failing to cross the organ. Is the liver losing synthetic power. Is portal blood carrying more burden than the organ can buffer. That is the habit. Mechanism first. The diagnosis comes later, and when it comes it is usually cleaner because the organ has already told you what kind of failure is unfolding. This matters because the same liver can be fatty, inflamed, cholestatic, scarred, congested, and synthetic-poor in different combinations, and the bloodstream will often blur those states into a few overlapping laboratory signals. The only way to think clearly is to force the problem back into tissue, flow, chemistry, and route.

The liver is especially unforgiving of slogan-thinking because its major jobs are transitional. It takes one form of chemistry and turns it into another. Portal glucose becomes buffered blood sugar or glycogen. Ammonia becomes urea. Unconjugated bilirubin becomes conjugated bilirubin and then bile pigment. Cholesterol becomes bile acid or lipoprotein cargo. A drug becomes a metabolite, then a conjugated waste, or a reactive intermediate if the sequence goes wrong. Hormones are bound, converted, weakened, or cleared. Iron is released or withheld. Bile is formed and moved. Every one of those tasks is a sequence. So when disease appears, the question is never only what is high or low. The question is where in the sequence the traffic failed. If you train yourself to ask that first, the liver stops looking mysterious. It starts looking like disrupted order.

This is why pattern reading is more valuable than number worship. A high ALT does not mean "liver disease" in any final sense. It means hepatocytes are leaking. Then you ask why. Viral burden. Drug toxicity. Steatohepatitic injury. Ischemia. Autoimmune attack. A high bilirubin does not mean jaundice and stop. It means pigment is not moving through its normal route. Then you ask where the route broke. Too much arriving from heme breakdown. Failed uptake. Failed conjugation. Failed secretion. Duct obstruction. A low albumin does not simply mean weak liver. It means the protein architecture of plasma is thinning, and then you ask whether the cause is hepatic synthetic decline, inflammation, protein loss elsewhere, or prolonged malnutrition. Mechanism-first interpretation protects you from treating a clue as if it were the whole case.

The same discipline has to be applied to symptoms. Itching is not just discomfort. It tells you retained biliary chemistry is circulating. Pale stool tells you pigment and bile did not reach the intestine properly. Asterixis tells you the brain is struggling inside blood the liver failed to clean. Ascites tells you portal pressure, albumin decline, renal compensation, and vascular misplacement are already working together. Splenomegaly tells you portal blood is backing up upstream. Gynecomastia and altered hair patterning tell you steroid handling and transport have changed. Once symptoms are read this way, the patient stops sounding like a scattered set of complaints and starts sounding like one organ failing through specific routes. The liver always announces its logic. The reader has to be disciplined enough to hear it.

Mechanism-first interpretation also means learning to distinguish the primary lesion from the downstream consequence. A person may present with renal failure, confusion, edema, or thrombocytopenia, but the first broken architecture may still be hepatic. Another person may present with abnormal liver tests while the original problem began in the gut, heart, muscle, blood, or immune system, and the liver is absorbing damage from elsewhere. So the expert question is not simply, "What is wrong with the liver?" It is, "What kind of burden is the liver carrying, and did that burden begin here or arrive from upstream?" That one habit prevents enormous diagnostic confusion. The liver sits in the middle of too many systems to be interpreted lazily.

Over time this way of thinking becomes almost physical. You stop reading hepatology as a set of memorized diseases and start seeing blood entering, bile leaving, proteins being made, toxins being transformed, pressure backing up, scar thickening the exchange space, and the wrong molecules reaching the wrong tissues. Once that happens, the liver becomes easier, not because it is simple, but because its complexity stops being shapeless. That is the goal. Do not start with the label. Start with the mechanism, and let the label earn its place afterward.

Why Expert Reasoning Begins Before the Diagnosis Label

A diagnosis label is the end of a line of thought, not the beginning of one. The liver punishes anyone who reverses that order. If you begin with the name, hepatitis, cholestasis, cirrhosis, fatty liver, drug injury, portal hypertension, you will start forcing the organ into a familiar box before you have asked what kind of failure is actually occurring. The expert starts earlier. What tissue is being hit. What route has slowed. What compound is being retained. What synthetic output is falling. What part of the traffic system is still intact and what part has already broken. Those questions come before the label because the liver does not fail as a word first. It fails as architecture, flow, chemistry, and sequence. The name only becomes useful after those have been read.

This matters because the same diagnosis label can hide very different tissue realities. "Hepatitis" can mean a brief acute inflammatory strike with preserved architecture and strong regenerative potential, or a chronic portal-interface inflammatory process already teaching the liver to scar. "Jaundice" can mean too much bilirubin arriving from hemolysis, failed hepatocyte conjugation, or a blocked biliary tree. "Cirrhosis" can describe an organ with some remaining compensation or one already spilling its failure into ascites, encephalopathy, renal vasoconstriction, and variceal risk. If the reader begins with the label, all of this gets flattened. If the reader begins with mechanism, the label becomes more precise instead of more misleading. The expert wants to know what the organ is doing wrong before deciding what chapter name the error belongs under.

This is why expert reasoning usually begins with a much more physical question. Is the main problem hepatocyte injury, bile retention, vascular distortion, synthetic decline, or metabolic overload. Is blood still meeting hepatocytes properly. Is bile still leaving the organ in the right direction. Is portal flow being resisted. Are proteins still being written into plasma at the rate the body needs. Has the liver become the first organ to absorb a systemic mismatch arriving from adipose tissue, gut barrier failure, drug exposure, or cardiac congestion. These questions sound more complicated than the diagnosis label, but they are actually simpler because they are closer to what the tissue is physically experiencing. A liver can only fail in a few deep ways. The expert learns to read those deep ways first and let the vocabulary follow afterward.

The same discipline applies to laboratory data and imaging. A novice sees a high ALT and wants the disease name. An expert sees a high ALT and asks what kind of hepatocyte injury would leak this pattern into blood and whether the rest of the liver still appears synthetically competent. A novice sees a high bilirubin and says jaundice. An expert asks whether the problem is upstream pigment load, failed conjugation, failed secretion, or failed outflow. A novice sees a nodular liver on imaging and says cirrhosis. An expert asks what kind of architecture remains functionally available inside those nodules, what portal pressure is doing, whether bile flow is compromised, and whether regeneration is still occurring inside scar. The label is not wrong. It is just late. Expert reasoning starts before it because the organ starts before it.

There is another reason this matters. Labels can become mental blinders. Once a reader decides "fatty liver," he may stop asking why the liver became the storage site for the body's failed fuel handling. Once he decides "drug-induced injury," he may stop asking whether the pattern is hepatocellular, cholestatic, or mixed, and what host susceptibility made that dose dangerous. Once he decides "autoimmune hepatitis," he may stop asking how much fibrosis the chronic inflammation has already fed and what the tissue future now looks like. Mechanism-first reasoning protects against this. It keeps the mind moving. It reminds the reader that the diagnosis is not the physiology. It is a shorthand summary for physiology that still has to be understood in real tissue terms if the summary is going to stay useful.

So expert reasoning begins before the diagnosis label because the liver itself begins before the label. Blood arrives. Bile leaves. Proteins are made. Toxins are transformed. Pressure builds or does not build. Cells die or compensate. Scar accumulates or does not. The organ tells its story in those events first. The expert listens there first. Only then does he allow the label to enter, not as a substitute for understanding, but as the name given after understanding has already begun.

Chapter 58

Symptoms Revisited Through Mechanism

Learn to Sort Symptoms by Compartment

A symptom becomes much easier to understand once the reader stops treating the liver as one undifferentiated organ and starts asking which compartment is speaking. The liver has workers, routes, vessels, pressures, and outputs. Hepatocytes do the main blood chemistry labor. Bile ducts and canaliculi carry the outgoing stream. Sinusoids and portal vessels determine how blood reaches the tissue. Hepatic veins determine how it leaves. The liver also writes proteins, hormones, and transport signals into the circulation, which means some "liver symptoms" are really blood symptoms first. If you do not sort by compartment, every complaint starts sounding the same. Fatigue, jaundice, itching, edema, confusion, dark urine, pale stool, abdominal distention, low platelets, bruising, sleep reversal, all of it becomes one blurred category called liver disease. That blur is exactly what expert reasoning tries to prevent. Symptoms follow the failing structure. The task is to ask which structure that is.

If the hepatocyte is the main injured compartment, the symptom pattern tends to look metabolic, inflammatory, and synthetic. Fatigue is common because the organ is no longer handling fuel, ammonia, and inflammatory traffic cleanly. Nausea, malaise, poor appetite, right upper quadrant discomfort, and later confusion can appear because the working cell is being stressed, dying, or falling behind on ammonia control, bilirubin handling, and protein synthesis. Jaundice can occur here too, but it comes from failed uptake, conjugation, or secretion at the cellular level rather than from a blocked duct first. So when the worker cell is the main problem, think of symptoms that suggest chemistry is no longer being stabilized properly. The body feels that as low energy, poor tolerance, mental slowing, and broader metabolic instability before it necessarily feels it as a mechanical drainage problem.

If the biliary compartment is the one speaking, the symptoms change tone. Now the liver may still be making bile, but bile is not moving. Dark urine appears because conjugated bilirubin is backing into blood and reaching the kidney. Pale stool appears because less bilirubin pigment is reaching the intestine. Itching appears because bile-related compounds are being retained in the bloodstream instead of leaving through the gut. Fat malabsorption, greasy stool, and fat-soluble vitamin deficiency appear because bile acids are not arriving where the meal needs them. This symptom set feels more directional. Something is not getting out. The route out is under pressure or erased, and the body starts revealing that by changing pigment, skin sensation, and the nutritional result of a meal. So when the symptom language is itch, pale stool, dark urine, and fat intolerance, think drainage first.

If the vascular compartment is the first one failing, the symptoms look hemodynamic and mechanical. Portal hypertension produces abdominal distention through ascites, enlarged spleen, collateral veins, varices, and fluid misplacement rather than only cell death. A person may bruise or bleed not only because the liver is making fewer clotting proteins, but because splenic sequestration has lowered platelets and varices have formed as thin-walled collateral exits under portal pressure. Edema means fluid has left the vascular space. Ascites means portal pressure, albumin decline, renal sodium retention, and lymphatic overload have converged. Dizziness, renal decline, and hepatorenal syndrome belong here too, because the circulation has changed enough that the kidney is now living in an underfilled arterial state despite total body fluid excess. So when the symptom pattern looks like swelling, distention, bleeding from varices, splenomegaly, or renal vasoconstriction, think flow and pressure before you think isolated hepatocyte injury.

The brain reveals another compartmental logic. If confusion, slowed thinking, sleep reversal, irritability, asterixis, and coma appear, the liver is now failing as a blood-clearing organ strongly enough that the nervous system is displaying it. This is not primarily a duct symptom or a local inflammatory ache. It is the symptom pattern of failed nitrogen handling, portosystemic shunting, and neuroactive burden escaping hepatic control. The gut keeps generating ammonia and other compounds. The liver fails to neutralize them or blood bypasses the liver. The brain then becomes the tissue forced to reveal the error. So mental-status changes in a liver patient should not be treated as generic "sick behavior" by default. They belong to the compartment of blood chemistry reaching the brain unedited.

Then there is the synthetic compartment, where the symptoms often look systemic before they look specifically hepatic. Low albumin produces edema and ascites because plasma can no longer hold water as effectively. Reduced clotting factor synthesis increases bleeding tendency. Lower thrombopoietin can contribute to thrombocytopenia. Reduced IGF-1 output can flatten anabolic tone. Altered steroid handling can shift libido, body hair, or breast tissue. These symptoms tell you the liver is no longer only injured. It is no longer writing the same bloodstream. When the synthetic compartment fails, the body does not necessarily feel liver pain first. It feels unstable plasma, unstable coagulation, unstable hormone transport, and unstable volume control.

This is why sorting symptoms by compartment is one of the fastest ways to stop thinking like a novice. Dark urine is not just "a liver symptom." It is a bilirubin-routing symptom. Itching is not just "a liver symptom." It is a retained bile chemistry symptom. Asterixis is not just "a liver symptom." It is a neurotoxic blood chemistry symptom. Ascites is not just "a liver symptom." It is a portal flow, albumin, and renal compensation symptom. Gynecomastia is not just "a liver symptom." It is a steroid-handling symptom. Once you train yourself to hear symptoms this way, the diagnosis label becomes less magical. The organ is already telling you which part of itself is failing. The expert advantage is simply that he knows how to listen compartment by compartment instead of as one blurred mass of "liver disease."

Itching

Itching in liver disease is usually a bile problem before it is a skin problem. The skin is where the person feels it, but the chemistry begins in a liver that is no longer moving bile cleanly. In cholestatic states, bile acids and other biliary compounds that were supposed to leave through the canalicular and ductal system are retained instead, first in liver tissue and then in the circulation. That matters because bile is not a passive fluid. It is a detergent-rich outgoing stream built to work in the intestinal lumen. When that stream stalls, the same chemistry that belonged in bile begins appearing in the wrong compartment, the bloodstream, and the body starts experiencing that error as pruritus, the urge to scratch. Cholestatic pruritus is a hallmark symptom of impaired bile flow and can be severe enough to dominate quality of life even before the patient looks deeply jaundiced or decompensated.

The mechanism is more complicated than the old shorthand that "bile salts get into the skin." That phrase points in the right direction but is too crude to explain what is actually happening. Current hepatology sources treat cholestatic itch as a multifactorial signal disorder rather than as one molecule simply irritating the epidermis. Retained bile acids may contribute, but they are probably not the whole explanation. Other circulating mediators, including pathways involving autotaxin and lysophosphatidic acid, opioid tone, and broader cholestatic signaling, appear to matter as well. The important principle is that the bloodstream has changed. The liver failed to clear and route biliary chemistry properly, and the nervous system is now reading that altered blood through itch pathways. The patient is not scratching because the skin became the primary diseased organ. The patient is scratching because bile has escaped its proper route and become systemic sensory chemistry.

This is why itching so often travels with the rest of the cholestatic pattern. Dark urine, pale stool, fat malabsorption, and elevated alkaline phosphatase all belong to the same failed stream. Bile did not reach the intestine in the amount it should have, so digestion changes. Conjugated bilirubin backed into blood, so urine darkens and stool loses pigment. Retained biliary compounds circulate, so itching appears. The liver is therefore speaking through multiple compartments at once. The gut shows loss of bile delivery. The kidney shows redirected pigment excretion. The skin shows retained bile-related signaling. Itching is one face of that same architectural failure.

The severity of the itch also teaches an important clinical lesson. Cholestatic pruritus can be profound even when some other signs of liver failure seem modest, which means symptom intensity does not always track neatly with how "sick" the liver looks on casual inspection. A patient may be relatively compensated in some respects and still be tormented by pruritus because bile flow and bile signaling are already disordered enough to keep the sensory system under constant attack. That is why itching should never be dismissed as a minor comfort issue in liver disease. It is one of the clearest signals that the outgoing stream has stopped behaving like an outgoing stream.

Ascites

Ascites is not simply water collecting in the abdomen. It is the abdomen becoming the place where failed liver architecture, failed circulatory control, and failed plasma protein support all finally meet. A healthy liver is supposed to be a soft low-resistance exchange organ. Portal blood enters, crosses an open sinusoidal bed, and drains onward without forcing the upstream splanchnic circulation to live under chronic backpressure. In cirrhosis that changes. Scar tissue replaces compliant exchange architecture, intrahepatic resistance rises, and portal hypertension develops. Fluid is then pushed out of the congested splanchnic vascular bed more easily, and the peritoneal cavity becomes one of the main places that displaced fluid can collect. Ascites is therefore a flow problem before it is a belly problem.

But portal pressure is only half the story. The other half is albumin. The cirrhotic liver often makes less albumin, and albumin is one of the main plasma proteins generating oncotic pressure, the inward pull that helps keep water inside the vascular compartment. Once portal and splanchnic hydrostatic pressure are pushing fluid outward and albumin is no longer pulling it back strongly enough, the circulation begins surrendering water to the wrong space. That is why ascites is not just excess fluid in the body. It is fluid misplacement. Blood volume may be present in total, but too much of that volume is no longer staying where circulation needs it.

The kidney then makes the problem larger while trying to solve it. Because the arterial tree is effectively underfilled in advanced cirrhosis, the body activates renin, angiotensin, aldosterone, sympathetic tone, and antidiuretic hormone. Sodium retention increases. Water follows. This is the same rescue program the body would use in dehydration or hemorrhage, but in cirrhosis it becomes maladaptive because the retained volume is poured back into a circulation still distorted by portal hypertension, splanchnic vasodilation, and weak oncotic restraint. More retained sodium means more retained fluid. More retained fluid means more ascites. The body is trying to defend perfusion and instead keeps feeding the leak.

There is another layer too. Hepatic and splanchnic lymph formation rise in portal hypertension, and once lymphatic drainage is outpaced, the excess fluid has to go somewhere. The abdomen becomes that reservoir. This is why ascites is such an important clinical threshold. It means the liver is no longer failing quietly as a biochemical organ only. It has become a hemodynamic problem for the whole body. A patient with ascites is not merely carrying fluid. That patient is declaring that portal flow, plasma protein architecture, renal compensation, and lymphatic handling are all now being forced to operate inside a cirrhotic circulation they can no longer fully control.

That is why ascites is one of the clearest visible signs of decompensated cirrhosis. It tells you the liver has stopped being an organ that blood can cross cheaply enough and that plasma can trust to hold itself together. The abdomen fills because the architecture failed first. The fluid is only the final shape that failure takes.

Fatigue

Fatigue in liver disease is often treated as if it were the least interesting symptom because it is common, subjective, and easy to dismiss. Mechanistically it is one of the most revealing. The liver sits at the center of fuel timing, plasma protein production, inflammatory signaling, bile flow, hormone conversion, and detoxification. When that organ begins losing control, the body does not usually fail first through one dramatic collapse. It often fails first through lowered usable energy, lower stability of blood chemistry, and a bloodstream that is becoming harder for tissues to live in. Fatigue is what that feels like from the inside. A person with liver disease may still be eating, still sleeping, still walking around, and yet the organism is no longer distributing fuel, clearing inflammatory burden, and maintaining endocrine tone with the same quiet precision. The body begins spending more effort to maintain the same output, and eventually the output falls.

One reason is metabolic timing. The liver is one of the main organs that keeps blood glucose from becoming too feast-driven after meals and too scarce between them. It stores glycogen when portal glucose is high, releases glucose when the intestine goes quiet, shifts into gluconeogenesis when glycogen falls, and later helps distribute ketones as alternate fuel during longer fasting. A damaged or congested liver loses some of that smooth control. Blood sugar handling becomes less graceful. Glycogen reserve may be weaker. Fuel transitions become rougher. The brain and muscle then stop receiving the same stable substrate environment they are used to, and what appears clinically as fatigue can begin as a fuel-availability problem long before overt encephalopathy or decompensation arrives. The person does not necessarily feel "hypoglycemic" in a dramatic way. The person feels less durable, less steady, less able to sustain mental and physical work across the day.

Inflammation adds another layer. A chronically injured liver is often a chronically signaling liver. Kupffer cells are more active, cytokines are higher, acute-phase proteins are being written into plasma, and the body shifts from a purely performance-oriented physiology toward a defense-oriented physiology. This matters because inflammation changes behavior on purpose. It lowers drive, narrows activity, increases sleep need, reduces appetite, and alters motivation. The liver participates directly in that shift by changing plasma composition in response to inflammatory signals from the gut, injured tissue, and the liver itself. So fatigue in liver disease is often not just weakness. It is sickness behavior, the nervous system reading an inflammatory bloodstream and adjusting the person downward. The symptom can look nonspecific from the outside, but at the tissue level it means the body has moved away from open metabolic confidence and toward guarded conservation.

Bile flow, hormonal handling, and blood architecture also matter. Cholestatic disease can make the body feel profoundly unwell even before failure becomes dramatic, because retained bile-related compounds circulate in the wrong compartment and because fat digestion and fat-soluble vitamin absorption begin falling behind. At the same time the liver helps convert thyroid hormone, shape steroid metabolism, make IGF-1, and maintain carrier proteins such as albumin and thyroid-binding proteins that determine how endocrine signals move through plasma. Once hepatic handling weakens, the person may live inside flatter thyroid signaling, altered sex steroid balance, lower anabolic tone, poorer nutrient extraction, and a plasma environment less able to hold fluid and transport molecules efficiently. Fatigue then becomes the shared output of multiple smaller losses. The bloodstream is less supportive, the hormones are less coherent, the meal is less usable, and the tissues begin feeling that combined decline before the diagnosis sounds dramatic enough to explain it.

This is why fatigue in liver disease should not be read as a soft symptom attached to a hard organ problem. It is often the first full-body expression of hepatic disorder. The liver is one of the organs that decides whether blood arrives at tissues as stable fuel, inflammatory burden, endocrine signal, or some unstable mixture of all three. When that decision-making weakens, the person feels it as reduced capacity before the scan, the biopsy, or the jaundice may fully display what is happening. Fatigue is the body saying that the liver is no longer keeping the internal environment as easy to work inside.

Bruising

Bruising in liver disease is not usually a skin problem first. It is a blood architecture problem becoming visible through the skin. A bruise appears when small vessels leak blood into tissue and the body does not contain that leak cleanly enough or quickly enough. The liver matters here because it makes most of the soluble clotting factors that allow blood to stabilize an injured vessel, and it also makes thrombopoietin, the main hormone supporting platelet production. In cirrhosis and other advanced liver disorders, PT and aPTT can become prolonged because hepatocytes are making less of the coagulation machinery, and thrombocytopenia can develop because thrombopoietin falls and because portal hypertension enlarges the spleen and traps platelets there. The result is not just "thin blood." It is a circulation with less reserve for sealing small vascular injury.

That is why bruising has to be read as a compartmental symptom. If the liver's synthetic side is failing, the plasma loses part of its clot-building protein structure. If portal hypertension is advanced, the spleen becomes a larger platelet reservoir and fewer platelets remain available in circulation. If cholestasis is present, fat-soluble vitamin K absorption may also be impaired, which can further weaken vitamin K-dependent clotting factor function. So the bruise on the skin is often the final surface expression of several deeper failures acting together, reduced clotting factor synthesis, lower platelet support, splenic sequestration, and in some patients vitamin K-related impairment on top of that.

This is also why bruising should not be treated as a trivial cosmetic sign in liver disease. It often means the liver is no longer only inflamed or fatty. It is now losing one of its whole-body construction jobs, keeping the bloodstream able to contain everyday vascular wear without spilling too easily into tissue. The bruise is small. The physiology behind it is not.

Confusion

Confusion in liver disease is not a vague sign that a person is generally unwell. It is one of the clearest signs that the liver has stopped protecting the brain from the chemistry of the gut. In hepatic encephalopathy, ammonia produced in the gastrointestinal tract is no longer being cleared and converted into urea efficiently enough, either because hepatocytes are failing or because portal blood is bypassing them through shunts. That ammonia and other gut-derived neuroactive compounds then enter the systemic circulation and reach the brain. This is why confusion belongs to the liver-brain axis, not just to the symptom list of advanced illness. The liver failed first as a blood-clearing organ. The brain becomes the organ forced to reveal that failure.

The brain responds to this burden through chemistry before it responds through obvious structural damage. Ammonia crosses into the brain and is taken up by astrocytes, the support cells that help maintain the neural environment. Astrocytes convert ammonia into glutamine, but glutamine accumulation draws water into the cell and promotes swelling. That swelling interferes with neurotransmission, energy handling, and network timing. The person may still be awake, still speaking, still physically present in the room, but thought loses sharpness. Attention drifts. The sequence of ideas becomes less reliable. Processing slows. That is why confusion in hepatic encephalopathy often looks less like dramatic madness and more like a brain trying to work through chemically thickened blood.

This is also why confusion often arrives with slowed thinking, poor concentration, sleep-wake reversal, and subtle personality change before coma ever enters the picture. Hepatic encephalopathy is a clinical syndrome, not a single lab value, and its cognitive features can fluctuate because the blood chemistry driving it fluctuates as well. A person may be coherent in one moment and tangential, inattentive, or delayed in the next. AASLD guidance emphasizes that hepatic encephalopathy is a clinical diagnosis and that ammonia levels alone do not reliably stage or confirm the syndrome, which means the confused patient has to be read through the whole pattern, gut-derived burden, liver failure, shunting, and neurologic behavior together.

The confusion also matters because it proves the liver is a necessity for consciousness, not just for digestion or detoxification in the casual sense. The gut keeps generating ammonia every day. The liver is supposed to make that chemistry livable. When it no longer can, or when blood bypasses it, the brain begins operating in an environment it was never meant to tolerate. Confusion is therefore one of the most direct demonstrations of hepatic necessity. It is the mental face of failed blood editing. The person seems disoriented because the bloodstream reaching the brain is no longer the kind of bloodstream a healthy liver would have allowed to exist.

Right Upper Quadrant Pain

Right upper quadrant pain becomes a liver symptom only when structure is being stretched, inflamed, or blocked enough to recruit tissue around the liver that can actually signal pain. The liver parenchyma itself is not a tissue that complains loudly when chemistry alone is disturbed. A person can have severe enzyme elevation, marked steatosis, or even substantial chronic fibrosis and feel little or nothing over the liver for a long time. Pain tends to appear when the liver capsule is distended, when the organ becomes enlarged and tender, or when the biliary system is inflamed or obstructed. Merck's clinical guidance states this directly: right upper quadrant pain from liver disorders usually results from distention, such as passive venous congestion or tumor, or from inflammation of the liver capsule.

That is why the quality of the pain matters. A dull constant ache or pressure in the right upper abdomen often points toward hepatomegaly, congestion, or inflammatory enlargement stretching the capsule rather than toward a purely luminal biliary event. Merck notes right upper quadrant discomfort from hepatic congestion as a consequence of stretching the liver capsule, and NIDDK lists mild pain or discomfort over the liver as one possible symptom of cirrhosis. This is pain from tension and swelling, not usually from the hepatocyte itself "hurting."

A sharper colicky right upper quadrant pain tells a different structural story. That pattern makes you think first about the gallbladder or bile ducts, especially transient obstruction by stones. NIDDK describes gallbladder attacks as sudden right upper abdominal pain when gallstones block the bile ducts, and Merck's jaundice guidance distinguishes colicky right upper quadrant pain from the more constant discomfort seen in other hepatobiliary states. In other words, not all right upper quadrant pain in liver work is hepatocellular. Much of it is biliary route pain.

This is why right upper quadrant pain has to be sorted by compartment. If the liver is enlarged, congested, acutely inflamed, or rapidly swollen, the capsule may be stretched and the person feels a more constant tender pain or discomfort. If the bile ducts or gallbladder are obstructed or inflamed, the pain may be more sudden, colicky, and mechanically dramatic. NIDDK and Merck both list right upper abdominal pain or discomfort among possible symptoms in conditions such as acute viral hepatitis, cirrhosis, PSC, PBC, autoimmune hepatitis, and alcohol-related liver injury, but the pattern and context decide what the pain is actually saying.

So right upper quadrant pain is not a generic "liver pain." It is a mechanical clue. Something in the hepatobiliary system is being stretched, inflamed, congested, or blocked enough that the body can finally feel the architecture failing.

Dark Urine and Pale Stool

Dark urine and pale stool belong together because they are two ends of the same broken pigment route. Bilirubin begins as the breakdown product of heme, mostly from aging red blood cells. The liver is supposed to take that bilirubin up from blood, conjugate it into a more water-compatible form, and secrete it into bile so it can be carried into the intestine. Once it reaches the gut, bacterial metabolism converts it through downstream steps that eventually help generate the brown pigment of normal stool. That is the healthy sequence. When bile flow is impaired, or when conjugated bilirubin cannot move forward through the biliary tree, the sequence splits. Less pigment reaches the intestine, and more of it backs up into blood. The stool loses color because the bowel is no longer receiving the pigment stream it needs. The urine darkens because the bloodstream is now carrying a water-soluble bilirubin form the kidney can filter.

The dark urine comes first from chemistry, not from dehydration in the ordinary sense. Conjugated bilirubin is different from unconjugated bilirubin. Unconjugated bilirubin is hydrophobic, albumin-bound, and poorly suited to renal filtration. Conjugated bilirubin has already been processed by the hepatocyte and made more water-compatible for biliary excretion. If that conjugated pigment is trapped by cholestasis or obstruction and refluxes back into blood, the kidney can now remove part of it. The urine becomes tea-colored, cola-colored, or simply much darker than expected because bilirubin is being rerouted into the wrong exit. This is an important clinical clue. Dark urine in a jaundiced patient points strongly toward conjugated hyperbilirubinemia and a biliary or hepatocellular secretion problem rather than a purely unconjugated burden arriving from hemolysis. The pigment has already passed through the liver's conjugation machinery. The route out failed afterward.

Pale stool tells the other half of the story. Normal stool color depends heavily on bilirubin making it into the intestine and then being metabolized by gut bacteria into downstream pigments, especially stercobilin-related compounds. If bile flow is blocked or severely reduced, the intestine is deprived of that pigment load. Stool then shifts toward pale brown, clay colored, gray, or frankly light depending on how complete the interruption is. This is not a cosmetic curiosity. It means the liver's outgoing stream is no longer reaching the bowel in the way it should. The same failed flow that is trapping bilirubin also means bile acids are reaching the meal less effectively, which is why pale stool may travel with fat malabsorption, greasy stool, bloating, and fat-soluble vitamin problems. The missing pigment and the missing detergent are both consequences of the same failed outflow.

That is why dark urine and pale stool should always be read as a paired mechanistic signal rather than as isolated symptoms. The blood is carrying conjugated bilirubin it should not still contain. The kidney is being forced to clear pigment that should have left through bile. The intestine is receiving too little bilirubin to make normal stool pigment and too little bile acid to support fully normal fat handling. One failed stream produces all three consequences at once. The patient does not merely look jaundiced. The body is revealing exactly where the pigment route broke.

Each Symptom Explained by the Structure That Failed

The cleanest way to understand liver symptoms is to stop treating them as a loose symptom list and start treating them as the exposed edges of broken structure. The liver is not one generic tissue producing one generic illness feeling. It is a composite organ built out of compartments that do different work. Hepatocytes govern blood chemistry, store glycogen, synthesize albumin and clotting factors, convert ammonia into urea, and secrete the first version of bile. Canaliculi and bile ducts carry the outgoing stream. Sinusoids and portal vessels determine how blood reaches the parenchyma. Hepatic veins determine how it leaves. Stellate cells and extracellular matrix determine whether the exchange field remains soft and open or becomes thick, resistant, and scarred. Once a reader starts from that architecture, symptoms stop looking random. They begin sounding like the compartment that failed.

Jaundice is what happens when the bilirubin route fails. Dark urine and pale stool belong to the same broken stream. Itching belongs there too.

The liver is supposed to take bilirubin out of blood, conjugate it, and move it into bile. If hepatocytes cannot do that well, or if bile cannot leave through ducts and canaliculi, conjugated pigment begins backing into blood instead of reaching the bowel. The kidney is then forced to clear what the biliary tree was meant to carry, and the urine darkens. The intestine receives less pigment, so stool pales. Bile acids and related biliary compounds circulate where they should not, and the skin begins to register that failure as pruritus. The symptom is not a generic "liver complaint." It is the voice of the outflow system failing.

Confusion, slowed thinking, sleep reversal, asterixis, and coma belong to another compartment entirely. They are the neurologic face of failed blood editing. The gut keeps producing ammonia and other neuroactive burden. The liver is supposed to intercept that portal chemistry, convert ammonia into urea, and keep the brain from bathing in intestinal nitrogen. When hepatocytes fail or when portal blood bypasses them through shunts and collateral routes, the brain becomes the organ forced to display the mistake. Edema and ascites belong to the vascular and synthetic compartments at the same time. Portal resistance rises because scarred architecture no longer allows cheap inflow. Albumin falls because synthetic function weakens. More fluid is pushed out of vessels. Less protein force pulls it back in. The abdomen fills. The tissues swell. So even the fluid symptoms are not vague markers of severe disease. They are the physical expression of failed flow and failed plasma construction.

Bruising, bleeding, and low platelets speak through the blood-making side of the liver and through the spleen. The liver makes most of the soluble clotting machinery, and it also makes thrombopoietin, which helps support platelet production. Portal hypertension enlarges the spleen and increases sequestration of circulating platelets. So the patient bruises not because the skin became fragile on its own, but because the blood beneath it is no longer carrying the same clotting reserve. Fatigue speaks through multiple compartments at once, which is why it is so often underestimated. A tired liver does not simply make a tired person in some vague way. It destabilizes fuel timing, raises inflammatory signaling, alters endocrine conversion, weakens nutrient extraction when bile flow is poor, and changes the composition of plasma itself. Fatigue is often the body's earliest confession that the bloodstream has become less supportive.

Right upper quadrant pain belongs to mechanics more than chemistry. The liver parenchyma does not complain loudly until something stretches the capsule, swells the organ, congests the venous side, or inflames the biliary system enough to recruit tissue that can signal pain. Splenomegaly, varices, and abdominal distention belong to portal pressure and rerouted blood rather than to primary hepatocyte leakage. Gynecomastia, altered libido, menstrual irregularity, and other endocrine-looking symptoms often belong to steroid transport and metabolism falling behind rather than to a primary gonadal failure. Once you start sorting symptoms this way, the organ becomes easier to read. The symptom is no longer just something the patient feels. It is a clue to which structure, route, or synthetic job stopped holding.

That is the deeper discipline. Do not memorize symptoms as if they hover over "liver disease" in a cloud. Ask which failed structure would make this symptom inevitable. Was the worker cell injured. Was bile trapped. Was blood rerouted. Was protein synthesis weakened. Was gut-derived chemistry no longer filtered before reaching the brain. Each symptom becomes clearer when it is forced back into the architecture that produced it. That is how symptoms stop being vague and start becoming mechanism.

Chapter 59

What the Healthy Liver Is Quietly Doing All Day

Post-Meal Buffering

The healthy liver spends much of the day preventing the bloodstream from becoming a violent mirror of the last meal. That is one of its quietest jobs and one of its most constant. Food is digested in pulses. Glucose can rise quickly after carbohydrate absorption. Amino acids can flood portal blood after protein digestion. Fructose can arrive in a form the rest of the body was never meant to see first. Lipid-derived traffic returns through a different route but still has to be sorted once it reaches hepatic handling. If the intestine were allowed to hand all of this directly to the general circulation without interception, the chemistry of eating would become the chemistry of instability. The liver sits between the gut and the rest of the body so that does not happen. It receives portal blood first, samples the surge, and decides how much of it will be stored, transformed, delayed, released, or redirected.

Glucose shows this most clearly. After a meal, hepatocytes take up a substantial fraction of the incoming portal glucose load and phosphorylate it so it stays inside the cell. Some is burned immediately. Some is linked together into glycogen, the liver's quick-access carbohydrate reserve. That storage step is not just an act of saving calories for later. It is a way of reducing the force of the post-meal rise before the rest of the circulation has to absorb it. The liver is flattening the curve in real time. Instead of allowing absorbed carbohydrate to strike every tissue as raw intestinal output, it converts part of that output into reserve and meters part of it onward more slowly. The bloodstream that leaves the liver is therefore already edited. Blood sugar after a meal is not simply what the intestine absorbed. It is what remains after hepatic buffering took its share.

The same buffering logic applies to amino acids, and this is where the liver's role becomes broader than "blood sugar control." Protein digestion sends a large nitrogen burden into portal blood. The liver receives that burden first and decides what will be used for protein synthesis, what will be oxidized, what carbon skeletons will be preserved for later metabolic work, and what nitrogen must be trapped and eventually moved toward urea. Without that buffering step, the meal would not simply look protein-rich in blood. It would become chemically disorderly. The liver prevents that by taking the incoming amino acid surge and translating it into synthesis, temporary holding, and controlled nitrogen disposal. The healthy liver is therefore not just handling calories after a meal. It is preventing the circulation from becoming flooded with half-processed substrate.

Fructose reveals the same principle in a sharper way. Much of absorbed fructose is handled first by the liver because peripheral tissues are not designed to be the main first-pass site for that carbon load. A healthy liver can absorb that burden, convert it, store part of it, and keep the rest of the body from seeing the full force of the incoming molecule. This is why the liver is such a powerful post-meal shock absorber. It does not merely store what is left over. It decides what kind of chemistry the rest of the body will be allowed to experience after the meal has already entered the bloodstream.

This buffering does not feel dramatic because when it works, nothing happens that a person notices. Thought stays clear. Blood sugar stays within a usable range. The next few hours remain metabolically livable instead of chemically erratic. That is the deeper point of a healthy liver. It is not only working in crisis. It is quietly absorbing ordinary excess, ordinary timing mismatch, and ordinary digestive surges all day long so the rest of the body can live in smoother blood than the gut alone could ever provide.

Fasting Support

The healthy liver does not go quiet when the gut goes quiet. It changes jobs. In the fed state it absorbs the shock of incoming substrate and keeps the bloodstream from becoming the raw extension of the last meal. In the fasting state it becomes the organ that keeps the body from discovering too quickly that the intestine has stopped supplying anything at all. That transition is one of the liver's most important daily tasks. The brain still needs fuel. Red blood cells still need glucose. The kidney medulla still depends on a stable blood chemistry. Skeletal muscle, heart, and other tissues begin shifting their preferences, but none of them can simply take over the whole problem. The liver has to keep the internal environment usable while intake has dropped to zero. Fasting support is the name for that quiet rescue.

The first line of support is glycogen. During the fed state the liver stored glucose in this compact branched polymer precisely so it could be reopened when the meal disappeared. Glycogenolysis, the breakdown of glycogen into glucose units, lets the hepatocyte release glucose back into blood during the early hours of fasting. This is why glycogen is not just storage. It is delayed circulation. A person who sleeps through the night without becoming immediately confused or weak is depending on this reserve, whether he knows it or not. The gut has gone silent, but the bloodstream has not. The liver is using what it stored yesterday to support what the body still needs today.

As glycogen begins to thin, the liver shifts again. Now the task is no longer only to release previously stored carbohydrate. It is to make new glucose. This is gluconeogenesis, the synthesis of glucose from noncarbohydrate substrates such as lactate, glycerol, and certain amino acids. Lactate returns from tissues that are producing it as part of glycolysis. Glycerol arrives from adipose tissue as stored triglyceride is broken down. Amino acid carbon skeletons contribute when needed. The liver gathers these fragments and rebuilds glucose for the circulation. This is one reason the organ is so central to fasting. It is not merely opening a reserve. It is manufacturing the continuation of blood sugar after the reserve begins to run low.

At the same time the liver changes what fuel it uses for itself. In fasting, adipose tissue releases fatty acids into blood, and the liver takes up much of this arriving fat. Instead of relying heavily on incoming dietary carbohydrate, hepatocytes begin oxidizing fatty acids in mitochondria, generating ATP and acetyl-CoA. ATP, adenosine triphosphate, is the immediately usable energy currency of the cell. This matters because gluconeogenesis is expensive. The liver has to burn one fuel so it can make another available. Fatty acid oxidation gives the hepatocyte the energy to keep glucose support running for tissues that still require it.

As fasting deepens further, the liver adds one more layer of support by making ketones. These are water-soluble molecules derived from fatty acid carbon that can circulate to other tissues, especially heart, muscle, and eventually the brain, as alternative fuel. This reduces how much glucose the liver has to keep supplying and spares some of the pressure that would otherwise fall on protein breakdown and gluconeogenesis alone. The liver therefore does not simply hold the fort during fasting. It reorganizes the fuel economy of the whole organism. It shifts the body from meal dependence to reserve dependence without letting the transition become a crisis.

This is what the healthy liver is quietly doing all day when no one is eating. It opens glycogen. It makes new glucose. It burns fat. It exports ketones. It keeps the blood from telling the rest of the body that the intestine has gone missing. That is fasting support.

Nitrogen Disposal

Nitrogen disposal is one of the clearest examples of why the liver is not optional to the rest of the body. Protein is constantly being digested, amino acids are constantly being broken down and rebuilt, and gut bacteria are constantly acting on nitrogen-containing material in the intestinal lumen. All of that chemistry produces ammonia and related nitrogen burden. Ammonia is not just another waste product drifting harmlessly through plasma. It is chemically small, diffusible, and dangerous enough that blood cannot be allowed to carry much of it for long, especially if the brain is expected to keep working normally. The body therefore has to solve a very specific problem every day. It has to keep using protein while preventing nitrogen from remaining in its most toxic circulating form. That is the job the liver quietly performs.

The sequence begins in the gut and in the tissues, not in the liver itself. Dietary protein is digested into amino acids. Enterocytes and other tissues metabolize some of those amino acids. Intestinal bacteria generate additional ammonia from luminal nitrogen sources. Portal blood then carries this burden forward to the liver. Under healthy conditions, hepatocytes intercept it and run the urea cycle, a series of mitochondrial and cytosolic reactions that convert ammonia into urea, a much safer and more water-soluble molecule. That conversion is not decorative biochemistry. It is the difference between a bloodstream that can carry nitrogen quietly to the kidney and a bloodstream that starts threatening the nervous system. The liver is not merely participating in nitrogen handling. It is changing the chemical form of nitrogen so the rest of the body can survive its own protein metabolism.

What makes this more important is that the liver is not working alone, but it is still the central organ in the sequence. Skeletal muscle can temporarily buffer ammonia by incorporating nitrogen into glutamine. Other tissues can contribute to local nitrogen handling. But those are support strategies, not final disposal. The major body-wide solution is still hepatic conversion of ammonia into urea followed by renal excretion of that urea. That is why liver failure and portosystemic shunting are so dangerous to the brain. The gut keeps producing nitrogen burden. The liver either fails to convert enough of it or blood bypasses the hepatocyte field through collateral routes. The bloodstream then begins carrying a chemistry it was not meant to carry in that form. The brain becomes one of the first organs forced to reveal that failure through slowed thinking, confusion, sleep reversal, asterixis, and coma.

So nitrogen disposal should be understood as a liver-centered transformation problem rather than as simple waste removal. The body is not just throwing nitrogen away. It is taking a dangerous product of ordinary protein use, converting it into a safer transportable form, and only then asking the kidney to finish the job. The gut creates the burden. The liver neutralizes its chemistry. The kidney removes the final product. When that middle step fails, the whole body begins discovering how much of everyday mental clarity depended on the liver all along.

Bile Secretion

Bile secretion begins as a membrane event, not as a fluid already sitting in a reservoir waiting to be drained. The hepatocyte makes bile by taking compounds out of blood on one side of the cell, modifying them internally, and then actively exporting them across the opposite membrane into the canaliculus, the tiny bile channel formed between neighboring hepatocytes. This only works because the hepatocyte is polarized. One surface faces sinusoidal blood. Another faces the bile space. The cell is therefore built to live between two traffic systems at once, portal and arterial blood arriving from one direction, bile being written outward in the other. Bile secretion is the point where that polarity becomes visible. The liver is not just processing chemistry inside the cell. It is deciding what chemistry must leave the body or enter the intestine and then physically pushing it into a separate stream.

That stream is assembled component by component. Bile acids are secreted into canaliculi by transporters that spend ATP, adenosine triphosphate, the cell's immediately usable energy currency, to move them against concentration gradients. Bilirubin is taken up from blood, conjugated into a more water-compatible form, and then secreted into the same canalicular space. Cholesterol and phospholipids are added as well. These are not random solutes drifting out of a leaky membrane. They are selected cargo. Bile acids are needed because the intestine requires detergent chemistry to emulsify fat and form micelles. Bilirubin needs an exit because it is a heme-derived pigment burden the body cannot leave in circulation. Cholesterol needs a route into bile because hydrophobic sterol cannot simply be dumped into urine in any meaningful way. So bile secretion is both digestive preparation and excretory decision. The same outgoing stream is being loaded with tools for the next meal and with burdens the body has already decided must go.

Water follows because secretion creates osmotic force. Once bile acids and other solutes are pumped into the canalicular space, water and electrolytes move with them and turn that chemical export into a flowing fluid rather than a paste of detergent and pigment. This is important because bile is not useful as a list of ingredients. It has to move. The canaliculus has to receive enough solute and water to create a stream that can pass into ductules, then larger ducts, then the gallbladder or intestine. The fluid phase is therefore not empty background. It is part of the physics of secretion. A hepatocyte that can make bile acids but cannot sustain the transporter work, membrane integrity, or osmotic conditions required for flow is not really secreting bile normally. It is only holding biliary chemistry inside itself a little longer.

The ductal system then modifies what hepatocytes began. Cholangiocytes, the epithelial cells lining bile ducts, are not the original makers of bile, but they do alter the fluid by adjusting bicarbonate and water content as it moves through the biliary tree. This matters because bile is not finished the moment it enters the canaliculus. It is initiated there. The hepatocyte writes the core secretion. The duct epithelium refines the stream as it travels outward. That two-stage design tells you something important about the liver. Secretion is not simply release. It is a coordinated program of hepatocyte export, osmotic flow, and ductal modification. If the hepatocyte side fails, bile cannot be generated properly. If the duct side fails, bile may still be formed but become abnormal in alkalinity, volume, or delivery.

This is why bile secretion is one of the clearest expressions of liver work and one of the easiest places for liver failure to become visible. A healthy hepatocyte spends energy to preserve polarity, take up bilirubin, synthesize bile acids, move cholesterol and phospholipids, and push all of it into a canalicular stream that the intestine will later depend on. A stressed hepatocyte loses ATP, loses transporter precision, loses membrane order, and begins failing on the bile-facing side at the same time it may still be receiving blood on the sinusoidal side. Cholestasis then appears, not because bile ceased to matter, but because secretion was always active work. The liver does not simply contain bile. It writes it continuously, and the body pays immediately when that writing starts to fail.

Hormone Conversion

The liver does not simply receive hormones after the endocrine glands are finished speaking. It changes the form in which those hormones circulate, the duration of their activity, the fraction that remains free versus protein-bound, and the route by which they eventually leave the body. That makes hormone conversion one of the liver's quieter but more consequential jobs. A gland can release a signal, but that signal still has to survive in blood, reach tissues in a usable form, avoid remaining active too long, and finally be cleared without the body drowning in yesterday's chemistry. The liver stands in the middle of that entire sequence. It binds, converts, weakens, activates, conjugates, and routes endocrine molecules so the bloodstream carries not just what the gland secreted, but what the liver decided that secretion should become.

Thyroid hormone shows this clearly. The thyroid gland releases a large amount of thyroxine, T4, but much of the body's meaningful hormonal work depends on how much of that precursor is converted into triiodothyronine, T3, the more active form. The liver is one of the major tissues participating in that peripheral conversion. It is also involved in clearing thyroid hormone through conjugation reactions such as glucuronidation and sulfation, which change solubility and direct the hormone toward excretion. So the liver is not only modifying excess thyroid signal after the fact. It is helping determine how much active thyroid instruction the body will actually experience. A person does not live only on what the thyroid secreted. A person lives on what the liver and other tissues allowed that secretion to become in circulation.

Vitamin D follows the same broader logic but in a different biochemical sequence. Vitamin D entering from skin or diet is not yet the major circulating reserve form the body uses to judge availability. The liver performs 25-hydroxylation, creating 25-hydroxyvitamin D, the main blood-borne intermediate measured clinically. That step is easy to underestimate because the kidney performs the later activation step that generates the more potent hormonal form. But the kidney can only refine what the liver first provided. Without hepatic conversion, the body does not build the stable circulating pool that the rest of the endocrine system depends on. The liver therefore is not a side station in vitamin D physiology. It is the organ that turns a precursor into a transportable endocrine resource.

Steroid hormones make the same point in a more dynamic way. Estrogens, androgens, progesterone-related compounds, and glucocorticoids do not simply circulate until tissues are done with them and then vanish. They are continuously being transformed. Hepatic enzymes oxidize them, reduce them, hydroxylate them, and then conjugate them through glucuronidation or sulfation so they become less active, more water-compatible, and more fit for biliary or urinary elimination. This is how the body prevents steroid signals from becoming chemically endless. The gonads and adrenals may release the hormone, but the liver decides whether that hormone remains active, becomes a metabolite, or is pushed toward exit. This is one reason liver disease can produce endocrine-looking symptoms without any primary lesion in the gland. The signal may still be secreted normally. The editing and clearance of the signal are no longer normal.

Transport proteins deepen this control. Hormones do not float through plasma in a simple free state. They ride in a carefully built transport environment shaped partly by the liver. Albumin, thyroxine-binding globulin, transthyretin, and sex hormone-binding globulin all influence how much hormone is buffered in circulation and how much remains immediately free to enter tissue. That means hepatic dysfunction can alter endocrine reality even when glandular output is relatively unchanged. Total hormone levels may shift because binding capacity changed. Free hormone may become less stable. A steroid or thyroid pattern may look "off" not because the gland spoke differently, but because the bloodstream is carrying the message differently. The liver is not only converting hormones chemically. It is constructing the fluid environment in which those hormones move.

This is why hormone conversion belongs in liver physiology at the deepest level. The liver is one of the organs that turns endocrine secretion into endocrine timing. It decides how long the signal lasts, how strongly it circulates, how much of it is transformed into a more active or less active form, and when the molecule is finally removed from the body. A damaged liver therefore does not only produce liver-specific symptoms. It can flatten thyroid tone, distort sex hormone balance, weaken anabolic signaling, alter vitamin D status, and change the entire feel of the endocrine environment because the bloodstream is no longer being edited with the same precision. The gland writes the first message. The liver determines how that message is carried, changed, and finally silenced.

Plasma Protein Manufacture

The liver does not simply clean the blood after other organs have finished using it. It helps build the blood itself. Plasma is not just water with nutrients and waste drifting through it. It is a protein-structured fluid whose behavior depends on what hepatocytes are constantly making and secreting into circulation. Albumin, clotting factors, transport proteins, complement proteins, acute-phase proteins, metal-binding proteins, hormone-carrying proteins, all of these are part of what gives blood its carrying power, its water-holding force, its clotting readiness, and its immunologic tone. The liver is therefore not merely standing beside the bloodstream as a processor. It is one of the organs writing the bloodstream into existence as a usable medium.

This matters because proteins are not passive cargo. Albumin helps determine whether water stays in the vascular compartment or escapes into tissue and body cavities. Binding proteins determine how thyroid hormone, sex steroids, fatty acids, bilirubin, metals, and many drugs move through plasma and how much of each remains free at any given time. Fibrinogen, prothrombin, and the broader coagulation proteins determine whether blood can seal a wound without turning the whole circulation into clot. Complement proteins and acute-phase proteins help turn plasma into an immune-active fluid when danger rises. Once the liver changes what it is making, blood itself changes character. The circulation becomes less stable, less buffered, less controlled, or more inflammatory depending on which proteins are falling and which are being prioritized instead.

Inside the hepatocyte this is a directional act of construction. Proteins are synthesized, processed, folded, sorted, and then secreted across the blood-facing side of the cell into the sinusoidal space. That means the same hepatocyte taking substances out of blood for metabolism is also returning newly manufactured proteins back into blood as part of normal physiology. This is one of the most important ways the liver differs from the crude public image of a detox organ. A detox organ sounds subtractive. Plasma protein manufacture is additive. The liver is not only removing burdens. It is supplying the structural molecules that make blood capable of transport, restraint, defense, and repair.

That is why chronic liver disease becomes visible in plasma long before the organ is physically absent. Albumin falls because synthetic power is weakening. PT and INR shift because clotting factor production is thinning. Platelet support weakens because thrombopoietin output falls. Hormone behavior changes because the transport environment is no longer the same. The body is not simply losing some reserve hidden inside the liver. It is losing part of the protein architecture of the circulation itself. A failing liver does not only leave toxins behind. It stops making the molecules that kept blood coherent.

So plasma protein manufacture is one of the clearest ways to understand what the healthy liver is quietly doing all day. It is building the fluid the rest of the body depends on. When that work is strong, blood holds volume, carries signals, transports fuel, controls clotting, and supports immunity without drawing attention to itself. When that work weakens, the whole body begins discovering how much of the bloodstream had been hepatic all along.

Nutrient Storage

The liver stores nutrients because the body cannot afford to live only on what entered the gut in the last hour. Digestion is intermittent, absorption is uneven, and demand changes constantly across feeding, fasting, infection, repair, growth, and sleep. A useful organ in that setting cannot simply pass nutrients through as they arrive. It has to catch some, hold some, reshape some, and release some later under different conditions. That is what storage means in the liver. It is not passive accumulation. It is controlled timing. Glycogen is the clearest short-horizon example. Portal glucose rises after a meal, hepatocytes take some of it up, phosphorylate it, and pack it into a branched polymer that can sit inside the cell without the osmotic disruption free glucose would create if it were left loose in large quantity. Later, when the intestine goes quiet and blood still needs glucose, the liver opens that reserve and returns part of it to circulation. The same organ that softened the meal now softens the fast. Storage is the hinge between those two states.

But the liver is not only a carbohydrate warehouse. It stores nutrients and metals whose importance is too high and whose chemistry is too dangerous to leave circulating freely. Vitamin A is a good example because it is not merely a nutrient to be counted on an intake chart. It is a signaling-active molecule involved in epithelial integrity, immune behavior, visual chemistry, and cell differentiation. The body cannot let it drift indefinitely as free active compound. So the liver stores it, mainly in stellate cells, as retinyl esters packed into lipid droplets. Vitamin B12 is stored heavily there as well, which is one reason B12 deficiency can take a long time to declare itself after absorption has already failed. Iron follows a different logic. It is stored because the body needs reserve for erythropoiesis and oxidative metabolism, but iron is also chemically reactive enough that it cannot be left free in tissue or plasma without risk. Ferritin solves that by buffering the metal in a more controlled intracellular form. Copper is similar. It is useful, enzymatically important, and potentially destructive if left loosely available, so the liver binds it, routes it, stores some of it, and decides when it leaves through bile. Nutrient storage in the liver is therefore always partly about future need and partly about present containment.

This is what makes hepatic storage so biologically revealing. The liver does not store one class of molecules for one reason. It stores fuel because blood glucose must remain stable across time. It stores vitamins because intake and need do not always line up and because some compounds are too biologically potent to leave ungoverned. It stores metals because they are both indispensable and chemically risky. The form of storage changes with the nature of the molecule. Glycogen is built for rapid access. Retinoids are esterified into reserve droplets. Iron is packed into ferritin. Copper is buffered by binding systems and handled with an eye toward biliary exit. The chemistry of the reserve tells you what kind of problem the body is solving. A useful molecule may need to be kept dense, quiet, nontoxic, or quickly releasable. The liver chooses the appropriate form and holds it until the body asks again.

This is also why nutrient storage creates both safety and diagnostic delay. A person can lose absorption, reduce intake, or shift into inflammatory iron restriction and still look biochemically stable for a while because the liver is spending stored capital to hide the shortage. Bloodwork may remain deceptively acceptable. Symptoms may lag far behind the real start of depletion. The reserve is protective, but it can also conceal decline. The same principle explains why storage disease becomes so destructive when storage stops being governed. Iron overload, copper overload, retained abnormal proteins, and severe steatosis all show what happens when a reserve function turns into a trapping function. The liver that normally holds useful material in a controlled way becomes the first organ injured by what it can no longer release, export, or buffer safely. Storage, then, is never just a quiet backup system. It is a controlled truce between immediate chemistry and future need.

So nutrient storage in the liver should be understood as strategic delay written into tissue. The meal arrives now. Need may come later. A useful compound may be harmless in one compartment and dangerous in another. A metal may need to be held tightly until the marrow asks for it. A vitamin may need to be kept in reserve for months. A sugar surge may need to be turned into a polymer before blood can live with it. The liver is the organ that catches these mismatches and turns them into order. That is what storage really is. It is not extra. It is one of the reasons the body can survive the gap between what was just absorbed and what will be needed next.

Microbial Filtering

The liver filters microbial burden because the intestine is never sending it a sterile stream. Portal blood arrives carrying the chemical remains of digestion, but it also carries low-level evidence of the microbial world living inside the gut. Bacterial fragments, endotoxin, DNA, cell wall debris, fermentation products, and countless smaller metabolites cross or approach the intestinal barrier every day. Some of that traffic is ordinary physiology. A colonized intestine will always generate a certain amount of microbial chemistry. The body does not survive by pretending that burden is absent. It survives by placing the liver directly downstream from it. The liver is the first major organ asked to decide how much of that microbial reality remains local to the gut and how much will be allowed to become systemic blood chemistry.

This filtering begins in the sinusoids, where portal blood slows enough to become readable. Kupffer cells sit there as resident macrophages, already positioned inside the stream rather than arriving later as reinforcements. That location is everything. They do not have to chase a signal through tissue after it has spread. They meet it inside the moving blood itself. Endotoxin, bacterial fragments, damaged-cell debris, and other particulate burdens are sampled, engulfed, and interpreted there. Some are cleared quietly. Some trigger cytokine release if the burden is too high or too inflammatory to be tolerated as background. Liver sinusoidal endothelial cells participate in this filtering environment as well, shaping what crosses, what is presented, and how immune tone is maintained at the blood-tissue interface. The liver is not straining microbes out like a mechanical sieve. It is biologically reading the stream and deciding whether the traffic belongs to ordinary portal life or to emerging threat.

That decision has to be exquisitely controlled, because overreaction would be as dangerous as underreaction. If the liver treated every small amount of bacterial evidence as if the bloodstream had become septic, digestion itself would turn into a daily inflammatory injury. The intestine is too biologically active and too densely populated with microbes for that kind of hypersensitivity to be survivable. So much of hepatic microbial filtering is really disciplined non-escalation. Kupffer cells clear without always sounding the alarm. Hepatocytes continue buffering glucose, handling amino acids, making bile, and synthesizing plasma proteins while this low-grade microbial traffic is being neutralized in the background. Tolerance here is not a failure to detect. It is a controlled refusal to convert every encounter into tissue-damaging immune noise. The liver keeps the rest of the body from having to feel the full inflammatory cost of living downstream from a colonized gut.

When the upstream burden changes, the filtering burden changes with it. A damaged intestinal barrier, dysbiosis, alcohol exposure, bile acid disruption, slowed transit, infection, or inflammatory bowel stress can all increase the amount or character of microbial material reaching portal blood. Now the liver receives a different stream. More endotoxin. More inflammatory metabolites. More bacterial structural residue. More reasons for Kupffer cells to shift from quiet scavenging into cytokine-driven activation. The liver then stops behaving like a mostly tolerant checkpoint and starts behaving like a defensive organ under siege. Acute-phase proteins rise. Hepatocyte stress increases. Stellate cells become easier to activate if the signal persists. The same organ that once filtered microbial traffic quietly begins converting that traffic into inflammation, fibrosis, and altered blood chemistry because the upstream burden no longer fits within the normal range of quiet clearance.

This is why microbial filtering belongs near the center of liver physiology rather than off to the side as an immune detail. The liver protects the body not only by processing nutrients and drugs, but by keeping the microbial chemistry of the intestine from becoming the chemistry of the whole organism. The gut makes the burden. Portal blood delivers it. The liver decides what stays local, what is neutralized, and what becomes systemic alarm. When that filtering is strong, the body barely notices how much microbial evidence is arriving from the bowel every day. When it weakens or is overwhelmed, the whole organism begins feeling the intestinal world much more directly. That is the deeper meaning of microbial filtering. The liver is one of the organs that keeps a colonized gut compatible with a clear bloodstream.

Why Liver Health Is Mostly Invisible Until Reserve Starts to Fail

The healthy liver is hard to feel because most of its work is preventive. It does not announce itself by doing spectacular things. It announces itself by making sure spectacular things do not happen. Blood sugar does not swing as hard after a meal because the liver absorbed part of the surge and stored it as glycogen. Ammonia does not rise into a neurologic burden because the liver converted it into urea. Bilirubin does not remain in circulation because the liver conjugated it and pushed it into bile. Fat reaches the intestine in a form the meal can use because bile acids were made and delivered. Plasma holds its water because albumin was synthesized. Blood clots when it should because clotting proteins were written into circulation in advance. Iron moves or stays trapped according to need because hepcidin was released. Drugs and foreign compounds are altered before they spread too freely through the body. None of this feels like a sensation. It feels like ordinary life because the liver is constantly preventing chemistry from becoming crisis.

This is why liver health is mostly invisible. The organ has enormous reserve and a talent for compensation. A person can lose hepatocytes, accumulate fat, carry significant fibrosis, distort portal flow, and still wake up, eat, think, work, and move through the day because enough functional tissue remains to hold the internal environment together. The bloodstream may already be rougher than it should be, the architecture may already be stiffer than it should be, and regeneration may already be occurring inside low-grade inflammation, yet the body can continue operating because the liver still has margin. It can still buffer enough glucose, clear enough bilirubin, make enough albumin, convert enough ammonia, and process enough portal blood that the rest of the organism does not yet feel the full price of what has been lost. A healthy-looking day can therefore sit on top of a liver already spending reserve to keep the illusion intact.

That reserve also hides time. Many organs declare failure quickly because they have less room to absorb decline before function becomes obviously impaired. The liver can cover a great deal of ground before symptoms become unmistakable. Nutrient stores can buffer poor intake or poor absorption for months or years. Synthetic capacity can thin slowly before edema, bruising, or prolonged clotting become obvious. Portal resistance can rise gradually before ascites or variceal bleeding reveal the new hemodynamic reality. The brain can remain clear until ammonia control and shunting cross a threshold. This delay is protective, but it is also deceptive. The person often believes the liver was healthy until the day symptoms appeared. More often the liver was compensating until the day compensation was no longer enough.

Once reserve starts to fail, the invisibility ends quickly because the liver's jobs are so widely distributed across the body. Fluid leaves the vasculature more easily because albumin and portal flow control are no longer sufficient. Bruising appears because clotting reserve and platelet support are thinning. Dark urine and pale stool appear because bilirubin and bile are no longer moving through the right route. Itching appears because biliary chemistry is now circulating instead of exiting. Confusion appears because the brain is now seeing what the liver used to intercept. Fatigue deepens because fuel timing, inflammatory signaling, hormone handling, and plasma stability have all become less reliable at once. The person does not suddenly become sick because the liver suddenly became important. The person becomes visibly sick because the liver has stopped being quiet enough to hide how important it always was.

That is the deeper truth. Liver health is mostly invisible not because the liver is doing little, but because it is doing so much of the body's background organizing so well that normal life feels normal. The organ remains hidden while it still has enough reserve to keep blood chemistry smooth, bile flowing, proteins supplied, toxins processed, and portal burden contained. When that reserve begins to fail, the body starts revealing how much of its everyday stability had been hepatic all along.

Chapter 60

The Liver as a Master Organ

Why the Liver Deserves a Central Place in Physiology

The liver deserves a central place in physiology because it sits where the body's major streams become reality. Blood from the intestine does not go first to the heart in its raw post-digestive form. It goes to the liver. Bile does not appear in the intestine by accident. The liver makes it, loads it with bile acids, phospholipids, cholesterol, and conjugated waste, and sends it outward so fat can be handled and certain burdens can leave the body. Plasma does not become a stable carrying medium on its own. The liver writes albumin, clotting factors, complement proteins, transport proteins, acute-phase proteins, thrombopoietin, hepcidin, and a long list of other molecules into circulation, changing what blood can carry, how long it can hold water, how it clots, how it signals, and how it defends. Nitrogen from protein turnover does not become safe because the body willed it. The liver converts ammonia into urea. Hormones do not simply circulate unchanged from gland to tissue. The liver binds them, converts them, conjugates them, and helps decide when their signal ends. This is what makes the organ central. Too many essential transitions pass through it for it to be treated as a side organ devoted only to detoxification.

A master organ is not merely one that does many things. It is one whose work changes the conditions under which other organs can do theirs. The liver fits that definition cleanly. The brain depends on it for glucose stability, ketone export during fasting, ammonia control, and protection from gut-derived neuroactive burden. The kidney depends on it for urea generation, plasma protein architecture, circulatory order, and a blood chemistry that has already been partly neutralized before filtration begins. The gut depends on it for bile flow, bile acid recycling, and the first-pass handling of what crosses the intestinal barrier. The marrow depends on it for iron traffic through hepcidin and for platelet production signals through thrombopoietin. The endocrine system depends on it for thyroid handling, vitamin D conversion, steroid metabolism, IGF-1 output, and the protein environment that determines how hormones travel. The circulation depends on it because the liver helps define whether blood remains a coherent transport fluid or becomes a chemically unstable mixture of excess substrate, retained waste, poor protein support, and inflammatory burden. This is why liver disease so rarely stays confined to the liver. Once this organ begins losing order, every other organ starts living inside a different bloodstream.

The liver also deserves this central status because it constantly solves competing demands without announcing itself. After a meal it must absorb the first glucose surge without starving the rest of the body of fuel. During fasting it must release stored carbohydrate, make new glucose, and later export ketones without letting the brain discover too quickly that the intestine has gone quiet. It must tolerate low-grade microbial traffic from the gut without turning digestion into systemic inflammation, yet still remain capable of strong immune defense when the portal stream becomes dangerous. It must hold iron tightly enough to prevent oxidative injury and microbial access, yet release enough of it for erythropoiesis when needed. It must process alcohol, drugs, hormones, bilirubin, and countless foreign compounds without letting the transitional chemistry of that processing destroy the cell that performed it. It must repair itself after injury while avoiding the trap of scar-driven architectural collapse. The liver is therefore central not only because it has many jobs, but because so many of those jobs are balancing acts between opposite risks. Storage versus circulation. Tolerance versus activation. Fuel buffering versus fuel release. Detoxification versus self-injury from the chemistry of detoxification itself. Few organs carry this much controlled contradiction as part of ordinary daily work.

This is also why the liver remains so misunderstood. Much of what it does is invisible precisely because it succeeds quietly. A healthy person does not feel his bilirubin being conjugated, his clotting proteins being synthesized, his bile acids being recycled, his steroid hormones being metabolized, his portal ammonia being trapped into urea, or his post-meal glucose rise being flattened before it becomes a problem. The organ's greatness is hidden inside the absence of crisis. People begin noticing it only when reserve starts to fail, when the bloodstream grows rougher, when bile stops reaching the meal, when edema appears because albumin and flow control have weakened, when confusion appears because ammonia escaped, when the abdomen fills because architecture can no longer hold fluid where it belongs, when the skin itches because retained bile chemistry is circulating, when bruising appears because the blood is no longer properly built. At that point the body starts revealing how much quiet order had been hepatic all along.

So the liver deserves a central place in physiology because it is one of the organs that turns raw biological traffic into livable internal order. It takes the chemistry of the gut, the burden of protein turnover, the instability of feeding and fasting, the danger of free metals, the volatility of endocrine signals, the risk of foreign compounds, and the demands of blood itself, and it reshapes all of them into forms the rest of the body can survive inside. A body without a functioning liver does not simply lose one organ among many. It loses one of the main organs that makes the others compatible with one another. That is what a master organ is. The liver earns the title every hour.

The Liver as Mediator Between Environment, Diet, Intestine, Blood, and Tissue Survival

The liver is the organ that stands between what comes from outside and what the body is willing to let become itself. That is the deepest way to understand it. The environment supplies light, food, microbes, alcohol, drugs, toxins, minerals, and countless chemical exposures the body did not make. Diet turns some of that environment into digestible matter. The intestine breaks that matter down, absorbs part of it, ferments part of it through microbial life, rejects part of it, and sends the resulting chemistry into portal blood. The liver receives that stream before the rest of the body does and decides, molecule by molecule and burden by burden, what will be stored, what will be transformed, what will be neutralized, what will be excreted, what will be returned to the blood, and what the tissues will finally be allowed to receive. That is mediation in the strict physiologic sense. The liver is not merely downstream from life. It is the organ that turns exposure into internal compatibility.

This matters because the intestine does not send the body finished nutrition. It sends a mixed report. Glucose arrives, but so do microbial metabolites. Amino acids arrive, but so does nitrogen burden that can become ammonia. Lipid digestion products arrive, but so do bile acid return signals, endotoxin, bacterial fragments, alcohol, drugs, plant compounds, and other xenobiotics the body did not make for its own structural use. The liver receives all of this in one moving stream. It buffers glucose before the brain and muscle are forced to live with the full post-meal surge. It sorts amino acids toward synthesis, oxidation, or urea production. It turns ammonia into urea so nitrogen metabolism does not become brain injury. It makes bile so dietary fat can become absorbable rather than chemically stranded in a watery lumen. It packages lipids into transport particles so hydrophobic fuel can move through plasma without becoming chaos. It conjugates bilirubin and sends it into bile so the pigment burden of heme turnover does not remain in blood. It transforms foreign compounds into forms more suitable for excretion. The environment enters the intestine as complexity. The liver turns part of that complexity into blood the tissues can survive inside.

This is why the liver is also the mediator between diet and tissue fate. A meal does not nourish the body simply because it was swallowed. It nourishes only to the degree that digestion can liberate it, the barrier can admit it, the portal stream can deliver it, and the liver can render it usable. A carbohydrate load that would otherwise produce a violent glucose swing is softened by hepatic uptake and glycogen storage. A fasting interval that would otherwise expose the brain to unstable fuel is covered by hepatic glycogenolysis, gluconeogenesis, and later ketone export. A fat-soluble vitamin in the meal does not simply cross into the body by virtue of its label. It depends on bile flow, micelle formation, intestinal absorption, hepatic handling, and proper storage or transport afterward. Iron does not become tissue iron because it was eaten. It becomes tissue iron because the liver decides through hepcidin whether the body is permitted to move it. The same is true in the other direction. A compound can be natural and still become a burden if it arrives in excess, resists safe handling, or imposes a cost on bile flow, redox balance, or nitrogen disposal that the body cannot carry cheaply. The liver is the organ that reveals that difference. It judges diet not by ideology but by net biological yield.

The liver also mediates between the intestine and the blood in a defensive sense. The gut is colonized. The barrier is selective but never absolute. Small amounts of microbial fragments, lipopolysaccharide, bacterial DNA, and other immune-active chemistry reach portal blood every day. The body survives that arrangement because the liver filters, samples, clears, and interprets the burden before the general circulation has to absorb it. Kupffer cells read the incoming stream. Hepatocytes adjust acute-phase output, bile acid synthesis, and metabolic priorities according to what the portal blood now means. If the intestinal barrier weakens, the microbial burden rises, or bile flow changes the luminal environment, the liver becomes the first organ forced to absorb that error. It is therefore not only a metabolic mediator. It is an immunologic mediator. It keeps the microbial reality of the gut from becoming the chemistry of the whole organism every time a meal passes through.

This same mediating role explains why the liver is so central to tissue survival far from the abdomen. The brain depends on the liver for glucose stability, ketone support during fasting, and protection from ammonia and other neuroactive gut-derived compounds. The kidney depends on the liver to convert ammonia into urea and to maintain a plasma environment with enough albumin, circulatory order, and hemodynamic stability that filtration can remain effective. The endocrine system depends on the liver to bind, convert, buffer, and clear hormones so glands do not become permanent unedited signals in the bloodstream. The marrow depends on it for thrombopoietin and for iron traffic control. The vasculature depends on it for albumin, clotting proteins, and the broader composition of plasma itself. A failing liver therefore does not only injure itself. It withdraws a layer of mediation the whole body had been relying on without noticing. The blood becomes rougher, less filtered, less stable, less protein-supported, less well buffered against the environment it is carrying.

That is the deeper reason the liver deserves to be seen as a mediator rather than as a detox organ, a bile organ, or a storage organ alone. Those names each capture one job. Mediation captures the principle behind all of them. The outside world becomes diet. Diet becomes intestinal chemistry. Intestinal chemistry becomes portal blood. Portal blood becomes systemic blood only after the liver has done its work. Tissue survival depends on that sequence being governed well enough that no organ is forced to live directly in the raw aftermath of the environment. The liver is where that governance happens. When it is healthy, the body experiences the world in a form made livable. When it fails, the body begins discovering how much of ordinary survival depended on one organ standing in the middle and saying yes, no, later, store this, release that, clear this, and do not let the tissues see the rest.

Why Almost Every Major System Depends on the Liver Being Organized

Almost every major system depends on the liver being organized because the liver is one of the few organs that sits at the crossing point of circulation, digestion, immunity, metabolism, endocrine signaling, and waste handling all at once. A tissue can be injured and still leave the rest of the body relatively untouched if its job is narrow enough. The liver does not have that luxury. It receives portal blood from the intestine carrying nutrients, microbial products, drugs, alcohol, ammonia precursors, bile acid return, and foreign compounds. It writes albumin, clotting factors, transport proteins, acute-phase proteins, hepcidin, thrombopoietin, and hormone-related intermediates into plasma. It converts ammonia into urea, conjugates bilirubin, packages lipids, builds bile, stores glycogen, shapes fasting fuel delivery, and edits endocrine signals before they move onward through the body. That means liver organization is not just a local structural preference. It is one of the conditions under which the rest of the organism remains livable. When the liver is soft, open, polarized, and well supplied, the blood leaving it is more stable than the blood entering it. When that organization begins to fail, every system downstream starts living in a rougher internal world.

The brain depends on this more than most people realize. It does not only need oxygen and circulation. It needs blood that has already been stripped of gut-derived neuroactive burden, buffered against violent glucose swings, and supported through fasting by glycogen release and ketone production. The liver provides all of that. The kidney depends on it as well. Urea does not appear by itself. The liver has to convert ammonia into a form the kidney can excrete, and the plasma reaching the nephron has to remain protein-supported and hemodynamically coherent enough for filtration to stay useful. The marrow depends on the liver for thrombopoietin and for iron traffic control through hepcidin. The endocrine system depends on it for steroid metabolism, thyroid hormone handling, vitamin D conversion, IGF-1 output, and the transport proteins that determine how much hormone remains free versus buffered in blood. The gut depends on it for bile flow and bile acid recycling, while the liver depends on the gut for the chemistry delivered through portal blood. None of these relationships are decorative. The liver is not helping around the edges. It is making the internal environment usable for each system in a different way.

This is why the bloodstream is one of the first places hepatic disorder becomes visible. A damaged liver does not merely stop detoxifying in the vague popular sense. It stops writing the same blood. Albumin falls and fluid begins leaving the vascular space more easily. Clotting factors thin and bruising or unstable coagulation appear. Bilirubin rises because pigment is no longer being moved through the right route. Ammonia rises because nitrogen is no longer being neutralized before reaching the brain. Bile acids fail to reach the intestine and fat handling changes. Hormone transport and clearance become less disciplined, so the person begins living in a different endocrine environment. Portal hypertension reroutes blood away from normal hepatic exchange, and now the spleen enlarges, varices form, ascites develops, and the kidney begins constricting inside a circulation it reads as underfilled. The liver has not simply become sick in one corner of the abdomen. It has stopped organizing the plasma, pressure, and chemical traffic the whole body was depending on.

Architecture is the hidden reason this dependence is so broad. The liver works because blood can cross a low-resistance sinusoidal bed, because hepatocytes can face blood on one side and bile on the other, because the space of Disse stays thin enough for exchange, because bile ducts remain open enough to carry the outgoing stream, and because regeneration occurs inside a scaffold still capable of supporting function. Once fibrosis thickens that exchange field, once nodules distort flow, once bile is retained, once portal blood is shunted, the organ may still be present anatomically but functionally unavailable in whole or in part. Blood no longer meets hepatocytes cleanly. Bile no longer leaves cleanly. Oxygen gradients worsen. Synthetic reserve falls. The rest of the body is then forced to adapt to chemistry that was supposed to be corrected before it reached them. Almost every system begins compensating for a liver that has stopped holding its position in the network. That is what advanced liver disease really is. It is not just injury to one organ. It is the collapse of an organizing hub.

So almost every major system depends on the liver being organized because the liver is one of the main organs that turns incoming burden into usable blood, turns intermittent intake into stable fuel, turns dangerous intermediates into excretable products, and turns hormonal signals into forms the body can actually live under. The heart can keep blood moving, but the liver helps decide what kind of blood is moving. The kidneys can filter plasma, but the liver helps decide what solutes reach them and in what form. The brain can think only inside the chemistry it is given, and the liver is one of the organs that decides how clean that chemistry will be. The bowel can absorb, but the liver decides what absorption is allowed to become. That is why liver organization matters so much. It is not one more physiologic detail. It is part of the hidden order that lets the rest of physiology remain possible.

The Final Framework: The Liver Keeps Blood Usable

The cleanest way to understand the liver is to stop asking what organ it sits beside and start asking what kind of blood would exist without it. That question clarifies almost everything. The intestine would still absorb. Muscle would still release lactate. Adipose tissue would still release fatty acids. Red blood cells would still age and shed heme burden. Microbes in the gut would still produce ammonia, endotoxin, and metabolites. Hormones would still be secreted. Drugs and alcohol would still enter the portal stream. But none of that chemistry would yet be fit for general circulation. It would be raw, unstable, mistimed, poorly buffered, or actively dangerous. The liver exists in the middle of that flood and changes it before the rest of the body has to live inside it. It does not merely add or subtract a few molecules. It decides whether blood is chemically civilized enough for tissues to survive in.

That is why so many different liver functions are actually one function seen from different angles. Post-meal glucose buffering is the liver keeping blood usable. Glycogen storage and fasting release are the liver keeping blood usable across time. Urea synthesis is the liver keeping blood usable for the brain. Bilirubin conjugation and biliary secretion are the liver keeping blood usable by removing pigment burden before it accumulates in plasma. Bile acid secretion is the liver keeping the gut usable so the next meal can become absorbable blood rather than wasted chemistry. Albumin synthesis is the liver keeping blood usable as a fluid compartment that can hold its own water. Clotting factor and anticoagulant synthesis are the liver keeping blood usable as a material that can remain fluid in motion and still solidify where a vessel breaks. Hepcidin is the liver keeping blood usable by deciding when iron may circulate and when it must be withheld. Hormone binding, conversion, and clearance are the liver keeping blood usable as an endocrine medium rather than a reservoir of unedited signals. These are not separate jobs in the deepest sense. They are one governing principle repeated across different classes of molecules. The liver keeps blood from becoming too much gut, too much meal, too much waste, too much inflammation, too much hormone, too much metal, too little protein, too little fuel, or too much danger.

This is also why liver disease so quickly becomes whole-body disease. The liver is not an isolated processor that can fail privately. When it begins losing architecture, the blood changes with it. Portal flow becomes more resistant and the upstream circulation begins backing up. Bile flow becomes less reliable and digestion loses one of its main tools. Albumin falls and fluid leaves the vasculature more easily. Clotting becomes unstable because the bloodstream is no longer being built with the same reserve on either side of the coagulation balance. Ammonia escapes control and the brain begins reading the chemistry of the bowel more directly than it should ever have to. Hormone handling loses precision, so endocrine symptoms appear in tissues that seem far from the liver. Iron traffic changes. Platelet support falls. The kidney begins constricting inside a circulation the scarred liver has already made unstable. This is why the liver is so easy to underestimate when healthy and so hard to ignore when failing. It is one of the organs that made the internal environment livable, and once it stops, every other organ begins discovering how dependent it was on a blood chemistry it never had to think about before.

Seen that way, the liver is not mainly a detox organ, a bile organ, a storage organ, a glucose organ, or a clotting organ, though it is all of those. It is the organ that stands between biological input and biological consequence. The environment becomes diet. Diet becomes intestinal chemistry. Intestinal chemistry becomes portal blood. Portal blood becomes systemic blood only after the liver has judged what may stay, what must go, what must wait, what must be stored, what must be transformed, and what must never be allowed to reach the wider circulation unchanged. The same is true for endogenous burden. Tissue turnover creates pigments, nitrogen, hormones, and metabolic residues that the body cannot simply leave as they are. The liver receives those too and decides whether they become urea, bile pigment, excretable metabolites, stored reserve, or circulating signal. It is the editor of internal consequence. Other organs contribute important chemistry. The liver decides how much of that chemistry becomes survivable blood.

So the final framework is simple enough to remember and broad enough to organize almost every chapter that came before it. The liver keeps blood usable. It keeps the bloodstream from becoming a raw extension of the gut. It keeps feeding from becoming instability and fasting from becoming collapse. It keeps nitrogen from becoming neurotoxicity, bile from becoming retention, fluid from leaving the vascular space too easily, and endocrine signals from remaining active longer than the body can tolerate. When it is healthy, this work is almost invisible because normal blood feels like no event at all. When it fails, the body begins revealing the truth in every direction. The skin yellows. The abdomen fills. The spleen enlarges. The kidneys constrict. The mind slows. The blood bruises, leaks, pools, and carries the wrong molecules to the wrong tissues. That is the proof. The liver was never just one organ among many. It was one of the main reasons the rest of the body could share one bloodstream and still remain a body.

Why This Organ Is Best Understood as a Continuity Machine

The liver is best understood as a continuity machine because its deepest work is not the production of one substance or the clearance of one toxin. Its deepest work is preventing biological discontinuity. The outside world enters in bursts, meals, microbes, alcohol, drugs, minerals, hormones, light-driven vitamin precursors, environmental chemicals. The inside of the body cannot afford to live in bursts. Blood cannot become a different fluid every time food crosses the intestine. The brain cannot think clearly only while glucose is arriving from the gut. The circulation cannot clot only after the liver remembers to make factors. The intestine cannot absorb fat only when bile happens to appear by accident. Iron cannot move freely one day and disappear the next. Nitrogen cannot be harmless after breakfast and neurotoxic by late afternoon. The body survives because the liver takes discontinuous, irregular, risky chemistry and converts it into something temporally smoother, chemically quieter, and structurally survivable. That is continuity. The organ stands between event and consequence and keeps one from arriving at the body with its full original violence.

This is why so many seemingly unrelated liver functions are actually the same function viewed through different substrates. After a meal, glucose arrives in portal blood as a surge. The liver absorbs part of it, stores part of it as glycogen, burns part of it, and meters the rest onward so the bloodstream does not become a violent reflection of the intestine. Hours later, when the intestine is quiet, the same liver opens glycogen and then makes new glucose so the brain and red blood cells do not experience the absence of feeding as immediate fuel collapse. In longer fasting, it shifts again and sends ketones outward so the body can reduce its glucose demand without disassembling itself too quickly. This is not just metabolism. It is continuity of fuel across time. The meal is episodic. The bloodstream must remain usable in between. The liver makes that possible.

The same logic governs waste. Ammonia is generated continuously from protein turnover and gut microbial activity, but the brain cannot tolerate a bloodstream that reflects that chemistry directly. The liver takes the nitrogen burden and turns it into urea, a safer excretory form the kidney can handle. Bilirubin arises from heme breakdown, but plasma cannot remain a permanent pigment reservoir. The liver conjugates it, pushes it into bile, and hands it onward to the intestine so the bloodstream is not left carrying what should have already left. Steroid hormones are secreted in pulses, but the body cannot remain trapped under every signal indefinitely. The liver transforms, binds, conjugates, and clears them so endocrine messaging has duration without becoming permanent occupation. Iron is needed constantly but dangerous when freely mobile, so the liver uses hepcidin to decide whether iron moves or stays trapped. Bile acids are expensive to synthesize and indispensable for fat handling, so the liver sends them out, receives them back, and reuses them rather than remaking the whole detergent system for every meal. In each case the principle is the same. A molecule with disruptive potential is prevented from becoming a permanent discontinuity in the blood. The liver either stores it, converts it, exports it, restrains it, or recycles it until the bloodstream can remain coherent.

This is also why the liver is inseparable from the gut. The intestine is a pulsed absorptive border sitting on top of a microbial ecosystem. It sends the liver nutrients, but also endotoxin, microbial metabolites, ammonia precursors, bile acid return, food antigens, alcohol, drugs, and countless foreign compounds. The bloodstream beyond the liver cannot be allowed to feel that raw stream directly. So the liver filters microbial traffic without overreacting to every meal, buffers substrate without letting each meal become a metabolic event for every tissue, and converts a chemically mixed portal inflow into a more disciplined systemic outflow. In that sense the liver is continuity not only across time, but across boundaries. It keeps the inside of the body from becoming too much like the chemistry of the gut. A colonized digestive tract and a clear bloodstream can coexist only because the liver stands between them and performs that translation constantly.

The circulatory side reveals the same truth in another form. Plasma is not self-maintaining. It needs albumin to hold water in the vascular space, clotting factors to seal injury, anticoagulant proteins to keep sealing from becoming thrombosis, transport proteins to carry hormones and lipids, acute-phase proteins to reshape defense during inflammation, and endocrine signals such as thrombopoietin and IGF-1 to tell other organs how to behave. The liver writes much of that into blood every day. That is continuity of blood as blood. Without it, the circulation becomes chemically thinner, hydraulically less stable, less able to hold volume, less able to transport with precision, less able to defend without overshooting, and less able to maintain timing across other organ systems. The liver is not just modifying blood contents. It is preserving the continuity of the medium through which every organ must live.

This is why liver failure feels so total when reserve starts to fail. A continuity machine hides its own importance while it is working because what it produces is the absence of disruption. When it weakens, the body does not lose one isolated task. It loses the smoothing force that kept one state connected to the next. Meals become less buffered. Fasts become less tolerable. Nitrogen becomes more neuroactive. Bile stops reaching the intestine and starts backing into blood. Fluid leaves the vasculature more easily. Portal blood reroutes around the organ instead of through it. The kidney receives a worse circulation. The brain receives worse chemistry. The gut sends a dirtier stream into a liver less able to absorb the burden quietly. Hormone signals flatten, linger, or drift because conversion and clearance no longer preserve their timing. The person then experiences not just liver disease, but discontinuity. Thought, energy, fluid balance, digestion, bleeding control, and metabolic stability all begin losing smoothness at once.

So this organ is best understood as a continuity machine because it takes inputs that are intermittent, excessive, microbial, foreign, reactive, hormonally unstable, or chemically dangerous and turns them into a bloodstream the tissues can inhabit without constantly noticing the violence of what just entered. It preserves continuity between meal and fast, gut and blood, burden and excretion, signal and silence, storage and release, survival and ordinary life. Other organs do essential work. The liver is one of the organs that keeps that work from being torn apart by the timing and chemistry of everything else.

Closing Synthesis for the Reader Who Now Sees the Whole Map

By this point the liver should no longer look like a red organ tucked under the diaphragm with a few famous jobs attached to it. It should look like a governing surface stretched between the world and the bloodstream. Food enters the gut as matter from outside. The intestine breaks that matter down, absorbs part of it, ferments part of it through microbial life, and sends the resulting chemistry into portal blood. The liver receives that stream first and decides what the rest of the body will be allowed to experience. Glucose is buffered. Amino acids are sorted. Nitrogen is trapped into urea. Lipids are packaged or stored. Bile is written outward so fat can become usable and waste can leave. Bilirubin is conjugated and excreted. Hormones are bound, converted, softened, or cleared. Iron is permitted to move or ordered to stay trapped. Plasma proteins are written into blood so fluid can remain in the vascular space, wounds can seal, and signals can travel in a stable medium. That is the map. The liver is not one more organ helping with metabolism. It is one of the organs that turns raw biological input into internal order.

Once you see that, most liver disease stops looking random. Steatosis is no longer just fat in the liver. It is fuel traffic failing. Cholestasis is no longer just bile backed up. It is the outgoing stream breaking and turning digestion and excretion into the same injury. Portal hypertension is no longer just pressure in a vein. It is blood discovering that the organ built to receive it has become too hard to cross. Ascites is no longer just abdominal fluid. It is circulation losing its grip on its own water after flow and protein architecture both fail. Hepatic encephalopathy is no longer mysterious confusion. It is the brain receiving blood the liver should have edited first. Even fibrosis and cirrhosis become clearer. They are not simply scar and end-stage disease. They are the point at which repair stops preserving function and starts replacing function with structure. The liver tries to save the body and, under chronic burden, begins rebuilding itself into the wrong shape.

That is also why the liver connects so naturally to every other major system. The gut depends on it for bile and for first-pass review of what crosses the barrier. The brain depends on it for glucose continuity, ketone support, and ammonia control. The kidney depends on it for urea generation, plasma stability, and a circulation that has not yet collapsed into portal and splanchnic disorder. The endocrine system depends on it for hormone conversion, hormone transport, hormone timing, and hormone clearance. The blood depends on it because plasma itself is partly a hepatic construction project. The immune system depends on it because the liver is one of the main organs deciding how much of the gut's microbial reality becomes systemic alarm. This is why liver failure never stays local for long. Once the organ that keeps blood usable begins losing order, every other organ starts living in a different internal world.

The deeper lesson is that physiology makes more sense when it is read through sequences instead of labels. Do not ask only what molecule is present. Ask what form it is in, what compartment it is in, what route it is taking, what tissue wrote it there, and what organ is supposed to change its fate next. Do not ask only whether the liver is inflamed. Ask whether the main problem is hepatocyte injury, bile retention, vascular distortion, synthetic decline, or metabolic overload. Do not ask only whether bilirubin is high. Ask whether too much is arriving, whether conjugation failed, or whether the route out is blocked. Do not ask only whether the patient is confused. Ask what gut-derived burden escaped hepatic control and why the brain is now paying for it. This is how the whole map stays coherent. The liver is an organ of transitions, and the only honest way to read an organ of transitions is to think in transitions.

If there is one framework worth carrying forward after all of this, it is simple. The liver keeps blood usable. That sentence is large enough to hold the rest. It keeps the bloodstream from becoming the raw extension of the intestine. It keeps the meal from becoming a glucose shock and the fast from becoming a fuel collapse. It keeps ammonia from becoming thought disorder. It keeps bile moving toward the gut and bilirubin moving out of the body. It keeps plasma full enough of the right proteins that water stays in vessels, hormones travel in buffered form, and coagulation remains possible without constant thrombosis. It keeps dangerous useful things, iron, copper, steroids, drugs, microbial products, in a state that tissues can survive. When it works, the body experiences that labor as normal life. When it fails, normal life starts breaking into compartments.

So the final synthesis is not that the liver does everything. It is that the liver stands at too many crucial junctions for any major system to remain fully normal once hepatic order is lost. It is the organ that absorbs irregularity, edits burden, stores reserve, writes plasma, moves bile, filters the gut, and translates exposure into survivable blood. If you can now see the liver that way, not as a detox organ, not as a bile organ, not as a glucose organ, but as a master continuity organ turning discontinuous, risky chemistry into a livable internal environment, then you have the right map. The details from here on only make that map sharper.

Visual Guide

LOCATION OF THE LIVER

The liver is located in the right upper abdomen, just beneath the diaphragm and under the right rib cage, extending slightly toward the center.

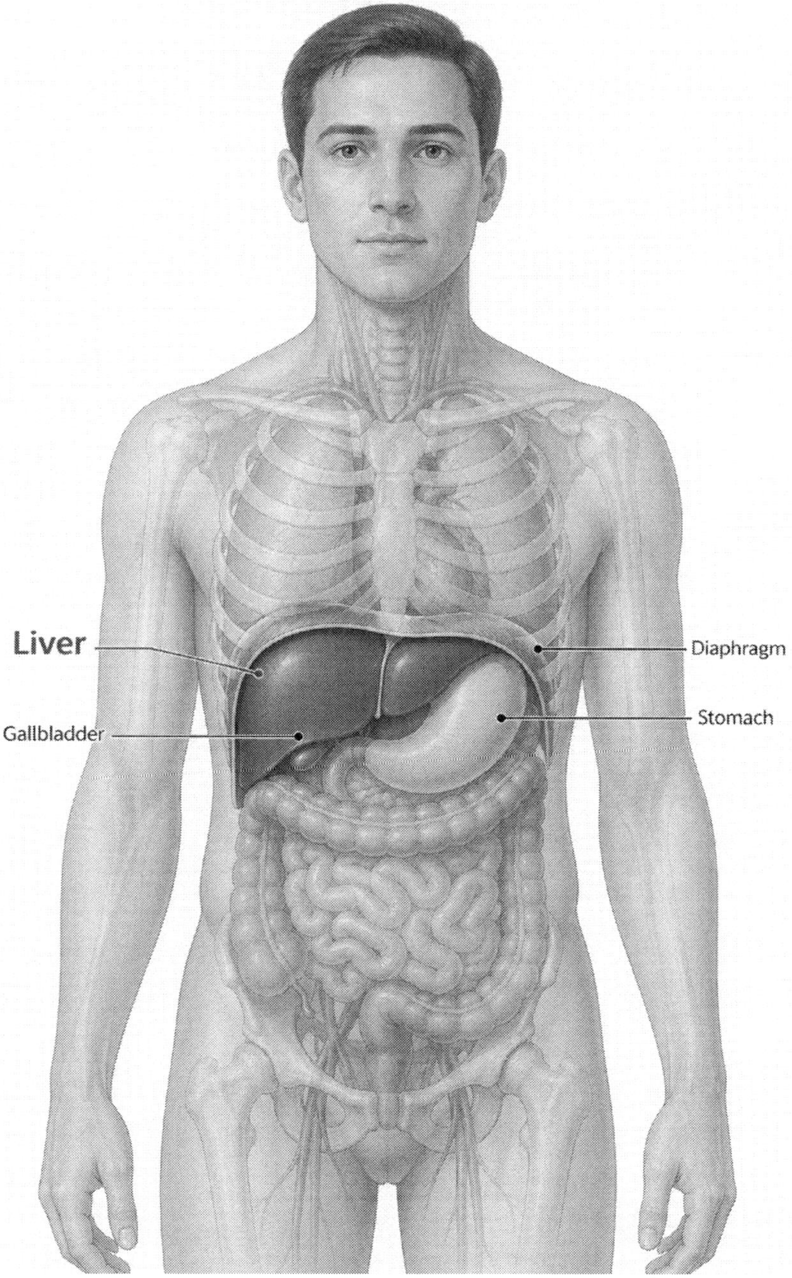

Liver Gross Anatomy

ANTERIOR VIEW

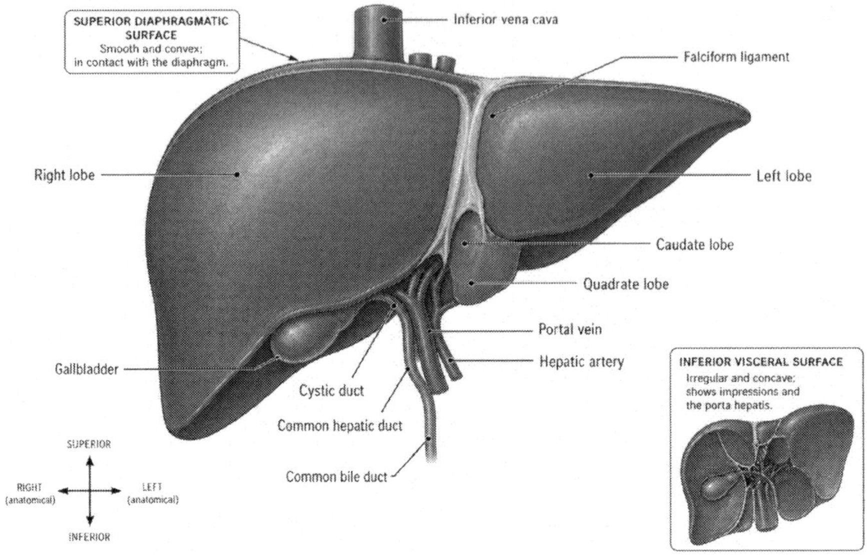

Portal Triad

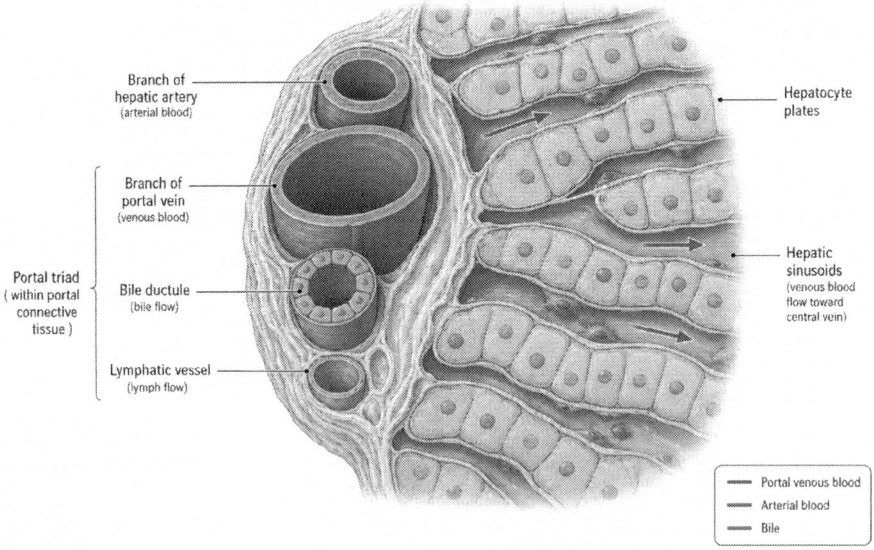

Lobule and Acinus

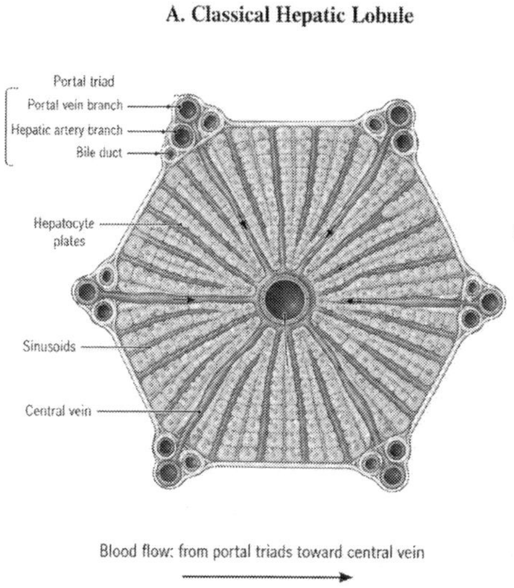

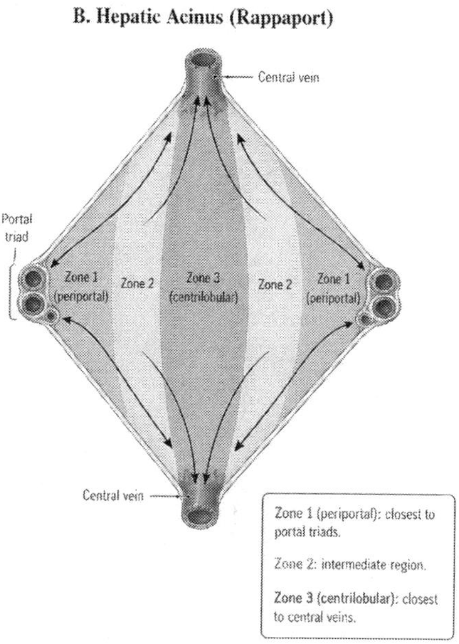

Blood Flow Versus Bile Flow

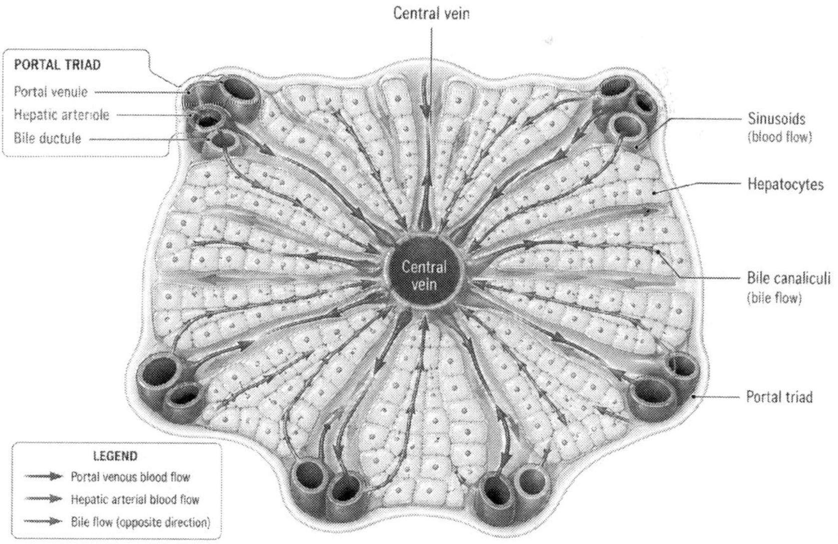

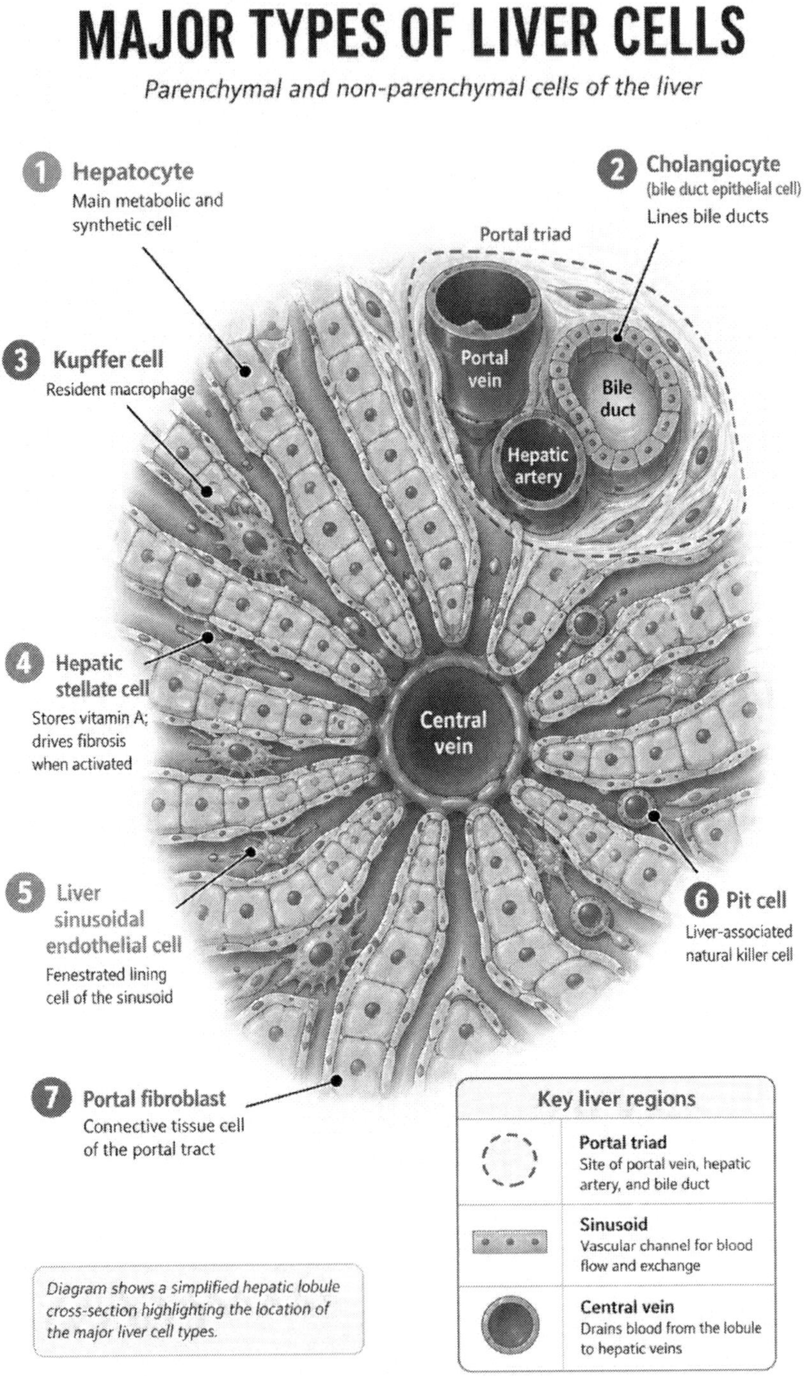
MAJOR TYPES OF LIVER CELLS
Parenchymal and non-parenchymal cells of the liver
1 Hepatocyte
Main metabolic and synthetic cell
2 Cholangiocyte
(bile duct epithelial cell)
Lines bile ducts
Portal triad
Portal vein
Bile duct
Hepatic artery
3 Kupffer cell
Resident macrophage
4 Hepatic stellate cell
Stores vitamin A; drives fibrosis when activated
Central vein
5 Liver sinusoidal endothelial cell
Fenestrated lining cell of the sinusoid
6 Pit cell
Liver-associated natural killer cell
7 Portal fibroblast
Connective tissue cell of the portal tract
Key liver regions
Portal triad
Site of portal vein, hepatic artery, and bile duct
Sinusoid
Vascular channel for blood flow and exchange
Central vein
Drains blood from the lobule to hepatic veins
Diagram shows a simplified hepatic lobule cross-section highlighting the location of the major liver cell types.

FIBROSIS AND CIRRHOSIS PROGRESSION

How chronic liver injury remodels liver tissue over time.

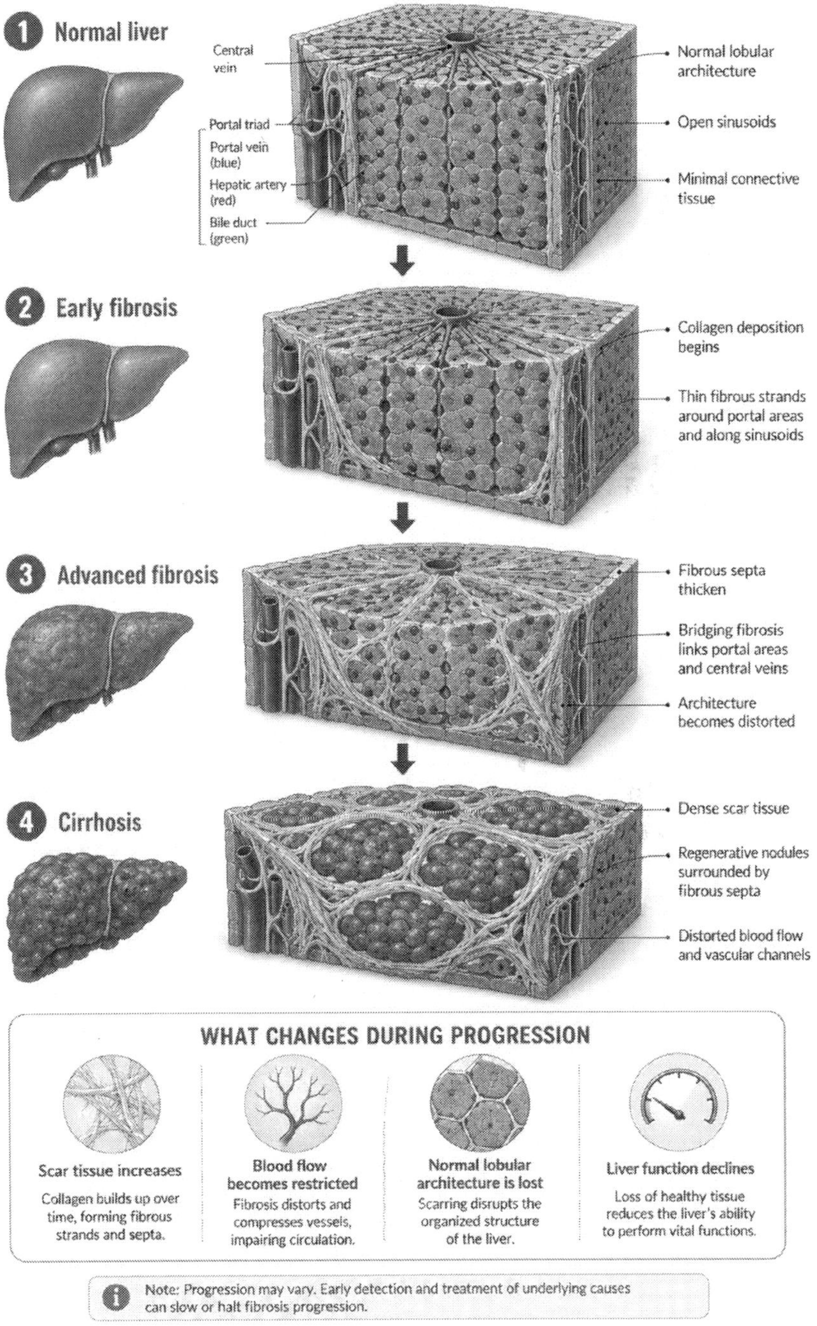

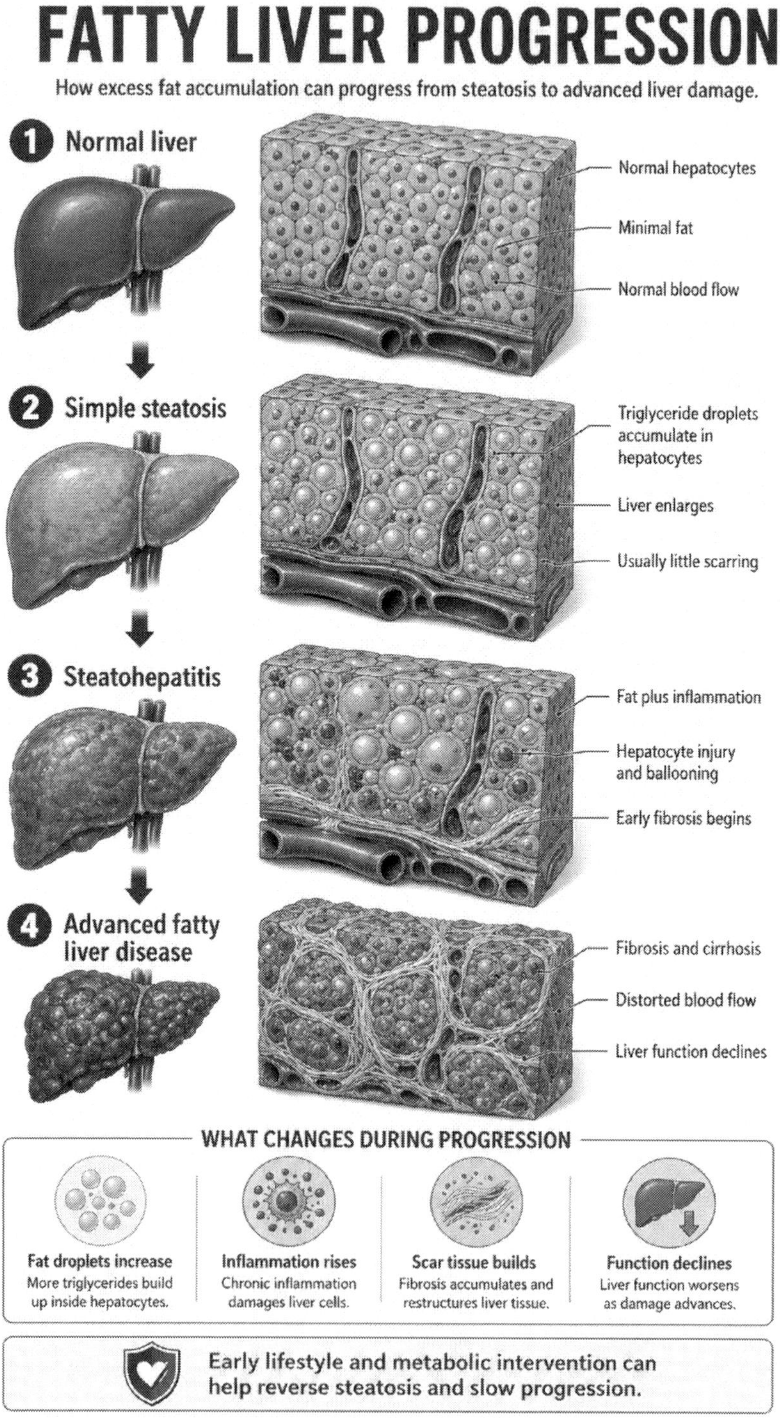
FATTY LIVER PROGRESSION
How excess fat accumulation can progress from steatosis to advanced liver damage.
1 Normal liver
Normal hepatocytes
Minimal fat
Normal blood flow
2 Simple steatosis
Triglyceride droplets accumulate in hepatocytes
Liver enlarges
Usually little scarring
3 Steatohepatitis
Fat plus inflammation
Hepatocyte injury and ballooning
Early fibrosis begins
4 Advanced fatty liver disease
Fibrosis and cirrhosis
Distorted blood flow
Liver function declines
WHAT CHANGES DURING PROGRESSION
Fat droplets increase
More triglycerides build up inside hepatocytes.
Inflammation rises
Chronic inflammation damages liver cells.
Scar tissue builds
Fibrosis accumulates and restructures liver tissue.
Function declines
Liver function worsens as damage advances.
Early lifestyle and metabolic intervention can help reverse steatosis and slow progression.

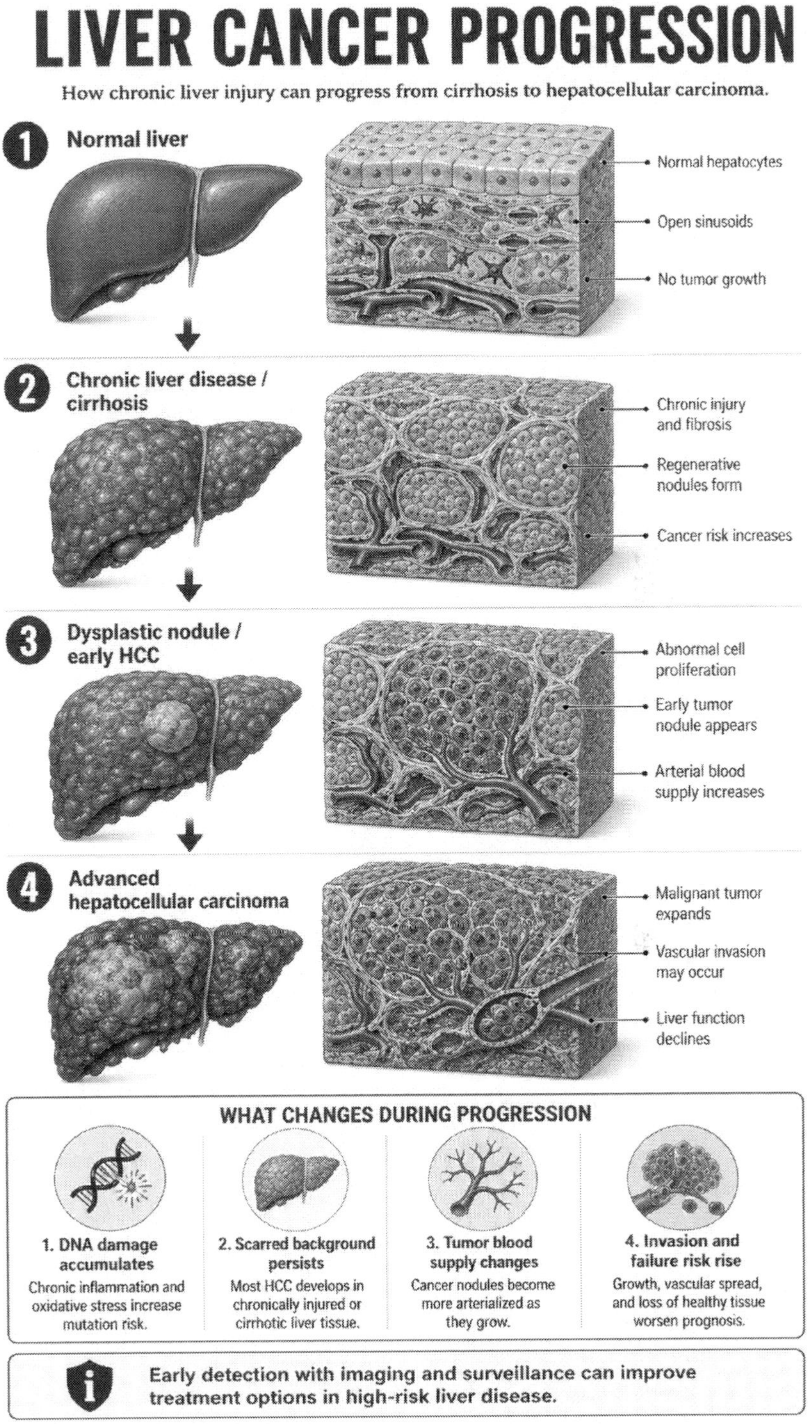
LIVER CANCER PROGRESSION
How chronic liver injury can progress from cirrhosis to hepatocellular carcinoma.
1 Normal liver
Normal hepatocytes
Open sinusoids
No tumor growth
2 Chronic liver disease / cirrhosis
Chronic injury and fibrosis
Regenerative nodules form
Cancer risk increases
3 Dysplastic nodule / early HCC
Abnormal cell proliferation
Early tumor nodule appears
Arterial blood supply increases
4 Advanced hepatocellular carcinoma
Malignant tumor expands
Vascular invasion may occur
Liver function declines
WHAT CHANGES DURING PROGRESSION
1. DNA damage accumulates
Chronic inflammation and oxidative stress increase mutation risk.
2. Scarred background persists
Most HCC develops in chronically injured or cirrhotic liver tissue.
3. Tumor blood supply changes
Cancer nodules become more arterialized as they grow.
4. Invasion and failure risk rise
Growth, vascular spread, and loss of healthy tissue worsen prognosis.
Early detection with imaging and surveillance can improve treatment options in high-risk liver disease.

Made in United States
Orlando, FL
04 November 2026

90218971R00411